T0335789

Biology and Evolution of the Mexican Cavefish

Biology and Evolution of the Mexican Cavefish

Alex C. Keene

Masato Yoshizawa

Suzanne E. McGaugh

AMSTERDAM • BOSTON • HEIDELBERG • LONDON
NEW YORK • OXFORD • PARIS • SAN DIEGO
SAN FRANCISCO • SINGAPORE • SYDNEY • TOKYO
Academic Press is an imprint of Elsevier

Academic Press is an imprint of Elsevier
225 Wyman Street, Waltham, MA 02451, USA
525 B Street, Suite 1800, San Diego, CA 92101-4495, USA
125 London Wall, London, EC2Y 5AS, UK
The Boulevard, Langford Lane, Kidlington, Oxford OX5 1GB, UK

© 2016 Elsevier Inc. All rights reserved.

No part of this publication may be reproduced or transmitted in any form or by any means, electronic or mechanical, including photocopying, recording, or any information storage and retrieval system, without permission in writing from the publisher. Details on how to seek permission, further information about the Publisher's permissions policies and our arrangements with organizations such as the Copyright Clearance Center and the Copyright Licensing Agency, can be found at our website: www.elsevier.com/permissions.

This book and the individual contributions contained in it are protected under copyright by the Publisher (other than as may be noted herein).

Notices
Knowledge and best practice in this field are constantly changing. As new research and experience broaden our understanding, changes in research methods, professional practices, or medical treatment may become necessary.

Practitioners and researchers must always rely on their own experience and knowledge in evaluating and using any information, methods, compounds, or experiments described herein. In using such information or methods they should be mindful of their own safety and the safety of others, including parties for whom they have a professional responsibility.

To the fullest extent of the law, neither the Publisher nor the authors, contributors, or editors, assume any liability for any injury and/or damage to persons or property as a matter of products liability, negligence or otherwise, or from any use or operation of any methods, products, instructions, or ideas contained in the material herein.

Library of Congress Cataloging-in-Publication Data
A catalog record for this book is available from the Library of Congress

British Library Cataloguing in Publication Data
A catalogue record for this book is available from the British Library

ISBN: 978-0-12-802148-4

For information on all Academic Press publications
visit our website at http://store.elsevier.com/

Typeset by SPi Global, India
www.SPi-global.com

Printed in USA

Contents

Part III
Morphology and Development

Part IV
Behavior

Preface

Evolution has created a spectacular assortment of diversity that has intrigued naturalists for centuries, and more recently, has been used by biologists to investigate basic principles of life on earth. In the same light that biomedical research uses dysfunction such as cancer or neurological disease to better understand function, the extreme cases of evolutionary processes can be used to study the basic principles that govern adaption in response to a changing environment. Perhaps one of the most distinct shifts in environment seen in nature is one of moving from surface to subterranean life.

Throughout the world are examples of cave animals, ranging from salamanders to small insects that have evolved cave-like traits that include albinism and eye loss. The Mexican cavefish, *Astyanax mexicanus*, provides a particularly striking system, because fish evolved in 29 geographically isolated caves over the last ~5 million years. While these fish look dramatically different from their river-dwelling counterparts, they remain interfertile, providing biologists with a tool to investigate the genetic basis for developmental, anatomical, and behavioral evolution.

The interest in cavefish extends well beyond the scientists using this system to those interested in cave exploration, biology, zoology, and evolution. The book is written to provide both historical perspective and a current snapshot of research on these fish. The first investigation of these caves, dating back to the early 1920s, included the heroic attempts by early speleologists to characterize the geology and biology of cave life. This book takes readers from the initial discovery of these caves to early experiments classifying fish through recent advances in genomics and neuroscience. As such, the diverse authors share a variety of perspectives that are pertinent to some of the ongoing discussions and debates about the biology of *Astyanax*.

The decision to write this book largely stems from the unique state of the research community. The recent era of genomics has provided powerful tools for investigating the evolutionary and population history of these fish. A genome for Mexican cavefish was published only last year, and the advent of genome-editing tools may allow for the identification of genes regulating behavior and developmental processes at a resolution previously thought possible only in genetically amenable systems, such as mice, zebrafish, and fruit flies. Therefore, we believe this is an excellent time to review the history of investigation in this field, as opportunities and interest in this system are likely to expand greatly in the future.

Many of the contributors to this book are the titans of the field and responsible for some of the most important discoveries in this system. Included are contributions from Bill Elliot, part of a small team that explored many caves for the first time; Bill Jeffery (University of Maryland) and Cliff Tabin (Harvard), who led work describing biology underlying albinism and eye loss in cavefish; Richard Borowsky (New York University), who has used genomics to trace the evolution of these fish; and Sylvie Rétaux (CNRS, France), who has identified many factors governing changes in brain development and behavior in cavefish.

Also included are contributions from more junior researchers that have recently started their independent research careers. As editors, we fall into this category and are grateful for the support we have received from our senior colleagues. We hope that this book serves as a captivating read and conveys the history and promise of this fascinating biological system.

Alex C. Keene
Florida Atlantic University

Masato Yoshizawa
University of Hawai'i

Suzanne E. McGaugh
University of Minnesota

Introduction: The Emergence of the Mexican Cavefish as an Important Model System for Understanding Phenotypic Evolution

Clifford J. Tabin
Department of Genetics, Harvard Medical School, Boston, Massachusetts, USA

Ever since the Modern Synthesis integrated population genetics with evolutionary change through natural selection (Huxley, 1942), evolutionary biologists have endeavored to understand the genetic architecture underlying phenotypic evolution. As new fields of molecular biology, developmental biology, and genomics have been added to the armament of evolutionary biologists, questions have become accessible that eluded previous generations of researchers. For example, is phenotypic evolution based upon many genetic changes each of small effect, or from a smaller number of genetic changes of large effect? Are the relevant changes generally in regulatory sequences affecting the amount and/or timing of gene expression, as has become the dogma in evolutionary genetics, or are there classes of phenotypic changes where coding mutations affecting protein specificity or activity playing the dominant role? When similar phenotypic alterations are observed in independent populations, are the same genetic pathways responsible for the morphological or behavioral changes, or can completely distinct molecular mechanisms be selected to yield the same phenotypic outcome? Under what circumstances is strict parallelism more likely to occur? Are the answers to these questions the same for regressive and constructive traits? Are they the same for morphological, behavioral, and metabolic traits? As organisms adapt within their environment, or to a new environment, to what extent do they rely on preexisting standing genetic variation, and when does genetic evolution rely on *de novo* mutation?

The molecular and genetic tools now available to ask such questions are extraordinarily powerful and also extremely versatile, allowing genetic and genomic integration of almost any species. Nonetheless, it requires a large community effort to develop a new evolutionary system, and thought must be given

Biology and Evolution of the Mexican Cavefish. http://dx.doi.org/10.1016/B978-0-12-802148-4.09999-0
© 2016 Elsevier Inc. All rights reserved.

in choosing a specific animal to develop into a new model. To establish a new model, experimental embryology, behavioral assays, and physiological assays need to be established. Required genetic infrastructure includes isolating genetic markers, constructing a genetic map, obtaining an assembled and annotated genome sequence (ideally from multiple phenotypically distinct populations), and a database of transcriptionally expressed sequences. For functional analyses, one also needs the construction of a bacterial artificial chromosome (BAC) library, transgenic technology, and genome editing. In this context, it is thus essential that the system chosen holds the promise of leading to a wealth of new insights into evolutionary problems on a genetic level. Ideally, one wants to start with a species for which there is already an extensive literature providing an ecological context for the questions one will ask, where the selective pressures are understood and the direction of evolutionary change is known (that is, there is a level of certainty regarding the ancestral versus the derived phenotypes). Ideally, the animal being studied would also exhibit a large number of distinct derived morphological and behavioral traits amenable to study. Moreover, one would like to choose an organism where similar traits have evolved in independent populations, isolated from one another, so that questions of convergence and parallelism can be addressed. For genetic studies to be carried out, it is critical that the animal of choice be amenable to being raised in a lab setting. Advantages in this context include being relatively small and having a rapid life cycle with large numbers of offspring. Most importantly, for carrying out genetic analyses, individuals from phenotypically divergent populations would ideally be interfertile.

Among vertebrates, fish have obvious advantages in these respects. Constitutively aquatic organisms are easy to maintain and can be grown in large numbers in a relatively small space. Fish also generally have large numbers of offspring per generation, compared with most terrestrial vertebrate systems. These advantages led to the development of zebrafish as a model for studying developmental processes. The system moved to the forefront of the field of embryology with the introduction of wide-scale genetic screens (reviewed in Vascotto et al., 1997) and as other tools, such as gene knock-down and whole-genome sequencing have been developed, work with zebrafish has led to major advances in understanding developmental mechanisms. These studies have laid a foundation for understanding morphogenesis in other species, as well as important genetic resources for comparative analyses. The zebrafish itself, however, exhibits little morphological variation in wild populations, making it a less than ideal system for evolutionary studies. Happily, the availability of new genetic tools that can be applied to nonmodel organisms have allowed the strength of genetic manipulation to be applied in evolutionarily relevant systems.

In the last few years, three groups of fish, in particular, have stood out as new, important models for evolutionary genetics: the sticklebacks, the cichlids, and the cavefish. All three systems have been extremely well studied on an ecological level, and each brings distinct advantages for evolutionary genetic analyses. The first to be exploited in this manner was the three-spine stickleback (reviewed in Peichel, 2005). The stickleback exists in two forms, an ancestral marine fish and multiple derived freshwater river and lake fish of the same species. The great advantage of the stickleback system is the extremely large number of independently evolved freshwater benthic populations, the ancestral marine form having invaded many freshwater inlets along both the Pacific and Atlantic rims at the end of the last ice age. A second evolutionary genetic model, with an equally strong history of ecological study, is the cichlid group from East Africa (reviewed in Henning and Meyier, 2014). While there are manifold examples of convergence in this setting as well, the great strength of this system is the diversity of different forms that have evolved. Indeed, there are more than 1500 different species just within the three major lakes studied in East Africa, displaying extremely diverse adaptive radiations, making it an ideal setting for genetic analysis of phenotypic diversification and speciation. The third emerging evolutionary genetic system is the Mexican cave tetra, *Astyanax mexicanus*, the subject of this volume. Perhaps the most important unique aspect of this system is the extreme environment in which it evolved.

An organism faces intense selection pressure when it enters a totally new environment, and the transition from life in the rivers to being entrapped in a cave is about as extreme a change as an organism is likely to encounter in nature. In addition to (and largely as a consequence of) being dark, caves are typically nutrient-poor environments with simplified ecosystems. The good news for an invading species is that there are likely to be few predators; the bad news is that there is very little to eat. Other parameters such as humidity, conductivity, pH, and temperature are also likely to be different from the invader's former home. When placed in an extremely different environment, many traits that were adaptive for an organism in its prior environment will no longer be helpful, and conversely over time, new traits evolve that increase fitness. In these conditions, cave-inhabiting creatures, or troglobites, have evolved an identifiable set of characteristics, including regressive traits such as reduced pigmentation, smaller or absent eyes, and loss of vision; constructive traits such as heightened sensitivity of nonvisual sensory systems and longer appendages, as well as lowered metabolic rate and a range of behavioral adaptations. This same convergent suit of phenotypes is seen in a broad range of phylum, including arthropods, mollusks, chordates, and various worms (Culver, 1982).

While caves are without question extremely harsh environments from the perspective of a newly invading species, less commonly discussed is the fact that caves can also serve as sanctuaries for species fortunate enough to adapt to them. The cave environment, shielded from many transient ecological fluctuations suffered on the surface, can be relatively stable. As such, cave species can outlast their sister surface taxa from which they were derived. Thus, in general, the ancestral species from which modern cave animals descended are no longer in existence. For example, the Olm cave salamander, the only chordate troglobite in Europe, is not just the only extant species in its genus, *Proteus*, it is also the only living European species in the entire Proteidae family.

The olm has a very long history in the scientific literature. It was first described in 1689 in a comprehensive compendium of the geography, fauna, flora, history, folklore, religion, culture, administration, and military exploits of the Duchy of Carniola (in present-day Slovenia) (Valvasor, 1689). While this early treatise misclassified the olm as a baby dragon, it was correctly catalogued as a species within Amphibia by the early herpetologist Josephus Lauranti (Laurenti, 1768). One hundred years later, it was the olm that led Charles Darwin to ponder what one now refers to as regressive evolution, the loss or reduction of structures in the cave environment or phenomena that he attributed to disease:

> *Far from feeling surprise that some of the cave-animals should be nearly anomalous...as is the case with blind Proteus with reference to the reptiles of Europe, I am only surprised that more wrecks of ancient life have not been preserved, owing to the less severe competition in which the scant inhabitants of those dark abodes will have been exposed.... As it is difficult to imagine that eyes, although useless, could be in any way injurious to animals living in the darkness, I attribute their loss wholly to disuse.*
>
> Darwin (1859)

To test these ideas, and to more broadly study the evolution of novel traits that evolved to allow survival in the unique cave environment, one would need a model system where, unlike the olm, the free-swimming ancestral surface morph still exists, where in fact the model is still interfertile with the cave forms, allowing genetic analysis. Moreover, ideally one would want a system where multiple isolated, independently invaded caves exist to allow study of parallelism and convergence. And, of course, one would want an animal easily reared in a laboratory setting. In short, one would want *A. mexicanus*, the Mexican cave tetra.

Astyanax has been bred and studied in the laboratory since 1947 (Breder and Rasquin, 1947). This volume illustrates how far the system has come since that time, to the point where it has now emerged as one of the most important vertebrate evolutionary genetic systems. Among its strengths is the fact that *A. mexicanus* is the most well studied cavefish system on an ecological level. The first section of this book provides that critical context, summarizing the geology, ecology, and biodiversity of the cave system in the Sierra de El Abra mountains where *Astyanax* resides. As a further background, this section closes with a chapter on the complex evolutionary history of *Astyanax* itself. The second section of the book focuses on genetics and genomics, aspects that have dramatically opened in the last decade through the advent of tools, such as plentiful genetic markers, a genetic map, a genome sequence, transcriptional profiling, methods for misexpression, and genome editing. Genetic studies have been complemented by detailed developmental "evo-devo" studies. These are reviewed in the third section, focusing on key morphological traits, such as loss of pigmentation and vision, craniofacial changes in both skeletal and sensory structures, and neuronal adaptations. Most recently, behavioral traits have been investigated in addition to ongoing work on morphological traits. The final major section of this book examines the range of cave-specific behavioral traits that have been studied thus far.

Taken together, this book provides a comprehensive look at the state of research on this important model system. It is hoped that it will provide a foundation upon which current and future generations of *Astyanax* researchers can build to gain deeper insight into adaption to the cave environment, and in so doing, provide a deeper understanding of the genetic architecture of evolutionary change.

REFERENCES

Breder, C.M., Rasquin, P., 1947. Comparative studies in the light sensitivity of blind characins from a series of Mexican caves. Bull. Am. Mus. Nat. Hist. 89, 323–351.

Culver, D., 1982. Cave Life Evolution and Ecology. Harvard Press, Cambridge.

Darwin, C., 1859. On the Origin of Species by Means of Natural Selection, or the Preservation of Favoured Races in the Struggle for Life. John Murray, London.

Henning, F., Meyier, A., 2014. The evolutionary genomics of cichlid fishes: explosive speciation and adaptation in the post-genomic era. Ann. Rev. Genomics Hum. Genet. 15, 417–441.

Huxley, J., 1942. Evolution: The Modern Synthesis. Allen and Unwin, London.

Laurenti, J.N., 1768. Specimen medicum: exhibens synopsin reptilium emendatam cum experimentis circa venena et antidota reptilium austriacorum. John Murray, London.

Peichel, C.L., 2005. Fishing for the secrets of vertebrate evolution in threespine Sticklebacks. Dev. Dyn. 234, 815–823.

Valvasor, J.V., 1689. The Glory of the Duchy of Corniola, Book III. Erazem Francisci, Neuremberg.

Vascotto, S.G., Beckham, Y., Kelly, G.M., 1997. The zebrafish's swim to fame as an experimental model in biology. Biochem. Cell Biol. 75 (5), 479–485.

Part I

Ecology and Evolution

Chapter 1

Cave Exploration and Mapping in the Sierra de El Abra Region

William R. Elliott
Association for Mexican Cave Studies, Missouri Department of Conservation (retired), Jefferson City, Missouri, USA

INTRODUCTION

Here I explore the worlds of biology and caving, and summarize what is known about the Mexican cavefish and its habitat. I will discuss the history of discovery, exploration, and mapping of caves in the Sierra de El Abra region (Figures 1.1 and 1.2). Much of this fieldwork was driven by an interest in the cavefish by about 200 biologists, geologists, and cavers (speleologists), who often worked together. Biologists and geologists made the first cavefish discoveries in the region. Only a few of the fish caves can be accessed on foot—vertical caving techniques and training are required in most. Many of the pit caves proved to be too challenging for academics. The cavers were younger explorers and adventurers, some of them graduate students excited by large, deep caves. Some of the professors became proficient in vertical caving, and some of the cavers became cave biologists. Americans, Canadians, Europeans, and Mexicans sometimes worked together in the field and laboratory. It also was a cultural phenomenon; the northerners learned more Spanish, fell in love with Mexico, and worked to create international goodwill. The teams found over 200 caves in the El Abra region, but just 29 of them are known to contain the Mexican cavefish.

Mexico is home to at least seven known species of cavefishes. These cavefishes have reduced or nearly absent eyes and pigment, and they have evolved from six families from widely separated areas: Characidae, Ictaluridae, Pimelodidae, Poeciliidae, Bythitidae, and Synbranchidae (Reddell, 1981). In this book, we refer to cavefishes of the species *Astyanax mexicanus*, which include the obsolete genus, *Anoptichthys*, as the "Mexican cavefish" (Figure 1.1). Whatever the Latin name may be, the Mexican cavefish is an evolving new species that is separating from its river form. The cave form can be purchased in aquarium shops and is easy to keep and breed. The aquarium breed came

Biology and Evolution of the Mexican Cavefish. http://dx.doi.org/10.1016/B978-0-12-802148-4.00001-3
© 2016 Elsevier Inc. All rights reserved.

FIGURE 1.1 *Asytanax mexicanus*, the Mexican cavefish from Cueva de El Pachón. *By Jean Louis Lacaille.*

FIGURE 1.2 Aerial photo looking south along the crest of the Sierra de El Abra. *By Robert W. Mitchell.*

from La Cueva Chica, described below; it is a hybrid between the river and cave forms (see the chapter on ecology and biodiversity).

A large technical literature exists in biology and speleology (caving) about caves and cavefishes in the Sierra de El Abra region of northeastern Mexico, also referred to as the "Huastecan Province." (The Huastecs are a group of native Americans in that area, whose language is related to Mayan.) Over 530 papers and reports have been published on the Mexican cavefish since 1936. A monograph on the cavefish was published by Mitchell et al. (1977). Another important study was John Fish's dissertation (1977, 2004) on karst (limestone cave) hydrology of the region. These monographs are available from the Association for Mexican Cave Studies (AMCS), Austin, Texas at http://www.mexicancaves.org, where thousands of cave maps from throughout Mexico are also available.

I can only present a few maps here and in the next chapter. For additional information and many maps, see *The Astyanax Caves of Mexico* (Elliott, in press).

Figures 1.3 and 1.4 depict the northern and southern parts of the Sierra de El Abra region, about 200km long and 60km wide. See Table 1.1 for a listing of the 29 known fish caves, and Table 1.2 for a list of the larger nacimientos (large springs or resurgences). Another focus of cavefish evolution is in the state of Guerrero, about 400 km to the south of the El Abra, with two populations of *Astyanax aeneus* (Espinasa et al., 2001).

Physiography and Hydrogeology

Mexico is a land of complex geology and many rock types. About 7500 caves have been recorded by the AMCS, ranging through six major karst areas and lava flows with lava tubes (Mejía-Ortíz et al., 2013). Karst is a landscape formed by the groundwater dissolution of soluble rocks such as limestone, dolomite, and gypsum, with underground drainage systems, caves, sinkholes, dolines, and springs. The subject of this book is located in the karstic Sierra Madre Oriental of northeastern Mexico.

During the late Jurassic to early Cretaceous period about 146-100mya (million years ago), a thick series of gypsum, anhydrite, and carbonate beds were deposited in shallow, warm seas in what is now northeastern Mexico. In the middle Cretaceous period, a widespread carbonate platform, or reef complex, grew on top, becoming what is now the El Abra limestone. The Sierra de El Abra is an elongated range along the eastern margin of that platform (Figures 1.3 and 1.4). During the late Cretaceous period (about 100-66mya), the region was covered by thick deposits of shale, impermeable to infiltrating water, unlike limestone. During the early Tertiary period (starting 66mya), the area was folded, uplifted, and subjected to erosion. The shales began to erode away, and the exposed limestone developed into a high-relief karst terrain, formed by the dissolving action of slightly acidic groundwater moving along joints (vertical fractures) and horizontal bedding planes (Fish, 2004). Later volcanic activity in the Gómez Farías area in the north created a ridge (Sierra Chiquita) that guided development of swallet (stream-capturing) caves in the karst valley immediately to the west.

Elevations in the region vary from 35 m above sea level at the Nacimiento del Río Choy in the south on the Gulf coastal plain, to 800m in the Sierra Tanchipa portion of the Sierra de El Abra, and 269 m at Gómez Farías to over 2000m in the Sierra de Guatemala. Annual rainfall in the region varies from 250 to 2500mm and is strongly concentrated from June through October, when large tropical storms come in from the Gulf of Mexico. Hydrogeological studies carried out in the Sierra de El Abra show that large conduits (caves carrying water) have developed, and that large fluctuations of the water table occur because of precipitation. The ancient caves on the eastern crest of the range were part of deep phreatic (below the water table) flow systems that circulated at least 300m below ancient water tables and discharged onto ancient coastal plains that were much higher

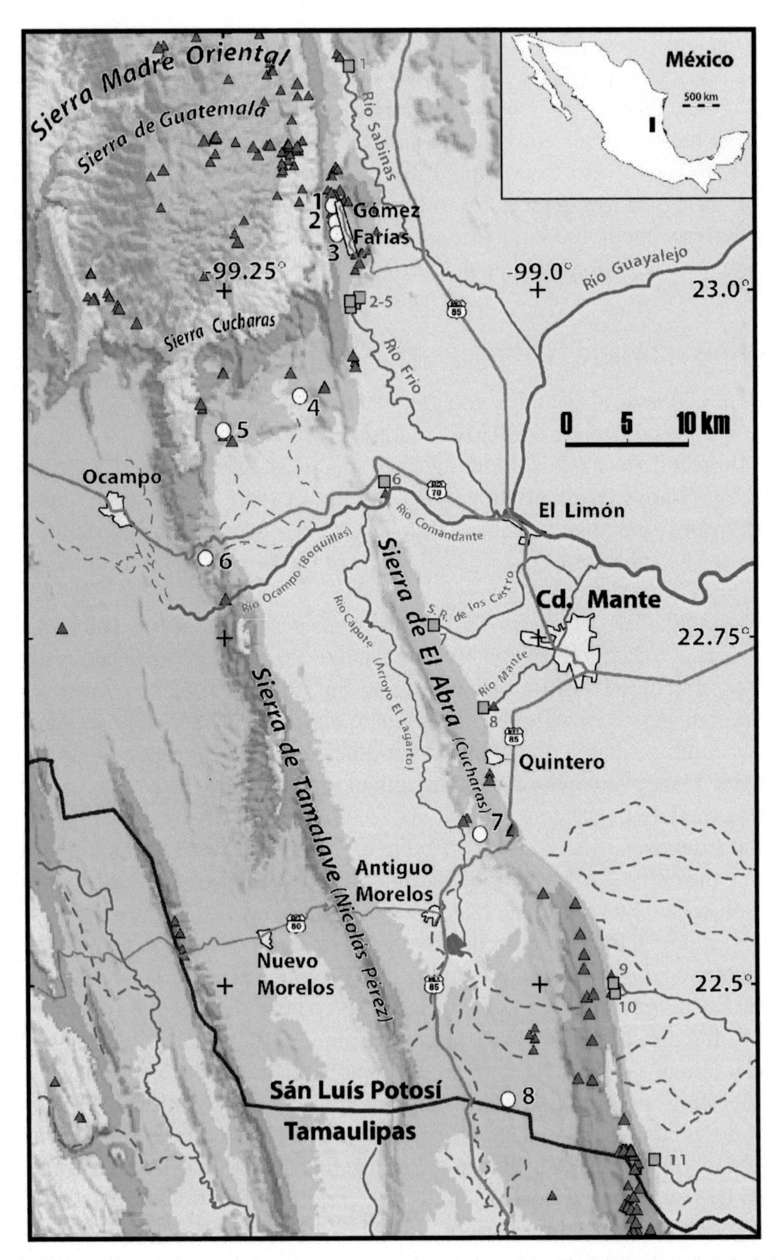

FIGURE 1.3 The Sierra de El Abra Region, northern map. Numbers for fish caves and nacimientos (springs) are in Tables 1.1 and 1.2. North is up, white dots are fish caves, triangles are other caves, and squares are nacimientos. By William R. Elliott based on INEGI 1:1,000,000 topographic map (San Luís Potosí sheet) and AMCS data. *Copyright © 2016 William R. Elliott. All rights reserved.*

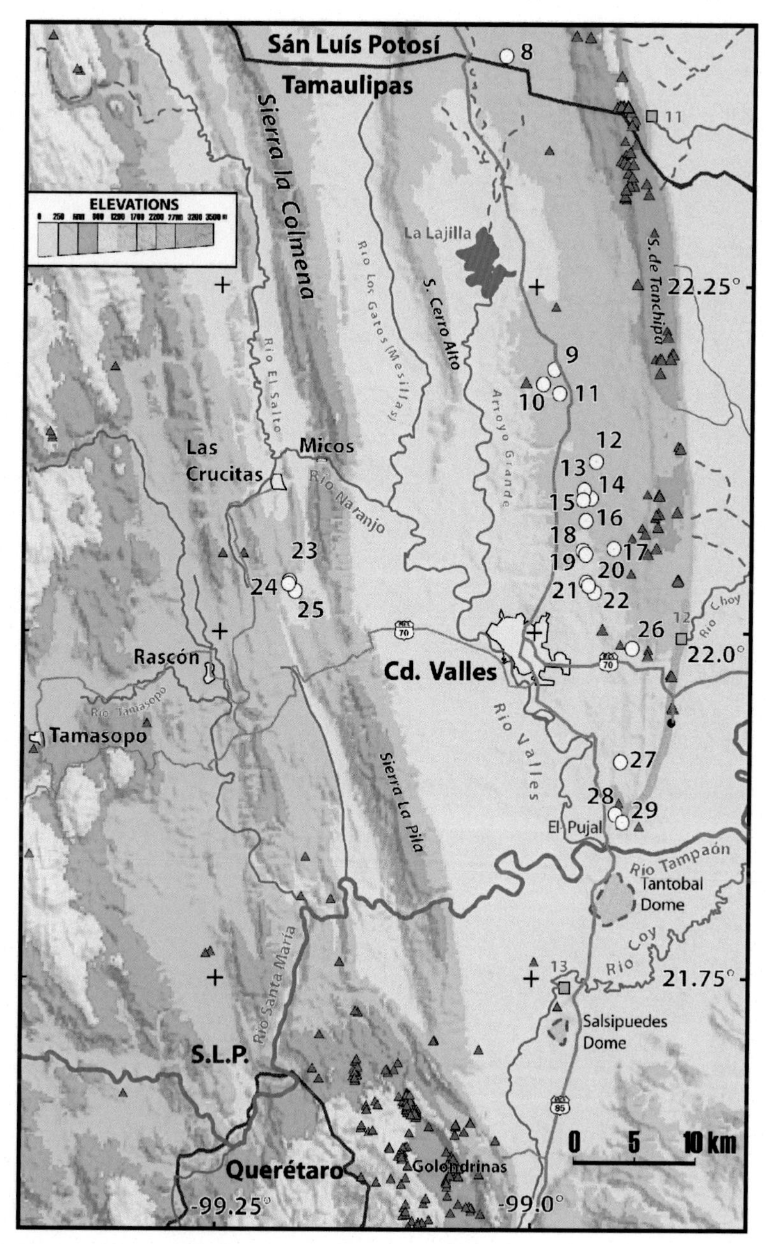

FIGURE 1.4 The Sierra de El Abra Region, southern map. See Tables 1.1 and 1.2. *By William R. Elliott.* *Copyright © 2016 William R. Elliott. All rights reserved.*

TABLE 1.1 The 29 Known *Astyanax* Caves of the Sierra de El Abra Region, with Label Numbers from Figures 1.3 and 1.4 and Dimensions in Meters

Label	Cave	Length	Elevation	Depth	Bottom
1	Sótano (Resumidero) de Jineo	302	292	144	148
2	Sótano del Molino	658	269	138	131
3	Sótano Escondido	100	303	150	153
4	Bee Cave	245	249	119	130
5	Sótano del Caballo Moro	285	320	211	110
6	Sótano de Vásquez	1500	422	277	145
7	Cueva de El Pachón	1000	211	8	203
8	Sótano de Venadito	4419	312	206	106
9	Sótano de Yerbaniz	2027	242	97	145
10	Sótano de Matapalma	1722	242	86	156
11	Sótano de Japonés	4500	243	140	104
12	Sótano del Tigre	3000	246	162	85
13	Sótano de la Roca	20	241	42	199
14	Cueva de los Sabinos	1502	239	96	144

15	Sótano del Arroyo	7202	192	134	58
16	Sótano de la Tinaja	4502	166	82	84
17	Sótano de Soyate	206	293	234	59 (6)
18	Sotanito de Montecillos	1741	190	92	99
19	Sótano de Pichijumo	1330	158	82	76
20	Sótano de Jos	338	176	85	92
21	Sótano de Las Piedras	405	145	47	99
22	Sótano de la Palma Seca	164	152	53	100
23	Cueva de Otates	269	220	15	205
24	Cueva del Río Subterráneo	475	239	32	207
25	Cueva del Lienzo	225	236	23	213
26	Cueva de la Curva	214	132	19	113
27	Sótano del Toro	66	92	5	88
28	Cueva Chica	320	68	19	49
29	Los Cuates (Cueva del Prieto)	400	62	22	40

The elevation at the entrance is in meters msl (above mean sea level), and is based on survey altimeter readings by Mitchell et al. (1977), at 27 caves. The elevations are within ±1 m of current INEGI topographic map elevations. Other cave elevations are based only on topographic maps and may be ±5 m. Cave depth is calculated from compass, clinometer and tape surveys. The bottom elevation (elevation-depth), is usually to the surface of the bottom-most pool.

TABLE 1.2 Thirteen Important Nacimientos (Springs) in the Sierra de El Abra Region, with Label Numbers from Figures 1.3 and 1.4 and Dimensions in Meters

Label	Nacimiento	Mean Flow (m^3/s)	Elevation
1	Nacimiento del Río Sabinas		160
2	Nacimiento del Río Nacimiento		85
3	Cueva del Nacimiento del Río Frío[a,b]		125
4	Nacimiento Florida		85
5	Nacimiento del Río Frío[c]	6	90
6	Nacimiento Riachuelo		110
7	Nacimiento de San Rafael de Los Castro[b]		95
8	Nacimiento del Río Mante	13	80
9	Nacimiento del Río Santa Clara		80
10	Nacimiento del Arroyo Seco		80
11	Nacimiento del Río Tantoán		80
12	Nacimiento del Río Choy	3	35
13	Nacimiento del Río Coy	19	31

The larger springs respond quickly to large storms, and water levels can rise by many meters.
[a]*May flow no longer.*
[b]Prietella lundbergi *site.*
[c]*Mean flow of springs #2, 4, and 5 gauged downstream.*

than the present one. These old caves may have formed by sulfuric acid speleogenesis (cave development), caused by hydrogen sulfide from petroleum deposits ascending and mixing with fresh groundwater, forming dilute sulfuric acid, a phenomenon now known in other karst areas (Palmer and Hill, 2005). Later the geochemistry evolved to the conventional mode of the dissolution of limestone, caused by CO_2 mixing with rain and groundwater to form weak carbonic acid.

The western margin of the El Abra contains younger swallets of the floodwater type (Table 1.1). Stream capture began to occur wherever the overlying San Felipe and Mendez shales eroded to where streams could invade the underlying El Abra limestone at prominent joints. The El Abra limestone probably was exposed first along high ridges before the present-day swallets formed in the lowlands near Ciudad Valles (Fish, 1977, 2004). Stream capture dramatically isolated colonizing fish populations underground while eliminating them from surface arroyos (wet weather streambeds) at the same time, and this occurred repeatedly in different places over a long period. We do not know where the first Mexican cavefish evolved, and the original caves probably eroded away, but the fishes probably spread through subterranean connections to other sites. Many of the fish caves lie under arroyos that may have been perennial streams long ago, but are now subterranean floodwater conduits.

Large springs, or nacimientos (birthplaces), are located along the east face of the El Abra, which discharge huge amounts of groundwater from caves and even from longer connections to the higher ranges in the Sierra Madre Oriental to the west (Table 1.2). Through geologic time, the subterranean connections have grown in size and volume, causing some nacimientos to increase their discharge while others shrank. Karst is three-dimensional, even four-dimensional when one considers the dimension of time. Older, higher elevation connections may have ceased to carry flow except during very large storm events. Some cavefish populations may reconnect with each other during flood times, which can cause groundwater to rise into upper air-filled cave passages. When the water levels drop again, this can strand cavefishes in pools perched as much as 100 m above the usual water table. Some of these perched pools or lakes may become permanent bodies of water, like natural cisterns, such as in Cueva de El Pachón.

The Mexican cavefish is distributed over large distances in 29 known caves that are semi-isolated from each other, but it has not been found in the nacimientos. By semi-isolated, I mean that many caves may only have temporary hydrological connections during and after large storms. It is important to note that cavers and biologists have explored hundreds of caves in the region, so we have a good idea of where the cavefish are absent. So far, they do not occur in waters at elevations above 300 m above sea level, even when suitable habitat is found. As yet, none of the fossil caves on the eastern crest reach water, so they are not cavefish habitat either. Cave divers have not seen the cavefish in the nacimientos on the eastern face of the Sierras. A small, blind catfish, *Prietella lundbergi* (Walsh and Gilbert, 1995), was found in two springs on the eastern face by Hendrickson et al. (2001) (Table 1.2), hinting at a different history of isolation and evolution than *Astyanax*, which is found only in the western, swallet caves or in large sinkhole caves that penetrate to groundwater.

History of Exploration and Mapping

One might say that there have been three generations of cavers and biologists involved in the study of *Astyanax* cavefish. The first generation was from 1936 to 1954, and the second from 1963 to 1998. After 1989, it became increasingly difficult for cavers to access parts of the region with increasing private land development and the establishment of two large bioreserves, Reserva de la Biósfera El Cielo in the Sierra de Guatemala, and Reserva de la Biósfera Sierra de El Abra Tanchipa. These reserves are beneficial for wildlife, flora, and the preservation of many karst features. The former contains several fish caves near Gómez Farías. The latter does not include any caves housing *A. mexicanus*.

In the 1990s, the only field work was mapping in Sótano de Venadito and cave diving in the nacimientos. The *Astyanax* International Meeting (AIM) started in 2009. We are currently in the third generation of Mexican cavefish studies with the advent of modern DNA analysis and the consolidation and interpretation of cave mapping and karst studies.

The first Mexican cavefish was described by Hubbs and Innes (1936) as *Anoptichthys jordani*, based on specimens collected earlier that year by Salvador Coronado in Cueva Chica, a cave located about 1 km north of the village of El Pujal, about 12 km southeast of Ciudad Valles, San Luís Potosí. Álvarez (1946) described a second species, *A. antrobius*, from Cueva de El Pachón, located near the village of El Pachón (Praxedis Guerrero), Tamaulipas. Álvarez (1947) described a third species, *A. hubbsi*, from a large cave, Cueva de Los Sabinos, located 11 km northeast of Ciudad Valles, San Luís Potosí (Mitchell et al., 1977). It was these three "species" to which so much study was devoted by Breder and many others until the late 1960s. Now most biologists consider the Mexican cavefish to be part of the species *A. mexicanus* or *A. fasciatus*.

Cueva Chica probably was not the original site of cavefish evolution in the region, but initial work suggests it represents a younger cave that already contained cavefishes when it was intersected by the Río Tampaón (Mitchell et al., 1977). Originally mapped by Breder in 1940, Elliott and others remapped the cave more accurately from 1970 to 1974, and surveyed overland to locate the nearby tinajas (waterholes), the Los Cuates cave, and Cueva El Mante. More details are in the cave descriptions below, and in my chapter on ecology and biodiversity in this volume.

Sótano del Arroyo and Sótano de la Tinaja were located in 1946 by Benjamin Dontzin and Edwin Ruda, who were commissioned by the American Museum of Natural History (Breder and Rasquin, 1947) to collect additional eyeless characins. These two caves are located near the previously known Cueva de Los Sabinos (see Elliott, in press, and Fish, 1977, 2004, for maps).

Early fieldwork also was done by Mexican scientists like Bonet, Bolívar y Pieltain, Osorio Tafall, Peláez, Álvarez, and American biologists. Although some of the caves were known to local residents and some biologists, scientists were not equipped to explore the vertical caves that require single-rope techniques and training. In the mid-1960s, as a result of exploration and mapping by the Texas-based AMCS, new sightings of cavefishes were reported. These reports came at the same time that Robert W. Mitchell's interest grew in the Sierra de El Abra cave fauna.

Then cavers and biologists from the University of Texas at Austin, Texas Tech University, and other parts of the United States began visiting Mexico. A trip to Xilitla, San Luís Potosí in 1958 inspired Robert W. Mitchell and his associates, followed by others. They were intrigued by Federico Bonet's 1953 papers on the Sierra de El Abra caves and the Xilitla area. Following a trip to the Tequila, Veracruz area in 1962, T.R. Evans organized the Speleological Survey of Mexico, which soon became the AMCS. The emphasis was on publications to inform the world of the cavers' discoveries.

The *Association for Mexican Cave Studies Newsletter* began in 1965. The AMCS Bulletin series began in 1967 with the influential Bulletin 1, *Caves of the Inter-American Highway*, a general guide to caves of northeastern Mexico (Russell and Raines, 1967). In 1967, Sótano de las Golondrinas near Aquismón, the world's deepest pit at that time, was explored and mapped by Evans and others (Figure 1.4). The AMCS work was done mostly by American and, later, Canadian cavers at their own expense. Today, many Mexican cavers are proficient in cave exploration and mapping, and groups from overseas, notably France, Italy, England, and Australia, have made significant discoveries. Some expeditions are multinational.

By 1965, Ed Alexander, David McKenzie, John Fish, Terry Raines, and others were discovering, exploring, and mapping large caves like Sótano del Arroyo, Sótano de la Tinaja, Sótano de Pichijumo, and Bee Cave. Sótano del Arroyo, the most extensive fish cave at 7202 m long, required about 50 cavers to map from 1961 to 1971. John Fish, William R. Elliott, Don Broussard, Neal Morris, and many American and Canadian cavers worked intensively in the El Abra from 1967 to 1974, mapping many caves and studying hydrology and biology. In total, about 150 cavers were involved in mapping the fish caves and assisting scientists. This work culminated in Fish's dissertation at McMaster University, Ontario, Canada (Fish, 1977, 2004), and Mitchell, Russell, and Elliott's monograph on cavefishes (1977).

Robert W. Mitchell's research group at Texas Tech University worked closely with the AMCS (Figure 1.5). Supported by grants, in 1969, Mitchell, Richard Albert, William H. Russell, Francis Abernethy, Don Broussard, Tom Albert, and others made an extensive aerial survey of the Sierra de El Abra region, discovering seven new fish caves. This aerial reconnaissance ended when Albert's airplane crashed in the Sierra Cucharas (foothills of the Sierra de Guatemala). He and his two passengers, Tom Albert and Don Broussard, survived, but it took 2 days to find their way out of the jungle.

Bill Russell, David McKenzie, and other AMCS cavers located many caves by logging back roads, hiking through the thorn forest and the arroyos, and talking with locals. The AMCS and Mitchell's group discovered a total of 23 new fish caves, most of which were explored and mapped over the next 12 years. I was involved as a graduate student in this work from 1969 to 1974. Later, I independently focused on the Sierra de Guatemala from 1978 to 1981. Altogether, I mapped or drafted maps for 17 of the 29 known fish caves.

In 1970, Horst Wilkens and Jakob Parzefall found Cueva del Río Subterráneo near Micos about 16 km west of Ciudad Valles, based upon information from rabies control workers who were looking for bat caves. They visited a few other caves, finding aquatic troglobites, but no cavefishes. Mitchell and Russell found two more fish caves near Micos, Cueva de Otates, and Cueva del Lienzo, which Elliott and others mapped in 1974. These caves contain interesting "half-cavefishes" that are at an early stage of evolution to a cave-adapted fish (Wilkens and Burns, 1972). The Micos area has not been adequately explored for caves

FIGURE 1.5 Robert W. Mitchell's research group at Rancho del Cielo, January 10, 1971. Left to right: Masaharu Kawakatsu (Fuji Women's College), Suzanne Wiley, Mel Brownfield, Jerry Cook, Robert W. Mitchell, William H. Russell, James R. Reddell, Virginia Tipton, William R. Elliott, and Ann Sturdivant. *By Robert W. Mitchell.*

since then. One small pit cave north of the Micos caves, entered by Elliott in 1974, had dangerous "bad air" (high CO_2 and low O_2), and could not be explored.

Explorations in light airplanes over the crest of the El Abra range revealed many large, deep pits. In the mid-1970s, a campaign of "jungle chops" began in which teams of cavers used machetes to chop their way through the jungle to the pits. Long streamers were sometimes dropped from an airplane to mark a route. The opening of new back roads aided in this work, and many new caves were explored and mapped. But no fish caves were found on the crest of the El Abra.

In 1977, cavers completed a 4-year campaign to "bottom" Cueva de Diamante on the El Abra crest. It had taken five expeditions and over 1500 man-hours to reach a depth of 621 m through an extreme obstacle course of tight canyons and razor-lined pits (Atkinson, 2004). At the time, Diamante was the fourth deepest cave known in the Western Hemisphere. The extreme difficulty of the cave has deterred anyone from returning, though several leads remain, including a major passage at −430 m that was taking water. Hoya de Zimapan, 320 m deep, is another major pit cave on the crest that does not go to water.

By the 1980s, the jungle chop era was over, although major pits are still unexplored; one can see them on Google Earth® today. Morris (1989) published an El Abra cave map folio. In 1989, the "Mexspeleo" caver convention was held at the Hotel Covadonga south of Ciudad Valles, and significant mapping was done in Sótano del Tigre (Figures 1.6 and 1.7). The only other significant expeditions since then were the trips led by Don Broussard to Sótano de Venadito (Figures 1.8 and 1.9). This extensive cave was mapped by 13 teams of 22 AMCS

FIGURE 1.6 Sótano del Tigre, Bill Farr on rope at the bottom of the entrance shaft, 1989. *By David Bunnell.*

cavers in 1968-1969, and then in 1989-1998, the longest effort for any cave in the El Abra. The Tigre and Venadito notes were recently drawn up for publication.

Cave divers explored nacimientos in the area from 1978 to 1989. Major dives by Bill Stone, Sheck Exley, Paul DeLoach, Jim Bowden, and others achieved depths to 55 m in Nacimiento del Río Sabinas, 264 m in Nacimiento del Río Mante (a world record at the time, Figure 1.10), 43 m in the Nacimiento del Río Choy against strong current (Figure 1.11), and 76 m in the Nacimiento del Río Santa Clara. These cutting-edge deep dives sometimes required special gas mixtures and long decompression times, and they demonstrated Fish's surmise that the springs had deep circulation under the Sierra, with some "B waters" gaining unusual chemistries from great depths (Atkinson, 2004). None of the divers reported seeing *Astyanax* cavefish in these places, but research dives in the 1990s found the new catfish species, *P. lundbergi*, "the phantom blindcat," at shallow depths in two springs on the eastern face (Table 1.2).

FIGURE 1.7 Jerry Broadus rigging rope in Sótano del Tigre, February 1, 1968. *By Robert W. Mitchell.*

MAPPING AND CARTOGRAPHY METHODS

Cave cartography is basic descriptive science. To explore a cave well, one must map it as one goes. A cave map is essential for route finding and understanding the hydrology, geologic history, and potential connections to other caves. A cave map is an important document in establishing a cave as an actual natural resource. Some of the early maps of fish caves were freehand sketches or compass and pace surveys, which are quite inaccurate in larger caves. Cave maps are used to note the locations of many points of scientific interest, including bat roosts and habitat. The special problems of speleology and cave biology are found in an extensive literature, but a good introduction may be found in Elsevier's 2007 *Encyclopedia of Caves and Karst*, in which Elliott discusses cave protection.

FIGURE 1.8 Sótano de Venadito aerial photo, with the shadow of the wing of Richard Albert's airplane, 1969. *By Robert W. Mitchell.*

FIGURE 1.9 James Reddell and John Fish at the low side drop of 46 m into Sótano de Venadito, 1969. *By Robert W. Mitchell.*

FIGURE 1.10 Nacimiento del Río Mante, 1969. *By Robert W. Mitchell.*

FIGURE 1.11 Nacimiento del Río Choy, March 2013. *By William R. Elliott.*

The 1940 expedition to Cueva Chica made the first published cave map from the region (Bridges, 1940; Breder, 1942). Charles Breder used a small plane table with a drawing sheet on a tripod, and an instrument for drawing lines and measuring distances optically to a stadia rod, which he sighted through a telescopic alidade. This method works well for outdoor surveys, but is not suited for dark caves with rugged terrain. Their map became increasingly inaccurate as errors in the drawing accumulated toward the lower end of the cave. The author remapped the cave from 1970 to 1974 using standard cave surveying equipment at that time: a steel survey tape and a Brunton pocket transit, which is a small magnetic compass with sights and a clinometer for measuring slopes. Similar equipment is still used by geologists, the military, and other field personnel. In the 1970s and 1980s, most cavers changed to the Suunto or similar liquid-filled compass and clinometer, which makes point-to-point, handheld surveying easier. In either case, meticulous notes are kept in a book with taped distance, magnetic azimuth (bearing), inclination, and LRUD (left, right, up, down) distances at each numbered station. Carefully scaled drawings are made in plan, profile, and cross-section views. Three or four cavers comprise a survey team, and multiple teams or trips may be needed to map the entire cave.

In the 1970s, cave survey data were still "reduced" to rectangular coordinates (x, y, z) using pencil, paper, and trigonometric tables, or plotted with a drafting machine. A slide rule or a calculator was used for calculations. AMCS cavers converted to the metric system in the 1980s, and cavers developed sophisticated mainframe computer programs for processing survey data from large caves, and plotting it on large plotters. Profiles, longitudinal sections of a cave, became easier to plot, and it became standard to depict caves in plan and profile views, especially pit caves requiring rope. Cave maps became better models of the caves, allowing scientific inferences to be made about hydrogeologic history.

Today the author and many cavers use the *Walls* cave survey program, one of several that allows one to plot the plan and profile survey lines along with LRUD data, and send it to a vector drawing program like Adobe Illustrator for the final drawing

GEOGRAPHIC INFORMATION SYSTEMS

Seeing caves in their true relationship in a geographic information system (GIS) is important to understand how they formed and where groundwater may flow. Groundwater connections are usually elucidated with dye tracing, but there has been only one water trace in the Sierra de El Abra from Sótano de Japonés (Fish, 1977, 2004). Instead, we have relied on elevations, hydrographs, and water chemistry, but now from cave maps we can see structural trends of cave clusters and flow directions. Until the late 1970s, we had no topographic maps of Mexico, so we surveyed overland between caves and landmarks to accurately locate caves. Mitchell et al. (1977) used a precision surveying altimeter tied to Mexican government benchmarks to measure elevations at cave entrances. By the 1980s, DETENAL topographic maps from the Mexican government became publicly available, which improved our efforts to record cave locations long before the GPS (global positioning system) was available. In recent years, some cavefish researchers have used GPS receivers to obtain satellite-based coordinates for *Astyanax* caves. Even these precision instruments can get wrong data if the geographic datum and the position error are not recorded. Today we can download Mexican topographic maps from the Instituto Nacional de Estadística, Geografía e Informática (INEGI), the federal mapping agency, but they must be converted to a different datum and projection for use in most GIS programs.

Initially, I developed cave location data using Google Earth, then it was converted to *WallsMap* format. I developed cave area maps using INEGI base maps in *WallsMap* with shape files from the *Walls* program or *Quantum GIS®* (2014). Researchers may obtain limited cave location and GIS data from the author and AMCS. Such data are not posted in public because of potential conservation problems.

CAVE DESCRIPTIONS

Numbers correspond to labels in Figures 1.3 and 1.4. Maps may be found at AMCS online, in Fish (2004), Elliott (in press), and in other publications cited below.

GÓMEZ FARÍAS AREA (SIERRA DE GUATEMALA)

The northern group of fish caves in the Sierra de Guatemala, Sierra Cucharas, and Sierra Tamalave may be separated hydrologically from the southern Sierra de El Abra fish caves by the large resurgences just south of Gómez Farías—Nacimiento del Río Nacimiento, Nacimiento del Río Frío, Nacimiento Florida, or at other resurgences farther south (Table 1.2). Some authors have debated whether or not the Cañon de la Servilleta of the Río Ocampo (Boquillas), which passes through the Sierra Cucharas, is a barrier to cavefishes. From the INEGI Loma Alta 1:50,000 topographic map, I found that the bottom of this Cañon runs west to east from 130 to 90 m msl (above mean sea level). Most of the nearby nacimientos are still below parts of the Cañon, but some are above it. An old cave, now dry, Cueva del Cañon de la Servilleta, is in the north wall of the Cañon. It is possible that water-filled caves still pass beneath the Cañon. The Cañon, per se, may not be much of a barrier in karst, which has three-dimensional flow paths that can change as water rises and falls after storms. Dye tracing studies during low and high water times might settle this problem, but it will be complex, owing to the many nacimientos in this area.

1. *Sótano* (*Resumidero*) *de Jineo* (Table 1.1, Figure 1.12) is the northernmost of three deep fish caves at Gómez Farías. Jineo corkscrews clockwise at −120 m, then bifurcates. The bottom pools are perched at 144 m deep (148 m msl), about 63 m above base level, which is about 85 m msl at the closest resurgence, Nacimiento del Río Nacimiento. Water flow apparently is west, and the end of the cave lies only 105 m from the end of Sótano de Molino, and I believe the two caves to be part of one system. A 1971 map showed a west-trending, muddy terminal passage, but a 2005 resurvey found mud blocking this route and a lower pool in another branch. An arroyo, lacking fishes, about 100 m long, drains a small area into the cave during wet weather. Some other nearby pit caves, checked in dry weather, had no cavefish pools, because they bottom above 200 m msl.
2. *Sótano de Molino* is the most extensive cave at Gómez Farías, with pools on four levels. Molino was mapped in 1971 (Elliott, 2002). Multiple trips by caving groups found cavefishes at the bottom, but sometimes cavefishes

FIGURE 1.12 Bill Elliott rappels into Sótano de Jineo at Gómez Farías, Mel Brownfield looks on, May 18, 1971. *By Francis Abernethy.*

occurred in debris-laden upper level pools at −73 to −77 m. This indicates the dynamic nature of the aquifer, with cavefish carried at least 70 m upwards by rising groundwater during wet times. The arroyo to the cave receives runoff from fields and the western side of the Sierra Chiquita. The arroyo is several hundred meters long and is the largest captured arroyo in the area, but it is not inhabited by fishes. Erosion in the arroyo produced an entrance headwall 10 m high. The Molino-Jineo system probably flows south under the valley toward the nacimientos about 8 km away.

3. *Sótano Escondido's* entrance is divided by a natural arch and is overgrown, hence the name Escondido (hidden). The 150-m-deep cave has seven pitches requiring equipment, with a counter-clockwise corkscrew leading to an east-trending terminal cavefish pool. This pool is perched well above the local base level at 153 m msl. Most of the fishes seen here were small and of uniform size, indicating that a few were transported upward by rising waters to be trapped in the pool, where they gave rise to the population. The shallow arroyo takes runoff, but it has no surface fishes. The cave runs to the east, but it could easily turn west again below the sump and join with the flow from the Molino-Jineo system.

Chamal-Ocampo Area (Sierra Cucharas and Sierra Tamalave)

4. *Bee Cave* (Sótano de las Colmenas) is a large collapse pit with a small pool at the bottom, with few cavefishes visible at any time.
5. *Sótano del Caballo Moro* is in a dolina (large sinkhole) in a saddle between two hills. The bottom is about 20 m above the Río Frío resurgences 14 km away to the northeast. The lake and stream at the bottom carries a large volume of water into a sump and toward the northeast. Sunlight hits a portion of the stream, where small-eyed fishes segregate from cavefishes (Mitchell et al., 1977; Elliott, 2002; Espinasa and Borowsky, 2000).
6. *Sótano de Vasquez* has the highest elevation entrance and is the deepest fish cave in the region. It has six levels, indicating a great age relative to other caves (Mitchell et al., 1977; Mothes and Jameson, 1984).

Northern El Abra (Sierra de El Abra or Sierra de Tanchipa)

7. *Cueva de El Pachón* is one of the three classic cavefish sites. It is the type locality, the place from which a new species is described, of Alvarez's *Anoptichthys antrobius*. The relatively high elevation and perched lake, 120 m above local base level at the Río Mante, may indicate a more ancient origin than most fish caves in the region. Additional upper-level water passage was mapped by Luis Espinasa et al. in 2003 (Espinasa, 2009).
8. *Sótano de Venadito* (Figures 1.8 and 1.9) is a large system developed on four levels at or near the groundwater divide between the northern and southern El Abra region. It flows southwest toward the La Lajilla lake and the headwaters of Arroyo Grande. It may also overflow into a north passage during high water, and thence possibly to small resurgences 12 km northeast, or to the Nacimiento del Río Mante on the east face of the range. A new map of the cave will appear in Elliott (in press).
9. *Sótano de Yerbaniz* (Figures 1.13–1.18) has three levels and floods violently. It has the largest catchment basin (area that captures runoff) of all the El

FIGURE 1.13 Sótano de Yerbaniz entrance, 1969. *By Robert W. Mitchell.*

FIGURE 1.14 Francis Rose examines flood debris in Sótano de Yerbaniz, 1969. *By Robert W. Mitchell.*

Abra caves at 16 km^2. Eyed fishes are sometimes found on Levels 1 and 2, cavefishes and eyed fishes on Level 3. The cave has a rich fauna.

The Yerbaniz cluster contains three hydrologically related caves: Yerbaniz Matapalma, and Japonés. Downstream are the older Sótano de Matapalma (1.4 km away, straight line) and Sótano de Japonés (2 km away). Yerbaniz is the youngest and probably the most hydrologically active of the three fish caves. Heretofore, authors have assumed that the Yerbaniz cluster flows only to the east. I can hypothesize groundwater flow paths from these caves based upon known flow paths on the cave maps, cave morphology, and elevations (Figure 1.18). They may flow to unknown wet-weather resurgences or risings on the Arroyo Grande 4-8 km southwest of the cluster. Dye tracing is needed to delineate these flow paths. Other overflow paths could also connect to the Sabinos cluster about 5-8 km to the southeast.

The name, *yerbaniz*, refers to an herb, the flowers of which are used in religious ceremonies on the Dia de Los Muertos. It is also spelled yerbanís or hierba anís.

Sótano de Yerbaniz is in the Arroyo Yerbaniz, about 22 km north of Ciudad Valles, San Luís Potosí (S.L.P.) Yerbaniz is described in detail here, as it is representative of a swallet cave that has captured a stream and its surface fishes, and is still capturing them. The cave was described in detail by Mitchell et al. (1977), and Elliott (2014) published a multi-colored map of the cave. That map is available on Elsevier's companion site and in the e-book. Figures 1.15 and 1.16 are simplified versions of the map for the printed volume of this book. Figure 1.17 is a legend of AMCS cave map symbols.

The cave was discovered from pilot Richard Albert's airplane on January 25, 1969 by Robert Mitchell, Richard Albert, Francis Rose, and Tom Albert. It was an accidental discovery after scouting for cavefish caves, as they

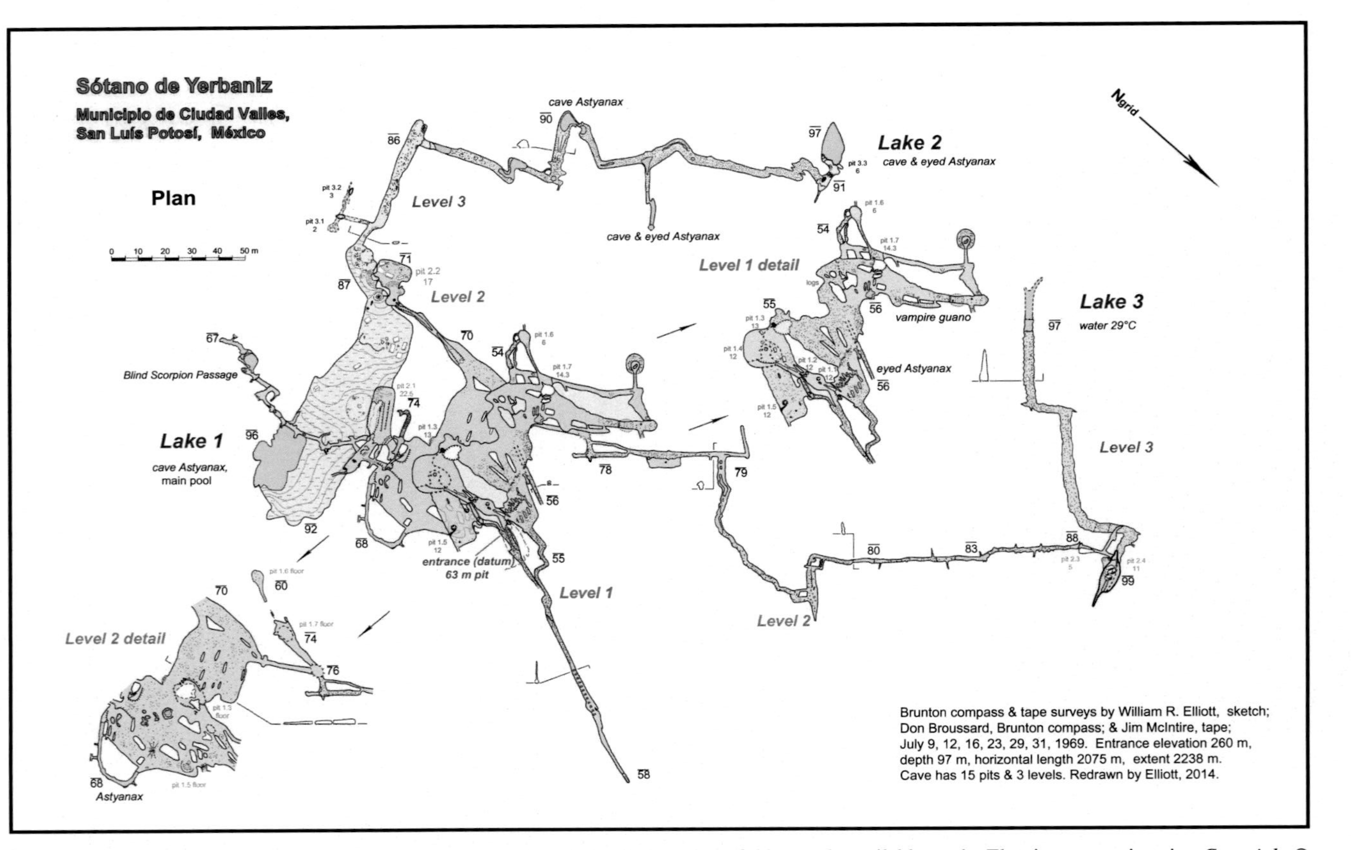

FIGURE 1.15 Sótano de Yerbaniz plan. By William R. Elliott. A detailed version of this map is available on the Elsevier companion site. *Copyright © 2015 William R. Elliott. All rights reserved.*

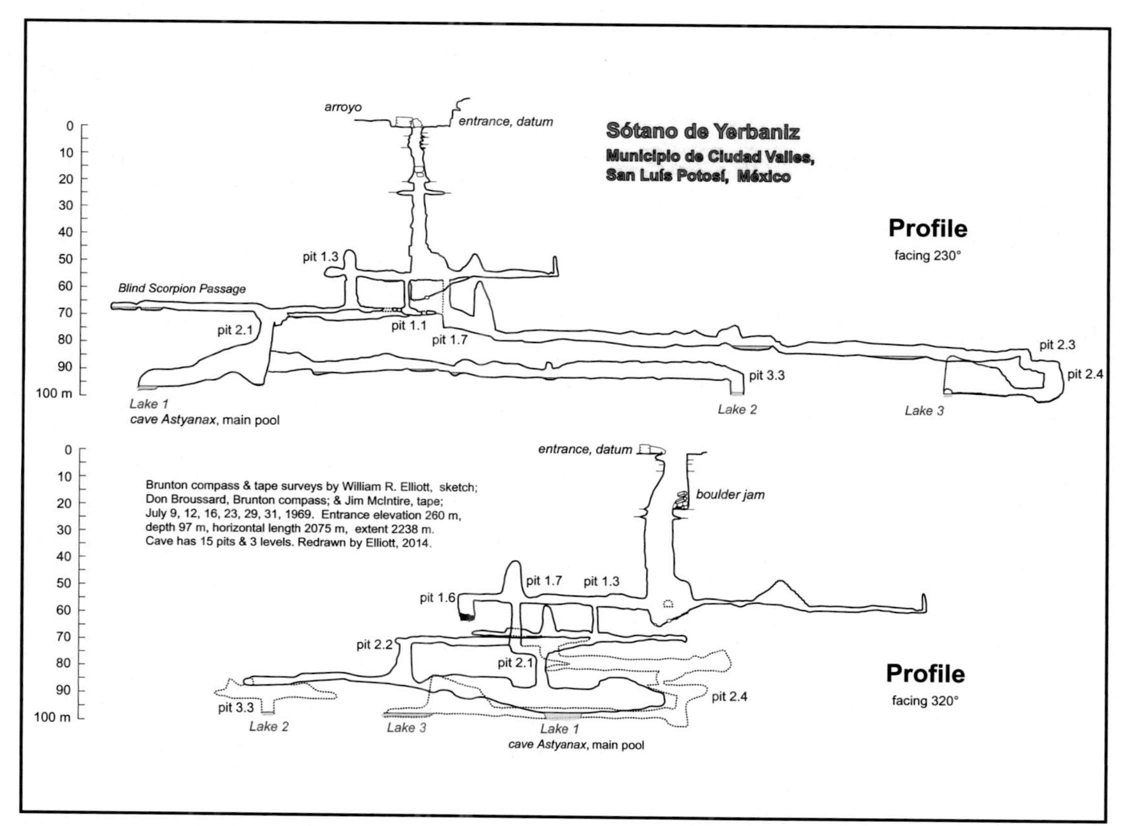

FIGURE 1.16 Sótano de Yerbaniz profile. *By William R. Elliott. Copyright © 2015 William R. Elliott. All rights reserved.*

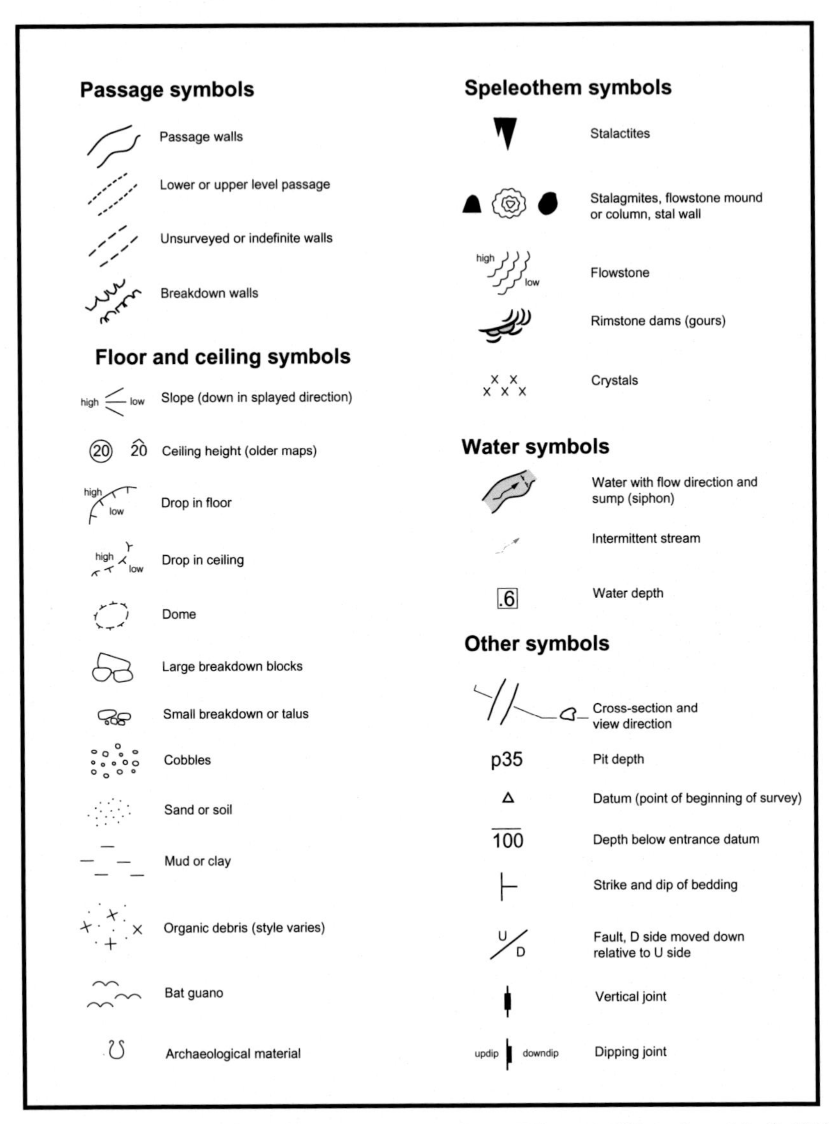

FIGURE 1.17 Legend of AMCS cave map symbols. *By William R. Elliott. Copyright © 2015 William R. Elliott. All rights reserved.*

searched at dusk for the landing strip at nearby Ponciano Arriaga. Little did they know they had found one of the most complex and biologically rich caves of Mexico.

Tony Mollhagen and Francis Rose, of Mitchell's research team from Texas Tech University, descended the 63-m entrance pit on January 28, 1969. On January 29, 1969, Mitchell, Rose, Richard, and his son Tom Albert entered the cave with more rope, explored parts of the first and second levels, and collected eyed, surface

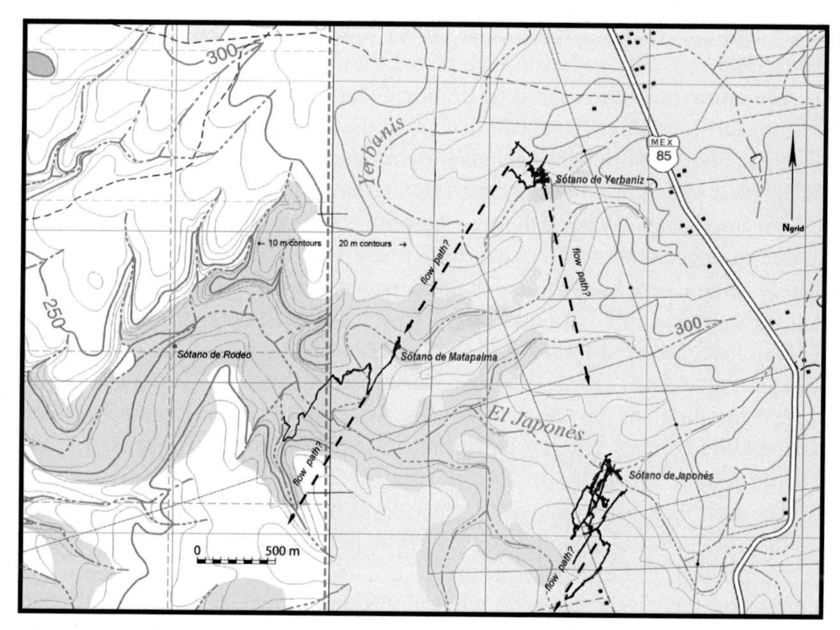

FIGURE 1.18 Map of the Yerbaniz cluster: Sótano de Yerbaniz, Sótano de Japonés, and Sótano de Matapalma, showing hypothesized groundwater flow paths based on cave morphology and elevations. *By William R. Elliott based on 1:50,000 INEGI topographic maps and AMCS data. Copyright © 2015 William R. Elliott. All rights reserved.*

Astyanax in pools on Level 1. On January 31, 1969, the cave was explored to Level 3 by Jerry Broadus, David Honea, Ann Lucas, Russell Harmon, Tony Mollhagen, and Joe Cepeda, who collected several cavefishes in Lake 1 on Level 3, about 96 or 97 m below the entrance. On February 2, 1969, a sizeable collection of cavefishes was made from the same lake by Bob Mitchell and Bill Russell.

The map of Yerbaniz went unpublished for 45 years, but it was worth the effort to redraw the July 1969 survey by Don Broussard, Jim McIntire, and me. I originally drafted the inked map on two large sheets in 1969-1970, but a planned series of papers on the *Astyanax* caves was not completed. Fortunately, I archived my survey notes and the two map sheets with the AMCS. Yerbaniz is the most complicated cave map I have ever drawn; it took 6 days to map and months to draw using a computer.

Surface fishes inhabit Arroyo Yerbaniz in abundance. The arroyo does not, however, support a permanently flowing stream, but in its course, there are pools in which the fish populations are maintained, some of which may be unseen deep within the bed of the arroyo.

Yerbaniz has 15 pits including the entrance, connecting three levels and three lakes on Level 3, which is in two parts. This is a challenge to represent on a two-dimensional map. I rendered the new map in three translucent colors to depict the overlapping levels, and I labeled the pits in three series. Pits from

Level 1 to Level 2 are numbered as pit 1.1 through pit 1.7. Level 2 pits are pit 2.1 through 2.4. Pits on Level 3 are pit 3.1 through 3.3.

The entrance of the cave is an elliptical slot about 10 m long by 1-4 m wide (Figure 1.13). It lies in the northern edge of the Arroyo Yerbaniz at an elevation of about 263 m. Although a relatively young opening, the entrance is of sufficient size to capture all the water flowing down the Arroyo Yerbaniz except at high flood stage. Water can move past the entrance at such times, as demonstrated in September 1969 when Bill Russell visited the cave entrance immediately after a very heavy rain. So much water was being carried in the arroyo that the entrance could not take all of it. Water about 1 m deep was flowing by the entrance, where a large whirlpool took water down, and at another point, mist shot up about 12 m as air was exhausted from the cave. The results of such violent flooding are seen within the cave, with cobble piles, large palm trunks wedged into corners, and log jams scattered in the floodwater mazes of Levels 1 and 2 (Figure 1.14).

The cave is a joint-controlled, three-level, floodwater maze with one major flow path (to Lake 1) and two overflow paths (to Lakes 2 and 3). The entrance pit drops to Level 1, which has joint sets trending at about 25° and 330°, witnessed by bedrock pillars in wide rooms and in narrow passages. Level 1, consisting of several large rooms, small intersecting passages, and one long northeast-trending passage, lies at 54-56 m beneath the datum (point of beginning of the survey at the entrance).

Level 2 at about −68 to −88 m has two large rooms (depending upon one's perspective), many small intersecting passages, one long northwest-trending passage to a second Level 3 and Lake 3, and a shorter south-trending passage (Blind Scorpion Passage). Joint sets trend at 0°, 45°, and 75°, with pillars trending along those joints.

Level 3, consisting of the Lake 1 Room the size of a football field and two long overflow passageways, lies about 91-97 m beneath the entrance. Joint sets trend at 45°, 75°, and 330°. The deepest points in the cave (surfaces of three lakes on the third level) lie 96-97 m beneath the entrance. Based upon mapping, these three lakes are all at about 145 msl and probably are connected via submerged passages. They are about 11 m above the bottom of Sótano de Matapalma, and probably are near local base level. The water in Lake 3 of Yerbaniz is warm, about 29 °C in July 1969. Sótano de Japonés, in a different arroyo, has a deeper bottom at 104 m msl.

I entered Yerbaniz 14 times over 3 years. On July 31, 1969, the last day of the survey, I was fortunate to discover a delicate blind scorpion on Level 2 (see Blind Scorpion Passage on map). Or rather the tiny, translucent scorpion found me, on my right thigh! I had been brushing off little *Brackenridgia* isopods as we surveyed the tubular passage, crouching against the wet walls. I was about to brush another one off when I looked down and saw what was to become the holotype specimen of a new species to science. I swore out loud and jumped up and down three times, then I collected it! Dr. Mitchell later described it as *Typhlochactas elliotti*. He found that it was similar to two other blind cave scorpions in Mexico, which he had published. Return trips found only three specimens total. Years later, it is still considered the world's most troglomorphic (cave-adapted) scorpion, and it has been moved to the genus *Sotanochactas* as the only known species in that genus. This was an exciting find, but especially since we did not know that blind scorpions could occur in the lowland tropics—the others were from montane areas. This began to change our ideas about cave animal evolution in the tropics, and now we understand that troglobites are often common in the tropics.

Some of us also found a new genus and species of cave schizomid, *Agastoschizomus lucifer*. Schizomids are tiny arachnids, relatives of the whip-scorpions, but this one is large by comparison. In 1970, Suzanne Wiley and I witnessed something strange on Level 2 while we were checking fish pools in the cave for her master's thesis. We found an *Agastoschizomus lucifer* sitting in a tiny rimstone pool on the wall. We thought it must have drowned, but when we gently poked it, the creature walked up out of the pool onto the wall. They are not supposed to be amphibious, but maybe in that environment, the scorpions and schizomids have become amphibious. The air is so saturated that it may not matter to the animal. On a return trip to the cave, Suzanne's carefully documented fish pools had been totally blown out by a flood. That told us how tough it is for fish in upper-level pools. Her thesis study was wrecked, so she ended up doing one on respiratory rates of *Rhadine* cave beetles in Texas.

10. *Sótano de Matapalma* (Figure 1.18) is a large cave with one level. It now lacks much arroyo runoff, because of the more recent upstream capture at Sótano de Yerbaniz, but it probably conveys deep groundwater to the southwest via a major unexplored stream passage.
11. *Sótano de Japonés* (Figure 1.18) is a large, complex cave on four levels. Its map shows that it conveys water to the southwest, and that may continue to Arroyo Grande. It captures runoff from the Arroyo Japonés, and the subterranean flow path would be on a similar, southwesterly flow path as Yerbaniz and Matapalma. John Fish conducted a *Lycopodium* spore water trace from Japonés with inconclusive results (few spores recovered) at the Nacimiento del Río Choy. Perhaps those results could be explained if most of the water flowed to the Arroyo Grande via risings in the streambed or wet-weather resurgences, although none are known to me at this time (Figure 1.18).

Sabinos Area

12. *Sótano del Tigre* (Figures 1.6 and 1.7) is a deep, biologically rich pit cave with three levels. Cavefish occur on Levels 2 and 3. Its direction of flow is to the south or southwest, and it probably feeds to the Los Sabinos cluster of caves 3 km to the south. The "Bats and Ricinuleids Passage" is a fossil stream passage that now contains bats and a thriving population of the rare arachnid, *Cryptocellus osorioi*, in copious bat guano. A new map will appear in Elliott (in press).

13. *Sótano de la Roca* is a small pit cave 800 m upstream of Sótano del Arroyo and the Sistema de Los Sabinos.
14. *Cueva de los Sabinos* is a classic cavefish cave and type locality of Álvarez's *Anoptichthys hubbsi*. It is a large, complex cave on one level, with a sump connecting to Sótano del Arroyo. It is part of the *Sistema de Los Sabinos*, comprising Cueva de los Sabinos, Sótano del Arroyo, and Sótano de la Tinaja.
15. *Sótano del Arroyo*, a major cavefish site and the longest cave in the El Abra region. A complex, dendritic cave on three levels, it receives flow via its arroyo pit entrance and via a sump from Sabinos, thence conducting flow to the east.
16. *Sótano de la Tinaja* may be the oldest cave in the *Sistema de Los Sabinos*, as it is farthest downstream and its surface pit has been eroded to a walk-in entrance. It is developed on two levels. The upper level Water Passage to the left of the entrance receives water flow from the west and transmits it to the north to a large lake passage. Another Lake Passage in a different branch may also connect. These two lake passages probably connect via sumps to the south end of Sótano del Arroyo. The lower level of Sótano del Arroyo, the Downstream Canyon, ends in a sump, the Lake Room Siphon, at −82 m, or about 84 m msl.
17. *Sótano de Soyate* is the deepest cavefish site in the southern Sierra de El Abra. Its 195-m entrance pitch leads to two short pitches to a large, clear lake at −234 m. In the lake resides a large population of cavefishes, studied by Elliott (Elliott, 1970; Mitchell et al., 1977). In Soyate, the surface of the lake was at 59 m msl, and the bottom of the lake was at 6 m msl, 29 m below the surface outlet of the Nacimiento del Río Choy. No flow direction was observed, but because of its deep bottom, the cave likely is a major conduit to the south and the Nacimiento del Río Choy. See further discussion of the cave's ecology in Chapter 2.
18. *Sotanito de Montecillos*, the northern half of the Sistema de Montecillos, connecting via a sump to Sótano de Pichijumo.
19. *Sótano de Pichijumo* (*Montecillos*), the eastern half of the Sistema de Montecillos. Three levels and three flow paths exist in the system. Floods captured at the sotanito entrance go two directions; the lower route flows north to a sump pointing east, meeting subterranean water that flows from the northwest. An overflow route connects to Pichijumo via another sump, and is carried east via the Lake Passage to the deepest sump at −82 m in Pichijumo, which points east.
20. *Sótano de Jos*, a young floodwater cave on one level, from its initial orientation, it turns 180° under the arroyo and drains to the north-northeast, possibly converging with flow from Soyate.
21. *Sótano de Las Piedras*, about 290 m downstream from Jos, is developed on two levels. Like Jos, it drains to the north-northeast.
22. *Sótano de la Palma Seca* is about 740 m from Piedras. It is developed on two levels, with a perched lake on Level 1 and a sump that is oriented to the east-southeast.

Micos Area (Sierra San Dieguito)

23. *Cueva de Otates*, one of the three known "Micos caves," was mapped in 1974 by Elliott, Prentice, and Walker. Its pit entrance is on the north valley margin, 250 m northeast of Cueva del Río Subterráneo. It probably is hydrologically connected to Subterráneo, coming underground within 160 m of its entrance.
24. *Cueva del Río Subterráneo* is about 250 m southwest of Otates entrance. It is the largest cave in the area, formed on two levels, but it needs to be re-mapped. It is the original site of the "Micosfish" (Wilkens and Burns, 1972), an evolving population with somewhat reduced eyes and pigment.
25. *Cueva del Lienzo* is a small pit cave on a hillside with a tiny cavefish pool at the bottom. It is across the valley, about 800 m southeast of Cueva del Río Subterráneo. Elliott, Prentice, and Walker mapped the cave in 1974.

Southern El Abra

26. *Cueva de la Curva*, near a curve in the rail line, is a shallow stream cave oriented north-south. It is just north of the ancient, dry river valley, the southern El Abra pass. The cave originally may have drained to an active river valley before the Nacimiento del Rio Choy captured this flow underground. The perched stream is 70-80 m above base level, and may rarely receive rising groundwater, if at all.
27. *Sótano del Toro* is a small fissure cave with a small cavefish pool near the north end. Its north-south orientation is like that of Cueva Chica and Los Cuates about 4 and 5 km, respectively, to the south.
28. *Cueva Chica* is an important fish cave, the original Mexican cavefish site from which Hubbs and Innes described *A. jordani* in 1936. It is an atypical fish cave, not far from the Río Tampaón, which backfloods into the lower cave pools via three pools near the river. The cave has a succession of pools stepping down from the entrance, with cavefish in Pool 1 and successively more hybrids with eyed fishes as one travels downstream to Pool 4. Eyed *Astyanax* enter the lower cave from the river during times of high water. See my chapter on ecology and biodiversity for a full discussion.
29. *Los Cuates* (the twins), also known as Cueva del Prieto, was mapped by John Fish et al. in 1971. Elliott, Morris et al. mapped it more extensively in 1974 when water levels were low. It has two entrances, and like Cueva Chica and Sótano del Toro, the cave is oriented north-south, following a joint along a structural fold associated with a ridge. Like Chica, the cave contains cavefishes and hybrids. The nearby Cueva El Mante only contains eyed *Astyanax*.

GLOSSARY OF MEXICAN, GEOGRAPHIC, AND GEOLOGIC TERMS

anticline a fold in geologic strata in the form of an arch
arroyo a wet weather streambed

Cañon de Servilleta the river canyon NW of Cd. Mante, through which flows the Río Ocampo (Boquillas), which becomes the Río Comandante
carbonate platform a flat-topped reef with limestone-forming marine life
Ciudad or Cd. Mante a city near the northern Sierra de El Abra and the Nacimiento del Río Mante
Ciudad or Cd. Valles a city near the southern end of the Sierra de El Abra
cueva a walk-in cave as opposed to a sótano or pit (vertical) cave
Cretaceous the geological period from about 145-100 mya (million years ago)
dolina or doline a large sinkhole
El Abra literally "the opening" or pass, refers to two, now-dry river passes at the northern and southern ends of the Sierra de El Abra
Gómez Farías a small town at the northern end of the region
gruta a large cave
karst landscapes formed by the dissolution of soluble rocks such as limestone, dolomite, and gypsum, with underground drainage systems, caves, sinkholes, dolines, and springs
joint a more-or-less vertical crack in bedrock, along which caves often form through dissolution by groundwater
Micos caves three fish caves south of Micos and Las Crucitas, S.L.P.
Huastecan Province synonymous with "Sierra de El Abra region"
msl or amsl above mean sea level
nacimiento literally "birth," a large spring (manantial) or resurgence (see Table 1.2)
ojo de agua literally "eye of water", synonymous with nacimiento or manantial
poza a pool
pozo a well or deep pit cave
resumidero a swallet cave that takes much runoff
río a river
San Luís Potosí or S.L.P. a northeastern Mexican state containing the southern half of the Sierra de El Abra
sierra a mountain range (literally "saw")
Sierra Cucharas literally "spoons," a local name for the northern Sierra de El Abra and the foothills of the Sierra de Guatemala
Sierra de El Abra a low range in the eastern Sierra Madre Oriental, between Cd. Mante and El Pujal
Sierra de El Abra region the subject of this chapter, including the El Abra, lower parts of Sierra de Guatemala (Cucharas), northern Sierra de Tamalave, and the Micos cave area
Sierra de Guatemala a higher range north of Ocampo and Gómez Farías, Tamaulipas
Sierra de Tamalave a range west of the Sierra de El Abra, also called Nícolas Pérez
Sierra de Tanchipa a local name for the Sierra de El Abra from south of the northern El Abra pass to the southern El Abra pass near Ciudad Valles
Sierra Madre Oriental the large mountain range in eastern Mexico, continuous with the Rockies
sima a chasm or abyss
speleogenesis cave development
sótano literally "cellar," commonly applied to a deep pit cave in Mexico
sumidero a sinkhole or pit cave that may not take runoff
swallet stream-capture cave
Tamaulipas or Tamps. the northeastern Mexican state containing the northern half of the Sierra de El Abra (see Figures 1.3 and 1.4)

Tertiary the geological period from 66-2.58 mya
tinaja a water jar or water hole
type locality the place from which a new species is described

ACKNOWLEDGMENTS

I am grateful to Sharon Mitchell, Linda Mitchell, David Bunnell, and the late professors, Robert W. Mitchell and Francis Abernethy for photographs. Thanks to Gayle Unruh, Gerald Atkinson, and Luis Espinasa for reviewing the manuscript, and to David McKenzie, who for many years has supported cavers with excellent programs such as *Walls* and *WallsMap* for cave cartography. Logan McNatt and Bill Mixon helped me many times in gathering up and scanning old survey notes and maps. Thanks to the AIM and the Association for Mexican Cave Studies for their support and for encouraging me to write these chapters.

REFERENCES

Álvarez, J., 1946. Revision del genero *Anoptichthys* con descripción de una especie nueva (Pisces, Characidae). An. Esc. Nac. Cien. Biol. Mexico 4, 263–282.

Álvarez, J., 1947. Descripción de *Anoptichthys hubbsi* caracinido ciego de La Cueva de Los Sabinos. S. L. P. Rev. Soc. Mex. Hist. Nat. 8, 215–219.

Association for Mexican Cave Studies (AMCS), 2014. http://www.mexicancaves.org.

Atkinson, G., 2004. Preface Pp. 9-11 in Fish Johnnie E 2004 Karst hydrology of the Sierra de El Abra, Mexico. AMCS Bull. 14, 186.

Breder Jr., C.M., 1942. Descriptive ecology of La Cueva Chica, with especial reference to the blind fish, *Anoptichthys*. Zoologica 27 (2), 7–16 fig. 1, pI. I-III.

Breder, C.M., Rasquin, P., 1947. Comparative studies in the light sensivity of blind characins from a series of Mexican caves. Bull. Am. Mus. Nat. Hist. 89, 325–351.

Bridges, W., 1940. The blind fish of La Cueva Chica. Bull. New York Zool. Soc. 43, 74–97.

Elliott, W.R., 1970. El Sótano de Soyate. Texas Caver 15, 63–66.

Elliott, W.R., 2002. Maps of Sótano del Molino and Sótano del Caballo Moro. AMCS Activities Newsletter 25 (96), 98Available at http://www.mexicancaves.org.

Elliott, W.R., 2007. Protecting caves and cavelife. pp. 458–468, In: Culver, D.C., White, W.B. (Eds.), Encyclopedia of Caves, second ed. Elsevier Academic Press, Amsterdam, The Netherlands, pp. 654.

Elliott, W.R., 2014. Sótano de Yerbaniz. AMCS Activities Newsletter 37, 125–130.

Elliott, W.R., in press. The Astyanax Caves of Mexico. AMCS Bull. 27.

Espinasa, L., 2009. Map of Cueva de El Pachón. P. 16, In: Astyanax International Meeting Program, March 15–18, 2009, Ciudad Valles, Mexico, pp. 49.

Espinasa, L., Borowsky, R., 2000. Eyed cave fish in a karst window. J. Cave Karst Studies 62 (3), 180–183.

Espinasa, L., Rivas-Manzano, P., Pérez, H.E., 2001. A new blind cavefish population of genus Astyanax: geography, morphology and behavior. In: The Biology of Hypogean Fishes. Springer, Netherlands, pp. 339–344.

Fish, J.E., 1977. Karst hydrology and geomorphology of the Sierra de El Abra and the Valles-San Luis Potosi region, Mexico. Ph.D. Dissertation. Hamilton, Ontario: McMaster University, xvii + 469 pp.

Fish, J.E., 2004. Karst hydrology of the Sierra de El Abra, Mexico. AMCS Bull. 14, 186.

Hendrickson, D.A., Krejca, J.K., Rodríguez-Martínez, M., 2001. Mexican blindcats genus *Prietella* (Siluriformes: Ictaluridae): an overview of recent explorations. Environ. Biol. Fish 62, 315–337.

Hubbs, C.L., Innes, W.T., 1936. The first known blind fish of the family Characidae: a new genus from Mexico. Occas. Papers Mus. Zool. Univ. Mich. 342, 7.

Instituto Nacional de Estadística, Geografía e Informática (INEGI), 2014. http://www.inegi.org.mx/.

McKenzie, D., 2014. Walls, Tools for Cave Survey Data Management and WallsMap, GIS Viewer and Shapefile Editor. Texas Speleological Survey, http://www.texasspeleologicalsurvey.org/.

Mejía-Ortíz, L.M., López-Mejía, M., Sprouse, P., 2013. Distribucion de los crustaceos estigobiontes de México. *Mundos Subterráneos*, Unión Mexicana de Agrupaciones Espeleológicas, A. C., 24, 20–32.

Mitchell, R.W., Russell, W.H., Elliott, W.R., 1977. Mexican eyeless characin fishes, genus *Astyanax*: environment, distribution, and evolution. Special Publ. Mus., Texas Tech. Univ. 12, 1–89.

Morris, N., 1989. Sierra de El Abra cave map folio. AMCS, 10 sheets.

Mothes, P., Jameson, R., 1984. Sótano de Vásquez. AMCS Activities Newslett. 14, 36–40.

Palmer, A.N., Hill, C.A., 2005. Sulfuric acid caves. pp. 573–581, In: Culver, D.C., White, W.B. (Eds.), Encyclopedia of Caves. Elsevier Academic Press, Amsterdam, pp. 654.

Quantum GIS, 2014. QGIS, A Free and Open Source Geographic Information System. http://www.qgis.org/en/site/.

Reddell, J.R., 1981. A review of the cavernicole fauna of Mexico, Guatemala, and Belize. Texas Mem. Mus. Bull. 27, 327.

Russell, W.H., Raines, T.W., 1967. Caves of the Inter-American Highway, a guide to caves of northeastern Mexico. AMCS Bull. 1, 126.

Walsh, S.J., Gilbert, C.R., 1995. New species of troglobitic catfish of the genus *Prietella* (Siluriformes: Ictaluridae) from northeastern México. Copeia 1995 (4), 850–861.

Wilkens, H., Burns, R.J., 1972. A new *Anoptichthys* cave population (Characidae, Pisces). Ann. Spéliol. 27 (1), 263–270.

Chapter 2

Hydrogeology of Caves in the Sierra de El Abra Region

Luis Espinasa[1] and Monika Espinasa[2]
[1]*School of Science, Marist College, Poughkeepsie, New York, USA*
[2]*SUNY Ulster, Stone Ridge, New York, USA*

PREFACE

Fresh water fish evolution is directly linked to the history of the drainages they inhabit (Waters et al., 2001). Rivers both at the surface and in caves change their paths through time and, as they do so, populations become isolated or united, dictating the direction for evolutionary processes, such as speciation, hybridization and colonization of new environments. The historical hydrogeology of the area is both the setting and the milieu in which *Astyanax* cavefish evolved. A thorough understanding of the hydrogeology of the El Abra region may ultimately provide an answer for the question "why" current cave populations are genetically structured as they are.

CURRENT SURFACE STREAMS AND SPRINGS

The El Abra region has several limestone mountain ridges (or sierras) where one typically does not find surface streams. To the north is the broad Sierra de Guatemala. From east to west is a series of long and narrow ridges: Sierra de El Abra, Sierra de Nicolas Pérez (or Sierra de Tamalave), Sierra de la Colmena, and some small hills that the *Astyanax* scientific community has referred to as the Micos area. Rivers either run through the valleys between these sierras or are born in springs (nacimientos) at the base of these mountains. In order to best interpret the description that follows, we recommend using Mitchell et al. (1977) map or Figures 1.3 and 1.4 of Chapter 1 from this book. Naming varies depending upon the source. Throughout the chapter we have maintained Mitchell et al. (1977) nomenclature, as it has been the main source of information for the *Astyanax* community. Other alternative names are in parentheses the first time they are named. The main rivers of the El Abra region are:

Biology and Evolution of the Mexican Cavefish. http://dx.doi.org/10.1016/B978-0-12-802148-4.00002-5
© 2016 Elsevier Inc. All rights reserved.

(1) Río Frío: Río Frío originates from Nacimiento del Río Nacimiento, Nacimiento de la Florida, and Nacimiento del Río Frío at the eastern base of the Sierra de Guatemala.
(2) Río Boquillas (Ocampo)/Río Comandante/Arroyo Lagartos (Capote): Río Boquillas is born in the mountains northwest of the El Abra area. It bisects the Sierra de Nicolas Pérez through a deep canyon and then divides the Sierra de Guatemala from the Sierra de El Abra through the fairly shallow (100 m) Servilleta (napkin) Canyon. Interestingly, the Servilleta Canyon has several cave entrances on its canyon walls at opposite sides facing each other, indicating that caves once connected the Sierra de Guatemala with the Sierra de El Abra before erosion formed this canyon (Espinasa et al., 2014). When it leaves this canyon, the Río Boquillas changes its name to Río Comandante.

    Before Río Boquillas enters the Servilleta Canyon, there is a tributary stream known as Arroyo Lagartos. This stream is of great significance, because it drains the northern portion of the Valle de Antiguo Morelos, west of Sierra de El Abra, where Pachón cave is located. The Pachón population is probably the most often used model in current studies. There is compelling evidence that there has been recent contact between Arroyo Lagartos and Pachón cave. In 1986, individual *Astyanax* with variable eye sizes and pigmentation were observed at Pachón. Prior and subsequent to this date, only albinotic and eyeless fish were observed. The presence of surface fish mitochondrial DNA (mtDNA) in Pachón cavefish is also thought to be the result of a fairly recent episode of introgression with surface fish (Langecker et al., 1991). This suggests that contact with surface fish is a relatively random event that occurs sporadically over time. Some have proposed that when a pipe was installed in Pachón to pump water out of the cave to the nearby town, somehow surface fish managed to travel through the pipe; however, a more likely alternative is that in very wet years, the Pachón entrance is a resurgence from which water pours out, creating a connection to Arroyo Lagartos. A branch of Arroyo Lagartos is only 1 km away. Although there is an approximate 50 m vertical climb between the cave's entrance and the stream, it is possible that surface fish can swim against the current and reach the cave's entrance, particularly when considering that *Astyanax* is routinely encountered above vertical inclines and waterfalls, such as is the case in the Micos waterfalls.
(3) Río Mante: It starts at Nacimiento del Río Mante at the eastern base of the northern Sierra de El Abra.
(4) Río Choy: Río Choy is born at the Nacimiento del Río Choy, on the eastern base of the southern portion of Sierra de El Abra.
(5) San Rafael de los Castros (Los Castro)/Río Santa Clara/Río Tantoán/Arroyo Seco: These four streams are born at comparatively small nacimientos on the eastern side of the Sierra de El Abra. Their small size relative to Mante and Choy indicates that they probably drain locally from a small surface and underground area. Nacimiento de San Rafael de los Castros is north of Mante, while the rest are between Mante and Choy.

(6) Río Naranjo/Río Valles/Río Puerco (Arroyo Grande)/Río Tampaón: The Río Naranjo is born in the mountains west of the El Abra region. After the town of Micos, the Naranjo River bisects the Sierra de la Colmena and changes its name to Río Valles, which eventually joins Río Tampaón. Río Valles receives as a tributary, Río Puerco. The relevance of Río Puerco is that it drains the southern portion of the Antiguo Morelos valley, west of Sierra de El Abra, where most caves with *Astyanax* populations are located. Undoubtedly in the past when the Antiguo Morelos valley was being eroded, portions of the Río Puerco were sequestered into some caves of the area.

(7) Río Santa María/Río Gallinas: Río Santa María is born on the mountains southeast of the El Abra region and it bisects the mountains south of the Micos area and the Sierra de la Colmena through a series of very deep canyons. After exiting the canyons, the river is known as the Río Tampaón.

Río Gallinas is included here due to its importance concerning surface populations. *Astyanax* cave populations originated from two waves of ancestral epigean forms from the south (Gross, 2012). These two waves derived from different "stocks" of surface fish, labeled as "old" or "new." Surface fish from the newer stock are still currently found throughout most of the rivers in the El Abra region. Remarkably, there is a single stream inhabited by surface fish of the old stock. This is the Gallinas River, where Ornelas-Garcia et al. (2008) collected old stock surface fish at Tamasopo and Rascón. Río Gallinas heads south and joins Río Santa María via the magnificent 105 m Tamul waterfall. This vertical drop seems to isolate the upriver stream from the lower drainages inhabited by the younger surface stock. Perhaps before erosion lowered the level of the valleys, the older wave of surface fish colonized and inhabited the entire area. Erosion then lowered the surrounding valleys, deepening the Río Santa María, leaving a perched population of isolated old stock *Astyanax* in the Gallinas River. Next, the old stock surface fish were replaced by the new stock in all the rivers of the area except the Gallinas River, which was inaccessible.

(8) Río Coy: This river is found south of Sierra de El Abra, is born at the Nacimiento del Río Coy, and drains into Tampaón. Nacimiento del Río Coy highlights a relevant phenomenon that will help understand the hydrogeological evolution of the El Abra region. This is one of the largest springs in the area, comparable in volume to Nacimiento del Río Choy and Mante. While these two springs are fed by about 769 km^2 of surface of the Sierra de El Abra (Mitchell et al., 1977), the Nacimiento del Río Coy is found in a diminutive 8 km^2 surface of the Salsipuedes dome (Figure 2.1(A)). Since this limestone dome is surrounded by impervious rock layers, underground water must come from elsewhere. Tectonic folding of the rock layers has forced the limestone layers of the Salsipuedes dome to dip under the impervious layers and then re-emerge about 10 km away in the area of the Aquismón mountains. The Nacimiento del Río Coy categorically shows that limestone caves and water conduits can pass under valleys with no

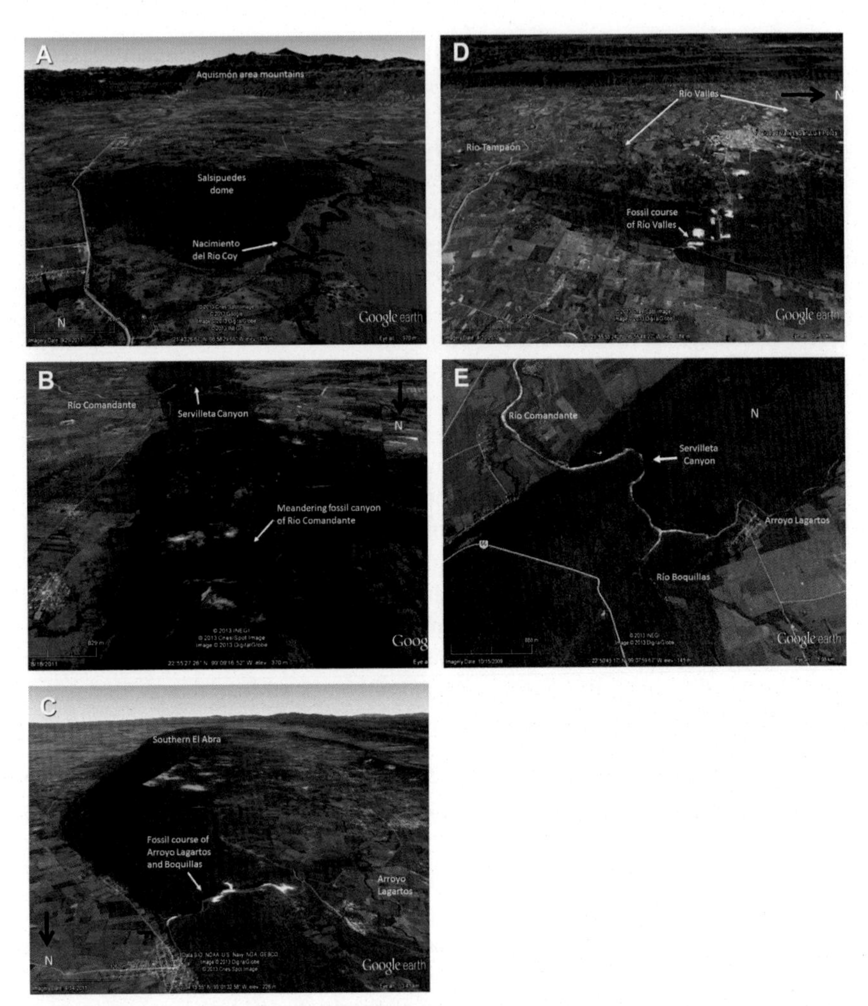

FIGURE 2.1 (A) The Nacimiento del Río Coy is in the small 8 km² surface of the Salsipuedes limestone dome. Tectonic folding of the strata makes the limestone layers dip under the surrounding impervious rocks and re-emerge in the Aquismón area mountains, seen in the distance. Water traverses under the valley through continuous underground conduits. (B) In the past, Río Comandante had a meandering path through a coastal plain farther north than its current path. About 9.1-2.3 mya, uplift caused it to erode a 100-m deep canyon. Further uplift and stream sequestration caused it to change its path south 6.3-1.6 mya, leaving as evidence a fossil canyon. (C) Río Lagartos also had a course farther south than its current path. Starting about 6.87-1.72 mya, it eroded an 80-m deep canyon. This canyon probably also sequestered Río Comandante for a time. The fossil canyon is currently used by Highway 85. (D) Río Valles had a more northern path in the past. About 4.4-1.1 mya, it started eroding a 60-m deep canyon. This abandoned pass is used by Highway 110 heading east from Ciudad Valles. (E) Río Comandante shifted its path north again to its current position about 3.3-0.8 mya, when it started cutting the 100-m deep Servilleta canyon. This canyon also has the diagnostic meandering path indicative of having crossed through an ancestral coastal plain when erosion first started.

surface limestone. It also explains why fossil resurgences such as the Hoya de Zimapán in Sierra de El Abra can be found at the top of the mountain ranges (Ford, 2000). Another Nacimiento that receives an influx from faraway mountain ranges is Nacimiento del Río Choy (Fish, 2004). In other words, distant sierras can be hydrologically connected.

Río Subterráneo and Lienzo caves are in separate karstic domes separated by a narrow valley (Mitchell et al., 1977), but cavefishes are clearly traversing under the valley and mixing. Canyons such as the ones between Sierra de Guatemala and Sierra de El Abra have also been traversed by troglobites (Espinasa et al., 2014) and there are at least four aquatic troglobites shared between the Sierra de El Abra and the Sierra de Guatemala: the entocytherid ostracod *Sphaeromicola cirolanae* Rioja, the cirolanid isopods *Speocirolana bolivari* (Rioja) and *S. pelaezi* (Bolivar), and the mysid *Spelaeomysis quinterensis* (Villalobos) (Reddell, 1981).

CURRENT UNDERGROUND DRAINAGES

There are five major underground drainages in the region with caves inhabited by *Astyanax*.

(1) Most of the southern region of Sierra de Guatemala and the portion of the Sierra de la Colmena north of Río Boquillas have an underground drainage that flows out through the closely situated Nacimientos of Florida, Río Frío, and Río Nacimiento. It is likely that the underground drainage flows via two main branches that join near the nacimientos. The first flows in a general northeast direction and starts at Sierra de la Colmena, where the terminal pool of Vásquez is at an elevation of about 145 m above mean sea level (msl). From here, water flows to Caballo Moro, whose base level lake is about 110 m msl. Caballo Moro must be one of the main hydrological conduits of the region due to the large amount of water that flows even during the dry season. Bee Cave, whose lower pool is at 130 m msl (Mitchell et al., 1977), joins this system. Water finally emerges at the aforementioned springs, which are at about 85-90 msl. The second branch flows south and drains the Gómez Farías region, whose caves (Jineo, Molino, and Escondido) are all thought to be related hydrologically. The lowest pool of these caves is at Molino, at about 131 m msl. Evidence of this drainage comes from geomorphology, as the small springs south of Caballo Moro are at 150 m msl, which is higher than the deeper pools in these caves, and through genetic studies of *Astyanax*. Espinasa and Borowsky (2000) showed that populations from Vásquez, Caballo Moro, and Molino caves are closely related.

(2 and 3) Sierra de El Abra appears to have two major underground drainages. Mitchell et al. (1977) pointed out that despite Nacimiento del Río Mante being 46.5 m higher than Nacimiento del Río Choy, its discharge was 2 1/3 times greater. This unusual circumstance was explained by suggesting a subterranean drainage divide between the southern and the central/northern sections of the Sierra de El Abra.

(4) In wet seasons, the Micos area valley has an ephemeral stream with surface fishes that goes underground at Subterráneo and Otates caves. The surface fishes enter the cave where they hybridize with the cave morph (Bibliowicz et al., 2013). To our knowledge, there has been no previous attempt to identify the flow direction of this underground system nor where its resurgence is located. An obvious hypothesis is that it flows north because the Micos caves are in a flat valley that very gently slopes toward Río Naranjo, near the town of Micos; however, an analysis using Google Earth and the reported cave depths (Mitchell et al., 1977) rejects this hypothesis; the caves' entrances are at an approximate altitude of 239-220 m msl, but their lowest pools are at about 205 m msl. Río Naranjo at Micos town has an altitude of 225 m msl. Underground flow in a northern direction would require the water to flow uphill. A better alternative lies to the south. Satellite images show a spring 5.2 km south (21°58′52.94″ N 99°11′34.76″ W) with an altitude of 181 m msl. This spring gives birth to a stream that joins Río Santa María at the town of Tanchachín, slightly downstream of the Tamul waterfall. One hopes that future studies will concentrate on this spring, as it may be a place where gene flow from cave to surface populations is significant and cave morph specimens may even be flushed out of the cave system.

(5) The Nacimiento de San Rafael de los Castros is located in the northern portion of the Sierra de El Abra between the Servilleta Canyon and Mante; therefore, the San Rafael de los Castros underground drainage does not include any recognized cave with *Astyanax*. The reason it is included in this discussion is that in the year 2000, while one of the authors of this chapter was collecting surface specimens, several showed eye degeneration similar to individuals at localities where hybridization between the cave and the surface morphs take place. This may be indicative that the area of the Sierra de El Abra north of Pachón has undiscovered caves inhabited by *Astyanax*.

Another locality where one of the authors of this chapter has found surface fishes with some degree of eye degeneration is in the small stream near the town of Coahuila, near the junction of Sierra de Nicolas Pérez and Sierra de Guatemala, close to Vásquez and Caballo Moro caves. There are no large springs in the area, but it is plausible that in wet seasons, the local subterranean hydrologic system overflows and underground water and fishes are flushed out into the surface. More research needs to be done at both localities to understand their significance.

GEOLOGY

The region of the El Abra was a calcareous reef complex during the middle Cretaceous era. The lower El Abra limestone was deposited during the early

Cenomanian around 93.5-99.6 mya and the upper limestone is dated to the late Turonian around 89.3-93.5 mya (Aguayo-Camargo, 1998). During the late Cretaceous period, the continental interior brought clastic materials and terrigenous sediments to the area, and by the end of the Cretaceous era, clastic deposits buried the entire reef and ended limestone deposition (Mitchell et al., 1977). The clastic deposits that overlaid the limestone are divided into three formations: the Agua Nueva, San Felipe, and Méndez. These formations determine the fate of water runoff in the El Abra region. The three formations are impervious to water and streams run on the surface; however, in places where limestone is exposed, water sinks underground through caves and there are no surface streams.

Uplift and warping of the layers started in the Late Cretaceous period, 70-80 mya during the Laramide orogeny, a period of mountain building in western North America. The Laramide orogeny occurred in a series of pulses with quiescent phases intervening. The major feature created by this orogeny was deep-seated, thick-skinned deformation of the strata. In the El Abra region, they are characterized by an east-to-west folding in which the limestone sierras are the anticlines (fold that is convex up), and the synclines (downward folds) form the valleys. Erosion in the sierras has progressively exposed the older limestone layers as the overlaying impervious layers are lost. The uplift of the area has continued up to recent times as indicated by Pleistocene/Recent lacustrine deposits that cap a small hill just south of Quintero, 6 km south of Nacimiento del Río Mante (Mitchell et al., 1977). This lacustrine limestone probably was deposited in an essentially sea-level lagoon, but it now lies at an altitude of 210 m msl and 110 m above Nacimiento del Río Mante and the surrounding coastal plain.

AGE OF THE CAVES

Chemical erosion of the limestone began even before uplift brought the area above sea level, resulting in extensive cavernous porosity while the limestone was still overlaid by other rock formations (Mitchell et al., 1977). Evidence of this comes from the El Abra limestone of the Faja de el Oro oil fields that are found to the east under the ocean and other rock layers. Here, oil has been encountered in cavernous porosities at depths of about 600-700 m, and the caves are so extensive that when an oil well is allowed to flow, there is an almost immediate pressure decline in other wells as far away as 5.5 km (Boyd, 1963).

Gross (2012) suggested that exposure of El Abra limestone started about 65 mya, but paleothermometry data indicate that the El Abra formation was buried by the Lower Tertiary sediments and uplifted about 30 mya (Yurewicz et al., 1997). Further evidence that limestone was not exposed until more recent times is that stream capture at Gómez Farías owes its presence to a narrow ridge formed by a Méndez base that persisted because of the protection afforded it by a Tertiary lava cap (Mitchell et al., 1977). Mitchell et al. (1977) state that "because uplift of the area occurred in early Tertiary and because overlying sediments were thin, some cavernous limestone surely was exposed by mid-Tertiary time," and "as removal of the original Méndez cover began… it is likely that such solution continued for a long period during Tertiary time, probably beginning at least as early as Miocene." The Miocene, in the mid-Tertiary, extends from about 23-5.3 mya.

Using molecular clocks, Ornelas-Garcia et al. (2008) have estimated that Mesoamerican colonization/expansion of *Astyanax* from South America occurred about 7.8-8.1 mya. Their data also show that the divergence between the old stock and the new stock of Mexican *Astyanax* occurred about 4.6 mya (estimated from their genetic distances using a Kimura K81uf unequal-frequency parameter of 5.74% at a 0.8% per million year rate). Finally, the old stock cavefishes diverged from their nearest surface relatives from Mezquital and Nazas—Aguanaval basins 1.12 ± 0.72 mya (mean of K81uf = $1.4\% \pm 0.9$); therefore, the first wave of colonization of the cave environment appears to have occurred sometime between 4.6 and 1.12 mya. As with many calibrations of molecular clocks, these dates should be considered approximate.

In terms of geologic time, erosion by rivers and streams is remarkably fast. For example, most geologists agree that much of the Grand Canyon was scoured by the Colorado River in only the last 5-6 million years (Pederson et al., 2008). Regrettably, the literature on *Astyanax* disregards erosion rates and is plagued with the idea that surface colonizers populated the subterranean waters of El Abra by way of the same caves we find the fish in today (for example, see Bradic et al., 2012). Considering the 4.6-1.12 millions of years that have passed since they first colonized, this is extremely unlikely. While the ages of individual caves may vary enormously, water-filled caves can increase their diameter up to 2-3 m in 1000 years (Palmer, 2002); thus, the youngest cave passages can have reached their present dimensions less than 100,000 years ago and older caves have largely been removed by continued surface lowering. For example, post-glacial diversion of water in New York state caves has formed passages up to a meter in diameter since the last glacial retreat about 13,000 years ago (Palmer, 2002). Relict caves millions of years old may survive in some buried limestone, but they are commonly found filled with younger sediments. This is certainly not the case of the active El Abra river caves with *Astyanax*, where mechanical erosion by the underground rivers is actively weathering and eroding the rock in these young caves.

Weathering and erosion is particularly fast in limestone rock. For example, Emmanuel and Levenson (2014) have measured up to more than 100 mm per thousand years weathered away by chemical erosion in limestone blocks. Twenty meters of rock, the approximate dimensions of a wide passage of El Abra caves, would erode away purely by chemical processes in about 200,000 years at this rates. Furthermore, mechanical erosion of a river is, again, many orders of magnitude faster than pure chemical erosion. When all this is taken into account, it is likely that not one of the caves that currently house *Astyanax* populations had been formed when the original wave of *Astyanax* colonization occurred 4.6-1.12 mya.

This does not mean there were no caves accessible for these fishes to colonize. There were other caves at higher levels of the limestone anticline. With uplift, the base level resurgences were abandoned and waters sought lower base levels. Erosion of newer passages and stream capture allowed the fishes to move into lower, younger passages. The older caves became fossil passages, or have since been filled with deposits and eroded away. The current caves where *Astyanax* is found are most likely the youngest caves in this often repeated process.

EVOLUTION OF THE HYDROGEOLOGY AT THE EL ABRA REGION

The hydrogeology of the El Abra region has been molded by two antagonistic geologic forces. On one side is the uplift of the area. On the other hand is erosion, which has created the valleys, stream courses, and caves. Dating erosion is challenging work and has not been done much for the greater part of El Abra. Uplift in the whole area seems to have been relatively uniform. For example, the anticline summit in the area of Gómez Farías and the anticline summit in Central Sierra de El Abra (east of Venadito cave) are both at a height of close to 800 m msl, so uplift has been more or less equivalent. Areas of slightly less uplift are in the northern and southern sections of Sierra de El Abra, which have lower altitudes above sea level. An area that has contrastingly different rates of uplift is the northern portion of Sierra de Guatemala, east of Gómez Farías, with heights of over 2000 m msl. This area could suffer earlier/higher rates of uplift, but they are in a different anticline than the one where El Abra and Gómez Farías caves are located and have no *Astyanax* caves, so this area is irrelevant to this discussion.

Regarding erosion, the present surface of the Sierra de El Abra is an almost perfectly stripped structural surface. The surface has been modified so little that in places, traces of drainage systems that once existed on the overlaying impervious strata may still be seen from the air (Mitchell et al., 1977). When the limestone was first exposed, the whole area was level in a coastal plain, as evidenced by the traces of the meandering paths of the rivers at the top of the sierras (Figure 2.1(B)–(D)). We would like to propose a simple model that provides a general means for dating geological structures and when they were first exposed by erosion of the overlaying impervious layers. To simplify matters, exposure at El Abra began at an undetermined time within the Miocene era; therefore, the model will use the lower Miocene (20 mya) and the upper Miocene (5 mya) as the high and low brackets for estimating dates. The highest point at Sierra de El Abra is at an altitude close to 800 m msl (as reported in Google Earth, 2014). The lowest base level in the region at Nacimiento del Río Choy and Río Tampaón is 39 m msl. The model assumes that in the southern tip of El Abra, where erosion rates have been the highest at close to 750 m, were eroded away in about 20-5 million years (my). The slowest erosion rate in the area occurred where the highest base level is at 300m msl, on the drainage divide between Arroyo Lagartos and Río Puercos. The model assumes that in this area, 500m has been eroded away since the Miocene period. To estimate the approximate date at which a particular feature, such as a cave, was first exposed and began to erode, we used the following formula: (width of sediments eroded to the feature X time of erosion to local base level)/width of sediments eroded to local base level.

For example, Soyate cave entrance is at an altitude of 293 m msl. In this area, the base level of Río Puerco is at an altitude of 100m msl. The ancestral level (800m msl) minus the current base level (100m msl) gives the width of sediments eroded (700m). Since 700m were eroded in about 20-5 my, the 507m of sediments that were eroded away to expose the area around the entrance of Soyate (800-293) took about 14.5-3.6 my; therefore, the Soyate entrance could only begin to form during the last 5.5-1.4 mya. Table 2.1 gives a list of the altitudes of the different features analyzed, the local base level near the feature, and the approximate age when the area around the feature was first exposed. Figure 2.2 provides the historical evolution of the hydrogeography of the area based upon this model.

This model attempts to incorporate differential rates of erosion in broadly defined, different zones. Nonetheless, uneven fluctuations in uplift and/or erosion are bound to result in some deviations and therefore all ages calculated by the model should be considered as rough estimates. Until direct dating is accomplished, these numbers provide a benchmark of the age of each individual cave inhabited by *Astyanax*. In time, we hope future research will improve the dating accuracy.

Surface rivers did not always run where they do now (Figure 2.2). Ancient paths eroded deep cuts throughout the Sierra de El Abra (Figure 2.1(B)–(D)). These canyons no longer have water, and the rivers were diverted to other areas (Mitchell et al., 1977). When the basal level was at our current altitude of 420 m msl, Río Comandante flowed farther north and eroded a 100-m deep canyon, called Puerto Chamalito, at a latitude of 22°56′, which is north of Caballo Moro and Bee caves. That the river flowed 9.1-2.3 mya through an ancestral coastal plain is evidenced by the diagnostic meandering path eroded in this fossil canyon (Figure 2.1(B)). As uplift continued and base levels reached 360m msl, Río Lagartos had a course farther south (Mitchell et al., 1977) where it started eroding 6.87-1.72 mya an 80-m deep canyon (Figure 2.1(C)) at latitude 22°36′, one km south of Pachón cave. This canyon eventually sequestered Río Comandante, which drastically changed its path south. Highway 85 follows this abandoned river path. Later, 4.4-1.1 mya when the base plain was at 240m msl level, Río Valles had a more northern path and started eroding a 60-m deep canyon (Figure 2.1(D)) at latitude 21°58′, between Curva and Toro caves. This abandoned pass is used by Highway 110 heading east from Ciudad Valles. The exit of the canyon is near the deviation toward the hotel Taninul. When the uplift set the base level at 215m msl, Río Comandante shifted its path north

TABLE 2.1 Altitude, Valley Base Level, and Approximate Estimated Age of Main Geologic Features

	Cave Entrance (a)	Valley Base Level (b)	Age (mya) max = ((a – b) × (20 mya))/(800 – b) min = ((a – b) × (5 mya))/(800 – b)
Caves			
Chica	49	39	0.26-0.07
Cuates	62	39	0.6-0.15
Toro	92	48	1.17-0.29
Curva	132	80	1.44-0.36
Palma, Piedras, Jos, Pichijumbo	145-176	100	2.17-0.32
Montecillo, Tinaja (canyon), Arroyo	190-192	100	2.63-0.64
Soyate	293	100	5.51-1.38
Sabinos, Roca, Tigre	239-246	120	3.71-0.88
Yerbaníz, Matapalma, Japonés	241-243	165	2.46-0.6
Venadito	312	270	1.58-0.4
Pachón	211	175	0.85-0.21
Gómez Farías (Molino)	269	130	4.12-1.03
Caballo Moro	320	190	4.26-1.07
Bee	249	150	3.05-0.76
Vásquez	422	160	8.19-2.05
Río Subterráneo, Otates, Lienzo	239-220	239	<0.1
Canyons	**Rim/ bottom**	**Valley base level**	**Start erosion/ abandonment**
North fossil Canyon/ Comandante	420/320	100	9.14-2.29/6.29-1.57
Central fossil Canyon/ Lagartos and Comandante	360/280	130	6.87-1.72/4.48-1.12
Servilleta Canyon/ Comandante	215/100	100	3.29-0.82/present
South Fossil Canyon/ Río Valles	240/180	80	4.44-1.11/2.78-0.69

Altitudes of multiple features are based upon data provided by Elliott within this book

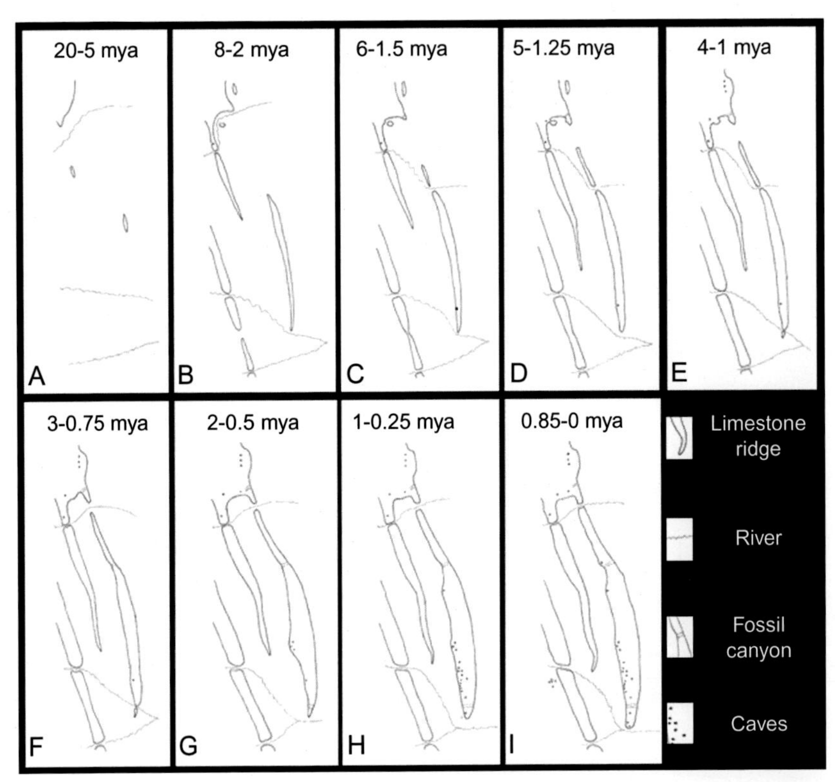

FIGURE 2.2 Hydrogeology of the El Abra region throughout time. Uplift of the Sierras changed the river paths and erosion of the overlaying impervious layers exposed the limestone, allowing for the development of caves. The following list of caves shows the earliest dates when the caves' entrances could have been formed, as well as the path of the surface streams at that time. (A) Emergence of El Abra limestone; (B) Vásquez cave. Río Comandante flowing farther north; (C) Caballo Moro. Río Comandante flowing farther south; (D) Soyate; (E) Molino, Jineo and Escondido. Río Valles flowing further north; (F) Río Comandante flows through the current path, forming Servilleta Canyon; (G) Yerbaníz, Matapalma and Japonés. Rio Valles changes path south to the current location; (H) All missing Astyanax caves exposed, except Pachón, Chica, Cuates, and Micos area caves; (I) Pachón and Cuates areas were exposed 0.85-0.15 mya, Chica 0.26-0.07 mya; Río Subterráneo, Otates and Lienzo less than 0.1 mya.

again to its current position 3.3-0.8 mya, where it started eroding the now 100 m deep Servilleta canyon. This canyon also has the diagnostic meandered path indicative of having flowed through an ancestral coastal plain when its erosion first started (Figure 2.1(E)). Finally, when the base level reached the current altitude of 39 m msl, Río Valles was sequestered by the faster eroding Tampaón and shifted its course south to its current path at el Pujal town, south of Cueva Chica.

This model also suggests that many of the caves currently housing *Astyanax* populations were overlaid with impervious layers at the time the first wave colonized the underground habitat; therefore, colonization had to occur at higher-level caves. The first area to be exposed was around Vásquez cave 8.2-2.0 mya (Figure 2.2(B)). The Río Comandante then changed its path from north of Caballo Moro cave to south of it by 6-1.5 mya, allowing for the entrance of Caballo Moro to be exposed later on (Figure 2.2(C)). This was followed by Soyate 5.5-1.4 mya (Figure 2.2D); Molino, Jineo and Escondido 4.1-1.0 mya (Figure 2.2(E)); Bee, Yerbaníz, Matapalma, Japonés, Tigre, Roca, and Sabinos 3.7-0.6 mya (Figure 2.2(G) and (H)); Venadito, Arroyo, Tinaja, Montecillos, Pichijumo, Jos, Piedras, Palma Seca, Curva and Toro 2.6-0.3 mya (Figure 2.2(H)); Pachón and Cuates 0.85-0.15 mya; Chica 0.26-0.07 mya; Río Subterráneo, Otates and Lienzo <0.1 mya (Figure 2.2(I)). It should also be remembered that these are the oldest possible ages when erosion could start; therefore, it is likely that in most cases, the entrance of the current caves is actually younger.

Do these conclusions contradict the ones from molecular phylogenetic studies (Bradic et al., 2012; Ornelas-Garcia et al., 2008; Strecker et al., 2003)? For example, is it incongruent that this study shows that Pachón cave is one of the youngest caves to be formed (0.85-0.15 mya) but molecular work indicates that Pachón's population is composed of the old stock and thus of the original colonization (4.6-1.12 mya)? We believe that the ages of the cave and its colonizing cavefish population can be different. The original colonization may have occurred in caves at a higher altitude in the nearby ridge, and later on the population migrated to its current location. As the zone uplifted and eroded, lowering the basal level, erosion and stream sequestration created new caves at the lower level. While this occurred, the ancestral populations could have migrated through underground passages to their current location by the capture and sequestration of the streams as deeper levels of cave passages formed. The original colonization might not have occurred at the current physical location of Pachón cave. Colonization would occur at a higher place along the Sierra, and the population has since been relocated by the evolutionary hydrogeology of the area. In conclusion, it is possible that an old stock of a cavefishes population inhabits a younger cave.

This is similar to the conundrum of the Galápagos tortoises. While the oldest Galápagos Island emerged 3-4 mya, molecular phylogenetic studies show that the Galápagos tortoise lineage divergence probably occurred 6-12 mya, before the origin of the oldest extant Galápagos Island (Parent et al., 2008). How can the date of island colonization be older than the islands themselves? These islands are formed by volcanoes that built up from the sea floor on top of a magma hot spot. As the tectonic plate moves west to east, new islands are formed and old ones are eroded away. The presence of several drowned seamounts east of the Galápagos suggests that earlier volcanic islands may have served as stepping stones for colonization by the Galápagos tortoise. The discovery of drowned seamounts extends the time the Galápagos have been available for colonization to at least 14 mya. In conclusion, both the Galápagos Islands and the caves at El Abra have a dynamic geologic history where colonization occurs at old islands or caves, which subsequently are eroded away, but the population migrates in a stepping stone fashion to their current location (Parent et al., 2008); thus, the apparent conflict between geologic age and phylogenetic age of inhabiting population can be non-controversial.

The major underground drainages within Sierra de El Abra have also changed drastically throughout this time. Nacimiento del Río Mante is 46.5 m higher than Nacimiento del Río Choy, but its discharge is 2 1/3 times greater. This unusual circumstance is the result of differential Pleistocene and possibly recent uplift of Sierra de El Abra that is higher in the north as indicated by Pleistocene lacustrine deposits of a sea-level lagoon that cap a small hill just south of Quintero, Tamaulipas (Mitchell et al., 1977) at an altitude of 210 m msl. Mitchell et al. (1977) hypothesized a subterranean drainage divide. The formation of this drainage divide during the Pleistocene/recent times postdates the colonization of the underground by *Astyanax* 4.6-1.12 mya.

By dividing the about 769 km^2 of infiltration area at Sierra de El Abra between the approximate 335 million cubic meters per year at Mante and the 141 at Choy, Mitchell et al. (1977) estimated the subsurface divide is located at 22°08′ or within that portion of outcrop lying between the Yerbaníz cave cluster and the Los Sabinos cave cluster (Mitchell et al., 1977). Genetic data do not fully corroborate with this estimate. Genetic structuring based upon microsatellites (Bradic et al., 2012) group instead the Yerbaníz and Japonés cave populations within the southern populations. As such, it would appear that the subterranean hydrological divide is farther north, so that only Pachón and Venadito are part of the subterranean hydrologic system that drains to Mante, while the rest of the Sierra de El Abra caves (Yerbaníz, Matapalma, Japonés, Tigre, Roca, Sabinos, Arroyo, Tinaja, Soyate, Montecillos, Pichijumo, Jos, Piedras, Palma Seca, Toro, Chica and Cuates) flow into Río Choy; however, further analysis by these same authors indicates that this may be an oversimplification. Gene flow data between caves appear asymmetric and these asymmetries seem related to relative altitudes. Pachón (203 m msl) sent more migrants to Yerbaníz/Japonés (145-104 m msl) than the reverse. The same was true with Yerbaníz/Japonés to Sabinos area caves (85-58 m msl). An asymmetric gene flow related to altitude between Pachón and Yerbaníz/Japonés is difficult to explain if we assume that the cave populations derive from independent colonizations from ancestral surface populations, with the cave populations being completely isolated from each other. This is easily explained if there are at least some sporadic migrants flowing downstream, which in this case suggest some type of underground hydrologic connection. We suggest that the inconsistencies between the genetic clustering, gene flow, and hydrological estimates could be resolved if the location of the drainage divide came about after the initial underground colonization by *Astyanax*, due to more recent geological changes, and the Yerbaníz cave cluster hydrologic system has been captured and sequestered towards the Mante and Pachón hydrological drainage. Genetic structure of the populations thus reflect a combination of past and present isolation/gene flow brought about by the dynamic history of the hydrogeology in the area.

DISCUSSION

The literature on cave *Astyanax* often purports the idea that surface colonizers populated the subterranean waters by way of the same caves we see today. It is assumed that the current topology of caves, as well as the hydrological systems, have remained unchanged from when the fishes first colonized the underground. Two of the first to express this idea were Breder and Rasquin (1947), who stated: "We may assume for purpose of discussion that river fishes entered La Cueva Chica, the while undergoing optical and pigmentary reduction." This assumes that at the time of original colonization, which could be as long ago as 4.6 mya, Cueva Chica had already been formed.

Mitchell et al. (1977) tried to correct this perception by giving a detailed review on how caves and stream capture develops through time and how erosion modifies the hydrologic system. They pointed out that through uplift, base level resurgences are abandoned and waters have sought lower base levels. They presented a dynamic process of cave development in which fishes "could then move into lower, integrating systems." Regrettably, at the time the book was written, their perception was that "cavefish evolution certainly began no earlier than the Pleistocene and probably not until the latter part of the Pleistocene or even later" (i.e., less than 0.012 mya instead of currently acknowledged 4.6-1.12 mya). They assumed the Sierra de El Abra drainage divide, which was caused by uplift of the northern portion of El Abra during the Pleistocene, had been a barrier present when *Astyanax* first colonized the cave environment. This has led to the idea that Pachón cave, for example, needed to be colonized independently from the southern populations, because there is a barrier, the current drainage divide (for example, Gross, 2012). We now know that cavefishes are older than the formation of this drainage divide and their original dispersal would not have been hindered by it. The dynamic nature of karst erosion since *Astyanax* cave colonization portrayed by Mitchell et al. (1977) is but a snapshot in time of what actually occurred over a period of 4.6-1.12 my of evolution.

It is somewhat surprising that later literature dealing with the question of how many independent colonizations occurred in the evolution of cave *Astyanax* has still maintained a static perspective. For example, Bradic et al. (2012) suggest that the first wave of surface fishes led to three independent subterranean invasion events (Pachón, central El Abra, and Chica) and that Cueva Chica population "was derived from an independent invasion of old stock. Because of its southernmost location, it may well be the earliest established of the cave populations." This assumes again that at the time of colonization, perhaps as early as 4.6 mya, Chica had already been formed, when in reality the limestone of the area may have still been overlaid by impervious strata until about 0.26-0.07 mya. It seems that the original colonization occurred at a higher, distant location, and as uplift/erosion created deeper river systems, cavefish migrated to the current locality at Chica cave. The second, younger wave of colonization that is currently creating the hybrid population is a recent, secondary phenomenon.

CONCLUSIONS

(1) We present a new possible model of the evolution of the hydrogeology of the El Abra region. A simple model for estimating the age of features by assessing rates of uplift/erosion is proposed. This suggests a much more dynamic process of cave development than is normally discussed by the scientific community studying cave *Astyanax*.

(2) Chemical erosion could have begun in the Cretaceous, while the limestone where we find caves today was still overlaid with impervious rocks, resulting in extensive cavernous porosity.

(3) Tectonic folding into anticlines and synclines allows for karstic sierras to be connected via continuous limestone strata that dip under impervious layers in the valleys. Water conduits can pass under the valleys and feed neighboring underground drainages. Fossil resurgences at the top of the mountain ranges indicate this process has been underway since El Abra limestone was first exposed.

(4) Uplift and erosion first exposed the limestone to active hydric erosion that further expanded the already formed extensive cavernous systems sometime between 23 and 5.3 mya. The first wave of underground colonization by *Astyanax* was estimated around 4.6-1.12 mya. This implies that the evolutionary history of the cavefish largely overlaps the erosional changes of El Abra; thus, extensive modifications of the underground drainages have occurred since colonization.

(5) Surface drainages have been modified extensively. Fossil canyons give evidence of the routes that rivers took at different times in geologic history. Stream capture and sequestering is an ongoing process. The surface drainages that divide the Arroyo Lagartos/Comandante and Río Puerco/Tampaón hydrographic networks are dynamic and could have not been effective barriers against *Astyanax* migration. This may explain why migration rate analysis (Bradic et al., 2012) show that surface fishes in the region form a metapopulation, with extensive exchange of genetic material among its component populations.

(6) It is likely that not a single one of the caves that currently house *Astyanax* populations had been formed when the original wave of colonization occurred 4.6-1.12 mya. Colonization had to occur at higher-level caves. As the zone uplifted and eroded, deeper-level cave passages formed. Capture and sequestration of the underground streams should have allowed the ancestral population(s) to migrate to their current localities.

(7) The current underground drainage divide of Sierra de El Abra could postdate the colonization of caves by *Astyanax*.

(8) Many species of troglobites other than *Astyanax* inhabit caves spanning from southern El Abra to Sierra de Guatemala. DNA sequences calibrated with a molecular clock indicate some were able to migrate among these areas in less than the last 12,000 years (Espinasa et al., 2014). Due to the extensive connectivity and the span of geological changes throughout the hydrogeological history, it is likely that there have been ample opportunities for all troglobites, including *Astyanax*, to migrate across most, if not all, the El Abra region.

(9) For the scientific community studying *Astyanax*, perhaps the most intriguing element of the proposed hydrogeology model is that the entrance of Pachón cave could be one of the youngest (0.85-0.15 mya) among Sierra de El Abra, but Pachón cavefishes are categorized as an old population (Strecker et al., 2003; Bradic et al., 2012) whose wave of colonization of the cave environment appears to have occurred sometime between 4.6 and 1.12 mya. According to our hydrodynamic model in conjunction with molecular phylogenetic data, it could suggest that the original colonization occurred at a higher, distant location, and as uplift/erosion created deeper river cave systems, cavefishes migrated to the current locality at Pachón cave. Regardless of all the imprecision of dating in the proposed model, the fact is that Pachón cave is only 36 m above the valley and as such, is probably a geologically extremely recent cave that postdates the colonization of the underground by *Astyanax*.

Therefore, we hereby propose that cave populations of El Abra region have undergone an underground migration from the colonization point(s) to the current cave locations as a result of a dynamic geohydrological evolution of the area.

REFERENCES

Aguayo-Camargo, J.E., 1998. The middle Cretaceous El Abra Limestone at its type locality (facies, diagenesis and oil emplacement), east-central Mexico. Rev. Mex. Cienc. Geol. 15, 1–8.

Bibliowicz, J., Alié, A., Espinasa, L., Yoshizawa, M., Blin, M., Hinaux, H., Laurent Legendre, L., Père, S., Rétaux, S., 2013. Differences in chemosensory response between eyed and eyeless *Astyanax mexicanus* of the Río Subterráneo cave. EvoDevo 4 (1), 25.

Boyd, D.R., 1963. Geology of the Golden Lane trend and related fields of the Tampico embayment. In: Geology of Peregrina Canyon and Sierra de El Abra. Corpus Christi Geological Society Field Trip Guidebook, Mexico, pp. 49–56.

Bradic, M., Beerli, P., Garcia-de Leon, F.J., Esquivel-Bobadilla, S., Borowsky, R.L., 2012. Gene flow and population structure in the Mexican blind cavefish complex (Astyanax mexicanus). BMC Evol. Biol. 12, 9.

Breder, C.M., Rasquin, P., 1947. Comparative studies in the light sensitivity of blind characins from a series of Mexican caves. Bull. Amer. Mus. Nat. Hist. 89, 325–351.

Emmanuel, S., Levenson, Y., 2014. Limestone weathering rates accelerated by micron-scale grain detachment. Geology 42 (9), 751–754.

Espinasa, L., Borowsky, R., 2000. Eyed cave fish in a karst window. J. Cave Karst Stud. 62, 180–183.

Espinasa, L., Bartolo, N.D., Newkirk, C.E., 2014. DNA sequences of troglobitic nicoletiid insects support Sierra de El Abra and the Sierra de Guatemala as a single biogeographical area: Implications for *Astyanax*. Subterranean Biol. 13, 35–44.

Field, M. S., 2002. A Lexicon of Cave and Karst Terminology with Special Reference to Environmental Karst Hydrology. National Center for Environmental Assessment. Washington Office, Office of Research and Development, US Environmental Protection Agency.

Fish, J.E., 2004. Karst hydrology of the Sierra de El Abra, Mexico. AMCS Bull. 14, 186.

Ford, D.C., 2000. Deep phreatic caves and groundwater systems of the Sierra de El Abra, Mexico. In: Speleogenesis: Evolution of Karst Aquifers. National Speleological Society, Huntsville, Alabama.

Gross, J.B., 2012. The complex origin of *Astyanax* cavefish. BMC Evol. Biol. 12, 105.

Langecker, T.G., Wilkens, H., Junge, P., 1991. Introgressive hybridization in the Pachón Cave population of *Astyanax fasciatusí* (Teleostei Characidae). Ichthyol. Explor. Freshw. 2, 209–212.

Mitchell, R.W., Russell, W.H., Elliott, W.R., 1977. Mexican Eyeless Characin Fishes, Genus Astyanax: Environment, Distribution, and Evolution. Texas Tech Press, Lubbock.

Ornelas-Garcia, C.P., Dominguez-Dominguez, O., Doadrio, I., 2008. Evolutionary history of the fish genus *Astyanax* Baird & Girard (1854) (Actinopterygii, Characidae) in Mesoamerica reveals multiple morphological homoplasies. BMC Evol. Biol. 8, 340.

Palmer, A.N., 2002. Speleogenesis in Carbonate Rocks. Evolution of Karst From Prekarst to Cessation. Inštitut za raziskovanje krasa, ZRC SAZU, Ljubljana. 43–59.

Parent, C.E., Caccone, A., Petren, K., 2008. Colonization and diversification of Galápagos terrestrial fauna: a phylogenetic and biogeographical synthesis. Phil. Trans. R. Soc. B: Biol. Sci. 363 (1508), 3347–3361.

Pederson, J., Young, R., Lucchitta, I., Beard, L.S., Billingsley, G., 2008. Comment on "Age and evolution of the Grand Canyon revealed by U-Pb dating of water table-type speleothems". Science 321 (5896), 1634.

Reddell, J.R., 1981. A review of the cavernicole fauna of Mexico, Guatemala, and Belize. Texas Mem. Mus. Austin Bull. 27, 1–327.

Strecker, U., Bernatchez, L., Wilkens, H., 2003. Genetic divergence between cave and surface populations of Astyanax in Mexico (Characidae, Teleostei). Mol. Ecol. 12, 699–710.

Waters, J.M., Craw, D., Youngson, J.H., Wallis, G.P., 2001. Genes meet geology: fish phylogeographic pattern reflects ancient, rather than modern, drainage connections. Evolution 55 (9), 1844–1851.

Yurewicz, D.A., Chuchla, R.J., Richardson, M., Pottorf, R.J., Gray, G.G., Kozar, M.G., Fitchen, W.M., 1997. Hydrocarbon generation and migration in the Tampico-Mislanta basin and Sierra Madre Oriental, east-central Mexico: evidence from an exhumed oil field in the Sierra de el Abra. In: Enos, P., Minero, C.J., Aracen, R.R., Yurewicz, D.A. (Eds.), Sedimentation and Diagenesis of Middle Cretaceous Platform Margins, East Central Mexico, vol. 4. Dallas Geological Society and SEPM Field Trip Guidebook, Dallas, pp. 1–24.

Chapter 3

Cave Biodiversity and Ecology of the Sierra de El Abra Region

William R. Elliott
Missouri Department of Conservation (retired), Association for Mexican Cave Studies, Jefferson City, Missouri, USA

INTRODUCTION

In this chapter, I discuss the general ecology and biodiversity of fish caves in the Sierra de El Abra region, and describe four important, but very different, caves within the El Abra region.

BIODIVERSITY

Over 200 caves have been found in the lowland Sierra de El Abra, with hundreds more in the adjacent areas. There are many distinct cave types within this region, including dry, fossil resurgence caves, deep collapse pits, and fissures on the eastern crest of the El Abra, and on the west flank of the El Abra hydrologically dynamic swallet caves, which contain most of the cavefishes. Between those areas a deep fissure cave, Sótano de Soyate, reaches to the base level of the regional aquifer, which drains to the large spring, Nacimiento del Río Choy (see Figures 1.3 and 1.4 in Chapter 1).

These environments contain more than just cavefishes—some are significant bat roosts with up to six species of bats, which contribute guano (droppings) and dead bodies to the cave's basic energy flow. Associated with the bat guano, flood debris, and moist organic films are rich invertebrate communities where new species have been found, including unusual arachnids, such as ricinuleids, schizomids, mites, a blind scorpion, and many insects. Lacking sunlight, cave communities depend upon decomposing organic materials transported in by animals, water, and gravity. These communities have short food chains compared to surface environments, with tiny springtails, thysanurans, mites, and millipedes at the base, and micropredators, such as arachnids and some beetles at the top. The aquatic and amphibious crustacean fauna is also rich in species. Nothing is known to prey on the cavefishes, which is the top predator/scavenger in the ecosystem.

Biology and Evolution of the Mexican Cavefish. http://dx.doi.org/10.1016/B978-0-12-802148-4.00003-7
© 2016 Elsevier Inc. All rights reserved.

Forty years ago, 265 species were documented in the caves of the lowland Sierra de El Abra (Reddell and Elliott, 1973a,b), but taxonomic work probably has increased this to nearly 300. Only six fish caves in the region have received thorough biological study: Cueva Chica (60 species identified including 4 troglobites), Sabinos (59 with 9), Tinaja (53 with 2 troglobites), Pachón (42 with 4 troglobites), Yerbaniz (35 with 10 troglobites, including a blind scorpion, Figure 3.1), and Tigre (29 species with 6 troglobites). Easy access on foot to the first four caves allowed teams of biologists to enter many times to collect and observe fauna. Sótano de Yerbaniz and Sótano del Tigre, although vertical, were visited repeatedly for mapping and biological study. Further studies are needed of the other caves. We still do not know exactly which species and energy sources nourish the cavefishes and other species the most, but bat guano appears to be important, which is often true in temperate caves as well.

There are several karstic biogeographic subregions in the area. The Sierra de El Abra subregion contains at least 25 species of troglobites (blind, usually albino, cave-adapted forms): 13 aquatic species and 12 terrestrial species (Reddell and Mitchell, 1971a,b; Reddell and Elliott, 1973a,b; Reddell, 1981), and it shares many species with the Sierra de Guatemala to the north (see the chapter on cave exploration and mapping). The Sierra de Guatemala subregion contains a remarkable assemblage with a higher proportion of troglobites, at least 38 known species, including 10 aquatic and 28 terrestrial species, some of them unique to this subregion. The highlands of the Sierra de Guatemala underwent more climatic changes and isolation of terrestrial species in caves than the tropical lowlands; however, the lowlands had a long history of marine forms invading subterranean habitats along the coast as the freshwater karst developed there, followed by the arrival of freshwater fishes and colonization by ancestral *Astyanax* in stream-capturing caves. The Micos area is another subregion.

FIGURE 3.1 *Sotanochactas elliotti* (Mitchell), a delicate troglobitic scorpion about 10 mm long from Sótano de Yerbaniz. *By Robert W. Mitchell (with permission of Sharon and Linda Mitchell).*

The nacimientos (large springs) may be considered yet another biogeographic subregion with some similarities to the El Abra and Guatemala caves, but with unique fauna. A tiny blind catfish, *Prietella lundbergi* (Walsh and Gilbert, 1995), inhabits two small nacimientos on the eastern face of the region, Cueva del Nacimiento del Río Frío and Nacimiento de San Rafael de los Castro, along with two new species of *Troglomexicanus* shrimp not found in the *Astyanax* caves (Hendrickson et al., 2001; Villalobos et al., 1999). Cave divers have not found *Astyanax* cavefish in the nacimientos, which receive "type-A waters" from the local ranges, as well as "type-B waters," with hydrogen sulfide and dissolved salts, from deeper circulation (Fish, 1977, 2004); therefore, these zoogeographic patterns suggest that ancestral cavefishes did not invade the karst through springs, but rather were isolated through stream capture.

The intensive stream capture that occurs in the El Abra region contrasts to other karst regions, such as the Yucatán peninsula, where marine-derived fishes, such as the bythitid, *Typhliasina pearsei*, and the synbranchid eel, *Ophisternon infernale*, invaded fresh groundwater in the karst from littoral habitats and perhaps through submarine karst springs (Reddell, 1981).

GENERAL CAVE ECOLOGY

Hazards to Cave Visitors

The following is a guide for field researchers or cavers who may want to explore caves in the Sierra de El Abra, or similar caves. Driving and hiking to a cave can be hazardous if one is not prepared with plenty of drinking water, proper clothing, maps, and planning. Heat exhaustion can occur in the humid tropical climate; temperatures in the El Abra region often exceed 40 °C in the spring and summer. Even the easy caves require caving gear: helmets with chin straps, three light sources per person, spare batteries, sturdy clothing, gloves, and good gripping boots. The vertical caves require rigorous training by experienced vertical cavers in single-rope technique and safety rules, such as in a tree or gymnasium, well before descending into a pit. A strong cable ladder with a belay line may be used for short pitches, but prior training is best. For pitches deeper than about 10 m, tough caving rope is much preferred over dynamic climbing rope, and it must be protected from rubbing, cutting, and excessive mud by various methods. Of course one should not enter the swallet caves during the rainy season (September to November), or anytime that rain threatens due to the possibility of flooding. Cave diving is quite dangerous and should not be attempted even for short distances without cave-diving certification. In some caves such as Arroyo, Tinaja, and Venadito, methane bubbles occasionally rise from stream bottoms when disturbed by wading. CO_2 at normal atmospheric level is about 0.035%, but it has been measured up to 3.8% in Arroyo, high enough to risk panting and exhaustion, and 0.08-0.10% in Tinaja, which is tolerable for fit people. While it has not been directly measured, it is likely that most of the caves in the Sierra de el Abra have elevated CO_2 in the summer and fall.

An additional and commonly encountered danger is histoplasmosis, a serious fungal infection of the lungs, but sometimes the eye, which often hits visitors from Europe, Canada, or the northeastern United States, where it is rare in the environment. It comes from microscopic spores growing on bat guano in caves or bird roosts. Most experienced cavers in Texas and Mexico are somewhat resistant to "histo," but it can result in hospitalization for those who have not been previously exposed. The symptoms are malaise, difficult breathing, and fever, but it can be effectively treated with specific antifungals upon diagnosis. Rabies virus may be found in some bats, and no one should handle bats unless they have been vaccinated and trained in bat study. A rabies infection is nearly always fatal. Bats are ecologically important and should not be unduly disturbed by visitors. Finally, be alert for the large, highly venomous fer-de-lance pit viper, *Bothrops asper*, which cavers have encountered in the bush or, rarely, at the bottom of entrance pits.

Temperatures

The caves of the region are warm, about 23 °C, approximating the average annual temperature of the region, as do most caves throughout the world. Breder (1942) thought that Cueva Chica had an unusually warm temperature, but follow-up studies found it to be normal for the area. Osorio Tafall (1943) noted that Ciudad Valles' annual average was 26 °C; today's more accurate value is 24.7 °C (Climate-data.org). Cooler temperatures may prevail in rainy season runoff, which has been noted to trigger reproductive behavior in the cavefishes. Pit caves may trap cold, dense air in winter, while air currents and limited evaporation may cool the cave slightly. The author has gleaned water temperatures from 20 measurements in 16 fish caves, from papers by several authors and unpublished cave maps. The mean temperature was 23.1 °C, the standard deviation 2.6 °C. The range was 10.5 °C, from 18.5 °C in winter in the upper pool of Sótano del Molino to 29 °C in summer at Lake 3 in Sótano de Yerbaniz. Probably these differences can be explained by winter versus summer storm runoff that occurred shortly before the temperatures were measured. The water temperatures would equilibrate to that of the bedrock given enough time.

Cavefish Food

Cavefish food includes bat guano, internal parasites in the guano, crickets, other fishes (cannibalism), and probably flood debris, flies, moths, floating dead bats, guano invertebrates, dead frogs, and perhaps free-swimming crustaceans. The cavefishes are known to be sensitive to disturbances and home in on objects in the water, sometimes schooling or swarming. Some authors have said they are bottom feeders and do not school, but they probably scavenge opportunistically anywhere in the water column, especially on floating objects.

Breder (1942) examined the gut contents of cavefishes in Cueva Chica and said, "Their stomach contents were found to consist of bat droppings and parts of other and smaller cave characins and their eggs. This would suggest that the only regular input of energy into the population for large parts of the year is bat dung. Their ability to thrive and reproduce on the ordinary foods supplied to small aquarium fishes also suggests the lack of any peculiar specialization in dietary requirements." Osorio Tafall (1943) made a thorough study of the invertebrates of Cueva Chica and Cueva de Los Sabinos. He found that the main source of cavefish food is bat guano. Plankton, though diverse and abundant, were not found in the cavefish gut, despite numerous examinations. His team found "murcielaguina" (bat guano), insect parts, and sometimes fragments of other cavefishes. In a 40 mm fish with a distended abdomen, they found a nearly intact gryllid cricket, *Paracophus apterus*. These wingless crickets usually are seen on walls and upside-down on ceilings, and they sometimes fall into the water. Bats and crustaceans are detailed below.

Wilkens and Burns (1972) studied the new population of "Micosfish" in Cueva del Río Subterráneo near Micos. The cave also has surface *Astyanax mexicanus* and *Poecilia sphenops* (Poeciliidae). *Astyanax* and Micosfish feed on bat guano, probably from the bat *Pteronotus parnellii* (field-identified from this cave by Elliott in March 2013), and the Micosfish also feeds on other fishes. The surface fishes appear to be undernourished and the Micosfish probably feeds on them as they succumb from competition in darkness.

Bats

Bat guano appears to be a critical food source for the cavefishes and other fauna. The richest fauna (highest number of species) occurs in caves with a large number of bat species. Having more species of bats probably indicates that there is more suitable bat roost habitat in that cave, such as many high domes at various temperatures, and therefore more total bat guano. Table 3.1 provides an overview of the known bats of the region. Most of the 29 fish caves occasionally have bats, but only a few have large bat colonies. The common vampire bat is often reported, but in low numbers; its guano often is a recognizable puddle of dark purple liquid with a strong ammonia odor and many flies and their larvae.

Crustaceans

The cavefishes could prey on several species of groundwater crustaceans: *Speocirolana* isopods, various shrimp, and undiscovered aquatic species. Five species of *Speocirolana* isopods have been found in the greater El Abra region. *Speocirolana bolivari* (Figure 3.2) crawls and swims in deep cave waters and could be a prey of *Astyanax*. The amphibious *S. pelaezi* is not a likely prey for cavefishes, as it inhabits smaller drip pools not usually inhabited by *Astyanax*; it has been identified in 15 caves in the region, three with no *Astyanax*. Amphibious isopods of the genus *Brackenridgia* also probably

TABLE 3.1 Nine Bat Species in 13 Fish Caves.

Label	Cave	Atja	Bapl	Dero	Diec	Glso	Mome	Nast	Ptda	Ptpa	Bats	Total
6	Vásquez										1	1
7	Pachón	1		1	1	1		1		1		6
8	Venadito										1	1
9	Yerbaniz			1								1
12	Tigre										1	1
14	Sabinos	1	1	1	1	1						5
16	Tinaja			1							1	2
18	Montecillos			1								1
21	Piedras										1	1
24	Subterráneo									1		1
26	Curva	1		1								2
28	Chica	1		1			1	1	1	1		6
29	Los Cuates			1								1

Five other bat species have been identified in nonfish caves of the region. Most data are from Mollhagen (1971). The identification of *Pteronotus parnellii* in Cueva del Río Subterráneo and some other data are by the author. Vampires usually are few, but are easily identified from their distinctive guano, so they are reported more often than other bats.

Abbreviations are as follows: Atja, *Artibeus jamaicensis*, Jamaican fruit bat; Bapl, *Balantiopteryx plicata*, Gray sac-winged bat; Dero, *Desmodus rotundus*, Common vampire bat; Diec, *Diphylla ecaudata*, Hairy-legged vampire bat; Glso, *Glossophaga soricina*, Common long-tongued bat; Mome, *Mormoops megalophylla*, Ghost-faced bat; Nast, *Natalus stramineus*, Mexican funnel-eared bat; Ptda, *Pteronotus davyi*, Davy's naked-backed bat; Ptpa, *Pteronotus parnellii*, Parnell's mustached bat; Bats, unidentified bats and guano indicated on cave maps.

FIGURE 3.2 *Speocirolana bolivari* (Rioja), about 20 mm long, from Grutas de Quintero. *By William R. Elliott.*

avoid cavefish predators. Swimming *Troglomexicanus* shrimp and mysid shrimp rarely have been seen by wading biologists, so we do not yet know their potential as prey for *Astyanax*.

ECOLOGY OF FOUR CAVES

Sótano de Yerbaniz

Sótano de Yerbaniz was discussed in detail by Elliott (2014) and in Chapter 1 as a typical swallet (stream-capture) cave, but with violent flooding. It has the largest catchment basin (area that captures runoff) of all the El Abra caves at 16 km^2. Yerbaniz, being a younger swallet than Matapalma, could have already contained cavefishes that arrived via groundwater before stream capture occurred. Surface fishes and hybrids are sometimes found in shallow pools in the cave. As there are few bats, most of the food input is from flood debris and dying surface fishes.

Sótano de Yerbaniz has 35 species with 10 troglobites:

1. *Sphaeromicola cirolanae* Rioja: an ostracod commensal on *Speocirolana pelaezi*
2. *Speocirolana pelaezi* (Bolívar): an amphibious isopod generally found in small pools
3. *Cylindroniscus* sp. nr. *vallesensis* Schultz: an amphibious isopod found on rotten wood
4. *Sotanochactas elliotti* (Mitchell): the world's rarest and most cave-adapted scorpion, about 10 mm long and translucent, known only from Yerbaniz (Figure 3.1)
5. *Agastoschizomus lucifer* Rowland: a small arachnid predator, but the largest species of the Order Schizomida, it is known only from three caves in the Sierra de El Abra

6. *Hoplobunus boneti* (Goodnight and Goodnight): a large harvestman with reduced eyes
7. *Newportia sabina* Chamberlin: a small, slender scolopendrid centipede known from Cueva de Los Sabinos, Sótano de la Tinaja, Sótano de Yerbaniz, and Bee Cave
8. *Anelpistina quinterensis* (Paclt, 1979): a rather large cave silverfish (8.5 cm long, antennae and caudal appendages included), which was first described from Grutas de Quintero near Ciudad Mante, Tamaulipas; when re-describing the species, Espinasa et al. (2007) reported it from Cueva de El Pachón and Yerbaniz
9. *Pseudosinella petrustrinatii* Christiansen: a tiny springtail (collembolan) that feeds on bacteria or fungi on organic material
10. *Astyanax mexicanus*

La Cueva Chica

La Cueva Chica was the first cave in which Mexican cavefishes were found, and the type locality of Hubbs and Innes' *Anoptichthys jordani* (1936), but it is the least representative of the known fish caves. Its entrance lies about 1 km north of El Pujal and 1.5 km north of the Río Tampaón.

Cueva Chica was originally mapped by Breder in 1940, but somewhat inaccurately (see Chapter 1). Elliott and others remapped the cave from 1971 to 1974 (Figure 3.3), and they surveyed overland to locate pools near the river, and the caves, Los Cuates and Cueva El Mante (Figure 3.4). Cueva Chica's entrance is at about 68 m above mean sea level (msl); its horizontal length is 573 m and its mapped extent is 591 m. The cave is 19 m deep from the entrance floor to the bottom of the sump at Pool 4. The cave follows a joint trend at 192° under a low ridge, which may be a structural fold in the El Abra limestone. The overlying San Felipe shale crops out just above the entrance ledge. The contact between the San Felipe and the El Abra can be seen in high domes in the first half of the cave, as depicted on the map.

A shallow arroyo from the north apparently downcut through the shale and was captured in the joint-controlled cave passage, resurging at risings under the Río Tampaón. As the terrain eroded and the river downcut, the cave was mostly drained, leaving a vadose (partially air-filled) conduit through which groundwater and captured floodwater flows to the sump, then to tinajas (water holes) near the river. The river backfloods into the lower cave pools via these three tinajas at about 48 m msl. The cave's south end is 1230 m from the river, and 970 m from the tinajas. The terminal sump bottom, at −19 m, is at 49 m elevation, about 7 m above the river's typical level of 42 m at the shoreline. Eyed *Astyanax* and even river prawns and crayfishes enter the lower cave system during high water times, fair proof of a connection to the river.

The cave has a succession of pools stepping down from the entrance, with cavefishes in Pool 1. Pools 1 and 2 receive a clear flow of subterranean water

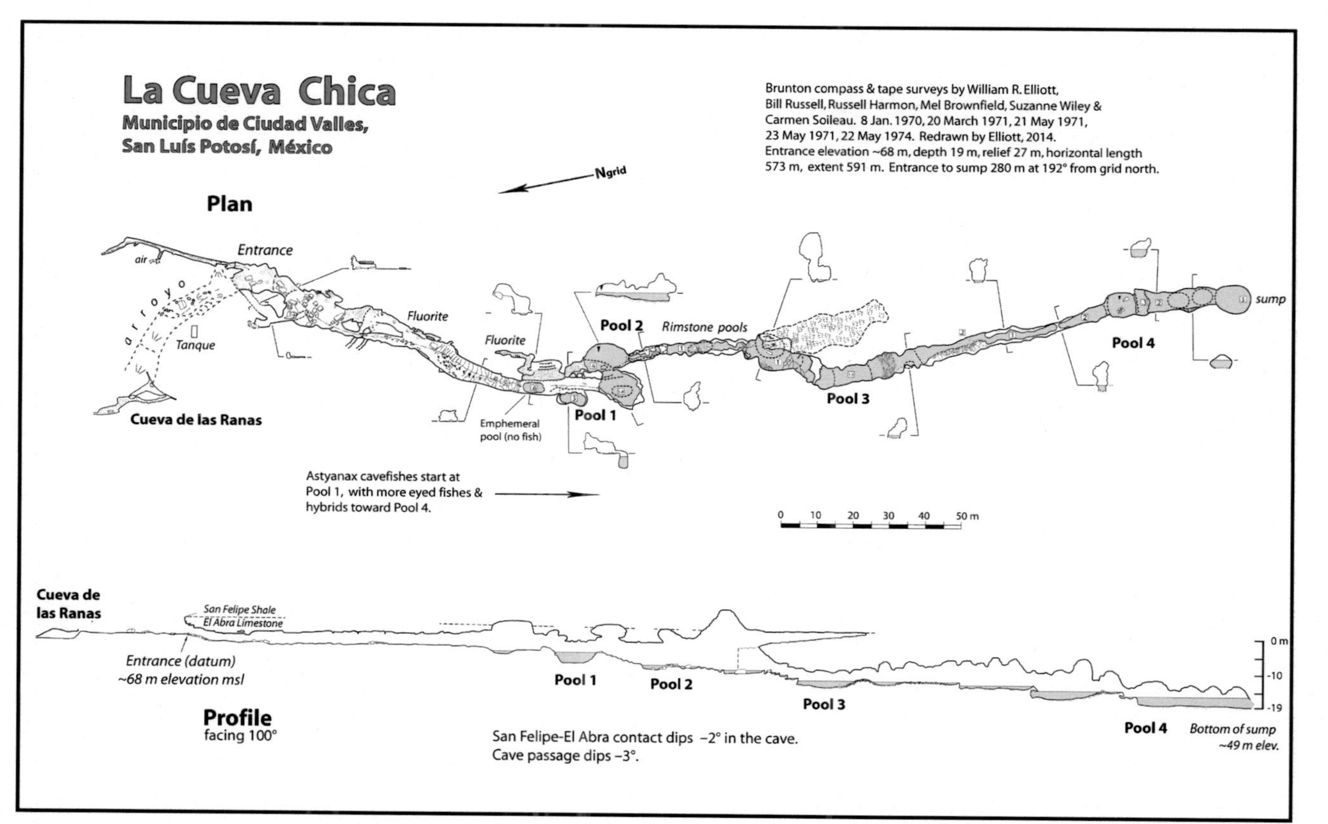

FIGURE 3.3 La Cueva Chica map, by William R. Elliott. See legend of AMCS cave map symbols in Chapter 1, Figure 1.17. *Copyright © 2015 William R. Elliott. All rights reserved.*

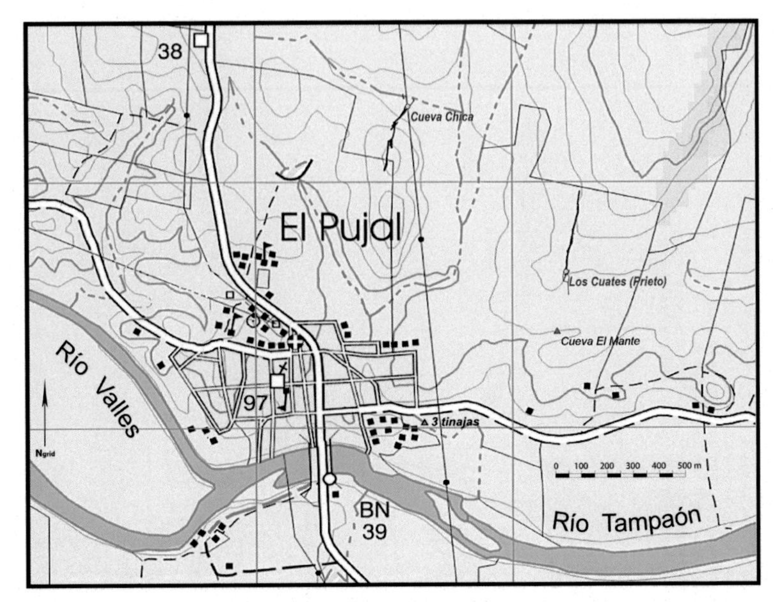

FIGURE 3.4 Cueva Chica area map. *By William R. Elliott based on INEGI 1:50,000 topographic map and 1974 surveys by the author. Copyright © 2015 William R. Elliott. All rights reserved.*

and, presumably, cavefishes from the regional aquifer to the north (Figure 3.5). There are successively more hybrids with eyed fishes as one travels downstream through cascades to Pools 3 and 4. This situation intrigued biologists for decades, as it was mistaken for a process of evolution from river to cave form within one cave. As more fish caves were found, it became clear that highly troglomorphic (cave-adapted) cavefishes occurred in most sites. A few caves, such as nearby Los Cuates, Sótano de Yerbaniz, Sótano de Matapalma, three Micos caves, Bee Cave, and Sótano del Caballo Moro have intermediate forms (Mitchell et al., 1977).

In the last half of Cueva Chica, most surfaces are covered with guano. Bridges (1940) characterized the last part of the cave as "a nightmare of slime and the stench of bats." The author can confirm Bridges' colorful observation, having swum through Pool 4 with a survey tape while beset with floating dead bats, vampire guano, and leeches.

The cave was known locally, but was "discovered" and its fishes collected in November 1936 by Salvador Coronado, who was employed by the Mexican fisheries agency, Dirección General de Pesca e Industrias Conexas, Secretaria de Marina de México (Mitchell et al., 1977). Specimens were sent to C. Basil Jordan of the Texas Aquaria Fish Company of Dallas, Texas, who soon introduced the cavefishes to the aquarium trade. Specimens were also sent to Carl Hubbs, the famous ichthyologist who described the new species and named it for Jordan (Hubbs and Innes, 1936). The original stock from Cueva Chica then became the common "commercial cavefish" in pet stores, which proved to be hardy and easy to raise. They are of reduced value to scientists because

FIGURE 3.5 Robert W. Mitchell collecting cavefishes in Cueva Chica, about 1969. *By Francis Abernethy and Robert W. Mitchell.*

of decades of selective breeding for troglomorphic traits, but they are a great educational tool. As far as is known, no additional cavefishes were taken from the cave for the aquarium trade, and it is not a conservation problem to possess commercial Mexican cavefishes.

Cueva Chica was studied in detail by a 1940 expedition of the American Museum of Natural History led by Charles M. Breder (Bridges, 1940; Breder, 1942). They concentrated on the ecology of the cavefish. Breder and his associates contributed greatly to our knowledge of the cavefishes in a series of some two dozen papers dealing with a wide variety of research. By 1946, additional caves were studied: Pachón, Arroyo and Tinaja (Breder and Rasquin, 1947). Overgeneralizations were sometimes made about the region's caves based upon Cueva Chica, which is an atypical sample of the region's fish caves. Cueva Chica probably was not the original site of cavefish evolution in the region. As we now know, most of the caves are not near a river, so surface fishes rarely get in, and if they do, they often perish or become fish food for the cave form; however, sometimes a few hybrids are seen. The dynamics of surface versus cavefish interactions have not been researched well *in situ*, as it is difficult to do (see the story of Suzanne Wiley's study in Yerbaniz in Chapter 1).

Cueva Chica and Cueva de Los Sabinos were also studied by Bibiano Osorio Tafall (1943) and a team of cave scientists from Mexico City's Escuela Nacional de Ciencias Biológicas: Candido Bolívar y Pieltain, Federico Bonet, and Dionisio Peláez, accompanied by three assistants, Cárdenas, Correa, and José Álvarez del Villar, who later described two more *Anoptichthys* (Álvarez, 1946, 1947). Their extensive sampling resulted in identifications of 37 aquatic species from the two caves and a description of a new species of troglobitic copepod. Many terrestrial species, including invertebrates, amphibians and bats, have been identified since then, giving a total of about 60 species (including four troglobites) in Cueva Chica and 59 (with nine troglobites) in Cueva de Los Sabinos. These remain among the richer known cave communities in the region.

Many authors have studied the Cueva Chica fishes, most frequently sampling from Pool 2, and noted that cave and surface fishes were hybridizing there (Breder, 1942, 1943; Sadoglu, 1957; Avise and Selander, 1972; Espinasa and Borowsky, 2001; Strecker et al., 2003; and others). A frequently overlooked study by Romero (1983) found that introgression between surface and cavefish had continued in Pool 2 since the original studies of 1936-1942, as measured using Breder's criteria for eye condition and pigmentation. Romero concluded that introgression (stabilization of backcross types) took place in Pool 2 in about 40 years. By 1983, there were fewer blind and fully eyed types, fewer pigmented types, and more intermediate types there. Strecker et al. (2003), working with microsatellite and mtDNA (mitochondrial DNA), and without knowledge of Romero's study, concluded that the Cueva Chica population evolved recently from a surface population, and subsequently hybridized with a phylogenetically older cave population. I have examined current topographic maps, and besides the connection to the river, it seems possible that nearby stock-watering ponds may be worth investigating as another source of surface fishes entering Cueva Chica via runoff into the cave entrance.

Troglobites in Cueva Chica:

1. *Diaptomus (Microdiaptomus) cokeri* Osorio Tafall: a tiny copepod found in pools of Cueva Chica and Cueva de Los Sabinos
2. *Speocirolana pelaezi* (Bolívar): a blind, albino, amphibious isopod known from 20 caves in the Sierra de El Abra region
3. *Brackenridgia bridgesi* (Van Name): a blind, amphibious isopod that inhabits nine caves in the Sierra de El Abra and one in the Sierra de Guatemala
4. *Astyanax mexicanus*

Cueva de Los Sabinos

Cueva de Los Sabinos is a classic cavefish cave and type locality of Álvarez's *Anoptichthys hubbsi* (1947). It is a large, complex cave on one level, with a southwesterly flowpath to a sump (submerged passage) connecting to Sótano del Arroyo. It is part of the Sistema de Los Sabinos, comprising Sabinos, Sótano del Arroyo, and Sótano de la Tinaja.[1]

This cave is located about 13 km north-northeast of Ciudad Valles and about 4 km east of the village of Los Sabinos, from which it takes its name.[2]

The entrance of the cave, about 30 m wide and 15.5 m high, is formed on the side of an indistinct branch of the arroyo captured by Sótano del Arroyo. The entrance probably no longer takes much floodwater. Even so, tree trunks and other organic debris make their appearance in the cave. A sump (submerged passage) unites the two caves, and floods from Sótano del Arroyo apparently invade Sabinos. Tinaja approaches close to the south side of Arroyo, and the three caves are hydrologically one unit with semi-isolation among them during dry times.

Sabinos was partly surveyed by the American Museum of Natural History expedition in 1946. About 900 m were surveyed to an indicated depth of 99 m. The resulting map was never published. Bonet (1953) provided a map of the cave apparently done by compass and pace, in which one uses a simple compass and paces off rough distances and directions while making notes. This method is useful for preliminary surveys or small caves only. The cave was resurveyed in 1972 by John Fish, Don Broussard, P. Thompson and others. This map (Fish, 1977, 2004) indicates a total length of 1502 m and a maximum depth of 95.5 at a bottom elevation of 144 m.

From the large entrance room, upper and lower levels trend northeast for 70 m to a large shaft where they rejoin. The lower level continues on for 150 m to the "Guano Room," inhabited by a large bat colony. Also from the entrance room is another passage that trends southeast for about 60 m to a point where a steeply sloping drop of 33 m is encountered. Beyond this drop, the passage enlarges to form a series of rooms extending about 150 m to a sump connecting to Sótano del Arroyo.

About 70 m past the sloping drop is a pool that leads through a small constriction into a large water passage, "Elliott's Swim," explored in 1970 by the author. After 250 m of northeast-trending pools 1 to 2 m in depth, walking passage is encountered, and the passage then continues about 200 m to a sump at the deepest point in the cave.

Troglobites in Cueva de Los Sabinos:

1. *Diaptomus (Microdiaptomus) cokeri* Osorio Tafall
2. *Sphaeromicola cirolanae* Rioja: this commensal ostracod crustacean, known from 15 caves in the El Abra and Sierra de Guatemala, has been taken from both *Speocirolana bolivari* and S. *pelaezi*
3. *Speocirolana pelaezi* (Bolívar): a 20 mm long, white, blind aquatic isopod

1. Maps of these caves may be seen in Fish (1977, 2004) and on the Association for Mexican Cave Studies (AMCS) website.

2. Its first exploration and the first blind-fish collection made in it were on April 3, 1942 by C. Bolívar y Pieltain, F. Bonet, B.F. Osorio Tafall, D. Peláez, M. Correa, and J. Álvarez (Osorio Tafall, 1943; Álvarez, 1946).

4. *Pseudosinella strinatii* Christiansen (det. K. Christiansen): a troglobitic collembolan (springtail) known from nine caves in the Sierra de El Abra
5. *Spherarmadillo cavernicola* Mulaik: an eyeless pillbug (terrestrial isopod crustacean) from the El Abra and Sierra de Guatemala
6. *Brackenridgia bridgesi* (Van Name): a sowbug (different type of isopod) from the El Abra and Sierra de Guatemala
7. *Hoplobunus boneti* (Goodnight and Goodnight): a large harvestman, a spider-like arachnid, with reduced eyes, this abundant species is frequently found on cave walls or silt banks, known from 15 caves
8. *Anelpistina quinterensis* (Paclt): a large, slender thysanuran or silverfish, known from many caves in the Sierra de El Abra and Sierra de Guatemala
9. *Astyanax mexicanus*

Sótano de Soyate

Sótano de Soyate (Figure 3.6) is the deepest cavefish site in the southern Sierra de El Abra. First explored by William R. Elliott, Jim McIntire, and Don Broussard in 1969, its 195-m entrance pitch leads to two short pitches to a large, clear lake at −234 m. This base level lake rises and falls in synchrony with the Nacimiento del Río Choy (Fish, 1977, 2004). In the lake resides a large population of cavefishes, studied by Elliott (Elliott, 1970; Mitchell et al., 1977), but this population has not been genetically analyzed. Live and preserved specimens were brought to Mitchell's laboratory in 1969, but they did not get into any genetics study. The population may represent an isolated, large, base-level population, or it may have episodic gene flow with the many perched-lake populations in different caves. The cave is oriented at 208° from true north, roughly north-northeast and south-southwest. No flow direction was reported, but the cave likely is a major conduit to the south and eventually to the Río Choy.

Soyate's deep fissure may have captured a shallow arroyo long ago, as there is a short headwall at the north end of the entrance, but there is no distinct arroyo remaining. Fish (1977, 2004) proposed that the cave is a phreatic void, a fissure that was enlarged by groundwater over time, followed by seepage waters depositing calcite on the upper shaft walls.

Troglobites from Sótano de Soyate:

1. *Hoplobunus boneti* (Goodnight and Goodnight): a large harvestman with reduced eyes, this abundant species is frequently found on cave walls or silt banks, known from 15 caves
2. *Anelpistina quinterensis* (Paclt): thysanuran
3. *Astyanax mexicanus*

Soyate has low food input. A few bats were seen by cavers in the upper shaft, but none at the bottom. There is little surface runoff. A few terrestrial invertebrates may fall into the lake. The Soyate cavefishes seemed to congregate at the surface of the deep lake, but they may only have been attracted

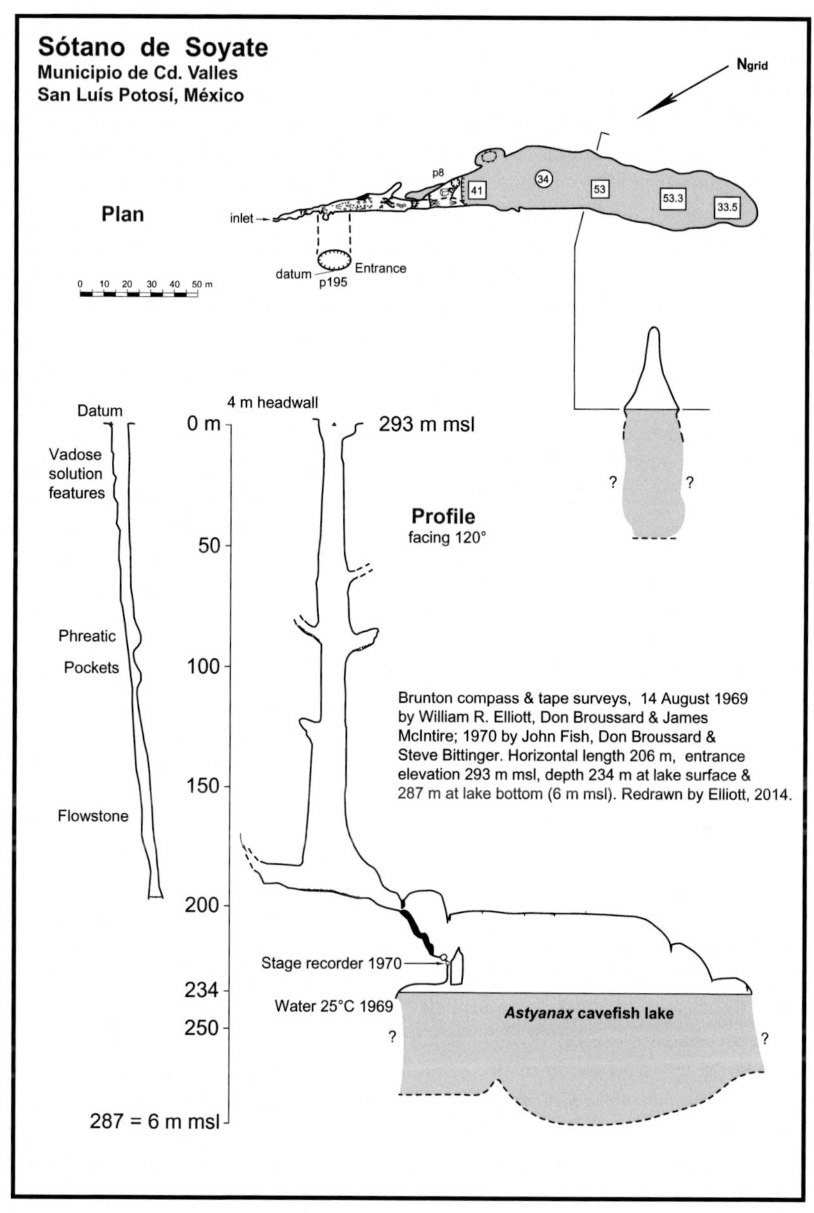

FIGURE 3.6 Sótano de Soyate map. *By William R. Elliott.* *Copyright © 2015 William R. Elliott. All rights reserved.*

to the disturbances caused by the explorers. Further studies of this population would yield information on their use of the water column, prey, and population biology. They may represent a large, base-level population differing from the perched-lake populations with which we are familiar. How such a large population subsists without much apparent food input is a mystery. I suggest that they may have an annual boom/bust cycle, in which the rainy season brings ample organic material into the deep groundwater, then they gorge, gain weight, and reproduce. This may be followed by a starvation period during which they cannibalize other cavefishes. Soyate deserves much more study, and its cavefishes should be sampled for genetic analysis.

CONCLUSIONS

Soyate and Yerbaniz both have large cavefish populations, but they may have opposite extremes of food input. We must remember, however, that humans can only observe for short periods of time, and not usually during and just after flood events, and therefore a detailed understanding of the ecology within the caves is lacking.

In March 1971, Mel Brownfield and I made cavefish population estimates in Sótano de Yerbaniz and Cueva del Pachón. Using a two-census, mark-recapture method (Lincoln Index), I statistically estimated the Lake 1 population in Yerbaniz at about 8700, with a wide 95% confidence interval of 1810-15,534 owing to the small number recaptured out of 201 marked. Pachón also had a large population, about 9800 ± 8502 (Mitchell et al., 1977).

There may be extremely large numbers of cavefishes in the El Abra region at base level, but not in upper pools. The bottom of Sótano de Soyate has a large population of cavefishes not yet estimated. This is interesting to cavefish biologists because of its implications for gene flow among semi-isolated populations. The vertical dimension of the caves must be kept in mind. Small, upper-level populations may sometimes die out, or experience genetic bottlenecking and fixation or loss of some alleles through inbreeding. Large populations could incorporate the results of these genetic drift "experiments," so to speak, when floods reconnect them temporarily to upper-level pools. The large populations could secondarily transmit some of these alleles to other caves through the base-level aquifer. Perhaps this is why some alleles are widespread while others are rare.

Finally, conservation efforts for the Mexican cavefish caves must begin. Land clearing and runoff of sediments, petroleum products, chemicals, and trash could adversely affect the cavefishes and groundwater (Elliott, 2007). The cavefishes are dependent upon bat colonies and other cave fauna for food; therefore, bat colonies should be maintained. A few fish caves already are within the Reserva de la Biósfera El Cielo in the Sierra de Guatemala. Perhaps some of the cave clusters and arroyos could be included as satellites of the Reserva de la Biósfera Sierra de El Abra Tanchipa. Further details and maps of 29 fish caves will be published in an AMCS Bulletin (Elliott, in press).

ACKNOWLEDGMENTS

Many thanks to Sharon Mitchell, Linda Mitchell, and the late professors, Robert W. Mitchell and Francis Abernethy, for photographs. I am grateful to Gayle Unruh and Luisa Espinasa for reviewing the manuscript and to James R. Reddell for help with literature.

REFERENCES

Álvarez, J., 1946. Revisión del género *Anoptichthys* con descripción de una especie nueva (Pisc., Characidae). An. Esc. Nac. Cienc. Biol. 4 (2-3), 263–282.

Álvarez, J., 1947. Descripción de *Anoptichthys hubbsi* caracinido ciego de La Cueva de Los Sabinos, S. L. P. Rev. Soc. Mex. Hist. Nat. 8, 215–219.

Association for Mexican Cave Studies (AMCS), 2014. http://www.mexicancaves.org.

Avise, J.C., Selander, R.K., 1972. Evolutionary genetics of cave-dwelling fishes of the genus *Astyanax*. Evolution 26, 1–19.

Bonet, F., 1953. Datos sobre las cavernas y otros fenomenos erosivos de las calizas de La Sierra de El Abra. Congreso Cientifico Mexicano Mem., (V.) 3, Ciencias Fisicas y Matematicas. Geologia 238–266.

Breder Jr., C.M., 1942. Descriptive ecology of La Cueva Chica, with especial reference to the blind fish, *Anoptichthys*. Zoologica 27 (2), 7–16, fig. 1, pI. I-III.

Breder Jr., C.M., 1943. Apparent changes in phynotypic ratios of the characina at the type locality of *Anopthichthys jordani* Hubbs and Innes. Copeia 1, 26–30.

Breder, C.M., Rasquin, P., 1947. Comparative studies in the light sensitivity of blind characins from a series of Mexican caves. Bull. Am. Mus. Nat. Hist. 89, 325–351.

Bridges, W., 1940. The blind fish of La Cueva Chica. Bull. N. Y. Zool. Soc. 43, 74–97.

Climate-data.org, December 12, 2014. http://en.climate-data.org/.

Elliott, W.R., 1970. El Sótano de Soyate. Tex. Caver 15, 63–66.

Elliott, W.R., 2007. Protecting caves and cavelife. In: Culver, D.C., White, W.B. (Eds.), Encyclopedia of Caves, second ed. Elsevier Academic Press, Amsterdam, pp. 458–468, 654 pp.

Elliott, W.R., 2014. Sótano de Yerbaniz. AMCS Act. Newslett. 37, 125–130.

Elliott, W.R., in press. The *Astyanax* caves of Mexico. AMCS Bull. 27.

Espinasa, L., Borowsky, R.B., 2001. Origins and relationship of cave populations of the blind Mexican tetra, *Astyanax fasciatus*, in the Sierra de El Abra. Environ. Biol. Fish 62, 233–237.

Espinasa, L., Flick, C., Giribet, G., 2007. Phylogeny of the American silverfish Cubacubaninae (Hexapoda: Zygentoma: Nicoletiidae): a combined approach using morphology and five molecular loci. Cladistics 23 (1), 22–40.

Fish, J.E., 1977. Karst hydrology and geomorphology of the Sierra de El Abra and the Valles-San Luis Potosi region, Mexico, (Ph.D. Dissertation), McMaster University, Hamilton, ON xvii + 469 pp.

Fish, J.E., 2004. Karst hydrology of the Sierra de El Abra, Mexico. AMCS Bull. 14, 186.

Hendrickson, D.A., Krejca, J.K., Rodríguez-Martínez, M., 2001. Mexican blindcats genus *Prietella* (Siluriformes: Ictaluridae): an overview of recent explorations. Environ. Biol. Fish 62, 315–337.

Hubbs, C.L., Innes, W.T., 1936. The first known blind fish of the family Characidae: a new genus from Mexico. Occasional Papers of the Museum of Zoology, vol. 342. University of Michigan, 7 pp.

Instituto Nacional de Estadística, Geografía e Informática (INEGI), 2014. http://www.inegi.org.mx/.

Mitchell, R.W., Russell, W.H., Elliott, W.R., 1977. Mexican eyeless characin fishes, genus *Astyanax*: environment, distribution, and evolution. Spec. Publ. Mus., Tex. Tech Univ. 12, 89.

Mollhagen, T., 1971. Checklist of bats in caves in the regions of the Sierra de Guatemala and Sierra de El Abra, northeastern Mexico. In: Reddell, J.R., Mitchell, R.W. (Eds.), Studies on the Cavernicole Fauna of Mexico, pp. 19–22, Assoc. Mex. Cave Stud. Bull. 4, 1–239.

Osorio Tafall, B.F., 1943. Observaciones sobre la fauna acuatica de las cuevas de la región de Valles, San Luís Potosí (Mexico). Rev. Soc. Mex. Hist. Nat. 4 (1-2), 43–71.

Reddell, J.R., 1981. A review of the cavernicole fauna of Mexico, Guatemala, and Belize. Tex. Mem. Mus. Bull. 27, 327.

Reddell, J.R., Elliott, W.R., 1973a. A checklist of the cave fauna of Mexico. IV. Additional records from the Sierra de El Abra, Tamaulipas and San Luis Potosi. In: Mitchell, R.W., Reddell, J.R. (Eds.), Studies on the Cavernicole Fauna of Mexico and Adjacent Regions, pp. 171–180, Assoc. Mex. Cave Stud. Bull. 5, 1–20.

Reddell, J.R., Elliott, W.R., 1973b. A checklist of the cave fauna of Mexico. V. Additional records from the Sierra de Guatemala, Tamaulipas. In: Mitchell, R.W., Reddell, J.R. (Eds.), Studies on the Cavernicole Fauna of Mexico and Adjacent Regions, pp. 181–190.

Reddell, J.R., Mitchell, R.W., 1971a. A checklist of the cave fauna of Mexico. I. Sierra de El Abra, Tamaulipas and San Luis Potosi. In: Reddell, J.R., Mitchell, R.W. (Eds.), Studies on the Cavernicole Fauna of Mexico, pp. 137–180, Assoc. Mex. Cave Stud. Bull. 4, 1–239.

Reddell, J.R., Mitchell, R.W., 1971b. A checklist of the cave fauna of Mexico. II. Sierra de Guatemala, Tamaulipas. In: Reddell, J.R., Mitchell, R.W. (Eds.), Studies on the Cavernicole Fauna of Mexico, pp. 181–215, Assoc. Mex. Cave Stud. Bull. 4, 1–239.

Romero, A., 1983. Introgressive hybridization in the *Astyanax fasciatus* (Pisces: Characidae) population at La Cueva Chica. NSS Bull., Natl. Speleol. Soc. 45, 81–85.

Sadoglu, P., 1957. Mendelian inheritance in the hybrids between the Mexican blind cave fish and their overground ancestor. Verh. Dtsch. Zoöl. Ges. 1957, 432–439.

Strecker, U., Bernatchez, L., Wilkens, H., 2003. Genetic divergence between cave and surface populations of *Astyanax* in Mexico (Characidae, Teleostei). Mol. Ecol. 12, 699–710.

Villalobos, J.L., Alvarez, F., Iliffe, T.M., 1999. New species of troglobitic shrimps from Mexico, with the description of *Troglomexicanus*, new genus (Decapoda: Palaemonidae). J. Crustac. Biol. 19 (1), 111–122.

Walsh, S.J., Gilbert, C.R., 1995. New species of troglobitic catfish of the genus *Prietella* (Siluriformes: Ictaluridae) from northeastern Mexico. Copeia 4, 850–861.

Wilkens, H., Burns, R.J., 1972. A new *Anoptichthys* cave population (Characidae, Pisces). Ann. Spéleol. 27 (1), 263–270.

Chapter 4

Phylogeny and Evolutionary History of *Astyanax mexicanus*

C.P. Ornelas-García and C. Pedraza-Lara
Departamento de Zoología, Facultad de Ciencias Naturales, Universidad Autónoma de Querétaro, Querétaro, Mexico

THE *ASTYANAX* GENUS

Patterns of diversity and fossil records support the general idea that the order Characiformes has a Gondwanic origin. The current distribution of most of the Characiformes includes the African continent and the Neotropical region (Otero and Gayet, 2001), with the only exception being the genus *Astyanax*, whose distribution reaches the Neartic region (Figure 4.1). The oldest fossil for the order in America has been recorded from the late Cretaceous (~68.4 mya) in Pajcha Pata formation (Bolivia), belonging to the subfamilies Tetragonopterinae and Serrasalminae (Gayet et al., 2001). Accordingly, the majority of diversity for the family Characidae is found at its southern distribution, at the Amazonian basin, with almost 220 species (Eschmeyer, 2014).

The genus *Astyanax* has the widest distribution for the family, occurring from Patagonia in Argentina to Texas and New Mexico in the United States (Lima et al., 2003). Currently, the genus comprises more than 170 species (Eschmeyer, 2014), although its taxonomic status continues to be under debate, as phylogenetic analyses have shown the non-monophyletic composition of *Astyanax* (Calcagnotto et al., 2005; Ornelas-García et al., 2008; Mirande, 2009; Valdez-Moreno et al., 2009; Javonillo et al., 2010; Oliveira et al., 2011).

Taxonomic classifications within the genus are under ongoing modification. In addition, a complete phylogeny for the genus *Astyanax* is still lacking. During the last two decades, several studies have tried to recover partial phylogenetic hypotheses for the genus and in particular in its northern distribution (e.g., Strecker et al., 2004; Ornelas-García et al., 2008), and the presence of recurrent morphological convergences in *Astyanax* (Ornelas-García et al., 2008) makes the delimitation of species difficult. In this context, delimitation of the species boundaries within the genus is far from an easy task and there is extensive discussion about validity of some species. Still, several efforts have been carried out in order to

Biology and Evolution of the Mexican Cavefish. http://dx.doi.org/10.1016/B978-0-12-802148-4.00004-9
© 2016 Elsevier Inc. All rights reserved.

FIGURE 4.1 Distribution pattern of Characiformes. The *Astyanax* is the only genus of the order in the Nearctic region (green pattern), in the figure the black line correspond to the Trans-Mexican Volcanic Belt (TMVB).

recover an integral taxonomic hypothesis within the genus in accordance with its evolutionary history (Strecker et al., 2004; Ornelas-García et al., 2008; Pie et al., 2009; Hausdorf et al., 2011; Coghill et al., 2014).

The troglobite populations (i.e., featuring cave-adapted traits) of the genus *Astyanax* inhabiting the Huasteca region in Mexico, located in the Sierra Madre Oriental region, in northeastern Mexico at the limits between San Luis Potosí and Tamaulipas states, is one of the most studied fish groups, because it represents an outstanding opportunity to study local adaptation and trait evolution (Avise and Selander, 1972; Gross, 2012; Gross et al., 2015). Similar to the rest of the genus, the taxonomic treatment of cave populations has been a subject of debate (Wilkens, 1988; Strecker et al., 2004; Ornelas-García et al., 2008; Hausdorf et al., 2011).

TAXONOMY OF TROGLOBITE *ASTYANAX* FROM THE HUASTECA REGION

The discovery by Salvador Coronado of the first troglobite characid population took place in 1936 at La Cueva Chica, in the southern portion of the Sierra de El Abra limestone, bringing to light an outstanding model to study evolutionary biology (Mitchell et al., 1977); however, this exceptional finding also represented a challenge to classification schemes that continues to be controversial. That same year, its obvious differing morphology (basically the lack of eyes and pigmentation) encouraged Hubbs and Innes to assign this fish to its own genus and species, *Anoptichthys jordani* (Hubbs and Innes, 1936). Approximately 10 years later, explorations made by groups of ichthyologists (led by J. Álvarez del Villar) discovered two more caves bearing blind fishes, and each population was described as a different species, *Anoptichthys antrobius* (Álvarez, 1946) from El Pachón cave and *Anoptichthys hubbsi* (Álvarez, 1947) from Cueva de Los Sabinos. At the time, the common thinking among taxonomists was that each cave population represented a different episode of evolution from epigean (surface) ancestors, and as such, corresponded to a different species. Further on, however, more populations were discovered from other caves, and the possibility that troglomorphic populations could represent only one taxonomic unit became apparent. As genetic studies involving interbreeding of troglobite and epigean *Astyanax* came into development (Şadoğlu, 1957), researchers realized how closely related the two forms were, and began assigning populations to one of two taxa, either *Astyanax fasciatus* or *Astyanax mexicanus* (Wilkens, 1971; Avise and Selander, 1972).

One idea regarding *Astyanax* taxonomy proposes that all populations of *Astyanax* from Brazil to the northern extreme of its distribution correspond to the species *A. fasciatus* (Cuvier, 1819) (Schuppa, 1984; Wilkens, 1988; Strecker et al., 2004). This assertion is based upon morphological variation (Schuppa, 1984), laboratory observations of interbreeding among populations, and in findings of gene flow between genetic clusters (Hausdorf et al., 2011; Strecker et al., 2012). In contrast, most taxonomic assertions coming from a variety of evidence agree that it is no longer acceptable to consider that the *A. fasciatus*, described from Brazil, is the same species that occurs as far north as Panama or Mexico (Miller, 2005; Ornelas-García et al., 2008). Evidence collected from karyotype studies showed differences in chromosome numbers and structure between *A. fasciatus* from Brazil and *A. mexicanus* (Klinkhardt et al., 1995). Evolutionary history of *Astyanax* populations from Mesoamerica inferred from mitochondrial and nuclear markers found at least 18 lineages corresponding to reciprocally monophyletic groups, in most of the cases with more than 2% of *p*-divergences in mitochondrial DNA (mtDNA), so that such diversity makes it difficult to explain the occurrence of only one taxonomically defined species (Ornelas-García et al., 2008). This observed diversity of evolutionary lineages has been confirmed by a number of studies, and has been asserted as deserving species recognition in other fish groups (Ornelas-García et al., 2008; Hausdorf et al., 2011).

Interbreeding among different species is a widely distributed phenomenon in a diversity of fish groups (Hubbs, 1955; Campton, 1987; Smith, 1992; Strecker, 2006; Mallet, 2008). In fact, it is thought to be a common process in several vertebrate groups (Campton, 1987; Allendorf and Waples, 1996; Mallet, 2005, 2008). Gene flow among fish taxa could be related to several factors, including: external fertilization, weak differentiation among mating behaviors, unequal density of parentals, competition for spawning habitat, decreased habitat complexity, and susceptibility to secondary contact (Hubbs, 1955; Campton, 1987; Cui et al., 2013). In spite of hybridization rates, several works have pointed out that divergence between lineages can be maintained even in the face of gene flow, and many hybrid populations and species remain genetically and ecologically distinct despite events of admixture (Hendry, 2009; Scascitelli et al., 2010; Vonholdt et al., 2010; Harrison and Larson, 2014; Schumer et al., 2014).

When prezygotic mechanisms are not responsible for speciation, it is highly probable that these mechanisms will not emerge until long after speciation has completed (Avise et al., 1984); thus, if secondary contact takes place, even after a long period of separation, divergent species may be able to interbreed, usually along limited contact zones of their distribution (Thorgaard and Allendorf, 1988). A reasonable treatment of such situations is to acknowledge and quantify the degree of gene flow in the context of the total gene pool of each species, and to estimate the extent each species is being affected, identifying the so-called genomic islands, or parts of the genome in which divergence is high among closely related species (Turner et al., 2005; Nadeau et al., 2012). Applying a biological species concept strictly (even more so if it is inferred under laboratory conditions) would drive to generalizations about diversity of certain groups, which may bias further comparative studies by ignoring a significant amount of variation among largely evolutionary, separate trajectories. We advise working toward a taxonomy that reflects the evolutionary relatedness accurately, as Mitchell and coworkers discussed several decades ago (Mitchell et al., 1977).

Following this idea, delimitation of species may be recognized along with the idea of evolutionary independent units (Simpson, 1944, 1953). Current nomenclatural treatment of surface and cave populations of *Astyanax* in most of northern Mexico considers *A. mexicanus sensu* (Miller, 2005), which includes the following general distribution range: Atlantic Slope, from Río Bravo basin in the northern tier of Mexican states east of Sonora and in New Mexico and Texas (Birkhead, 1980), southward into the Pánuco basin, Cazones system, and in uplands into Río Papaloapan drainage, lower Río Grijalva system, upper Río Balsas basin and upper Río Usumacinta, Guatemala, Maya Mountains of Belize and upper Río Polochic, Guatemala. Miller also recognized *A. aeneus* as valid, and that both species co-occur along the boundaries of their distribution ranges, mainly on the Atlantic slope (Figure 4.2).

Besides taxonomical controversies, *Astyanax* in Mexico is phylogenetically diverse. Dowling et al. (2002) questioned the monophyly of *A. aeneus* and *A. mexicanus* based upon sequences of the mitochondrial ND2 gene and inclusion of individuals of *A. aeneus* from southern Mexico and Costa Rica and *Astyanax bimaculatus* (Twospot Astyanax) from Perú (as outgroup). Mexican populations of *A. aeneus* were more closely related to *A. mexicanus* than to populations of *A. aeneus* from Costa Rica. Other phylogenetic studies recovered two well differentiated lineages (Figure 4.2), one distributed mainly northward of the Trans-Mexican Volcanic Belt (TMVB) and one to the south of this boundary (Strecker et al., 2004; Ornelas-García et al., 2008), with a mean divergence between them of $D_{K81uf}=2.3\% \pm 1.21$ (Cyt*b*). Appealing to a combination of monophyly and vicariant criteria, Ornelas-García et al. (2008) assigned these two lineages to *A. mexicanus* and *A. aeneus*, mainly coincident with Miller's

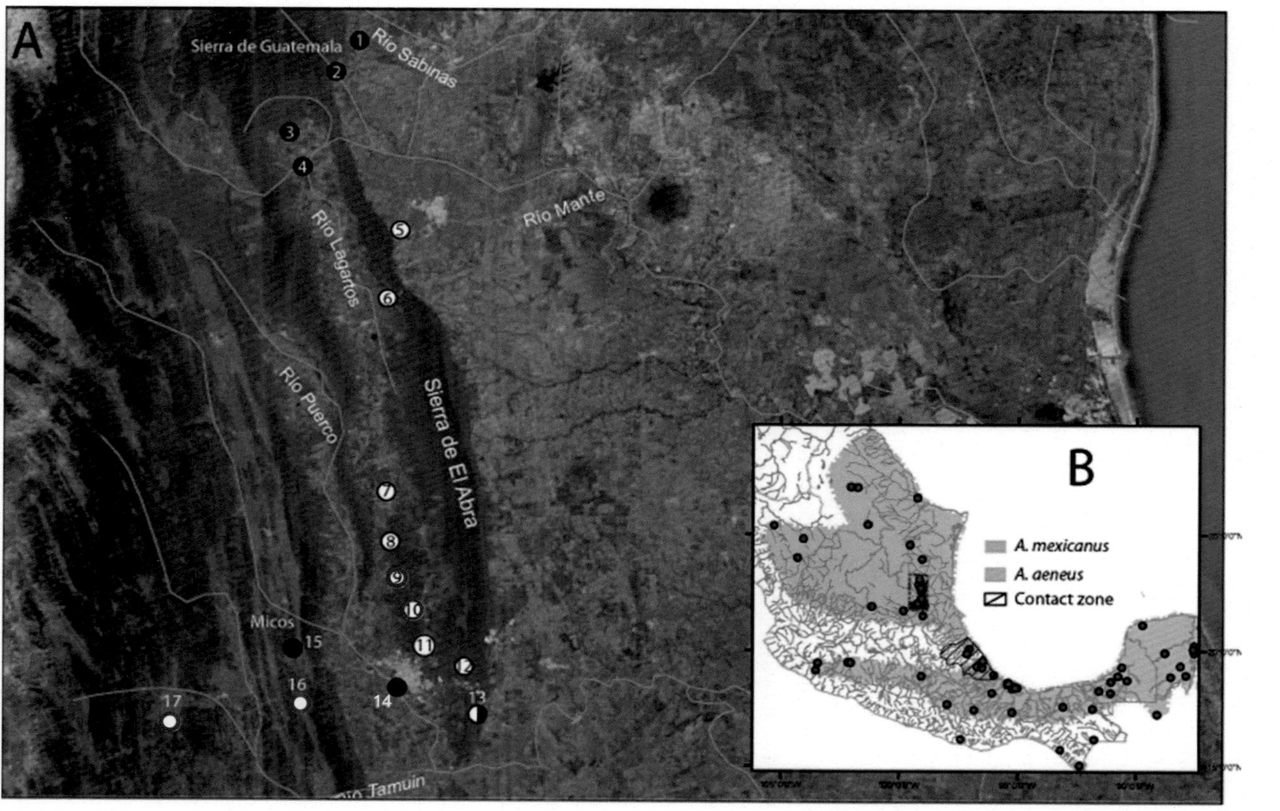

FIGURE 4.2 (A) Cave populations in the Huasteca region. Red circles correspond to the "old" colonization populations. (B) Distribution patterns of *A. mexicanus* and *A. aeneus;* area denoted by black stripes corresponds to the contact zone between the species. **Sierra de Guatemala:** 1 = Molino cave, 2 = Guayalejo river, 3 = Caballo Moro cave, 4 = El Limón river; 5 = Nacimiento del Río Mante; **Sierra de El Abra:** 6 = Pachón, 7 = Mante Dam, 8 = Yerbaníz, 9 = Sabinos, 10 = Tinaja, 11 = Piedras, 12 = Curva, 13 = Chica, 14 = Pte. La Raya; **Micos**: 15 = Micos cave; surface "old" lineages; 16 = Rascon, 17 = Tamasopo.

proposal (Figures 4.2 and 4.3). Despite finding similar divergences between these two groups (D_{K2P}=2.3%, *Cox*1), Hausdorf et al. (2011), provided arguments against the recognition of *A. mexicanus* and *A. aeneus* because of the observation of certain amount of gene flow between them in a contact zone and grouped all *Astyanax* from Mexico into one single taxonomic unit. In contrast, the same authors argue for parallel ecological speciation and recognize *A. jordani* for all cave populations, despite demonstrating that some cave populations obviously have hybridized with surface populations (Strecker et al., 2012). The intricate evolutionary history of *Astyanax* makes necessary the adoption of integrative strategies that accurately account for the diversity displayed within the clade, a task still requiring of intense work in the future.

Apart from the two aforementioned lineages (i.e., *A. mexicanus* and *A. aeneus sensu*, Ornelas-García et al., 2008), at least five highly divergent lineages have been found within Mexico (Dowling et al., 2002; Strecker et al., 2004; Ornelas-García et al., 2008; Bradic et al., 2012; Coghill et al., 2014). Divergence of those lineages is higher than that commonly observed among species in other groups of fishes (De la Maza et al., 2015; Silvia et al., 2010; Doadrio and Dominguez, 2004; Zardoya and Doadrio, 1999) and times for splitting among these five lineages and the rest of *Astyanax* from Mexico have been estimated around 6.7 mya (Ornelas-García et al., 2008) (Figure 4.4).

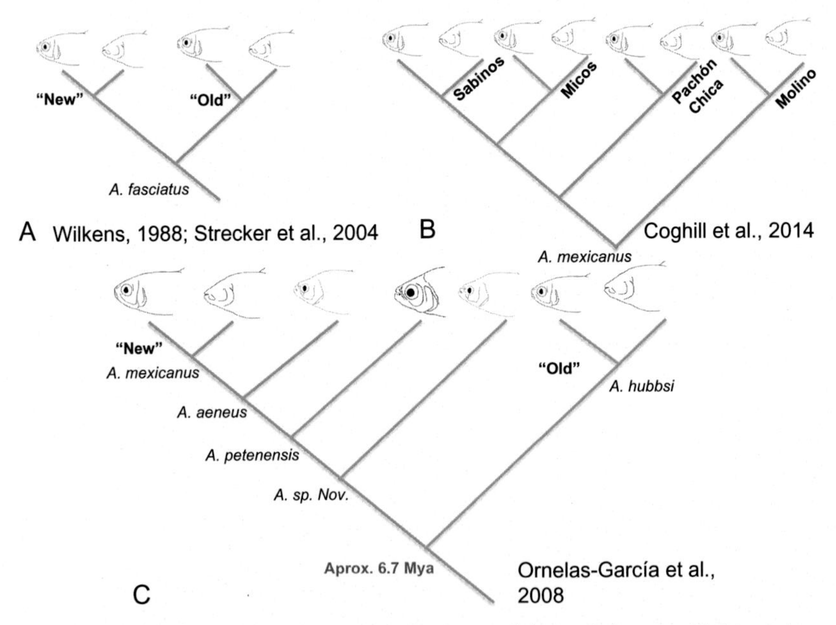

FIGURE 4.3 Different taxonomic proposals for the cavefish populations considering phylogenetic reconstructions. (A) *Astyanax fasciatus* based upon both morphological and molecular data (Wilkens, 1988; Strecker et al., 2004). (B) *A. mexicanus* based upon genomic information (Coghill et al., 2014). (C) *A. mexicanus* and *A. hubbsi*.

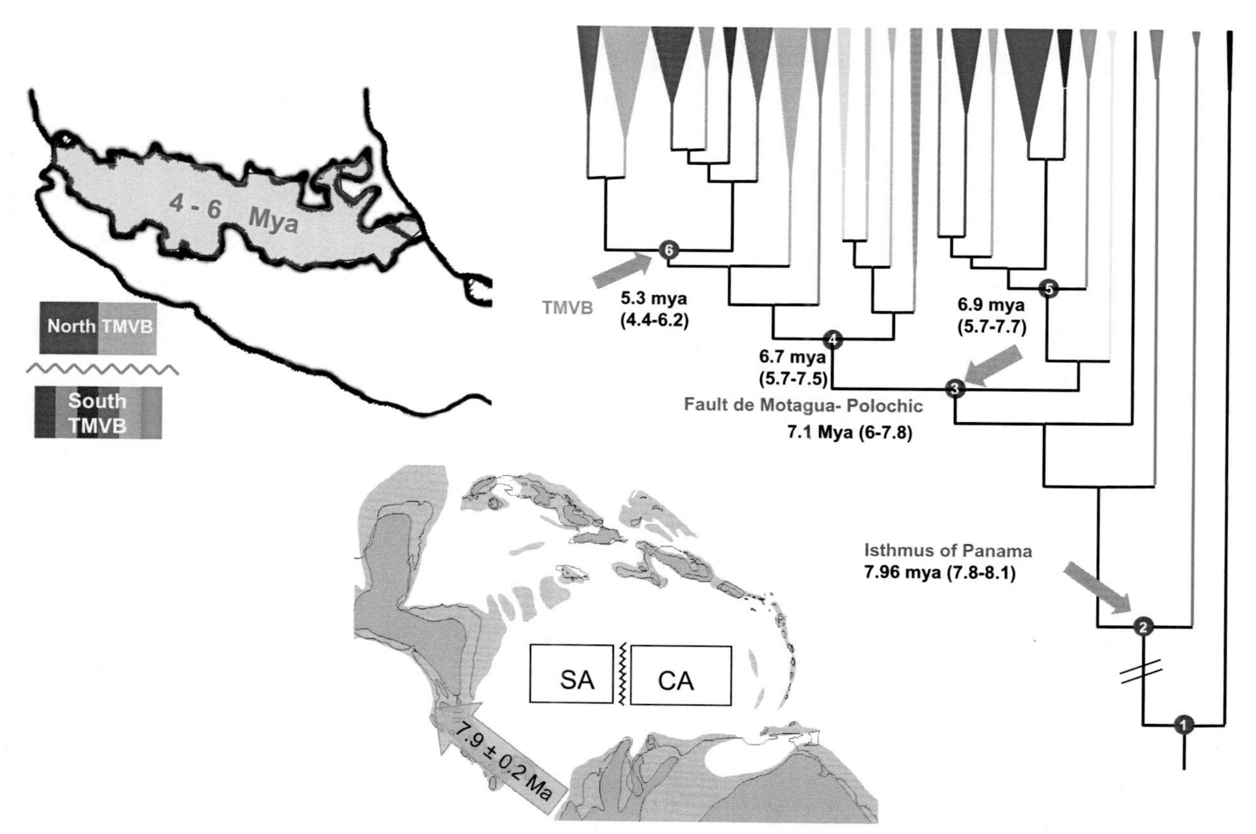

FIGURE 4.4 Biogeographic scenario based upon molecular clock previously proposed (Ornelas-García et al., 2008). The major vicariant and dispersion events occurred for the genus are shown.

Ornelas-García et al. (2008) suggest that these five lineages could correspond to different species. Considering mtDNA, mean genetic distance among these lineages and the rest of Mexican *Astyanax* is $D_{K81uf}=4.3\%$ (from 3.5% to 6.7%, Cyt*b*). Of these, at least 2-4 lineages have been found to occur in the caves from the Huasteca region (Strecker et al., 2004; Ornelas-García et al., 2008; Coghill et al., 2014).

Several studies have suggested that troglobite populations correspond to different episodes of invasion to cave environments from epigean forms. The "old" and "new" elements in caves have been suggested based upon levels of genetic diversity and demographic history (Avise and Selander, 1972), degree of morphological adaptation (Wilkens, 1988), evolutionary context (Dowling et al., 2002; Strecker et al., 2003, 2004; Ornelas-García et al., 2008), and genetic clustering (Bradic et al., 2012). In order to explain this phenomenon, we could make use of phylogeographic information of freshwater fauna of the region. Based upon their current distribution, Rosen, and similarly Myers, suggest that some groups of fish could colonize Mesoamerica from South America prior to the complete closure of the Panama Isthmus (Myers, 1966; Rosen, 1975).

This "old element" *sensu* (Rosen) could have passed through a proto Antillean bridge during the Cretaceous, and includes the following freshwater groups: Cichlids, Poecilids, and *Astyanax*. For several of these groups, this scenario has been confirmed (Darrin Hulsey et al., 2004; Concheiro Perez et al., 2007; Hrbek et al., 2007; Říčan et al., 2012); however, there is not a complete coincidence with the Cretaceous times. For Characids, and other primary freshwater fauna (i.e., *Rhamdia*), there is evidence of a Mesoamerican colonization during the Pliocene (Bermingham et al., 1998; Perdices et al., 2002, 2005). Ornelas-García et al. (2008) suggested that *Astyanax* colonized Mesoamerica during the Miocene (8 mya, see Figure 4.4), which would require a land connection previous to the final closure of the Panama Strait, which was complete around 3 mya (Ornelas-García et al., 2008).

According to phylogenetic structure, Ornelas-García et al. (2008) postulated a biogeographic scenario for the origin of populations of *Astyanax* in North America. Once Mesoamerican invasion took place, the genus *Astyanax* rapidly expanded in the region (Figure 4.4), and apparently reached the northern part of Mexico (i.e., Nazas-Aguanal basin, in Durango) during the Middle Miocene (about 7 mya). Multiple lineages from the lower part of Mexico expanded northward in at least one expansion event (although they could represent more than one such event). After this, different lineages experienced a new expansion event corresponding to *A. mexicanus* and *A. aeneus*, producing a lineage-turnover process and leaving the old lineages in different populations fragmented along the previous distribution as relicts of the older expansion episodes. In this way, the biogeography of the genus suggests the occurrence of independent lineages, which are known to occur in the troglobitic populations as a result of the arrival of such related evolutionary units to the Huasteca region.

CAVE INVASION BY *ASTYANAX* LINEAGES

The understanding of how lineages reached the caves is fundamental for studies of trait evolution, as it provides accurate spatial and temporal frameworks. This is especially important and nontrivial, as different studies have shown discordant results about the number of colonizations and genetic structuring among caves (Ornelas-García et al., 2008; Strecker et al., 2012; Bradic et al., 2012; Coghill et al., 2014), and it is unknown if this discordance comes from differences in geographic sampling coverage and markers used.

All the studies find evidence of at least two lineages inhabiting caves, and these have been interpreted as a convergent evolution from independent colonization events from epigean ancestors (Ornelas-García et al., 2008; Jeffery, 2009; Strecker et al., 2012; Bradic et al., 2012, 2013; Borowsky and Cohen, 2013); furthermore, there is evidence that regions of the genome, or even specific genes, have been modified independently among lineages, ultimately demonstrating convergent evolution (Protas et al., 2006; Bradic et al., 2013).

Two kinds of information are available to describe the observed genetic structure in caves: one coming from mitochondrial (maternal line), and other coming from the nuclear information. Mitochondrial information suggests two lineages present in the region (including cave populations). One identified as an old lineage included the central caves in El Abra limestone (Figure 4.2), specifically Sabinos, Tinaja, Piedras, and Curva (Dowling et al., 2002; Strecker et al., 2004; Ornelas-García et al., 2008) showing a mean divergence within these populations of $\bar{D}_{K81uf} = 2.13\%$ Cyt*b* (Ornelas-García et al., 2008). This lineage presents a low level of genetic differentiation within the surface-dwelling populations of *A. mexicanus* from the Mezquital and Nazas-Aguanaval basins (mean of $\bar{D}_{K81uf} = 1.4\% \pm 0.9$); therefore, this could be considered a sister linage of the first expansion episode of the genus, given previous to the closure of TMVB (Figure 4.4).

Also based upon mitochondrial information, the new lineage currently inhabits several caves in El Abra limestone: Pachón, Chica, and Yerbaníz caves, as well as Sierra de Guatemala: Molino and the Micos region, including Micos cave (Subterraneo). This new component includes most of the surface populations from the region (Dowling et al., 2002; Strecker et al., 2004; Ornelas-García et al., 2008).

Nuclear and genomic studies (single nucleotide polymorphisms [SNPs] and microsatellites) are mostly coincident, and found a minimum of three, and up to five, independent origins of cave populations (Strecker et al., 2012; Bradic et al., 2012; Coghill et al., 2014); however, some important differences exist. Strecker et al. (2012) proposed three invasions of the caves: one corresponding to most of the El Abra caves, another including Chica and Micos caves, and a third including Caballo Moro from Sierra de Guatemala. Bradic et al. (2012), who included a greater number of loci (26 microsatellite), found five genetic groups in the caves. The old lineage included most of the El Abra caves (i.e.,

O2-O7 populations). The Bradic paper proposed that subsequent to this, the surface old stock went extinct and was replaced by a new stock that gave rise to the cave populations in the two other regions, Guatemala (Molino y Caballo Moro caves) and Micos (Subterraneo cave). Finally, based on F_{ST} and common ancestry analysis, Bradic et al. proposed a fifth group (Pachón). Interestingly, Pachón and Chica corresponded to different clusters, and tentatively independent cave invasions, but potential gene flow from the surface, especially in Chica, makes this difficult to discern.

In contrast, Strecker et al. (2012) found Chica clustered in the same genetic group as Micos cave, while Bradic et al. (2012), found Micos and Chica as different clusters. In addition, Strecker et al. (2012) found Sabinos and Tinaja recovered as old lineages for both mitochondrial and nuclear markers and reported mitochondrial introgression in Pachón from the surface new lineage.

Finally, Strecker et al. (2012) also suggest that Chica, Micos, and Caballo Moro likely have limited gene flow with the surface populations, while no gene flow is observed between old caves (e.g., Sabinos, Tinaja, and Pachón) and surface populations. These results are interpreted as a confirmation of ecological speciation; however, Bradic et al. (2012) found high migration rates among surface and Chica, Pachón and Micos, as wells as migration among caves in El Abra limestone.

Coghill et al. (2014), using a phylogenomic analysis coming from SNPs, recover cave populations in four different lineages, which are interpreted as at least four independent invasions. Coghill et al. suggest that Chica and Pachón are derived from one lineage. Another independent lineage was recovered for Sabinos cave, which was closely related to lineages from far southern Mexico and Central America. Interestingly, this was previously suggested in phylogenetic studies of *Astyanax* (Dowling et al., 2002; Strecker et al., 2004; Ornelas-García et al., 2008). Coghill et al. suggest Sabinos to be a descendant of an extinct surface lineage. The Micos and Molino populations were recovered as two additional independent lineages, similar to previous studies (Bradic et al., 2012).

The occurrence of different lineages in the region, however, does not necessarily mean that the fishes have invaded the caves in different temporal "waves." Another scenario is possible, in which diversity of lineages could co-occur in the region previous to cave formation. It would be possible that formation of one cave would "capture" a specific lineage, inhabiting a certain part of the region in that particular moment. Action of evolutionary mechanisms on each genetic pool as those referred to before (Wilkens, 1988; Dowling et al., 2002; Protas et al., 2006) would result in the rising of troglobitic adaptation, but not necessarily related to the lineage age (Coghill et al., 2014). Support of this assumption is given by the current configuration of lineages, where old and new lineages are present in the surface populations (Ornelas-García et al., 2008).

As evidenced by the aforementioned studies to elucidate the history of cave invasion, understanding the demographic and phylogeographic history of the caves still represents a challenge for evolutionary biologists. Common patterns across studies identify that at least some of the cave invasions occurred from different epigean stocks. The specific number of invasions is still under discovery or debate, suggesting that future efforts to generate a more complete coverage of cave genetic diversity will be key to resolving these debates and providing a framework for understanding adaptation within the caves. As previously suggested, genetic drift could be an important factor when determining the current genetic composition (Avise and Selander, 1972) in cavefishes, where effective population sizes are lower in orders of magnitude compared to surface populations (Bradic et al., 2012). We advise those making future efforts to consider demographic stochasticity and migration as key factors in modeling allele composition at one specific moment of sampling (Avise and Selander, 1972).

REFERENCES

Allendorf, F.W., Waples, R.S., 1996. Conservation and genetics of salmonid fishes. In: Conservation Genetics. Springer, Massachusetts, USA, pp. 238–280.

Álvarez, J., 1946. Revisión del género Anoptichthys con descripción de una especie nueva (Pisc., Characidae). An. Esc. Nac. Cienc. Biol. Mex. 4, 263–282.

Álvarez, J., 1947. Descripción de *Anopichthys hubbsi* caracinindo ciego de La. Cueva de Los Sabinos. SLP Rev. Soc. Mex. Hist. Nat. 8, 215–219, 194.

Avise, J.C., Bermingham, E., Kessler, L.G., Saunders, N.C., 1984. Characterization of mitochondrial DNA variability in a hybrid swarm between subspecies of bluegill sunfish (*Lepomis macrochirus*). Evolution 38, 931–941.

Avise, J.C., Selander, R.K., 1972. Evolutionary genetics of cave-dwelling fishes of the genus *Astyanax*. Evolution 26, 1–19.

Bermingham, E., Coates, A., Cruz, G., Emmons, L., Foster, R.B., Leschen, R., Seutin, G., Thorn, S., Wcislo, W., Werfel, B., 1998. Geology and terrestrial flora and fauna of Cayos Cochinos, Honduras. Rev. Biol. Trop. 46, 15–37.

Birkhead, W.S., 1980. *Astyanax mexicanus* (Filippi), Mexican tetra. Atlas of North American freshwater fishes. North Carolina State Museum of Natural History, Raleigh, 139 p.

Borowsky, R., Cohen, D., 2013. Genomic consequences of ecological speciation in *Astyanax* cavefish. PLoS One 8. e79903.

Bradic, M., Beerli, P., Garcia-de Leon, F.J., Esquivel-Bobadilla, S., Borowsky, R.L., 2012. Gene flow and population structure in the Mexican blind cavefish complex (*Astyanax mexicanus*). BMC Evol. Biol. 12, 9.

Bradic, M., Teotónio, H., Borowsky, R.L., 2013. The population genomics of repeated evolution in the blind cavefish *Astyanax mexicanus*. Mol. Biol. Evol. 30, 2383–2400.

Calcagnotto, D., Schaefer, S.A., DeSalle, R., 2005. Relationships among characiform fishes inferred from analysis of nuclear and mitochondrial gene sequences. Mol. Phylogenet. Evol. 36, 135–153.

Campton, D., 1987. Natural hybridization and introgression in fishes: methods of detection and genetic interpretations. Population Genetics and Fishery Management. University of Washington Press, Seattle, pp. 161–192.

Coghill, L.M., Hulsey, C.D., Chaves-Campos, J., de Leon, F.J.G., Johnson, S.G., 2014. Next generation phylogeography of cave and surface *Astyanax mexicanus*. Mol. Phylogenet. Evol. 79, 368–374.

Concheiro Perez, G.A., Rican, O., Orti, G., Bermingham, E., Doadrio, I., Zardoya, R., 2007. Phylogeny and biogeography of 91 species of heroine cichlids (Teleostei: Cichlidae) based on sequences of the cytochrome *b* gene. Mol. Phylogenet. Evol. 43, 91–110.

Cui, R., Schumer, M., Kruesi, K., Walter, R., Andolfatto, P., Rosenthal, G.G., 2013. Phylogenomica reveals extensive reticulate evolution in *Xiphophorus* fishes. Evolution 67, 2166–2179.

Cuvier, G., 1819. Sur les poissons du sous-genre Hydrocyon, sur deux nouvelles espèces de Chalceus, sur trois nouvelles espèces du Serrasalmes, et sur l'Argentina glossodonta de Forskahl, qui est l'Albula gonorhynchus de Bloch. Mémoires du Muséum National d'Histoire Naturelle, Paris (N. S.) (Série A) Zoologie 5, 351–379, Pls. 26–28.

Darrin Hulsey, C., García de León, F.J., Sánchez Johnson, Y., Hendrickson, D.A., Near, T.J., 2004. Temporal diversification of Mesoamerican cichlid fishes across a major biogeographic boundary. Mol. Phylogenet. Evol. 31, 754–764.

De la Maza-Benignos, M., et al., 2015. Phylogeographic analysis of genus Herichthys (Perciformes: Cichlidae), with descriptions of Nosferatu new genus and *H. tepehua* n. sp. Hydrobiologia 748 (1), 201–231.

Doadrio, I., Domínguez, O., 2004. Phylogenetic relationships within the fish family Goodeidae based on cytochrome b sequence data. Mol. Phylogenet. Evol. 31 (2), 416–430.

Dowling, T.E., Martasian, D.P., Jeffery, W.R., 2002. Evidence for multiple genetic forms with similar eyeless phenotypes in the blind cavefish, *Astyanax mexicanus*. Mol. Biol. Evol. 19, 446–455.

Eschmeyer, W.N. (Ed.), 2014. Catalog of Fishes. Updated internet version, 18 June 2014. Catalog databases of CAS cited in FishBase (website).

Gayet, M., Marshall, L.G., Sempere, T., Meunier, F.J., Cappetta, H., Rage, J.C., 2001. Middle Maastrichtian vertebrates (fishes, amphibians, dinosaurs and other reptiles, mammals) from Pajcha Pata (Bolivia). Biostratigraphic, palaeoecologic and palaeobiogeographic implications. Palaeogeogr. Palaeoclimatol. Palaeoecol. 169, 39–68.

Gross, J.B., 2012. The complex origin of *Astyanax* cavefish. BMC Evol. Biol. 12, 105.

Gross, J.B., Meyer, B., Perkins, M., 2015. The rise of *Astyanax* cavefish. Dev. Dyn. http://dx.doi.org/10.1002/dvdy.24253.

Harrison, R.G., Larson, E.L., 2014. Hybridization, Introgression, and the nature of species boundaries. J. Hered. 105, 795–809.

Hausdorf, B., Wilkens, H., Strecker, U., 2011. Population genetic patterns revealed by microsatellite data challenge the mitochondrial DNA based taxonomy of *Astyanax* in Mexico (Characidae, Teleostei). Mol. Phylogenet. Evol. 60 (1), 89–97.

Hendry, A., 2009. Ecological speciation! Or the lack thereof? Can. J. Fish. Aquat. Sci. 66, 1383–1398.

Hrbek, T., Seckinger, J., Meyer, A., 2007. A phylogenetic and biogeographic perspective on the evolution of poeciliid fishes. Mol. Phylogenet. Evol. 43, 986–998.

Hubbs, C.L., 1955. Hybridization between fish species in nature. Syst. Zool. 4, 1–20.

Hubbs, C.L., Innes, W.T., 1936. The First Known Blind Fish of the Family Characidae: A New Genus from Mexico.

Javonillo, R., Malabarba, L.R., Weitzman, S.H., Burns, J.R., 2010. Relationships among major lineages of characid fishes (Teleostei: Ostariophysi: Characiformes), based on molecular sequence data. Mol. Phylogenet. Evol. 54, 498–511.

Jeffery, W.R., 2009. Regressive evolution in *Astyanax* cavefish. Annu. Rev. Genet. 43, 25–47.

Klinkhardt, M., Tesche, M., Greven, H., 1995. Database of Fish Chromosomes. Heidelberg, Germany, 237 p.

Lima, F.C.T., Malabarba, L.R., Buckup, P.A., Pezzi Da Silva, J.F., Vari, R.P., Harold, A., Benine, R., Oyakawa, O.T., Pavanelli, C.S., Menezes, N.A., Lucena, C.A.S., Malabarba, M.C.S.L., Lucena, Z.M.S., Reis, R.E., Langeani, F., Cassati, L., Bertaco, V.A., 2003. Genera Incertae Sedis in Characidae. In: Reis, R.E., Kullander, S.O., Ferraris, Jr., C.J. (Eds.), Checklist of the Freshwater Fishes of South and Central America. EDIPUCRS, Porto Alegre, Brasil, pp. 106–168.

Mallet, J., 2005. Hybridization as an invasion of the genome. Trends Ecol. Evol. 20, 229–237.

Mallet, J., 2008. Hybridization, ecological races and the nature of species: empirical evidence for the ease of speciation. Philos. Trans. R. Soc. Lond. B: Biol. Sci. 363, 2971–2986.

Miller, R., 2005. Freshwater Fishes of Mexico. The University of Chicago Press, Chicago.

Mirande, J.M., 2009. Weighted parsimony phylogeny of the family Characidae (Teleostei: Characiformes). Cladistics 25, 574–613.

Mitchell, R.W., Russell, W.H., Elliott, W.R., 1977. Mexican Eyeless Characin Fishes, Genus Astyanax: Environment, Distribution, and Evolution. Texas Tech Press. The Museum Special Publications 12, 89 p.

Myers, G.S., 1966. Derivation of freshwater fish fauna of Central America. Copeia 4, 766–773.

Nadeau, N.J., Whibley, A., Jones, R.T., Davey, J.W., Dasmahapatra, K.K., Baxter, S.W., Quail, M.A., Joron, M., Blaxter, M.L., Mallet, J., 2012. Genomic islands of divergence in hybridizing *Heliconius* butterflies identified by large-scale targeted sequencing. Philos. Trans. R. Soc. Lond. B: Biol. Sci. 367, 343–353.

Oliveira, C., Avelino, G.S., Abe, K.T., Mariguela, T.C., Benine, R.C., Ortí, G., Vari, R.P., Correa e Castro, R.M., 2011. Phylogenetic relationships within the speciose family Characidae (Teleostei: Ostariophysi: Characiformes) based on multilocus analysis and extensive ingroup sampling. BMC Evol. Biol. 11, 275.

Ornelas-García, C.P., Dominguez-Dominguez, O., Doadrio, I., 2008. Evolutionary history of the fish genus *Astyanax* Baird & Girard (1854) (Actinopterygii, Characidae) in Mesoamerica reveals multiple morphological homoplasies. BMC Evol. Biol. 8, 340.

Otero, O., Gayet, M., 2001. *Palaeoichthyofaunas* from the Lower Oligocene and Miocene of the Arabian Plate: palaeoecological and palaeobiogeographical implications. Palaeogeogr. Palaeoclimatol. Palaeoecol. 165, 141–169.

Perdices, A., Bermingham, E.A.M., Doadrio, I., 2002. Evolutionary history of the genus Rhamdia (Teleostei: Pimelodidae) in Central America. Mol. Phylogenet. Evol. 25, 172–189.

Perdices, A., Doadrio, I., Bermingham, E., 2005. Evolutionary history of the synbranchid eels (Teleostei: Synbranchidae) in Central America and the Caribbean islands inferred from their molecular phylogeny. Mol. Phylogenet. Evol. 37, 460–473.

Pie, M., Baggio, R., Boeger, W., Patella, L., Ostrensky, A., Vitule, J., Abilhoa, V., 2009. Molecular data reveal a diverse *Astyanax* species complex in the upper Iguaçu River. J. Fish Biol. 75, 2357–2362.

Protas, M.E., Hersey, C., Kochanek, D., Zhou, Y., Wilkens, H., Jeffery, W.R., Zon, L.I., Borowsky, R., Tabin, C.J., 2006. Genetic analysis of cavefish reveals molecular convergence in the evolution of albinism. Nat. Genet. 38, 107–111.

Říčan, O., Piálek, L., Zardoya, R., Doadrio, I., Zrzavý, J., 2012. Biogeography of the Mesoamerican Cichlidae (Teleostei: Heroini): colonization through the GAARlandia land bridge and early diversification. J. Biogeogr. 40 (3), 579–593.

Rosen, D.E., 1975. Vicariance model of Caribbean biogeography. Syst. Zool. 24, 431–464.

Şadoğlu, P., 1957. A Mendelian gene for albinism in natural cave fish. Cell. Mol. Life Sci. 13 (10), 394–394.

Scascitelli, M., Whitney, K., Randell, R., King, M., Buerkle, C., Rieseberg, L., 2010. Genome scan of hybridizing sunflowers from Texas (*Helianthus annuus* and *H. debilis*) reveals asymmetric patterns of introgression and small islands of genomic differentiation. Mol. Ecol. 19, 521–541.

Schumer, M., Cui, R., Powell, D.L., Dresner, R., Rosenthal, G.G., Andolfatto, P., 2014. High-resolution mapping reveals hundreds of genetic incompatibilities in hybridizing fish species. Elife 3. e02535.

Schuppa, M., 1984. Morphometrische und meristische Untersuchungen an verschiedenen Astyanax-Populationen (Characidae) Mexikos. Fachbereich Biologie. Universitiit Hamburg, Hamburg.

Silvia, P., et al., 2010. Phylogenetic relationships and biogeographical patterns in Circum-Mediterranean subfamily Leuciscinae (Teleostei, Cyprinidae) inferred from both mitochondrial and nuclear data. BMC Evol. Biol. 10 (1), 265.

Simpson, G.G., 1944. Tempo and Mode in Evolution. Columbia University Press, New York, Chichester, West Sussex.

Simpson, G.G., 1953. The Major Features of Evolution. Columbia University Press, New York.

Smith, G.R., 1992. Introgression in fishes: significance for paleontology, cladistics, and evolutionary rates. Syst. Biol. 41, 41–57.

Strecker, U., 2006. The impact of invasive fish on an endemic *Cyprinodon* species flock (Teleostei) from Laguna Chichancanab, Yucatán, Mexico. Ecol. Freshw. Fish 15, 408–418.

Strecker, U., Bernatchez, L., Wilkens, H., 2003. Genetic divergence between cave and surface populations of *Astyanax* in Mexico (Characidae, Teleostei). Mol. Ecol. 12, 699–710.

Strecker, U., Faundez, V.H., Wilkens, H., 2004. Phylogeography of surface and cave *Astyanax* (Teleostei) from Central and North America based on cytochrome *b* sequence data. Mol. Phylogenet. Evol. 33, 469–481.

Strecker, U., Hausdorf, B., Wilkens, H., 2012. Parallel speciation in *Astyanax* cave fish (Teleostei) in Northern Mexico. Mol. Phylogenet. Evol. 62, 62–70.

Thorgaard, G.H., Allendorf, F.W., 1988. Developmental genetics in fishes. Developmental Genetics of Animals and Plants. MacMillan, New York, pp. 369–391.

Turner, T.L., Hahn, M.W., Nuzhdin, S.V., 2005. Genomic islands of speciation in *Anopheles gambiae*. PLoS Biol. 3. e285.

Valdez-Moreno, M., Ivanova, N., Elías-Gutiérrez, M., Contreras-Balderas, S., Hebert, P., 2009. Probing diversity in freshwater fishes from Mexico and Guatemala with DNA barcodes. J. Fish Biol. 74, 377–402.

Vonholdt, B.M., Stahler, D.R., Bangs, E.E., Smith, D.W., Jimenez, M.D., Mack, C.M., Niemeyer, C.C., Pollinger, J.P., Wayne, R.K., 2010. A novel assessment of population structure and gene flow in grey wolf populations of the Northern Rocky Mountains of the United States. Mol. Ecol. 19, 4412–4427.

Wilkens, H., 1971. Genetic interpretation of regressive evolutionary processes: studies on hybrid eyes of two Astyanax cave populations (Characidae, Pisces). Evolution 530–544.

Wilkens, H., 1988. Evolution and genetics of epigean and cave *Astyanax fasciatus* (Characidae, Pisces). Evolutionary Biology. Springer, USA, pp. 271–367.

Zardoya, R., Doadrio, I., 1999. Molecular evidence on the evolutionary and biogeographical patterns of European cyprinids. J. Mol. Evol. 49 (2), 227–237.

Part II

Genetic Diversity and Quantitative Genetics

Chapter 5

Regressive Evolution: Testing Hypotheses of Selection and Drift

Richard Borowsky
Department of Biology, New York University, New York, New York, USA

INTRODUCTION

Cavefishes species that complete their life cycles entirely in the dark typically exhibit traits thought to have evolved in response to the environmental shift from surface to subterranean life. The cavefish system brings to evolutionary biology much of the power of the replicated experiment. Each independently derived species or population is the result of a similar experiment, one that examines the evolutionary consequences of the environmental shift to life in darkness and with limited food availability. Because we have multiple replicates of this experiment, we can test hypotheses formed from observations on one cavefish species by its predictions for others.

Augmentations of the non-visual senses and changes in metabolism and behavior have evolved repeatedly in cavefishes and are thought to be the result of selection for adaptations to life in perpetual darkness (Culver et al., 1995). The elaboration of neuromast cells in the lateral line system (Poulson, 1963), increased sensitivity to dissolved amino acids in the water (Bibliowicz et al., 2013; Protas et al., 2008), and attraction to vibration in the water (Yoshizawa et al., 2010) presumably all facilitate the detection and capture of food items in the absence of visual information. The increased sensitivity to dissolved amino acids is of particular interest, because it is known that many aquatic invertebrates release them into the water (Johannes and Webb, 1970) and when detected by vertebrates, they can elicit striking food search behavior (Ferrer and Zimmer, 2007). Similarly, the evolution of a more efficient metabolism (Moran et al., 2014; Poulson, 1963) and increased wakefulness (Duboue and Borowsky, 2012; Duboue et al., 2011) are likely adaptations for coping with food scarcity. Given the straightforward functional significances of these constructive changes, there is little disagreement that

Biology and Evolution of the Mexican Cavefish. http://dx.doi.org/10.1016/B978-0-12-802148-4.00005-0
© 2016 Elsevier Inc. All rights reserved.

the force driving their evolution is natural selection, although it should be noted that there is also little direct evidence supporting this view.

In contrast, the most obvious and seemingly consistent troglomorphic traits are losses of eyes and pigmentation; but as these are regressive changes, there is no consensus on how they evolved. While neither vision nor pigmentation is of any obvious, direct value in the dark, it is hard to imagine the selective advantage of their losses. This is especially the case for melanin pigmentation, because melanin production is down-regulated under conditions of constant darkness (Parker, 1948). In consequence, there is longstanding debate (Culver, 1985) as to whether the driving forces for these changes are simply random mutation and genetic drift (Wilkens, 1988, 2007), or selection, either direct (Protas et al., 2007) or indirect (Bilandzija et al., 2013; Borowsky and Wilkens, 2002; Yamamoto et al., 2009). It is likely that all three mechanisms have played roles in driving cave adaptation, and the challenge is to identify the driver or drivers in specific cases (Borowsky, 2013).

In this chapter, I examine current evidence related to three types of regressive changes—decrease in eye size, decrease in melanophore numbers, and albinistic loss of the ability to synthesize melanin—deriving hypotheses from observations and applying the principle of the replicated experiment to test them by their predictions.

EYES AND PIGMENTATION, HYPOTHESES AND THEIR TESTS

Protas et al. (2007) studied the genetic architecture of decreases in eye size and melanophore number in Pachón cavefishes relative to surface *Astyanax*. Genetic mapping and quantitative trait loci (QTL) analysis identified 12 locations in the genome where allelic variation was correlated with variation in eye size, and 13 regions in the genome where allelic variation was correlated with variation in melanophore density. While the numbers of QTL and distributions of effect sizes were similar for both traits, the patterns of substitution effects differed significantly. In every case, the eye size QTL had the same polarity, where homozygotes for the cave allele had smaller eyes than homozygotes for the surface allele. In contrast, the polarities for the melanophore QTL were mixed, with eight QTL in which homozygotes for the cave allele had fewer melanophores than homozygotes for the surface allele and five QTL having the opposite polarity.

The patterns of polarity of the two traits were significantly different, and it was hypothesized that this difference reflected biologically significant differences in the mechanisms driving their regressions. Specifically, it was hypothesized that the consistent polarity of eye QTL signaled that eye regression was driven by direct natural selection. That is, eyes are detrimental in the dark cave environment. On the other hand, it was hypothesized that the inconsistent polarity of melanophore QTL reflected the stochastic nature of another principal driver, genetic drift. The main theme of this chapter is that it is possible to test such hypotheses by their predictions.

The hypotheses make three testable predictions, which are:

1. The basic observation should be replicable. That is, if we map and characterize QTL for different, independently evolved cave populations, we should detect the same patterns of polarity differences. If not, this would seriously challenge the generality or validity of the hypotheses.
2. The second prediction is that eyes should regress much more quickly than pigmentation, because, given reasonable estimates of population sizes and selective coefficients, selection should drive changes in allelic frequency much faster than drift.
3. The third prediction is that the linkage of eye size QTL and regions bearing hallmarks of selection should be much tighter than the linkage of melanophore QTL with such regions.

Testing the First Prediction

The first prediction was tested by replicating the QTL analysis in independently evolved cave populations of *Astyanax mexicanus*. Previous complementation analyses of eye size in complex crosses among cave and surface populations showed that different sets of genes are responsible for eye regression in three cave populations of this species: Pachón, Tinaja, and Molino (Borowsky, 2008). That is, the three caves are independent replicates of very similar evolutionary experiments, and Molino and Tinaja cavefishes are appropriate choices to test the first prediction. The Molino/surface cross was previously described (Protas et al., 2006), the Tinaja/surface cross is described in section "Materials and Methods."

These replicate studies revealed eight new QTL for eye size in the Tinaja/surface cross and three new eye size QTL in the Molino/surface cross. All 11 QTL had the same polarity, with cave alleles causing smaller eyes than surface alleles. Two QTL affecting melanophore numbers were found in the Tinaja/surface cross. These two QTL opposed each other in polarities. These results are entirely consistent with the prediction of the hypothesis. Combining the original data with the new data shows the differences in patterns of polarity for the two traits are highly significant (23:0 vs. 9:6, $P=0.0018$, Fisher's exact test). One eye size QTL reported in the literature from the Pachón cave population (Borowsky, 2013) is unique, and its polarity is opposite that of all other known eye size QTL. Including this QTL in the analysis does not abolish statistical significance ($P=0.0082$). The newly described QTL and their substitution effects are summarized in Table 5.1.

Testing the Second Prediction

The second prediction is that the rate of eye regression should exceed that of melanophore loss, because evolutionary change is much faster when it is driven

TABLE 5.1 Properties of New Eye Size and Melanophore Number (E and M) QTL from the Tinaja and Molino Cave Populations. LG = Linkage Group; LOD = Log Odds Score After MIM Mapping (Kao et al., 1999); Position of Locus on the LG; Proportions of Explained (PEV) and Explained Additive Variance (PEVad); the Substitution Effect of a Cave Allele, and the Heterozygote Effect. Heterozygote Effect not Measurable in the Molino Backcross

Trait	LG	LOD	*P* Value	Locus Position	PEV	PEVad	Substitution Effect	Heterozygote Effect
QTL from Tinaja X surface F_2								
E	LGT10	5.7	0.001	3	0.048	0.047	−0.097	0.013
E	LGT13	4.7	0.001	10	0.058	0.004	−0.030	0.074
E	LGT14	3.3	0.003	103	0.027	0.012	−0.049	0.039
E	LGT19	6.9	0.001	2	0.057	0.057	−0.107	0.000
E	LGT2	3.5	0.003	16	0.037	0.028	−0.075	0.030
E	LGT21	26.9	0.001	0	0.260	0.255	−0.227	−0.023
E	LGT3	5.0	0.001	59	0.056	0.056	−0.106	0.000
E	LGT6	4.8	0.001	4	0.052	0.042	−0.093	0.031
M	LGT24	14.9	0.001	46	0.270	0.175	−9.071	4.724
M	LGT3	4.3	0.001	38	0.083	0.063	5.445	2.154
QTL from Molino X surface backcross								
E	LGM11	2.4	0.005	21	0.064	N/A	−0.122	N/A
E	LGM27	2.4	0.007	4	0.077	N/A	−0.134	N/A
E	LGM4	2.5	0.006	93	0.072	N/A	−0.130	N/A

by selection rather than drift. This prediction was tested by comparing regression of eyes and pigmentation in five cavefish species from an unrelated family, the Balitoridae, from southeast Asia. Cavefishes are generally stated to have both reduced eyes and pigmentation (Proudlove, 2004; Romero, 2004; Weber et al., 1998), but it is widely recognized that in some cases, reduced pigmentation may be a physiological or morphological response to life in the dark that can be reversed. As early as 1890, it was observed that the cave salamander, *Proteus anguinus*, pure white when maintained in darkness, darkened considerably when exposed to light in the laboratory (Poulton, 1890). By 1948 the bleaching of aquatic vertebrates when maintained in darkness was recognized as a general phenomenon (Parker, 1948); Vandel (1965) distinguished among cave animals that were stably depigmented and those that were in an "unstable state of depigmentation" and would darken when brought into the light (Vandel, 1965). The balitorid cavefishes examined in this study all belong to the second category; they exhibited little or no melanin pigmentation at the time of capture, but all showed a noticeable dusting of melanophores within days of being exposed to light, and they continued to darken over months. All had been maintained in the laboratory under conditions of 12:12 L:D for over 8 years prior to being scored for pigmentation for this study and were long stabilized in their pigmentation state.

Five cave species and three surface species were available for analysis. The three surface species were considered typical in coloration for surface balitorids based upon comparison with published figures in three monographs covering southeast Asia (Kottelat, 1990, 2001; Rainboth, 1996) and comparison with other balitorid species imaged on the Web (Loaches_online, 2014). As such, they were taken to be representative of the character states of the unknown ancestors of the cave species. Overall pigmentation of the eight species was determined from photographs of free-swimming individuals using a standard gray card in each photograph, which allowed for corrections for minor differences in lighting. Details are given in section "Materials and Methods." Eye development was scored on a scale of 0.0-1.0 based upon the following criteria: 0.33 for externally visible eyes; 0.33 for behavioral response to bright light; 0.34 for full visual function, as demonstrated by responses to moving objects, visually based feeding, etc. Populations that were polymorphic for externally visible eyes were rated as 0.33/2. This scoring system is arbitrary, but was designed to count as equally important the development of the dioptric apparatus (lens and anterior chamber), the sensory apparatus (retina), and the fully integrated function of the whole system.

A total of 25 individuals represented the eight species, with N ranging from one to five for each species. In spite of the small numbers, the results were clear. On average, the five species of cavefishes were no lighter in pigmentation than the three surface species. The most pigmented of the cave species was *Schistura kaysoniae*, which was significantly darker than two of the surface species, *S. similis* ($t_6=6.61$, $P<0.001$) and *S. mahnerti* ($t_5=4.56$, $P<0.01$).

Next most pigmented of the cavefishes was *S. jaruthannini*, which was also significantly darker than two of the surface species, *S. similis* ($t_7=4.89$, $P<0.01$) and *S. mahnerti* ($t_6=2.89$, $P<0.05$). Sample sizes were too small to test the significance of other differences, but the two least pigmented of the eight were the cave species *S. oedipus* and *Nemacheilus troglocataractus*. It is possible that these two cave species are significantly lighter in pigmentation than the three surface species because the three individuals that were phenotyped were the survivors of larger groups held in the lab at an earlier time (*S. oedipus*, total $N=5$; *N. troglocataractus*, total $N=4$), and images made earlier revealed no noticeable difference in coloration among individuals within groups. Thus, cave species were among the darkest and lightest of the group and, as a whole, showed little sign of overall pigmentation loss (Table 5.2).

In contrast, eyes were considerably reduced in all of the cave species. None had visual function, although two were responsive to bright light and two others variably exhibited external eye spots (Table 5.3). It is clear that vision has regressed considerably in the balitorid cavefishes, and much faster than any regression of pigmentation (Figure 5.1).

The rate of evolution driven by recurrent mutation and drift is equal to the mutation rate, μ, while the rate driven by selection is approximately $4N_e s\nu$, where N_e is the effective population size, s is the selective coefficient, and ν is the rate of favorable mutations (Kimura and Ohta, 1971). Because in this

TABLE 5.2 The Mean Reflectance of Balitorid Surface and Cavefishes Measured from Photographs Reduced to Gray Tone. Reflectance Values were Standardized Against 18% Gray Cards as in section "Materials and Methods." *N*=Number of Fish Measured

	Reflectance		
Species	**Mean**	**StDev**	***N***
Surface			
S. similis	24.5	3.3	4
S. mahnerti	21.7	4.7	3
S. moeiensis	15.0	N/A	1
Cave			
N. troglocataractus	36.0	N/A	1
S. kaysoniae	4.5	5.1	4
S. oedipus	40.0	18.4	2
S. jaruthannini	13.4	3.4	5
S. spiesi	22.6	14.2	5

TABLE 5.3 Visual System Status in Surface and Cave Balitorids (1 = Presence, 0 = Absence). Blind = Incapable of Forming Image; Reactive = Showing Behavioral Response to Bright Light; Eye Spots = External and Visible. (*S. jaruthannini* and *S. spiesi* are Polymorphic for Eye Spots and are Rated 0.5.) Visual System Overall is the Weighted Average of all Three Attributes

	Visual System			
	Blind	Reactive	Eye Spots	Overall
Surface				
S. similis	1	1	1	1
S. mahnerti	1	1	1	1
S. moeiensis	1	1	1	1
Cave				
N. troglocataractus	0	0	0	0
S. kaysoniae	0	1	0	0.33
S. oedipus	0	1	0	0.33
S. jaruthannini	0	0	0.5	0.165
S. spiesi	0	0	0.5	0.165

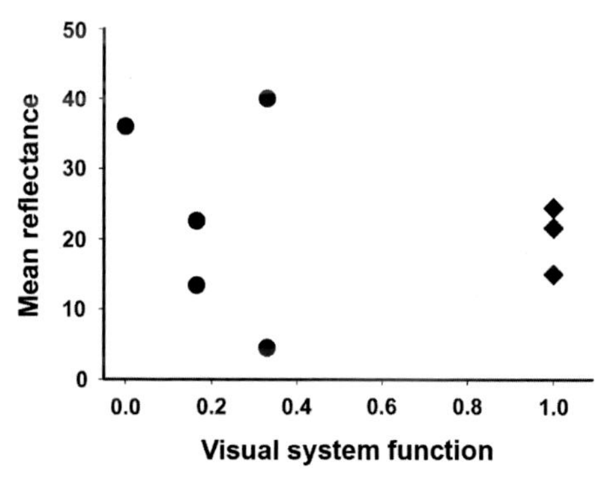

FIGURE 5.1 Mean reflectance of five cave (circles) and three surface (diamonds) balitorid species plotted against degree of visual system function. Visual function is more regressed than pigmentation in the cave species.

case we are interested in only one direction of change, regression of eyes or of pigmentation, only mutations with the same polarity, whether adaptive or neutral, are under consideration. Thus, for this subset of mutations, $\mu = \nu$. Previous studies estimated the population size of *S. oedipus* from direct counts to be in "the thousands" (Trajano et al., 2002). While population sizes for the other cavefishes have not been estimated directly, estimates of nucleotide diversity, a function of effective population size, have been made for *S. oedipus*, *S. speisi*, and *N. troglocataractus*; all are very similar to one another (approx. 10^{-3}) (Borowsky and Vidthayanon, 2001). Thus, a reasonable estimate of effective population size for these cave species is in the low thousands. Taking the number 2500 for the sake of illustration, allelic variants with selective coefficients smaller than $1/2N_e$ (= 2.0 e^{-4}) would be effectively neutral (Kimura and Ohta, 1971). Given an effective population size of 2500, regressive evolution would proceed 10 times faster under selection with even a small selective coefficient of 10^{-3} than it would by mutation pressure and drift.

Testing the Third Prediction

QTLs denote regions in the genome where allelic differences between cave and surface forms have phenotypic consequences. By themselves, however, they are silent on whether the allelic differences arose through selection or drift. It is possible, though, to identify regions in the genome with hallmarks of selection. Thus, the third prediction is that QTL for eye size will have closer association with such regions than will melanophore number QTL.

Hallmarks of selection can be detected through population genetic analyses that identify marker loci where allelic frequencies in cave populations are significantly more diverged from those in surface populations than would be expected on the hypothesis of neutral drift. Population genetic analyses of five different *Astyanax* cave populations and representative surface populations have identified over 70 such SNP (single nucleotide polymorphism) loci ("significant SNPs") that are significantly diverged, based on F_{ST} values, presumably because of directional selection in the cave populations (Bradic et al., 2013). Thus, to test the associations of eye and melanophore QTL with such SNPs, I remapped a previously studied F_2 progeny derived from a cross between a surface and a Pachón cavefish (Figure 5.1S). This allowed me to measure the distance in centimorgans (cM) from the center of each QTL to the nearest significant SNP.

Eye QTL were significantly closer than melanophore QTL to significant SNPs (median distances of 6.2 vs. 19.5 cM; Mann-Whitney $U=30$, $Z=1.76$, one-sided $P=0.039$; Figure 5.2). 6 of the 12 eye QTL versus 1 of the 10 melanophore QTL were closer than 5 cM to the nearest significant SNP (Table 5.4). Thus, the observations are in accord with the prediction.

To summarize this part, all three predictions of the hypotheses are supported by the data. The patterns of QTL polarities detected in the Tinaja and Molino populations are consistent with those in the Pachón population that first suggested the hypotheses. The prediction that eyes will regress faster than melanophore pigmentation is supported by the analysis of balitorid cavefishes. The prediction that eye size QTL will be more closely associated than melanophore number QTL with areas of the genome exhibiting hallmarks of selection is borne out by the analysis of the Pachón by surface fishes F_2. Thus, the evidence strongly supports the hypotheses that selection is the principal driver of eye regression while drift is the principal driver of melanophore regression.

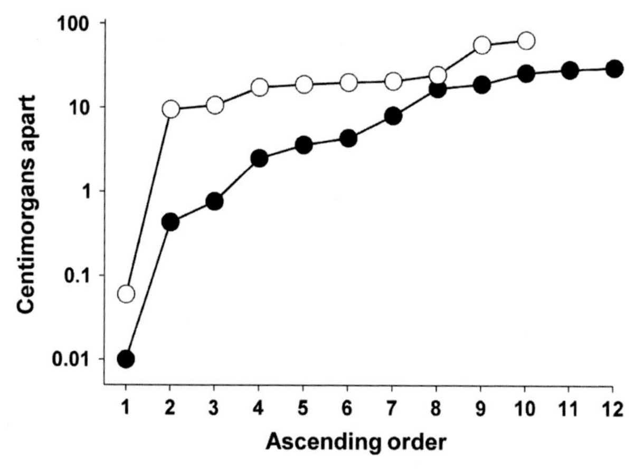

FIGURE 5.2 Distances between QTL and the nearest marker in the genome bearing hallmarks of selection ordered from closest to most distant. Solid circles = eye size QTL, hollow circles = melanophore number QTL.

TABLE 5.4 Distances in Centimorgans from QTL for Eye Size or Melanophore Number to Nearest Region in the Genome with Hallmarks of Selection. SNPs with Hallmarks of Selection are Shown in Red in Figure 5.1S (companion site). LG Numbers are Linkage Groups as Numbered in Figure 5.1S

Eye QTL		Melanophore QTL	
Distance	LG	Distance	LG
0.01	12	0.06	38
0.43	6	9.50	2
0.76	14	10.64	34
2.50	30	17.40	12
3.62	1	18.90	30
4.36	3	20.00	25
8.08	37	20.80	31
17.00	27	24.60	26
19.14	31	56.40	13
26.00	29	64.00	8
28.60	20		
30.00	14		
Median		Median	
6.22		19.45	

This does not mean that selection plays no role in loss of melanophores or that drift plays no role in eye regression. For example, the melanophore QTL on Lg38 is only 0.06 cM from a significant SNP (m204), suggesting that selection may have played a role in its evolution, either directly or indirectly through hitchhiking. A second example is that of the gene *mc1r* (Lg13), mutant alleles of which cause reduced numbers of melanophores and are found in high frequency in multiple independently evolved cave populations of *Astyanax* (Gross et al., 2009).

Indirect selection may also play a role in regression of eyes or pigmentation through pleiotropic effects of alleles that have principal effects on other traits (see O'Quin and McGaugh in Chapter 7). One example is the hypothesis that direct selection for more taste buds and more robust feeding apparatus in the cave forms of *A. mexicanus* indirectly drives eye regression. The zone of mid-line *hedgehog* expression is expanded in cavefishes compared to surface fishes with the result of decreasing the bilaterally symmetrical zones of *pax6* expression early in development. The smaller zone of *pax6* expression leads to decreased eye size. This hypothesis has considerable experimental support (Jeffery, 2005; Yamamoto et al., 2009) and is strengthened by the fact that similar changes in *hedgehog* expression occur in three independently derived populations of *Astyanax* cavefishes (Jeffery, 2009). Nevertheless, indirect selection would as often have positive as negative effects and would produce mixed polarities of pleiotropic QTL, the same as drift. Thus, it is unlikely to have been the principal driver of eye regression, because the polarities of eye QTL are consistent. In the case of pigmentation regression, the evidence points to slower change and inconsistent polarity of substitution effects. This suggests that mutation pressure and drift are at work, although it doesn't rule out a role for indirect selection through pleiotropic effects.

MELANIN TRAITS

Albinism

Albinism, the complete or near complete absence of melanin, is a regressive change in pigmentation that has been reported in about half of all troglomorphic cavefish species surveyed (47/86 = 55%) (Romero and Green, 2005). This figure may be an overestimate, because synthesis of melanin is down-regulated in constant darkness (Parker, 1948), presenting the possibility of misclassification of specimens fixed shortly after collection. So, while it is common in cavefishes, albinism is by no means close to universal. Because albinism is not more common in cavefishes, it would appear to be a neutral rather than advantageous trait in the cave environment. Nevertheless, in at least three different cave populations of *Astyanax*, albinism has arisen independently through mutations in the same gene, *oca2*. These mutations affect either the coding sequence or the control of the gene's expression, but the resulting loss of *oca2* function needed for melanin synthesis is the same (Protas et al., 2006). These independent convergences on the same phenotype via mutations affecting the function of the same key protein suggest selection; however, *oca2* is a large gene, 123 kb in length, and codes for a functional protein of 888 amino acids, and thus presents a large mutational target, suggesting that mutation and drift may also account for its repeated involvement in albinism of *A. mexicanus*.

In the Pachón X surface hybrid cross, a strong QTL for eye size nearly centers on the *oca2* locus (Figure 5.1S, companion site). Humans with oculocutaneous albinism almost universally have serious retinal defects (Fulton et al., 1978), so the association of an eye QTL with the *oca2* locus may reflect a functional relationship. Thus, the selective advantage of *oca2* mutations in the cave environment might simply reflect their indirect pleiotropic effects on eye structure.

A second way in which *oca2* mutants might be favored in the cave environment also hinges on pleiotropy. Both the melanin and the catecholamine synthesis pathways compete for L-tyrosine, a shared precursor. Thus, a plausible selective advantage to loss of melanin synthesis in cavefishes is that it could increase the availability of L-tyrosine for synthesis of the neurotransmitters that play roles in the numerous behavioral changes in cavefishes (Bilandzija et al., 2013). Indeed, catecholamine levels, including those for norepinephrine (NE), were significantly higher in the brains of adult cavefishes than in those of surface fishes raised in the laboratory (Bilandzija et al., 2013). Furthermore, knock-downs of *oca2* in surface fishes increased levels of L-tyrosine and dopamine (see also Jeffery et al. in Chapter 9).

There is a potential link between increased availability of NE and the increased wakefulness phenotype of the *Astyanax* cavefishes: the cave phenotype is abolished by treatment with propranolol, an inhibitor of β-adrenergic signaling (Duboue et al., 2012). Thus, it is tempting to hypothesize that positive selection for wakefulness drove the loss of *oca2* activity to silence the melanin synthesis pathway and increase the availability of NE.

Nevertheless, for the wakefulness trait, this link seems unlikely. Increased wakefulness of larvae was documented in three independently derived *Astyanax* cave populations, Tinaja, Pachón, and Molino (Duboue et al., 2011). Of the three, Tinaja fishes are most wakeful, yet they synthesize melanin while Molino and Pachón fishes do not (see also Duboué and Keene in Chapter 16). This is the opposite of the prediction of the hypothesis. Furthermore, in the F_2 progeny of a Pachón X surface cross, there was no significant correlation between pigment state and wakefulness. Finally, I note that all five balitorid cavefishes tested exhibited significant increases in wakefulness compared to surface balitorids (Duboue and Borowsky, 2012), yet all have intact and functional *oca2*.

It is also unclear how much of an advantage, if any, there would be to eliminating *oca2* function in the cave environment. Assuming that silencing the melanin synthesis pathway is beneficial in the cave environment, an *oca2* null phenotype is not needed to accomplish this. It has long been recognized that melanin synthesis is minimized in the dark in many species of fishes (Eigenmann, 1900; Parker, 1948), and the rule pertains to *Astyanax* surface fishes, as well (Romero and Green, 2005). Thus, a surface fish entering the cave environment would achieve the needed phenotypic state physiologically. That competition between the two biosynthetic pathways could drive the silencing of *oca2* is an intriguing hypothesis, but one that needs more evidence.

Brown Phenotype

Another melanin trait in *Astyanax* cavefishes is the brown phenotype, named for the color of the larval eye caused by changes in structure of the melanin in the retinal pigmented epithelium (RPE). Wild type color of the RPE is black, whereas it is brown in homozygotes for the gene. The brown phenotype has been reported from seven populations (Sadoglu and Mckee, 1969; Wilkens and Strecker, 2003), which represent at least three independent evolutionary origins (Bradic et al., 2012). Two different mutational changes in the gene *mc1r*, one a loss of function and the other a hypomorph, have been shown to be responsible for the phenotype (Gross et al., 2009). In addition to causing structural changes in melanin, the *mc1r* mutations in *Astyanax* cause reduced numbers of melanophores, a novel phenotype for this gene (Gross et al., 2009). As in the case with *oca2*, the convergence on the same phenotype through independent mutations in the same gene suggests a role for selection, although there is no direct evidence for this.

oca2 and *mc1r* interact epistatically

The *oca2* and *mc1r* loci are unlinked and their products are not known to interact directly, but there is evidence that they may interact epistatically. Hybrid F_1 males derived from a cross between a Pachón cave and a surface fish produce sperm cells with all four combinations of parental alleles at the two loci (mc1rC/oca2C, mc1rC/oca2S, mc1rS/oca2C, and mc1rS/oca2S). These four types of gametes should be transmitted in equal numbers to the next generation, but combinations that are homospecific (mc1rC/oca2C and mc1rS/oca2S) appear to be transmitted preferentially, in departure from independent assortment (Borowsky and Cohen, 2013). The mechanism for this transmission bias is unknown, but whatever it is, it could provide selective advantage to allelic variants in the cave population. Consider a situation in which an *oca2* mutant allele had reached high frequency in a cave population, through drift or selection, as discussed above. If an *mc1r* mutant allele were introduced into the population, it would have a major selective advantage over the wild type allele, because it would be transmitted with higher frequency than the wild type allele from individuals that were heterozygotes. The fitness advantage of the mutant allele estimated from data when the phenomenon was first discovered (Borowsky and Cohen, 2013) was greater than 50%, but this was an overestimate, due to ascertainment bias. In a subsequent experiment (unpublished) the advantage was 11%. This epistatic drive would decrease the probability of loss of the mutant allele when rare, and would speed fixation in the population. As a caveat, there is no direct proof that *oca2* and *mc1r* are the interacting entities and the epistasis may be between functional regions closely linked to the two loci.

CONCLUSION

In their evolutionary adaptation to subterranean life, cavefish species converge phenotypically on a suite of traits, including loss of eyes and pigmentation, augmentation of the non-optic senses, and physiological and behavioral changes related to food scarcity. Not every cavefish population exhibits every trait, but the convergences are consistent enough to warrant explanation. Increased sensitivity to vibration or dissolved chemicals in the water, as examples, are constructive changes that facilitate the detection of food in the dark. A more efficient metabolism, also constructive, increases survivorship and fecundity in an environment with low-energy input. It is generally agreed that the evolution of such constructive changes is driven by natural selection.

On the other hand, the forces driving regressive evolutionary changes, eye and pigmentation losses, are more difficult to identify and are still debated. What roles do selection and genetic drift play in regressive evolution? If selection is important, is it direct selection on the trait or is it indirect, due to the pleiotropic effects of selection for a constructive trait? The cavefish system is ideal to address such questions, because each separately evolved cavefish population or species is an independent replicate of a very similar evolutionary experiment. As such, hypotheses developed through observations on one cavefish taxon can be tested in others, based upon their predictions.

Analysis of the polarities of substitution effects of QTL related to eye size and numbers of melanophores led to the hypotheses that eye regression was driven by selection and melanophore loss was due to genetic drift. These hypotheses make specific predictions that were used to test them: (1) that the observations will be reproducible in independently evolved cave populations, (2) that the rate of eye loss will exceed that of melanophore loss in unrelated species of cavefishes, and (3) that QTL for eye size will be better associated with hallmarks of selection in the genome than QTL for melanophore numbers. All three predictions were supported by the data, leading to the conclusion that eye loss in cave populations is driven predominantly by selection and that this selection is primarily direct, rather than indirect. In contrast, losses of melanophores are driven primarily by drift, perhaps with a small component of indirect selection. The analyses illustrate the power of the approach, which is observation to hypothesis, hypothesis to prediction, and prediction to test, in replicate populations or species.

In addition to melanophore loss, losses of the ability to synthesize melanin or alterations in melanin structure are examples of pigmentation regressions that may be driven by selection. This conclusion is based upon the repeated, independent alterations in the *oca2* and *mc1r* genes that are the basis for albinism and the brown phenotype in *Astyanax* cave populations. It is suggested that epistatic interactions between these two loci could drive the fixation of mutant alleles at one or the other loci through transmission bias.

MATERIALS AND METHODS

Tinaja/Surface Cross

The F_2 mapping pedigree is designated Asty111 and was derived from two F_1 individuals that were half-siblings, from the same Tinaja cave female parent mated to two different surface males derived from the Arroyo Sarco population (Di Palma et al., 2007). A total of 285 Asty111 were genotyped for 235 markers including microsatellites, SNPs and genes previously mapped, using published methods (Protas et al., 2006, 2007, 2008). Individuals were phenotyped by measuring the eyes post-mortem with an eyepiece micrometer. Eye size was corrected for body length from regression against standard length and represented as relative eye size (= measured diameter divided by predicted diameter) for QTL analysis. Melanophore density was determined from photographic images of the fish. All melanophores in a 1.0 mm swath from the insertion of the dorsal to the insertion of the anal fin were counted. Total counts were converted to densities by dividing them by the area inspected.

Mapping and QTL Analysis

The linkage map was constructed using MultiPoint software (Ver. 2.2, MultiQTL Ltd., Haifa) and consisted of 36 linkage groups (haploid number = 25). QTL analysis was performed with MultiQTL (Ver. 2.5, MultiQTL Ltd., Haifa). In brief, we started by identifying linkage groups with significant or suggestive trait associations ($P=0.10$) using simple interval mapping (Lander and Botstein, 1989). These linkage groups were used as starting sets for multiple interval mapping (MIM) using the MIM function of MultiQTL. MIM assesses the significance of QTL iteratively, estimating the effects of all detected QTL on each other to minimize background variation and optimize estimation of QTL parameters (Kao et al., 1999). With each round of MIM analysis, some of the candidate linkage groups may be revealed as not being significant. When this occurred, they were discarded from further analysis. The process was repeated until a stable set of QTL was detected. For the final program output, the false detection rate was set at 0.10. Details of both the linkage analysis and QTL mapping procedures used are published (Protas et al., 2008).

Assessment of Pigmentation

To rate the relative darkness of individual fishes, they were photographed, free swimming, with their flanks parallel to the image plane. Included in each photograph was a standard black/white/gray card (WhiBal, PictureFlow LLC, Melbourne). Images were recorded in RAW format and opened in Photoshop (Adobe). They were then standardized by using the black and white standards to set the full range of luminance. The standard gray was used to adjust white balance and intermediate gray. The overall tone of the fish was obtained by selecting the full fish image and then using the blur function to average over the whole fish. The image mode was then converted to grayscale and the percentage luminosity was read from the image. Although the fishes were photographed only when they got close to the WhiBal card, there were variations in proximity from one image to the next, which is a source of variation in the estimates.

REFERENCES

Bibliowicz, J., Alie, A., Espinasa, L., Yoshizawa, M., Blin, M., Hinaux, H., Legendre, L., Pere, S., Retaux, S., 2013. Differences in chemosensory response between eyed and eyeless *Astyanax mexicanus* of the Rio Subterraneo cave. EvoDevo 4, 25.

Bilandzija, H., Ma, L., Parkhurst, A., Jeffery, W.R., 2013. A potential benefit of albinism in *Astyanax* cavefish: downregulation of the oca2 gene increases tyrosine and catecholamine levels as an alternative to melanin synthesis. PLoS One 8, e80823.

Borowsky, R., 2008. Restoring sight in blind cavefish. Curr. Biol. 18, R23–R24.

Borowsky, R., 2013. Eye regression in blind *Astyanax* cavefish may facilitate the evolution of an adaptive behavior and its sensory receptors. BMC Biol. 11, 81.

Borowsky, R., Cohen, D., 2013. Genomic consequences of ecological speciation in *Astyanax* cavefish. PLoS One 8. e79903.

Borowsky, R., Wilkens, H., 2002. Mapping a cave fish genome: polygenic systems and regressive evolution. J. Hered. 93, 19–21.

Borowsky, R.L., Vidthayanon, C., 2001. Nucleotide diversity in populations of balitorid cave fishes from Thailand. Mol. Ecol. 10, 2799–2805.

Bradic, M., Beerli, P., Garcia-deLeon, F., Esquivel-Bobadilla, S., Borowsky, R.L., 2012. Gene flow and population structure in the Mexican blind cavefish complex (*Astyanax mexicanus*). BMC Evol. Biol. 12, 9.

Bradic, M., Teotonio, H., Borowsky, R.L., 2013. The population genomics of repeated evolution in the blind cavefish *Astyanax mexicanus*. Mol. Biol. Evol. 30, 2383–2400.

Culver, D.C., 1985. Regressive evolution. In: Flurkey, A.J. (Ed.), National Speleological Society, Bulletin 47, no. 2.

Culver, D.C., Kane, T.C., Fong, D.W., 1995. Adaptation and Natural Selection in Caves. Harvard University Press, Cambridge, MA.

Di Palma, F., Kidd, C., Borowsky, R., Kocher, T.D., 2007. Construction of bacterial artificial chromosome libraries for the Lake Malawi cichlid (*Metriaclima zebra*), and the blind cavefish (*Astyanax mexicanus*). Zebrafish 4, 41–47.

Duboue, E.R., Borowsky, R.L., 2012. Altered rest-activity patterns evolve via circadian independent mechanisms in cave adapted balitorid loaches. PLoS One 7, e30868.

Duboue, E.R., Borowsky, R.L., Keene, A.C., 2012. Beta-adrenergic signaling regulates evolutionarily derived sleep loss in the Mexican cavefish. Brain Behav. Evol. 80, 233–243.

Duboue, E.R., Keene, A.C., Borowsky, R.L., 2011. Evolutionary convergence on sleep loss in cavefish populations. Curr. Biol. 21, 671–676.

Eigenmann, C.H., 1900. The Blind-Fishes; Eighth Lecture, Biological Lectures from the Marine Biological Laboratory of Wood's Hole. Ginn and Company, The Athenaeum Press, Boston.

Ferrer, R.P., Zimmer, R.K., 2007. Chemosensory reception, behavioral expression, and ecological interactions at multiple trophic levels. J. Exp. Biol. 210, 1776–1785.

Fulton, A.B., Albert, D.M., Craft, J.L., 1978. Human albinism. Light and electron microscopy study. Arch. Ophthalmol. 96, 305–310.

Gross, J.B., Borowsky, R., Tabin, C.J., 2009. A novel role for Mc1r in the parallel evolution of depigmentation in independent populations of the cavefish *Astyanax mexicanus*. PLoS Genet. 5, e1000326.

Jeffery, W.R., 2005. Adaptive evolution of eye degeneration in the Mexican blind cavefish. J. Hered. 96, 185–196.

Jeffery, W.R., 2009. Regressive evolution in *Astyanax* cavefish. Annu. Rev. Genet. 43, 25–47.

Johannes, R., Webb, K.L., 1970. Release of dissolved organic compounds by marine and freshwater invertebrates. In: Institute of Marine Sciences (Alaska) Occasional Publications, vol. 1. pp. 257–273.

Kao, C.H., Zeng, Z.B., Teasdale, R.D., 1999. Multiple interval mapping for quantitative trait loci. Genetics 152, 1203–1216.

Kimura, M., Ohta, T., 1971. Theoretical Aspects of Population Genetics. Princeton University Press, Princeton, NJ.

Kottelat, M., 1990. Indochinese Nemacheilines: A Revision of Nemacheiline Loaches (Pisces:Cypriniformes) of Thailand, Burma, Laos, Cambodia, and southern Viet Nam. Pfeil, München.

Kottelat, M., 2001. Fishes of Laos. WHT Publications, Colombo.

Lander, E.S., Botstein, D., 1989. Mapping Mendelian factors underlying quantitative traits using Rflp linkage maps. Genetics 121, 185–199.

Loaches_online, 2014. https://www.google.com/#safe=off&q=loaches+online.

Moran, D., Softley, R., Warrant, E.J., 2014. Eyeless Mexican cavefish save energy by eliminating the circadian rhythm in metabolism. PLoS One 9, e107877.

Parker, G.H., 1948. Animal Colour Changes and their Neurohumours, first ed. Bambridge University Press, London.

Poulson, T.L., 1963. Cave adaptation in amblyopsid fishes. Am. Midl. Nat. 70, 257–290.

Poulton, E.B., 1890. The Colours of Animals, second ed. Kegan Paul, Trench, Trübner, & Co. Ltd, London.

Protas, M., Conrad, M., Gross, J.B., Tabin, C., Borowsky, R., 2007. Regressive evolution in the Mexican cave tetra, *Astyanax mexicanus*. Curr. Biol. 17, 452–454.

Protas, M., Tabansky, I., Conrad, M., Gross, J.B., Vidal, O., Tabin, C.J., Borowsky, R., 2008. Multi-trait evolution in a cave fish, *Astyanax mexicanus*. Evol. Dev. 10, 196–209.

Protas, M.E., Hersey, C., Kochanek, D., Zhou, Y., Wilkens, H., Jeffery, W.R., Zon, L.I., Borowsky, R., Tabin, C.J., 2006. Genetic analysis of cavefish reveals molecular convergence in the evolution of albinism. Nat. Genet. 38, 107–111.

Proudlove, G.S., 2004. Pisces (Fishes), Encyclopedia of Caves and Karst Science. Fitzroy Dearborn, New York, pp. 593–595.

Rainboth, W.J., 1996. Fishes of the Cambodian Mekong. Food and Agriculture Organization of the United Nations, Rome.

Romero, A., 2004. Pisces (Fishes): Amblyopsidae, Encyclopedia of Caves and Karst Science. Fitzroy Dearborn, New York, pp. 595–597.

Romero, A., Green, S.M., 2005. The end of regressive evolution: examining and interpreting the evidence from cave fishes. J. Fish Biol. 67, 3–32.

Sadoglu, P., Mckee, A., 1969. A second gene that affects eye and body color in Mexican blind cave fish. J. Hered. 60, 10–14.

Trajano, E., Mugue, N., Krejca, J.K., Vidthayanon, C., Smart, D., Borowsky, R., 2002. Habitat, distribution, ecology and behavior of cave balitorids from Thailand (Teleostei: Cypriniformes). Ichthyol. Explor. Freshw. 13, 169–184.

Vandel, A., 1965. Biospeleology: The Biology of Cavernicolous Animals, first ed. Pergamon Press, Oxford, New York.

Weber, A., Proudlove, G.S., Parzefall, J., Wilkens, H., Nalbant, T.T., 1998. Pisces (Teleostei). In: Juberthie, C., Decu, V. (Eds.), Encyclopaedia Biospeologica. Société Internationale de Biospéologie. Moulis & Bucarest, pp. 1177–1213.

Wilkens, H., 1988. Evolution and genetics of epigean and cave *Astyanax fasciatus* (Characidae, Pisces)—support for the neutral mutation theory. Evol. Biol. 23, 271–367.

Wilkens, H., 2007. Regressive evolution: ontogeny and genetics of cavefish eye rudimentation. Biol. J. Linn. Soc. 92, 287–296.

Wilkens, H., Strecker, U., 2003. Convergent evolution of the cavefish *Astyanax* (Characidae, Teleostei): genetic evidence from reduced eye-size and pigmentation. Biol. J. Linn. Soc. 80, 545–554.

Yamamoto, Y., Byerly, M.S., Jackman, W.R., Jeffery, W.R., 2009. Pleiotropic functions of embryonic sonic hedgehog expression link jaw and taste bud amplification with eye loss during cavefish evolution. Dev. Biol. 330, 200–211.

Yoshizawa, M., Goricki, S., Soares, D., Jeffery, W.R., 2010. Evolution of a behavioral shift mediated by superficial neuromasts helps cavefish find food in darkness. Curr. Biol. 20, 1631–1636.

Chapter 6

Mapping the Genetic Basis of Troglomorphy in *Astyanax*: How Far We Have Come and Where Do We Go from Here?

Kelly O'Quin[1] and Suzanne E. McGaugh[2]
[1]*Centre College, Biology Program, Danville, Kentucky, USA*
[2]*Department of Ecology, Evolution, and Behavior, University of Minnesota, Falcon Heights, Minnesota, USA*

INTRODUCTION

The Mexican tetra, *Astyanax mexicanus*, is the premier model system for the study of troglomorphy, or cave adaptation. Troglomorphic animals live permanently in caves or subterranean pools and typically exhibit reduced eyes and pigmentation, as well as enhanced tactile and sensory organs (Mohr and Poulson, 1966; O'Quin et al., 2013). Like many troglomorphic animals—including over 80 other species of cavefish (CF) (Romero and Paulson, 2001)—*Astyanax* CF exhibit reduced eye size, pigment cell size and number, thoracic vertebrae and rib number, and schooling behaviors. But in addition to these regressive traits, *Astyanax* CF also exhibit increased jaw size, tooth and tastebud number, olfactory bulb size, cranial neuromast size, fat content, and food-searching behaviors (Jeffery, 2008). Within this long list of evolutionary changes, individual traits are also the result of numerous minute changes. For example, the reduced eyes of *Astyanax* are actually the end result of numerous independent changes to eye development, including reduced eye vesicle and ventral retina size, early lens and retinal apoptosis, failed photoreceptor differentiation, and arrested growth (Rétaux and Casane, 2013). But what makes *Astyanax* exceptional among cave animals is the continued existence of its surface-dwelling conspecific, or surface fish (SF), which can often be found living in streams and rivers directly outside of caves (Mitchell et al., 1977). Unlike CF, *Astyanax* SF retain eyes and pigmentation, school frequently, and have sensory systems akin to normal fish. More importantly, however, *Astyanax* CF and SF remain interfertile and can be crossed in the lab to produce viable hybrids (Sadoglu, 1957). This ability to perform

Biology and Evolution of the Mexican Cavefish. http://dx.doi.org/10.1016/B978-0-12-802148-4.00006-2
© 2016 Elsevier Inc. All rights reserved.

genetic crosses makes *Astyanax mexicanus* a powerful tool for dissecting the genetic underpinnings of troglomorphic traits.

In order to identify the genetic bases of eye loss or other traits related to troglomorphy, variation in genotypes is statistically associated with variation in phenotypes resulting from a genetic cross (Falconer and Mackay, 1996; Lynch and Walsh, 1998). The resulting associations are called quantitative trait loci, or QTL, and they highlight regions of the genome containing mutations responsible for some trait of interest (Figure 6.1). At its heart, QTL mapping can be thought of as an analysis of variation (ANOVA) with genotypes as the independent variable and phenotypes as the dependent variable. QTL mapping has been used extensively in *Astyanax* and other systems to identify QTL and mutations responsible for a variety of traits from crop yield to human disease. For methods and examples related to evolution, we refer readers to excellent articles and reviews by (Hughes, 2007; Stinchcombe and Hoekstra, 2007; Ellegren and Sheldon, 2008; Stern and Orgogozo, 2009; Nadeau and Jiggins, 2010; Elmer and Meyer, 2011; Vidal et al., 2011; Schielzeth and Husby, 2014). Thus far, over 200 QTL have been identified in *Astyanax*, which are the subject of this chapter (Figure 6.2). These QTL represent a diverse range of troglomorphic traits, including eye size, pigmentation, number of tastebuds, and even dark preference. Past QTL studies have contributed substantially to our understanding of troglomorphy in *Astyanax*, and their results continue to shape our current research in the species.

In this chapter, we discuss the success of QTL mapping strategies in *Astyanax*, and attempt to integrate the results of these various studies into a single picture of the genetic basis of troglomorphy in *Astyanax*, as it is currently understood. We also detail future directions to link genotype-to-phenotype in *Astyanax* and, most importantly, explain why we think the continued pursuit of genetic mapping is a worthy goal in this species (sensu Lee et al., 2014).

PART I: QUANTITATIVE GENETICS AND QTL MAPPING IN *ASTYANAX*

Mendelian vs. Quantitative Traits

The genetic basis of different phenotypes can either be Mendelian or quantitative (Falconer and Mackay, 1996). Mendelian traits exhibit two distinct phenotypes that are most commonly controlled by two alleles at just a single locus.

FIGURE 6.1—cont'd eye size (data from Yoshizawa et al., 2012). The genotypes of the F2 at several microsatellite or SNP markers are then used to construct a genetic linkage map of the *Astyanax* genome (not shown). Once constructed, the statistical association (log of odds [LOD] score) between each phenotype and marker is calculated and plotted along each linkage group. Genetic markers with LOD scores rising above some predetermined level (usually LOD 3 or 4, which approximates $P \approx 0.05$) are considered significantly associated with the trait of interest and are deemed QTL. True Mendelian traits will have only a single QTL, while quantitative traits will have many.

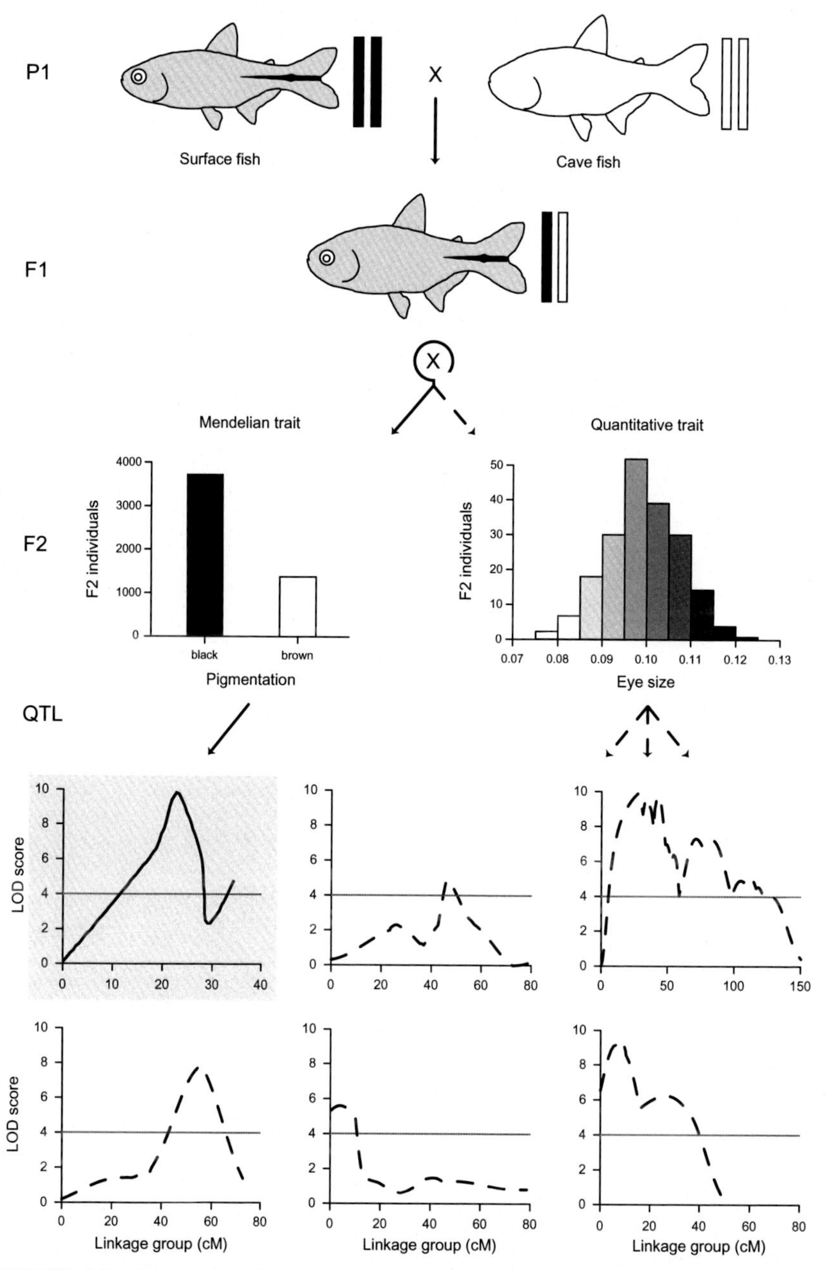

FIGURE 6.1 *Experimental Design for QTL Analysis of Mendelian and Quantitative Traits.* Phenotypically and genetically different *Astyanax* SF and CF (P1) crossed with each other to create a hybrid generation (F1). F1 progeny have a mix of CF and SF genotypes (illustrated by the chromosomes shown to the right of each fish). In the most common experimental design, the F1 progeny are then intercrossed with themselves to create a second hybrid generation (F2, not shown). These F2 progeny will exhibit a range of phenotypes. If the phenotype is Mendelian with complete dominance of one allele, the F2 will exhibit two alternate forms of the trait in a 3:1 ratio, as does "brown" pigmentation (data from Sadoglu, 1957; Gross et al., 2009). If the phenotype is Quantitative, the F2 will exhibit a diverse range of phenotypes, from completely CF-like to completely SF-like, as does

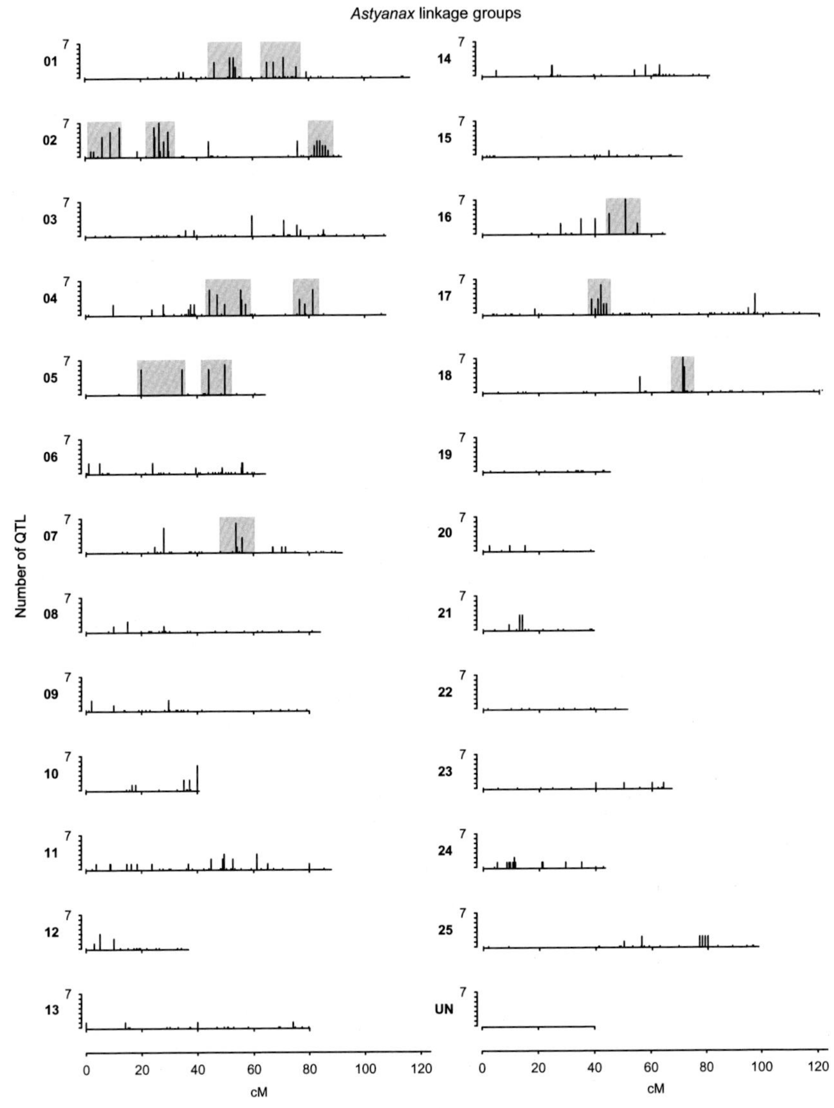

FIGURE 6.2 *Summary of QTL for troglomorphy in Astyanax.* So far, over 200 QTL for troglomorphy have been identified in *Astyanax* and represent a diverse range of phenotypes, from eye size to pigmentation and even behavior. These QTL are distributed across 25 chromosomes or linkage groups (LG 1-25), and one unordered region (UN) (Protas et al., 2008; O'Quin et al., 2013). Here, the linkage groups are numbered following O'Quin et al. (2013) and McGaugh et al. (2014). The most QTL that map to any one genetic marker is seven (shown on the *y*-axis), and there are three markers with that many QTL located on LG 2, 16, and 18. The remaining QTL are largely clustered within 13 regions (gray boxes) on just eight linkage groups: LGs 1, 2, 4, 5, 7, 16, 17, and 18. For the purpose of this review, we define a QTL cluster as any 20-cM region that contains markers associated with a total of 10 or more QTL; however, it is also worth nothing that many linkage groups contain just one or a few QTL, and none are found on LGs 19 or 22.

These simple traits produce a characteristic 3:1 ratio of phenotypes among the progeny of heterozygotes (Mendel, 1866). Classic examples of Mendelian traits in humans, for example, include brachydactyly (Bell, 1951) and wet versus dry earwax (Matsunaga, 1962; Petrakis et al., 1967). In contrast, quantitative traits exhibit a large number of continuous, intermediate phenotypes that are controlled by alleles at two or more genes, along with additional environmental factors. These complex, or multifactorial, traits characteristically produce a broad range of phenotypes resembling a bell-curve among a sample in the population (Falconer and Mackay, 1996; Lynch and Walsh, 1998). A classic example of a quantitative trait in humans is adult height, 80% of which is controlled by alleles at 50 or more genes, and 20% of which is influenced by environmental factors, such as diet or maternal effects (reviewed in McEvoy and Visscher, 2009). Although Mendelian traits are easier to study from a transmission genetics perspective, most traits likely have a complex genetic basis (Fisher, 1918; Rockman, 2012; Travisano and Shaw, 2013).

Early quantitative genetic studies in *Astyanax mexicanus* revealed the Mendelian or quantitative nature of many troglomorphic traits (see also Wilkens, Chapter 5, this volume). In particular, Sadoglu discovered that two aspects of *Astyanax* coloration, albinism and "brown," were both inherited in a simple Mendelian fashion (Sadoglu and McKee, 1969; Sadoglu, 1975). In contrast, Sadoglu (1975) and Wilkens (1971) both discovered that eye size was inherited in a complex fashion, with a broad range of intermediate eye sizes present among the hybrid progeny of SF and CF. Later, Wilkens (1988) estimated the minimum number of genetic factors underlying numerous other traits in *Astyanax*, including eye size, melanophore number, feeding behavior, and location of tastebuds on the head. As expected from theoretical expectations of Fisher's polygenic model of inheritance (Fisher, 1918), Wilkens (1988) found that most troglomorphic traits in *Astyanax* are indeed quantitative.

Quantitative Trait Locus Mapping: Complex Traits and QTL Clusters

Quantitative Trait Locus (QTL) mapping attempts to identify those regions of the genome containing mutations responsible for Mendelian and quantitative traits (Falconer and Mackay, 1996; Lynch and Walsh, 1998). To accomplish this, parental generations (F0) with different traits are crossed to produce a hybrid generation (F1). Then, depending upon the experimental design, these F1 can then be crossed back to one of the parental lines (backcross) or crossed with each other (intercross) to form an additional hybrid generation (F2). F2 intercross designs provide more power to detect QTL, and these have been the most common design used to study troglomorphy in *Astyanax*. Once the hybrid family or mapping cross has been created, alleles segregating within a family are identified and genotyped via microsatellite analysis or DNA sequencing, and then used to construct a genetic linkage map of the genome. Finally, the hybrid progeny are assayed for one or more phenotypes of interest, and then the genotypes are analyzed for their statistical association to these traits. The resulting associations are deemed QTL (Figure 6.1). Theoretically, if a large enough sample of individuals are genotyped and phenotyped, and the genetic basis of the trait is sufficiently simple, the resulting QTL can be used to estimate the genetic architecture of these traits (but see Rockman, 2012; Travisano and Shaw, 2013). The genetic architecture describes the number and location of all genetic factors underlying each trait, as well as the strength of their effects and any interactions among them. Thus, in practice with current technologies, understanding the complete genetic architecture of most traits is extremely challenging; however, QTL mapping is a first step toward identifying the genes and mutations responsible for troglomorphy. QTL mapping has been used extensively in *Astyanax* to identify large-effect genomic regions underlying numerous traits that differ between SF and CF (Table 6.1).

Borowsky and Wilkens (2002) were the first to map QTL for troglomorphic traits in *Astyanax*. Using a backcross of Pachón CF and SF, and alleles at randomly amplified polymorphic DNA (RAPD) markers, Borowsky and Wilkens (2002) identified multiple QTL for eye size, melanophore number, condition factor, and albinism. Additionally, these authors found that QTL for many of these traits were linked, or found together in the genome, hinting at the interesting pos-

TABLE 6.1 Catalog of All Known QTL Mapped Between CF and SF of *Astyanax Mexicanus*

Study	Populations	Type	Trait	QTL
Borowsky and Wilkens (2002)	Pachón × SF (?)	BC	Eye size	3
			Pigmentation	2
			Albinism	1
			Condition factor	2
Protas et al. (2006)	Molino × SF (?), Pachón × SF (RV)	BC, F2	Albinism	1
Protas et al. (2007)	Pachón × SF (RV)	F2	Eye size	8
			Lens size	6
			Jaw bone(s) length	7
			Maxillary tooth number	6
			Tastebud number	6
			Pigmentation	18

TABLE 6.1 Catalog of All Known QTL Mapped Between CF and SF of *Astyanax Mexicanus*—cont'd

Study	Populations	Type	Trait	QTL
Protas et al. (2008)	Pachón × SF (RV)	F2	Eye size	8
			Pigmentation	4
			Condition factor	9
			Weight loss	1
			Tooth number	5
			Peduncle depth	3
			Fin placement	9
			Anal fin rays	2
			SO3 width	6
			Rib number	6
			Length	5
			Chemical sense	4
Gross et al. (2009)	Molino × SF (?), Pachón × SF (RV)	BC, F2	Brown melanophore	1
Yoshizawa et al. (2012)	Pachón × SF (TX)	F2, F3	Albinism	1
			Eye size	5
			Vibration attraction	2
			Superficial neuromasts	2
O'Quin et al. (2013)	Pachón × SF (TX)	F2	Retinal thickness	4
Kowalko et al. (2013a)	Pachón × SF (?)	F2	Feeding Angle	2
			Jaw Angle	1
			Eye size	5
			Head depth	1
			Feeding Angle	1
			Tastebuds	2
			Body depth	1
			Eye orbit diameter	5

Continued

TABLE 6.1 Catalog of All Known QTL Mapped Between CF and SF of *Astyanax Mexicanus*—cont'd

Study	Populations	Type	Trait	QTL
Kowalko et al. (2013b)	Tinaja × SF (?)	F2	Schooling behavior	2
			Dark preference	1
			Eye size	1
			Pupil size	1
Gross et al. (2014)	Pachón × SF (RV)	F2	Albinism	1
			Eye size	2
			Craniofacial bones	12
			Sex	1
			Standard length	1

sibility that different troglomorphic traits could share a common genetic basis. Later, Protas et al. (2007) extended these results using a much larger mapping family of Pachón CF and SF F2 progeny and nearly 200 microsatellite loci. In addition to eye size, melanophore number, and condition, Protas et al. (2007) identified QTL for numerous other traits, including lens size, jawbone length, maxillary tooth number, and tastebud number. Protas et al. (2007) not only demonstrated that most traits are indeed complex and governed by multiple loci, but they also found a consistent directional pattern of genetic effects for the 12 QTL related to eye and lens size. CF alleles at all 12 eye or lens QTL resulted in a decrease in eye size. Although not tested formally, this pattern is strongly consistent with the evolution of eye reduction via natural selection (e.g., Orr's QTL sign test; Orr, 1998). In contrast, CF alleles at 13 pigmentation QTL did not show a consistent pattern of reduction in CF, a pattern that is expected due to neutral evolutionary divergence through drift (see also Borowsky, this chapter, this volume). Thus, Protas et al. (2007) provided the first genetic evidence that eye reduction in CF might be adaptive, possibly as a result of the high energetic costs of maintaining eyes in darkness (Laughlin et al., 1998; Okawa et al., 2008).

Protas et al. (2008) then extended their previous results by mapping QTL for multivariate suites of traits associated with troglomorphy, including reduced eye size and melanophores, enhanced chemical reception, tooth count, and relative condition. Since similar suites of traits are often found in other cave-dwelling fish, including the Amblyopsid CF of North America (Poulson, 1963) and the Cyprinid CF of China (Romero et al., 2009; Zhao et al., 2011), Protas et al. (2008) hypothesized that many troglomorphic traits could share a common genetic or pleiotropic basis. Considering 12 traits individually,

Protas et al. (2008) also found evidence that single-trait QTL for troglomorphy were clustered within the genome significantly more than would be expected by chance; indeed, many single-trait QTL mapped to within 5 cM of each other. They then employed a multiple-trait QTL mapping strategy and found 12 independent multi-trait QTL for nine traits clustered in a total of six different sets of 3-5 traits. These results, therefore, identified significant clustering of QTL in *Astyanax*. Similar QTL clusters have been observed in examples of rapid evolution or domestication, including jaw and craniofacial shape in African cichlid fish (Albertson et al., 2003), plant morphology in domesticated rice (Cai and Morishima, 2002), and wing morphology and coloration in *Heliconius* butterflies (Brown and Benson, 1974). In these and other systems, QTL clustering has variously been attributed to pleiotropy, genetic hitchhiking, the formation of "supergenes" or tightly linked clusters of coadapted genes, and inversions. Although, the results of QTL mapping alone cannot distinguish among these hypotheses, Protas et al. (2008) argued that such extensive QTL clustering for so many seemingly unrelated traits is most likely the result of pleiotropy, or a few alleles that affect multiple phenotypic traits in complex and sometimes counterintuitive ways (see also Yoshizawa et al., 2012).

The general conclusions of Borowsky and Wilkens (2002) and Protas et al. (2007, 2008) have been reaffirmed by subsequent QTL studies of numerous other troglomorphic traits. Some of the most interesting of these are QTL for behavioral differences observed between *Astyanax* SF and CF. For example, CF tend to be attracted to vibrating stimuli, while SF avoid these (Yoshizawa et al., 2010). Additionally, CF tend to feed at a 45-degree angle, while SF feed at a 90-degree angle (Schemmel, 1980), and CF also tend to avoid schooling (Parzefall and Fricke, 1989). Yoshizawa et al. (2012) and Kowalko et al. (2013a,b) quantified these behaviors in the F2 progeny of Pachón CF and SF, and Tinaja CF and SF and mapped the resulting QTL. Like Borowsky and Wilkens (2002) and Protas et al. (2007, 2008) before them, these authors found that behavioral traits in *Astyanax* are multifactorial and are clustered within the genome, in some cases with other functionally related traits. Yoshizawa et al. (2012) identified two QTL for vibration attraction (VAB) and found that these were significantly clustered with QTL for two other traits: eye size and the number of superficial neuromast cells found within the eye obit (see also Yoshizawa, Chapter 14, this volume). The experimental removal of these superficial neuromasts significantly decreased VAB in *Astyanax* hybrids, leading Yoshizawa et al. (2012) to speculate that pleiotropy or tight physical linkage could actually promote rapid adaptation in *Astyanax* by linking functionally related QTL necessary for cave life. Kowalko et al. (2013a) identified two QTL for feeding angle, consistent with a polygenic basis for this trait (Schemmel, 1980); however, these authors found no detectable clustering of QTL for feeding posture with any other troglomorphic traits. Conversely, Kowalko et al. (2013b) found only a single QTL for loss of schooling, but found that it overlapped considerably with a QTL for dark preference. Thus, the clustering of QTL remains a common and notable occurrence in *Astyanax*.

More recent advances in QTL mapping in *Astyanax* have employed comparative mapping to other model species, more extensive genetic sampling techniques, and increasingly finer measurements of troglomorphic traits. Gross et al. (2008) used the sequences of 155 microsatellite markers and candidate genes to construct an *Astyanax* linkage map and then used the program BLAST (Altschul et al., 1990) to anchor this linkage map to the zebrafish (*Danio rerio*) genome sequence. This comparative analysis revealed large regions of synteny between the *Astyanax* and *Danio* genomes. Importantly, this analysis also revealed that the sequence of *Danio* could be used to identify candidate genes within *Astyanax* QTL. Additionally, O'Quin et al. (2013) and Kowalko et al. (2013a,b) each employed a relatively new method of genetic marker identification based upon next-generation sequencing technology called restriction site associated DNA sequencing (RADseq; Baird et al., 2008). Using this method, O'Quin et al. (2013) constructed a high-density linkage map of nearly 700 RADseq and microsatellite markers. Comparative mapping to *Danio* using this high-density map also revealed extensive regions of synteny between *Astyanax* and *Danio*, and it confirmed the utility of this model organism for identifying candidate genes within *Astyanax*. Additionally, O'Quin et al. (2013) quantified retinal thickness in Pachón CF and SF F2 progeny. This study was the first to assay quantitative genetic differences in eye development finer than gross measurements of eye or lens size. As in previous studies, O'Quin et al. (2013) identified multiple QTL for retinal thickness, many of which clustered with previously identified QTL for eye and lens size. Finally, Gross et al. (2014) used Pachón CF and SF F2 progeny to quantify fine-scale differences in *Astyanax* craniofacial structure. Their results revealed that craniofacial structure is a multifactorial trait, and they also identified several "asymmetric QTL" that are responsible for changes to only one side of the head, either right or left (see also Gross and Powers, Chapter 11, this volume).

To summarize the results of these various studies, we have generated an integrated map of the *Astyanax* genome based upon the linkage map of O'Quin et al. (2013), while also including additional markers from Protas et al. (2007, 2008). Where genetic markers were present in Protas et al. but absent from O'Quin et al., we inferred their likely linkage group and cM position based upon the presence of nearby markers common to both maps. Markers that could not be integrated easily were grouped into a single unordered linkage group, UN. Once integrated, the positions of QTL for various eye, pigment, craniofacial, behavioral, sensory, physiological, and anatomical traits were denoted for each marker. Finally, the total number of QTL mapping to each marker was summed. Afterwards, we looked for QTL clusters by identifying those 20 cM regions associated with a total of 10 or more QTL. Figure 6.2 summarizes the location of most of the 204 troglomorphic QTL that have been identified so far in *Astyanax*. These QTL give a genome-wide view of the genomic positions responsible for troglomorphic differences between *Astyanax* SF and CF.

As Figure 6.2 shows, QTL for troglomorphic traits are distributed throughout much of the genome. The most QTL that map to any one genetic marker is seven. Three genetic markers on LG 2 (Am120c), LG 16 (Am227a), and LG 18 (Am226e) have markers that are each associated with seven QTL; furthermore, the QTL for one such marker, Am226e on LG 18, are for traits as diverse as eye size, pigmentation, craniofacial structure, sensory perception, physiology, and general anatomy (Figure 6.2). These three markers may therefore represent pleiotropic loci responsible for numerous troglomorphic traits in *Astyanax*. In addition to these three markers, we found several more examples of single markers or loci that are associated with multiple quantitative traits. These include 5 markers that are each associated with six QTL, 11 markers that are each associated with five QTL, 10 markers that are each associated with four QTL, 22 markers that are each associated with three QTL, 43 markers that are each associated with two QTL, and a whopping 81 markers that are associated with at least one QTL. We note, however, that these numbers do not account for epistasis or linkage, and several markers can be associated with the same QTL. Furthermore, Figure 6.2 illustrates that, while common, QTL clustering is not as pervasive as one might think. Thirteen clear QTL clusters are found on LG 1, 2, 4, 5, 7, 16, 17, and 18—indeed, three clear clusters are found on LG 2 alone (gray boxes in Figure 6.2); however, most LG are associated with only a few QTL here and there, while LG 19 and 22 are not associated with any QTL. Although the reasons for QTL clustering remain unclear, possible explanations include pleiotropy, hitchiking, large-scale structural variants between the SF and CF genomes (e.g., inversions), or even simply that these regions have very low recombination, making QTL easier to detect. In either case, our results surprisingly suggest that many phenotypic differences between *Astyanax* SF and CF may be explained by genotypes found on just eight chromosomes (Figure 6.2).

Quantitative Trait Nucleotide (QTN) Fine-Mapping: Success for Mendelian Traits

Underlying quantitative trait loci are genotypes, segregating within natural populations and mapping crosses, which produce phenotypic variation in some traits. Identifying these alleles, or QTN, is the goal of many researchers hoping to determine the molecular basis of phenotypic traits such as troglomorphy, domesticated crop yield, or human disease (Blangero, 2004; Ron and Weller, 2007; Rockman, 2012); however, some of the same features that increase the power to detect QTL in small mapping crosses, such as high linkage disequilibrium between adjacent genetic markers, can actually decrease the power available to detect individual QTN, especially when multiple QTN underlie a single QTL (Noor et al., 2001). Additionally, since most traits are presumably governed by numerous alleles of very small effect (Fisher, 1930), the number of hybrid offspring and genetic markers needed to unambiguously detect these QTN by conventional QTL mapping strategies is often prohibitive, if not impossible

(Rockman, 2012; Travisano and Shaw, 2013). As a result, the leap from QTL to QTN is a long one for all but the simplest traits with large-effect QTN. It is unsurprising, therefore, that the only traits that have been successfully mapped from QTL to QTN in *Astyanax* are those that are inherited in a Mendelian fashion: albinism (Protas et al., 2006) and "brown" pigmentation (Gross et al., 2009) (see also Jeffery et al., Chapter 9, this volume).

Using a Molino CF and SF backcross, as well as a Pachón CF and SF F2 intercross, Protas et al. (2006) identified a single QTL for albinism, consistent with Sadoglu's previous discovery that albinism is a Mendelian trait (Sadoglu, 1957). Interestingly, the albinism QTL was localized to the same linkage group and genetic marker in both mapping crosses, suggesting that the genetic basis of albinism is similar in both Molino and Pachón CF. To identify this gene, Protas et al. (2006) added several candidate genes to their genetic linkage map and discovered that one gene, *ocular and cutaneous albinism-2* (*Oca2*), mapped to the center of their QTL. The authors then sequenced *Oca2* directly in both Molino and Pachón CF and discovered that Molino CF were missing exon 21 and Pachón CF were missing exon 24. The authors then carried out a functional analysis by experimentally expressing these CF mutations in mouse cell lines, which showed that the CF mutations could not rescue pigmentation. The discovery of these two mutations confirmed the feasibility of QTL-to-causal-variant mapping in *Astyanax*, at least for simple traits of very large effect.

Gross et al. (2009) used the same strategy to identify the mutations responsible for "brown" pigmentation. Using a Pachón CF and SF F2 intercross, these authors identified a single QTL for "brown" pigmentation, again consistent with previous studies that demonstrated that "brown" is a Mendelian trait (Sadoglu and McKee, 1969). To identify the QTN responsible for "brown" pigmentation, Gross et al. (2009) took a different candidate gene approach by first anchoring their genetic markers to the genome of *Danio rerio* using BLAST. They identified a ~15 Mb region of synteny between *Astyanax* and *Danio* that flanked the "brown" QTL region; they then searched this region for appropriate candidate genes using the genome sequence of *Danio*. They found only two genes that would likely alter pigmentation, including *melanocortin 1 receptor* (*mc1r*). The authors genotyped a polymorphism at the *mc1r* locus within the mapping cross and discovered that it mapped to the center of the "brown" QTL. The authors then sequenced *mc1r* in several *Astyanax* SF and CF populations. They discovered a 2-base-pair frameshift deletion in Pachón CF relative to SF and *Danio rerio*. The authors also discovered a separate nonsynonymous mutation in *mc1r* in two other CF populations, Yerbaníz and Japonés. To examine the effect of cave and surface alleles, the authors experimentally knocked down *mc1r* expression in *Danio rerio* and found that treated individuals exhibited reduced melanin, similar to *Astyanax* individuals with the "brown" mutation. Further, both Pachón and Yerbaníz CF alleles failed to rescue pigmentation *Danio*, but the SF allele produced fish with pigmentation closer to normal in most cases. As in the previous study of albinism (Protas et al., 2006), Gross et al. (2009)

discovered that the same gene was affected in several CF populations with the "brown" phenotype, although the QTN responsible differed among populations.

Thus far, no QTN for complex traits such as eye size or lens degeneration have been identified in *Astyanax*; however, researchers have successfully identified several genes involved in these processes through studies of candidate genes. Most of these studies are described in more detail elsewhere in this volume (e.g., Yamamoto, Chapter 10). Below, we provide a brief summary of their results. First, Strickler et al. (2001) and Yamamoto et al. (2004) examined the effect of the genes *pax6* and *shh*, both of which are important in the control of animal eye development (Ekker et al., 1995; Gehring and Ikeo, 1999). Following the discovery that *pax6* expression is significantly reduced in the CF lens (Strickler et al., 2001), Yamamoto et al. (2004) analyzed *shh* expression, which regulates *pax6* (Ekker et al., 1995). Yamamoto et al. (2004) found that *shh* expression is actually expanded in CF, resulting in reduced *pax6* expression, increased lens apoptosis, and reduced eye size. Overexpression of *shh* in SF resulted in small eyes, mimicking the CF phenotype, and knock-down of *shh* in CF resulted in larger eyes, partially rescuing eye development and confirming that alteration of *shh* and *pax6* expression is partially responsible for eye degeneration in *Astyanax* (Yamamoto et al., 2004). Interestingly, however, no mutation responsible for reduced *pax6* expression or expanded *shh* expression has yet been identified, and no QTL for eye size has ever implicated *shh* as a possible candidate gene, though a QTL for eye diameter maps close to *pax6* (Yoshizawa et al., 2012).

Second, based upon the results of Yamamoto et al. (2004) as well as an earlier study that demonstrated that the expression of several signaling factors necessary for proper eye and brain development were altered in CF (Alunni et al., 2007), Pottin et al. (2011) examined the influence of expanded *shh* on two signaling genes required for eye and retinal patterning, fibroblast growth factor 8 (*fgf8*) and LIM homeobox 2 (*lhx2*). By manipulating *shh* pharmacologically, Pottin et al. (2011) demonstrated that the expanded *shh* expression of CF directly results in earlier expression of *fgf8* and *lhx2*, which modifies the fate of ventral retinal cells and results in the reduced development of the ventral retina. Finally, spurred by studies of canalization and the expression of the molecular chaperone HSP90 in laboratory model systems, Rohner et al. (2013) recently tested the influence of this protein on eye development in *Astyanax*. These authors determined that the chaperone HSP90 buffers standing genetic variation that influences eye size; when HSP90 function is pharmacologically inhibited, variation in eye size increases among SF as well as CF and SF F2 progeny (Rohner et al., 2013). Furthermore, the authors found that this variation could be selected to produce additional generations with increasingly smaller eyes, and that the HSP90 reduction could be environmentally triggered by mimicking the low conductivity environment of cave pools (see also Rohner, Chapter 8, this volume). This latter result is particularly interesting because it reveals an environmental source of variation in eye size that may have limited the power of past studies to detect and fine-map QTL and QTN for eye size and other troglomorphic traits.

PART II: LOOKING FORWARD: REMAINING QUESTIONS AND NEW APPROACHES

The Future of QTL Mapping in *Astyanax*

How does the future look for our attempts to map the causative mutations underlying QTL for troglomorphy in *Astyanax*? For traits that are Mendelian or that are controlled by just a few large-effect alleles, the outlook remains bright (Mackay et al., 2009; Stern and Orgogozo, 2009). New methods of genetic marker identification based upon next-generation sequencing will make large-scale QTL mapping even faster, and the recent completion of the first CF genome (McGaugh et al., 2014) has already made the search for candidate genes much easier. Additionally, the recent advent of new methods of *in vivo* transgenesis and gene knock-down (e.g., CRISPR-Cas, Hale et al., 2012; Hwang et al., 2013; Elipot et al., 2014) will make it easier to functionally test these candidate genes, greatly expanding our ability to make the jump from QTL to QTN. But it is not just Mendelian traits that will benefit. Many complex traits in *Astyanax* may be controlled in part by just one or two large-effect QTL (e.g., percent variance explained ~20%, (but see Beavis, 1994, 1998; Slate, 2013); including eye and lens size, the number of anal fin rays, and the number of ribs (Protas et al., 2007; Protas et al., 2008). These few large-effect QTL are also ripe for dissection with new QTL mapping and functional genomic tools. Below, we discuss several ways to aid the dissection and identification of the genes underlying complex traits related to troglomorphy. We also briefly cover the continued motivation for these studies and how to link laboratory-based mapping studies to field-relevant genetic bases and fitness.

Increasing the Accuracy and Precision of QTL Mapping

For the more complex traits that characterize CF evolution, increasing the number of genetic markers, the size and information content of mapping crosses, and the precision of phenotypic measurements can all increase the power to detect QTL and fine-map QTN. Both the advent of affordable methods of genetic marker identification using next-generation sequencing (e.g., RADseq; Baird et al., 2008) and the availability of a complete reference genome make it possible to identify and choose genetic markers that are more evenly distributed across the genome. Furthermore, ongoing efforts by the Beijing Genome Institute and our labs to produce a comparable SF reference genome will help identify a set of differentially fixed markers, which can increase the number of genotypes available for statistical analysis. The greater marker density and information content that results from these tools will help ensure that significant QTL are not missed due to a lack of markers linked to the causal variant (but see Stanton-Geddes et al., 2013). And, as evidenced by the results of O'Quin et al. (2013) and Gross et al. (2008), increased numbers of genetic markers can considerably aid comparative mapping and candidate gene identification;

however, the real benefit of greater marker densities—that is, the refinement of ever-narrower QTL peaks so that fewer candidate genes must be identified and analyzed—will only be possible if the size of our mapping crosses and number of recombination events increase as well.

Indeed, the power of most recent QTL studies is limited not by the number of genetic markers, but by the number of individuals sampled and the number of recombination events breaking up linkage between markers (Mackay et al., 2009; Slate, 2013). Larger mapping crosses can also increase the accuracy and precision of QTL mapping. For instance, for sample sizes commonly used in QTL studies (or even as large as 1000-3000 individuals depending upon the study design), it has been clearly shown that the locations of the QTL are often highly inaccurate, the percent variance explained is overestimated, and the number of QTL contributing to the trait is underestimated (Beavis, 1994, 1998; Slate, 2013). As the trait variance explained decreases and/or the number of QTL contributing to a trait increases, the number of individuals needed to provide meaningful estimates of these parameters increases exponentially (Mackay et al., 2009). Given that most QTL studies in *Astyanax* have used fewer than 500 hybrid individuals—and many have used considerably less—the studies described above have likely underestimated the number of genomic regions that contribute to troglomorphy. Thus, we urge caution in drawing strong conclusions regarding the total genetic architecture of troglomorphic traits, as most have sample sizes well below that needed to accurately and confidently map all QTL responsible for complex traits (Beavis, 1994, 1998; Rockman, 2012; Slate, 2013). We also encourage replicate mapping of phenotypes from past studies to confirm the location and effect size of current QTL.

Improving phenotyping can also increase the precision and accuracy of QTL studies (Falconer and Mackay, 1996). Many complex traits in *Astyanax* are actually a combination of numerous (potentially) simpler traits. For example, adult eye size, which is associated with at least eight different QTL (Protas et al., 2007, 2008) is actually the end result of numerous independent changes to eye development, including reduced eye vesicle and ventral retina size, early lens and retinal apoptosis, failed photoreceptor differentiation, and arrested growth (Jeffery and Martasian, 1998; Alunni et al., 2007; Wilkens, 2007; Pottin et al., 2011; Rétaux and Casane, 2013). Studied individually, these traits should prove more amenable to genetic mapping than their combined end result, e.g., "eye size." Wilkens (1971) argued that it would be impossible to separate these factors, since they were so tightly integrated into the developmental program of the eye; however, successful dissection of ventral retinal development by Pottin et al. (2011) and left/right asymmetry by Gross et al. (2014) suggest that this type of atomization may be possible. Additionally, although Sadoglu (1975) found that eye and lens size are each polygenic, she also found that alterations to these different optic structures sort independently of one another, and that fewer genes are probably responsible for variation in lens size than eye size. Furthermore, Sadoglu (1975) found that some specific aspects of lens and retinal degeneration, such as the formation of vacuolar lenses and the loss of photoreceptors, may even be Mendelian. These results suggest that the atomization of complex traits like eye development is possible and can yield simpler traits that are more amenable to genetic dissection via QTL and QTN studies. The use of even finer methods of phenotypic dissection, such as X-ray microtomography (micro-CT), is sure to aid in this respect.

Additional Methods: Population Scans, Genome Sequencing, and Sampling

QTL studies are increasingly being performed in conjunction with additional genome-wide analyses, such as scans for positive selection (e.g., Rogers and Bernatchez, 2005). The assumption of these studies is that by combining different analyses, it should be possible to refine the causal region for traits that have diverged due to natural selection or drift. Genome scans for positive selection require RADseq, transcriptome, or whole-genome sequence data from multiple individuals from each population ($N < 50$ in most cases) in order to compare the frequency of genetic variants. These scans can be done within and between the same populations or morphs used for QTL mapping, though an existing QTL map is not required (Jones et al., 2012; Roesti et al., 2012b). The scans are then used to identify alleles whose frequencies are significantly different between two populations or morphs and to determine whether those alleles lie within any known QTL (Nadeau et al., 2012; Roesti et al., 2012a). In addition to potentially providing narrower candidate regions, population scans can also complement QTL studies by linking them back to natural variation. QTL studies are generally conducted under laboratory conditions, and the genes that contribute to the trait in the lab may not always be important in shaping phenotypic variation in the wild (Weinig et al., 2002; Anderson et al., 2011). Further, population scans are not reliant upon a candidate gene approach, which is prone to overlook causal factors not specific to model organisms (Elmer and Meyer, 2011).

Based upon the unique evolutionary history of CF, we have several recommendations for applying scans for selection in *Astyanax*. First, while inflated F_{ST}, a measure of fixation between populations, is commonly used as evidence of outlier regions (Nosil et al., 2009; Kaeuffer, et al., 2011; Ellegren et al., 2012; Roesti et al., 2012a), inflated F_{ST} can also be the result of the region having low nucleotide diversity due to a reduced recombination rate (Charlesworth, 1998; Noor and Bennett, 2009; Cruickshank and Hahn, 2014). Thus, it is imperative to employ absolute measures of divergence (e.g., D_{xy}) when scanning for outlier regions instead of or in addition to relative measures such as F_{ST} (Noor and Bennett, 2009). In the case of suspected adaptive regions in CF, a convincing outlier caused by selection in CF would be one with high absolute divergence (D_{xy}) between CF and SF, allelic diversity in SF that is on par with the median for the SF genome, but low diversity in CF. Such a result would indicate a likely selective sweep in CF, but offer no other inherently obvious reason why that particular region would have exceptionally low diversity relative to the rest of the genome. Further evidence for an area of adaptive importance would be provided if such a region also had an overabundance of derived alleles in CF. We encourage more population genomic studies with comprehensive population sampling in *Astyanax* (Bradic et al., 2012, 2013; Gross, 2012; Coghill et al., 2014), as a tool to better understand the relationships among the various CF and SF populations, as well as to identify gene flow (including potentially cave-derived alleles) between them. An understanding of the relationships and patterns of gene flow among the various populations is essential to understand how evolution has proceeded in the cave environment (e.g., were key traits derived from standing genetic variation or *de novo* mutations?). One broad theme that has already emerged from genome-wide population sampling in other systems is that different parts of the genome can have very different evolutionary histories (e.g., Nosil et al., 2009; McGaugh and Noor, 2012). Given the ongoing gene flow inferred between several cave and surface populations (Bradic et al., 2012), we suspect *Astyanax* may be an exemplar of this interesting phenomena.

Whole-genome sequencing can also increase the power of genome-wide scans of selection or other population parameters. Whole-genome sequences from individually barcoded samples are preferred over reduced representation sequences, such as RADseq or pooled samples for several reasons. First, recent work has indicated that RADseq suffers from biases in inferring population genetic parameters (Gautier et al., 2012; Arnold et al., 2013; but see Davey et al., 2013) and is unlikely to lead to QTN discovery with genome scans (Stanton-Geddes et al., 2013; Tiffin and Ross-Ibarra, 2014). Second, whole-genome sequencing of multiple individuals provides data on the frequency of structural variants (e.g., indels, inversions, gene duplications), which are impossible to obtain with RADseq and likely difficult to obtain with pooled samples (but see Zhu et al., 2012). Structural variation is increasingly appreciated as a source of important genetic variation in ecotypic divergence (Manoukis et al., 2008; Lowry and Willis, 2010; Feulner, et al., 2012). Further, evidence continues to grow that evolution likely proceeds through genetic changes that regulate when, where, and how much a gene is expressed (King and Wilson, 1975; Jones et al., 2012), and genome-wide data can aid in detecting these evolutionarily important variants (Stapley et al., 2010; Jones et al., 2012). For example, in the QTN mapping of albinism in *Astyanax*, Protas et al. (2006) could only hypothesize that the *Oca2* mutation found in the Japonés CF population was regulatory, since the actual mutation could not be identified, because no genomic sequence was available. Recently developed long-read technology, such as the single-molecule sequencing system supplied by Pacific Biosciences, can enhance our ability to detect large structural variants in the future (Ritz et al., 2014). Additionally, this long-read technology can aid in the analysis and assembly of genomic regions that are too divergent, contain duplications of genes (which may be especially important in teleost fish that have undergone additional historic genome duplications comparing with a lobe-fin lineage, which includes mammals) or that contain repetitive DNA elements, which might normally be excluded from realignments to a reference genome (Tiffin and Ross-Ibarra, 2014). In fact, an exciting question remains as to whether these difficult-to-align regions, normally missed by standard resequencing strategies (e.g., Illumina), are fundamentally different in their diversity and evolutionary rate (Tiffin and Ross-Ibarra, 2014). If so, then these regions, which are accessible only by long reads, could provide a more nuanced story about the evolutionary history of *Astyanax* and other organisms.

Finally, there are several methodologies that are ripe for use in *Astyanax*, including expression QTL (eQTL) mapping, admixture mapping, and even the use of other CF and SF populations. Extensive reviews are written on the first two methods, so we only briefly touch on the goals of each here. First, eQTL studies treat gene expression as the trait of interest in QTL mapping in order to identify *cis*- and *trans*- regulatory factors underlying gene regulatory networks (Gilad et al., 2008; Rockman, 2008; Rockman et al., 2010; Tur et al., 2014). In some model systems, eQTL studies have even been extended to protein-expression analyses, as mRNA is often not a perfect correlate to translated products (Albert et al., 2014). Such eQTL studies would be an extension of recent next-generation transcriptome studies that have already contributed a great deal to our understanding of regulatory evolution in *Astyanax* (Gross et al., 2013; Hinaux et al., 2013; McGaugh et al., 2014). eQTL mapping might even identify the mutations responsible for altered *shh*, *pax6*, and *fgf8* gene expression, which partially explain eye and retinal reduction, but for which no causative mutation has yet been found (Yamamoto et al., 2004; Pottin et al., 2011). Second, admixture mapping, which is a special case of association mapping, takes advantage of natural variation among genotypes and phenotypes in hybrid populations (Reich et al., 2005; Smith and O'Brien, 2005; Winkler et al., 2010). This method could be extremely fruitful in caves with significant introgression with a known SF population. This method is particularly powerful, because mapped variants are putatively responsible for the phenotype under field conditions, as opposed to laboratory-based QTL studies. Admixture mapping might be particularly fruitful in the CF population found in Chica or Subterráneo caves, which contain both eyed and eyeless individuals (Mitchell et al., 1977). Finally, although the repeated evolution of troglomorphic traits among different CF populations is often touted as a strength of the *Astyanax* system, this evolutionary diversity has so far been underutilized with regard to QTL and QTN mapping. Most QTL studies in *Astyanax* have relied upon members of the "old" CF group, either Pachón or Tinaja, crossed with SF from either an "old" (Texas) or more recent invasion; however, at least 27 other populations are theoretically available for analysis (Mitchell et al., 1977). Some of the younger cave populations, such as those from the Micos or Guatemala regions, may have fewer mutations differentiating SF and CF. These younger CF populations may, therefore, be more amenable to genetic dissection and shed more light on the evolutionary process of cave adaptation than "old" CF. Additionally, many of these populations have evolved reduced eye size independently from Pachón and Tinaja (Borowsky, 2008) and may, therefore, reveal completely different genes and mutations following QTL and QTN mapping.

The Final Step: Genotype-to-Phenotype-to-Fitness

Simply identifying the genes responsible for troglomorphic traits is a laudable goal—but it is not sufficient in itself for understanding the evolutionary process (Rockman, 2012; Travisano and Shaw, 2013; Lee et al., 2014; Tiffin and Ross-Ibarra, 2014). Broadening our understanding of selection in wild populations, the prevalence of genetic drift, and population demography are also necessary to increase our understanding of the forces that have shaped the evolution of CF (Travisano and Shaw, 2013; Lee et al., 2014). Studies that document the strength of selection against SF traits or alleles in the cave environment are needed to understand the evolutionary forces in caves and among different caves as well, since slightly different conditions of light, temperature, and conductivity may exist in each cave. Possible studies include leveraging repeated population sampling in the field to measure selection against surface alleles in admixed caves (similar to Mathieson and McVean, 2013); or, alternatively, within a cave environment, one could obtain estimates of selection against surface alleles by setting up large enclosures containing advance intercrossed fish. These selection studies could then begin to answer some of the larger questions in CF evolution, such as whether iconic traits such as eye loss and albinism have occurred through positive selection or relaxed negative selection (Protas et al., 2007; Lahti et al., 2009; Bilandžija et al., 2013). Other recent studies of genotype-to-phenotype-to-fitness have used structural equation modeling to estimate the influence of QTL and environment on life history traits (Gove et al., 2012; Fournier-Level et al., 2013). In *Astyanax*, this may be a promising method to explore more about the evolutionary process that generated these unique phenotypes (see also Tabor and Burgess, Chapter 20, this volume).

The *Astyanax* community has made large and impressive strides in linking phenotype to genotype. As we look toward the future, we see room for many exciting advances and refinements in this area, as well as new tools to increase our understanding of how selection, drift, and gene flow interact to shape the evolution of this unique species.

REFERENCES

Albert, F.W., Treusch, S., Shockley, A.H., Bloom, J.S., Kruglyak, L., 2014. Genetics of single-cell protein abundance variation in large yeast populations. Nature 506, 494–497.

Albertson, R.C., Streelman, J.T., Kocher, T., 2003. Genetic basis of adaptive shape differences in the cichlid head. J. Hered. 94, 291–301.

Altschul, S.F., Gish, W., Miller, W., Myers, E.W., Lipman, D.J., 1990. Basic local alignment search tool. J. Mol. Biol. 215, 403–410.

Alunni, A., Menuet, A., Candal, E., Pénigault, J.B., Jeffery, W.R., et al., 2007. Developmental mechanisms for retinal degeneration in the blind cavefish *Astyanax mexicanus*. J. Comp. Neurol. 505, 221–233.

Anderson, J.T., Lee, C.-R., Mitchell-Olds, T., 2011. Life history QTLs and natural selection on flowering time in *Boechera stricta*, a perennial relative of *Arabidopsis*. Evolution 65, 771–787.

Arnold, B., Corbett-Detig, R.B., Hartl, D., Bomblies, K., 2013. RADseq underestimates diversity and introduces genealogical biases due to nonrandom haplotype sampling. Mol. Ecol. 22, 3179–3190.

Baird, N.A., Etter, P.D., Atwood, T.S., Currey, M.C., Shiver, A.L., et al., 2008. Rapid SNP discovery and genetic mapping using sequenced RAD markers. PLoS One 3, e3376.

Beavis, W., 1994. The Power and Deceit of QTL Experiments: Lessons from Comparative QTL Studies. ASTA, Washington D.C., pp. 250–266.

Beavis, W., 1998. QTL Analyses: Power, Precision, and Accuracyin *Molecular Dissection of Complex Traits*. CRC Press, Boca Raton, USA, pp. 145–162.

Bell, J., 1951. The Treasury of Human Inheritance: On Hereditary Digital Anomalie. On Brachydactyly and Symphalangisms. Cambridge University Press, Cambridge.

Bilandžija, H., Ma, L., Parkhurst, A., Jeffery, W.R., 2013. A potential benefit of albinism in *Astyanax* cavefish: downregulation of the *oca2* gene increases tyrosine and catecholamine levels as an alternative to melanin synthesis. PLoS One 8, e80823.

Blangero, J., 2004. Localization and identification of human quantitative trait loci: king harvest has surely come. Curr. Opin. Genet. Dev. 14, 233–240.

Borowsky, R., 2008. Restoring sight in blind cavefish. Curr. Biol. 19, R23–R24.

Borowsky, R., Wilkens, H., 2002. Mapping a cave fish genome: polygenic systems and regressive evolution. J. Hered. 93, 19–21.

Bradic, M., Beerli, P., León, F.G.D., Esquivel-Bobadilla, S., Borowsky, R., 2012. Gene flow and population structure in the Mexican blind cavefish complex (*Astyanax mexicanus*). BMC Evol. Biol. 12, 9.

Bradic, M., Teotónio, H., Borowsky, R.L., 2013. The population genomics of repeated evolution in the blind cavefish *Astyanax mexicanus*. Mol. Biol. Evol. 30, 2383–2400.

Brown, J.K.S., Benson, W.W., 1974. Adaptive polymorphism associated with multiple Müllerian mimicry in *Heliconius numata* (Lepid. Nymph.). Biotropica 1974, 205–228.

Cai, H., Morishima, H., 2002. QTL clusters reflect character associations in wild and cultivated rice. Theor. Appl. Genet. 104, 1217–1228.

Charlesworth, B., 1998. Measures of divergence between populations and the effect of forces that reduce variability. Mol. Biol. Evol. 15, 538–543.

Coghill, L.M., Hulsey, C.D., Chaves-Campos, J., García de Leon, F.J., Johnson, S.G., 2014. Next generation phylogeography of cave and surface *Astyanax mexicanus*. Mol. Phylogenet. Evol. 79, 368–374.

Cruickshank, T.E., Hahn, M.W., 2014. Reanalysis suggests that genomic islands of speciation are due to reduced diversity, not reduced gene flow. Mol. Ecol. 23, 3133–3157.

Davey, J.W., Cezard, T., Fuentes-Utrilla, P., Eland, C., Gharbi, K., et al., 2013. Special features of RAD Sequencing data: implications for genotyping. Mol. Ecol. 22, 3151–3164.

Ekker, S.C., Ungar, A.R., Greenstein, P., von Kessler, D.P., Porter, J.A., et al., 1995. Patterning activities of vertebrate *hedgehog* proteins in the developing eye and brain. Curr. Biol. 5, 944–955.

Elipot, Y., Legendre, L., Père, S., Sohm, F., Rétaux, S., 2014. *Astyanax* transgenesis and husbandry: how cavefish enters the laboratory. Zebrafish 11, 291–299.

Ellegren, H., Sheldon, B.C., 2008. Genetic basis of fitness differences in natural populations. Nature 452, 169–175.

Ellegren, H., Smeds, L., Burri, R., Olason, P.I., Backström, N., et al., 2012. The genomic landscape of species divergence in *Ficedula* flycatchers. Nature 491, 756–760.

Elmer, K.R., Meyer, A., 2011. Adaptation in the age of ecological genomics: insights from parallelism and convergence. Trends Ecol. Evol. 26, 298–306.

Falconer, D., Mackay, T., 1996. Introduction to Quantitative Genetics. Longman, Harlow.

Feulner, P.G.D., Chain, F.J.J., Panchal, M., Eizaguirre, C., Kalbe, M., et al., 2012. Genome-wide patterns of standing genetic variation in a marine population of three-spined sticklebacks. Mol. Ecol. 22, 635–649.

Fisher, R., 1930. The Genetical Theory of Natural Selection. Oxford University Press, New York.

Fisher, R., 1918. The correlation between relatives on the supposition of Mendelian inheritance. Trans. R. Soc. Edinburgh 52, 399–433.

Fournier-Level, A., Wilczek, A.M., Cooper, M.D., Roe, J.L., Anderson, J., et al., 2013. Paths to selection on life history loci in different natural environments across the native range of *Arabidopsis thaliana*. Mol. Ecol. 22, 3552–3566.

Gautier, M., Gharbi, K., Cezard, T., Foucaud, J., Kerdelhué, C., et al., 2012. The effect of RAD allele dropout on the estimation of genetic variation within and between populations. Mol. Ecol. 22, 3165–3178.

Gehring, W.J., Ikeo, K., 1999. *Pax 6*: mastering eye morphogenesis and eye evolution. Trends Genet. 15, 371–377.

Gilad, Y., Rifkin, S.A., Pritchard, J.K., 2008. Revealing the architecture of gene regulation: the promise of eQTL studies. Trends Genet. 24, 408–415.

Gove, R.P., Chen, W., Zweber, N.B., Erwin, R., Rychtář, J., et al., 2012. Effects of causal networks on the structure and stability of resource allocation trait correlations. J. Theor. Biol. 293, 1–14.

Gross, J.B., 2012. The complex origin of *Astyanax* cavefish. BMC Evol. Biol. 12, 105.

Gross, J.B., Borowsky, R., Tabin, C.J., 2009. A novel role for *Mc1r* in the parallel evolution of depigmentation in independent populations of the cavefish *Astyanax mexicanus*. PLoS Genet. 5, e1000326.

Gross, J.B., Furterer, A., Carlson, B.M., Stahl, B.A., 2013. An integrated transcriptome-wide analysis of cave and surface dwelling *Astyanax mexicanus*. PLoS One 8, e55659.

Gross, J.B., Krutzler, A.J., Carlson, B.M., 2014. Complex craniofacial changes in blind cave-dwelling Fish are mediated by genetically symmetric and asymmetric loci. Genetics 196, 1303–1319.

Gross, J.B., Protas, M., Conrad, M., Scheid, P.E., Vidal, O., et al., 2008. Synteny and candidate gene prediction using an anchored linkage map of *Astyanax mexicanus*. Proc. Natl. Acad. Sci. U. S. A. 105, 20106–20111.

Hale, C.R., Majumdar, S., Elmore, J., Pfister, N., Compton, M., et al., 2012. Essential features and rational design of CRISPR RNAs that function with the Cas RAMP module complex to cleave RNAs. Mol. Cell 45, 292–302.

Hinaux, H., Poulain, J., Da Silva, C., Noirot, C., Jeffery, W.R., et al., 2013. De novo sequencing of *Astyanax mexicanus* surface fish and Pachón cavefish transcriptomes reveals enrichment of mutations in cavefish putative eye genes. PLoS One 8, e53553.

Hughes, A.L., 2007. Looking for Darwin in all the wrong places: the misguided quest for positive selection at the nucleotide sequence level. Heredity 99, 364–373.

Hwang, W.Y., Fu, Y., Reyon, D., Maeder, M.L., Tsai, S.Q., et al., 2013. Efficient genome editing in zebrafish using a CRISPR-Cas system. Nat. Biotechnol. 31, 227–229.

Jeffery, W.R., 2008. Emerging model systems in evo-devo: cavefish and microevolution of development. Evol. Dev. 10, 265–272.

Jeffery, W.R., Martasian, D.P., 1998. Evolution of eye regression in the cavefish *Astyanax*: apoptosis and the *Pax-6* gene. Am. Zool. 38, 685–696.

Jones, F.C., Grabherr, M.G., Chan, Y.F., Russell, P., Mauceli, E., et al., 2012. The genomic basis of adaptive evolution in threespine sticklebacks. Nature 484, 55–61.

Kaeuffer, R., Peichel, C.L., Bolnick, D.I., Hendry, A.P., 2011. Parallel and nonparallel aspects of ecological, phenotypic, and genetic divergence across replicate population pairs of lake and stream stickleback. Evolution 66, 402–418.

King, M.-C., Wilson, A.C., 1975. Evolution at two levels in humans and chimpanzees. Science 188, 107–116.

Kowalko, J.E., Rohner, N., Linden, T.A., Rompani, S.B., Warren, W.C., et al., 2013a. Convergence in feeding posture occurs through different genetic loci in independently evolved cave populations of *Astyanax mexicanus*. Proc. Natl. Acad. Sci. U. S. A. 110, 16933–16938.

Kowalko, J.E., Rohner, N., Rompani, S.B., Peterson, B.K., Linden, T.A., et al., 2013b. Loss of schooling behavior in cavefish through sight-dependent and sight-independent mechanisms. Curr. Biol. 23, 1874–1883.

Lahti, D.C., Johnson, N.A., Ajie, B.C., Otto, S.P., Hendry, A.P., et al., 2009. Relaxed selection in the wild. Trends Ecol. Evol. 24, 487–496.

Laughlin, S.B., van Steveninck, R.R., Anderson, J.C., 1998. The metabolic cost of neural information. Nat. Neurosci. 1, 36–41.

Lee, Y.W., Gould, B.A., Stinchcombe, J.R., 2014. Identifying the genes underlying quantitative traits: a rationale for the QTN programme. AoB Plants 6, plu004.

Lowry, D.B., Willis, J.H., 2010. A widespread chromosomal inversion polymorphism contributes to a major life-history transition, local adaptation, and reproductive isolation. PLoS Biol. 8, e1000500.

Lynch, M., Walsh, B., 1998. Genetics and Analysis of Quantitative Traits. Sinauer Associates, Sunderland, Massachussets.

Mackay, T.F., Stone, E.A., Ayroles, J.F., 2009. The genetics of quantitative traits: challenges and prospects. Nat. Rev. Genet. 10, 565–577.

Manoukis, N.C., Powell, J.R., Touré, M.B., Sacko, A., Edillo, F.E., et al., 2008. A test of the chromosomal theory of ecotypic speciation in Anopheles gambiae. Proc. Natl. Acad. Sci. 105, 2940.

Mathieson, I., McVean, G., 2013. Estimating selection coefficients in spatially structured populations from time series data of allele frequencies. Genetics 193, 973–984.

Matsunaga, E., 1962. The dimorphism in human normal cerumen*. Ann. Hum. Genet. 25, 273–286.

McEvoy, B.P., Visscher, P.M., 2009. Genetics of human height. Econ. Hum. Biol. 7, 294–306.

McGaugh, S.E., Gross, J.B., Aken, B., Blin, M., Borowsky, R., et al., 2014. The cavefish genome reveals candidate genes for eye loss. Nat. Commun. 5, 5307.

McGaugh, S.E., Noor, M.A.F., 2012. Genomic impacts of chromosomal inversions in parapatric *Drosophila* species. Philos. Trans. R. Soc., B 367, 422–429.

Mendel, G., 1866. Versuche über Pflanzenhybriden. Verh. Naturforsch. Ver. Brunn 4, 3–47.

Mitchell, R.W., Russell, W.H., Elliott, W.R., 1977. Mexican Eyeless Characin Fishes, Genus Astyanax: Environment, Distribution, and Evolution. Texas Tech Press, Lubbock.

Mohr, C.E., Poulson, T.L., 1966. Life of The Cave. McgrawHill Book Company, New York.

Nadeau, N.J., Jiggins, C.D., 2010. A golden age for evolutionary genetics? Genomic studies of adaptation in natural populations. Trends Genet. 26, 484–492.

Nadeau, N.J., Whibley, A., Jones, R.T., Davey, J.W., Dasmahapatra, K.K., et al., 2012. Genomic islands of divergence in hybridizing *Heliconius* butterflies identified by large-scale targeted sequencing. Philos. Trans. R. Soc. B 367, 343–353.

Noor, M.A.F., Bennett, S.M., 2009. Islands of speciation or mirages in the desert? Examining the role of restricted recombination in maintaining species. Heredity 103, 439–444.

Noor, M.A.F., Cunningham, A.L., Larkin, J.C., 2001. Consequences of recombination rate variation on quantitative trait locus mapping studies: simulations based on the *Drosophila melanogaster* genome. Genetics 159, 581–588.

Nosil, P., Funk, D.J., Ortiz-Barrientos, D., 2009. Divergent selection and heterogeneous genomic divergence. Mol. Ecol. 18, 375–402.

O'Quin, K.E., Yoshizawa, M., Doshi, P., Jeffery, W.R., 2013. Quantitative genetic analysis of retinal degeneration in the blind cavefish *Astyanax mexicanus*. PLoS One 8, e57281.

Okawa, H., Sampath, A.P., Laughlin, S.B., Fain, G.L., 2008. ATP consumption by mammalian rod photoreceptors in darkness and in light. Curr. Biol. 18, 1917–1921.

Orr, H.A., 1998. Testing natural selection vs. genetic drift in phenotypic evolution using quantitative trait locus data. Genetics 149, 2099–2104.

Parzefall, J., Fricke, D., 1989. Alarm reaction and school in cave and river populations of *Astyanax fasciatus* (Pisces, Characidae) and their hybrids. Mem. Biospeleology 16, 177–182.

Petrakis, N.L., Molohon, K.T., Tepper, D.J., 1967. Cerumen in American Indians: genetic implications of sticky and dry types. Science 158, 1192–1193.

Pottin, K., Hinaux, H., Rétaux, S., 2011. Restoring eye size in *Astyanax mexicanus* blind cavefish embryos through modulation of the *Shh* and *Fgf8* forebrain organising centres. Development 138, 2467–2476.

Poulson, T.L., 1963. Cave adaptation in amblyopsid fishes. Am. Midl. Nat. 70, 257–290.

Protas, M., Conrad, M., Gross, J.B., Tabin, C., Borowsky, R., 2007. Regressive evolution in the Mexican cave tetra, *Astyanax mexicanus*. Curr. Biol. 17, 452–454.

Protas, M., Hersey, C., Kochanek, D., Zhou, Y., Wilkens, H., et al., 2006. Genetic analysis of cavefish reveals molecular convergence in the evolution of albinism. Nat. Genet. 38, 107–111.

Protas, M., Tabansky, I., Conrad, M., Gross, J.B., Vidal, O., et al., 2008. Multi-trait evolution in a cave fish, *Astyanax mexicanus*. Evol. Dev. 10, 196–209.

Reich, D., Patterson, N., De Jager, P.L., McDonald, G.J., Waliszewska, A., et al., 2005. A whole-genome admixture scan finds a candidate locus for multiple sclerosis susceptibility. Nat. Genet. 37, 1113–1118.

Rétaux, S., Casane, D., 2013. Evolution of eye development in the darkness of caves: adaptation, drift, or both? EvoDevo 4, 26.

Ritz, A., Bashir, A., Sindi, S., Hsu, D., Hajirasouliha, I., et al., 2014. Characterization of structural variants with single molecule and hybrid sequencing approaches. Bioinformatics 30 (24), 3458–3466. btu714.

Rockman, M.V., 2008. Reverse engineering the genotype-phenotype map with natural genetic variation. Nature 456, 738–744.

Rockman, M.V., 2012. The QTN program and the alleles that matter for evolution: all that's gold does not glitter. Evolution 66, 1–17.

Rockman, M.V., Skrovanek, S.S., Kruglyak, L., 2010. Selection at linked sites shapes heritable phenotypic variation in *C. elegans*. Science 330, 372–376.

Roesti, M., Hendry, A.P., Salzburger, W., Berner, D., 2012a. Genome divergence during evolutionary diversification as revealed in replicate lake-stream stickleback population pairs. Mol. Ecol. 21, 2852–2962.

Roesti, M., Salzburger, W., Berner, D., 2012b. Uninformative polymorphisms bias genome scans for signatures of selection. BMC Evol. Biol. 12, 94.

Rogers, S., Bernatchez, L., 2005. FAST-TRACK: integrating QTL mapping and genome scans towards the characterization of candidate loci under parallel selection in the lake whitefish (*Coregonus clupeaformis*). Mol. Ecol. 14, 351–361.

Rohner, N., Jarosz, D.F., Kowalko, J.E., Yoshizawa, M., Jeffery, W.R., et al., 2013. Cryptic variation in morphological evolution: HSP90 as a capacitor for loss of eyes in cavefish. Science 342, 1372–1375.

Romero, A., Paulson, K.M., 2001. It's a wonderful hypogean life: a guide to the troglomorphic fishes of the world. In: The Biology of Hypogean Fishes. Springer, New York, pp. 13–41.

Romero, A., Zhao, Y., Chen, X., 2009. The hypogean fishes of China. Environ. Biol. Fish 86, 211–278.

Ron, M., Weller, J., 2007. From QTL to QTN identification in livestock-winning by points rather than knock-out: a review. Anim. Genet. 38, 429–439.

Sadoglu, P., 1957. Mendelian inheritance in the hybrids between the Mexican blind cave fishes and their overground ancestor. Verh. Dtsch. Zool. Ges. Graz 1957. Zool. Anz. Suppl. 21, 432–439.

Sadoglu, P., 1975. Genetic Paths Leading to Blindness in *Astyanax Mexicanus*, in *Vision in Fishes*. Springer, New York, pp. 419–426.

Sadoglu, P., McKee, A., 1969. A second gene that affects eye and body color in Mexican blind cave fish. J. Hered. 60, 10–14.

Schemmel, C., 1980. Studies on the genetics of feeding behaviour in the cave fish *Astyanax mexicanus* f. anoptichthys. Z. Tierpsychol. 53, 9–22.

Schielzeth, H., Husby, A., 2014. Challenges and Prospects in Genome-Wide Quantitative Trait Loci Mapping of Standing Genetic Variation in Natural Populations. Annals of the New York Academy of Sciences, New York, USA.

Slate, J., 2013. From Beavis to beak color: a simulation study to examine how much QTL mapping can reveal about the genetic architecture of quantitative traits. Evolution 67, 1251–1262.

Smith, M.W., O'Brien, S.J., 2005. Mapping by admixture linkage disequilibrium: advances, limitations and guidelines. Nat. Rev. Genet. 6, 623–632.

Stanton-Geddes, J., Paape, T., Epstein, B., Briskine, R., Yoder, J., et al., 2013. Candidate genes and genetic architecture of symbiotic and agronomic traits revealed by whole-genome, sequence-based association genetics in *Medicago truncatula*. PLoS One 8, e65688.

Stapley, J., Reger, J., Feulner, P.G.D., Smadja, C., Galindo, J., et al., 2010. Adaptation genomics: the next generation. Trends Ecol. Evol. 25, 705–712.

Stern, D.L., Orgogozo, V., 2009. Is genetic evolution predictable? Science 323, 746–751.

Stinchcombe, J.R., Hoekstra, H.E., 2007. Combining population genomics and quantitative genetics: finding the genes underlying ecologically important traits. Heredity 100, 158–170.

Strickler, A.G., Yamamoto, Y., Jeffery, W.R., 2001. Early and late changes in *Pax6* expression accompany eye degeneration during cavefish development. Dev. Genes Evol. 211, 138–144.

Tiffin, P., Ross-Ibarra, J., 2014. Advances and limits of using population genetics to understand local adaptation. Trends Ecol. Evol. 29, 673–680.

Travisano, M., Shaw, R.G., 2013. Lost in the map. Evolution 67, 305–314.

Tur, I., Roverato, A., Castelo, R., 2014. Mapping eQTL Networks with mixed graphical Markov models. Genetics 198, 1377–1393.

Vidal, M., Cusick, M.E., Barabasi, A.-L., 2011. Interactome networks and human disease. Cell 144, 986–998.

Weinig, C., Ungerer, M.C., Dorn, L.A., Kane, N.C., Toyonaga, Y., et al., 2002. Novel loci control variation in reproductive timing in *Arabidopsis thaliana* in natural environments. Genetics 162, 1875–1884.

Wilkens, H., 1971. Genetic interpretation of regressive evolutionary processes: studies on hybrid eyes of two *Astyanax* cave populations (Characidae, Pisces). Evolution 25, 530–544.

Wilkens, H., 1988. Evolution and genetics of epigean and cave *Astyanax fasciatus* (Characidae, Pisces): Support for the neutral mutation theory. Evol. Biol. 23, 271–367.

Wilkens, H., 2007. Regressive evolution: ontogeny and genetics of cavefish eye rudimentation. Biol. J. Linn. Soc. 92, 287–296.

Winkler, C.A., Nelson, G.W., Smith, M.W., 2010. Admixture mapping comes of age*. Annu. Rev. Genomics Hum. Genet. 11, 65–89.

Yamamoto, Y., Stock, D.W., Jeffery, W.R., 2004. Hedgehog signalling controls eye degeneration in blind cavefish. Nature 431, 844–847.

Yoshizawa, M., Gorički, Š., Soares, D., Jeffery, W.R., 2010. Evolution of a behavioral shift mediated by superficial neuromasts helps cavefish find food in darkness. Curr. Biol. 20, 1631–1636.

Yoshizawa, M., Yamamoto, Y., O'Quin, K.E., Jeffery, W.R., 2012. Evolution of an adaptive behavior and its sensory receptors promotes eye regression in blind cavefish. BMC Biol. 10, 108.

Zhao, Y.H., Gozlan, R., Zhang, C.G., 2011. Out of sight out of mind: current knowledge of Chinese cave fishes. J. Fish Biol. 79, 1545–1562.

Zhu, Y., Bergland, A.O., Gonzalez, J., Petrov, D., 2012. Empirical validation of pooled whole genome population re-sequening in *Drosophila melanogaster*. PLoS One 7, e41901.

Chapter 7

Selection Through Standing Genetic Variation

Nicolas Rohner
Stowers Institute for Medical Research, Kansas City, Missouri, USA

Animals display a fascinating diversity in nature. The generation of this variation is tightly linked to the different environmental challenges to which animals have adapted over the course of evolution. New climates, varying ecological niches, and species interactions, as well as different environmental milieus are the selective agents for change. Heritable adaptations to these different environments are ultimately caused by changes in the genome. Thus, when surface fish are swept into a cave, they have to adapt to this new and extreme environment, and they do so through changes in their genomic composition. Mutations can be introduced into the genome, by either intrinsic (e.g., errors in DNA replication) or extrinsic means (e.g., physical damage to the DNA by UV light or mutagens). Those, in turn, lead to multiple different types of mutations (e.g., point mutations, insertions, inversions, duplications). Such mutations can subsequently also lead to multiple differing effects (e.g., affecting the protein structure, the regulation of a gene or its doses, just to name a few). Differences among environments can affect the type of mutations that arose. For example, in the case of the cave environment with its absence of (UV) light, mutations are less likely to result from pyrimidine dimers, a type of lesion generated via photochemical reactions; however, the purpose of this chapter will be primarily to deal with the timing of mutational events in adaptation.

DE NOVO VERSUS STANDING GENETIC VARIATION

When a species colonizes a new environment, an important and still unanswered question is if the changes that drive the adaptation to the new environment are primarily generated *de novo* or if they more commonly derive from standing genetic variation (Figure 7.1). By definition (compare also Box 7.1), "*de novo*" means that the changes have occurred after a selective change in the environment or a geographic isolation event. In the case of cavefish, *de novo* thus refers to changes following the event of surface fish being swept into the caves. In other

Biology and Evolution of the Mexican Cavefish. http://dx.doi.org/10.1016/B978-0-12-802148-4.00007-4
© 2016 Elsevier Inc. All rights reserved.

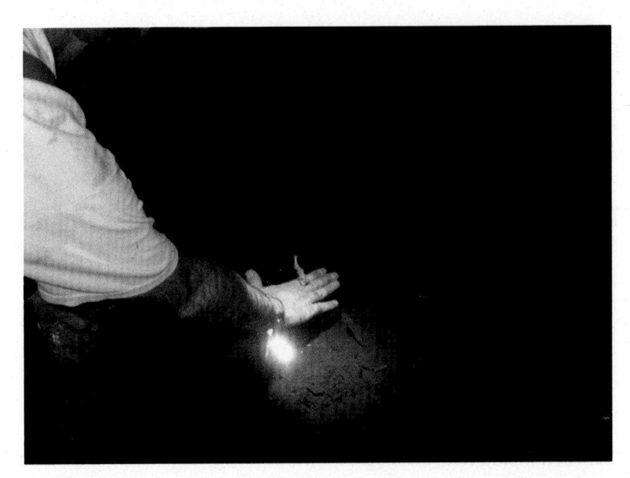

FIGURE 7.1 The cave environment. Cavefish had to adapt to the completely dark cave environment. One crucial question in the genetics of adaptation is where the mutations that help the fish in survival are coming from. Are they derived *de novo* in the cave after the fish are swept into the cave, or are they maybe already present in the surface fish populations? Latest results are pointing toward the role of standing genetic variation in adaptation of cavefish. *Picture taken at the Tinaja cave. Credit Nicolas Rohner.*

Box 7.1 Definition of Keywords

De novo mutation	Mutation occurring AFTER the change in environment; not present in the ancestral population
Standing genetic variation	Mutation already present in the ancestral population, usually in low frequencies; mutation occurring BEFORE the change in environment
Cryptic variation	Polymorphic loci that have no phenotypic effect until perturbed by unusual conditions
Canalization	Phenomenon that some genetic variation is silenced or the phenotype is buffered against genetic or environmental perturbations
HSP90 (heat shock protein 90)	Has a dual role of canalization and keeping proteins from misfolding during environmental stress
Genetic assimilation	Selection for the cryptic trait will ultimately lead to its fixation and the new trait will develop even under normal conditions
QTL (quantitative trait locus)	A region in the genome containing the genetic basis responsible for variation in a quantitative trait

words, in the *de novo* context, the change in the environment comes first, and subsequently, the isolated population acquires new mutations, which, if helpful for survival, will eventually be selected for and fixed within that population. In contrast, standing genetic variation refers to mutations that were already present at some level within the original population and were only selected after the change in the environment. In the case of *Astyanax mexicanus*, this would mean the beneficial mutations were present in the population of surface fish that were swept into the cave. The individuals carrying such mutations have a selective advantage in this new environment. In the extreme case, only those individuals will survive and thrive, ultimately leading to a fixation of the mutation in the cave population. Evolution from standing genetic variation may also proceed by the generation of novel alleles through recombination of multiple loci, which makes it highly unlikely for *de novo* mutations to occur, at least in the short term.

Each mode of genetic change leaves a distinct signature of selection in the genome, and each has particular advantages and disadvantages when it comes to adaptation. For instance, *de novo* mutation is generally considered to be a relatively slow process, as the population has to wait for the mutation to occur before individuals of higher fitness can be selected (Barrett and Schluter, 2008) (this is particularly true in smaller populations). Additionally, *de novo* mutations come with a bias against recessive mutations (a process called Haldane's sieve (Haldane, 1927), simply because the chance that a mutation will be lost before displaying an effect and experiencing selection is relatively high. The same seems to be true for small-effect mutations (Barrett and Schluter, 2008). So all in all, utilization of standing genetic variation is generally considered to be a faster process, leading to the spread of more alleles of small effect and more recessive alleles (Orr and Betancourt, 2001). Of course, this does not mean that evolution through *de novo* mutation does not occur. On a bigger macro-evolutionary scale, the enormous changes in evolution, such as the transition from land to water or from ape to human, would have been impossible without *de novo* mutation. But also from a micro-evolutionary point of view, there are indeed many examples where it is believed that *de novo* mutations underlie adaptation (e.g., Bersaglieri et al., 2004; Chan et al., 2010). Nonetheless, the above mentioned disadvantages of *de novo* mutation (slow, bias against recessive and small-effect alleles) make standing genetic variation a compelling model for the primary genetic substrate of evolutionary change and a target of rising interest in evolutionary genomics (e.g., Domingues et al., 2012; Colosimo et al, 2005); however, selection on standing genetic variation can be difficult to detect (see below), hence, often the signatures for standing genetic variation in the genome have been overlooked in the past.

The fact that evolution from standing genetic variation is faster than *de novo* mutation is of course especially important when dealing with quickly changing ecological conditions. Organisms need to adapt quickly to even survive, let alone optimally exploit their environments (as in the case of the cavefish *Astyanax mexicanus*, where surface forms are suddenly confronted with a completely new set of environmental parameters when swept into the cave);

however, adaptation through standing genetic variation faces an alternative problem—how is it possible for preadaptive mutations to be generated in the original population and, more importantly, how are they maintained in the absence of selection? Since there is no reason to believe that the mutations that provide a certain individual with an advantage to the new (and different) environment should per se be advantageous in the original environment (indeed, this may often be deleterious), how is it possible that a reservoir of mutations can be maintained in the original population and kept "silent until needed." Such variation where the selection coefficient is small in the current/ancestral environment or genetic background is called cryptic variation.

CRYPTIC GENETIC VARIATION AND CANALIZATION

Cryptic variation is genetic variation that is invisible, meaning it has no phenotypic effect under normal conditions but that can, under certain circumstances (such as stress, or in particular genetic backgrounds), have phenotypic effects. There has been a long debate about the extent to which cryptic variation exists and the role it plays in evolution. The concept of cryptic variation as a potentiality of evolution originated with Conrad Waddington in his famous paper in 1942 (Waddington, 1942) where he introduced the concept of canalization. Canalization is defined as a process during development that stabilizes or buffers a phenotype against genetic variation or certain environmental stresses. More a theoretical concept at the time, in a later paper (Waddington, 1953), Waddington provided actual experimental evidence for canalization using the fruit fly *Drosophila melanogaster*. He noticed that when a certain strain of these flies was exposed to a heat shock during development, some of the adult flies (roughly a third of the treated flies) displayed a specific wing phenotype (a somewhat unspectacular reduction of some of the veins in the wing), which he dubbed "crossveinless." This rather unremarkable phenotype, however, never occurred in untreated flies. When he selected for this trait over subsequent generations, the trait would reach almost 100% penetrance, meaning that all the flies that were treated with the heat shock displayed the crossveinless phenotype. Interestingly, when raising these flies under normal conditions without a heat shock, they still displayed the crossveinless phenotype. The trait became "genetically assimilated." Thus, he was able to conclude that there was cryptic genetic variation in this trait already present in the original population, but masked in the absence of heat shock stress during development.

Waddington used his observations to speculate that canalization of developmental processes underlying cryptic genetic variation could play a role in evolution. It seems paradoxical at first—how can a mechanism that prevents variation be involved in evolution?—but this brings us back to the question of standing genetic variation. With such a mechanism, it is conceivable that cryptic mutations can accumulate under mutation-selection-drift balance, and when an individual encounters a new environment (which may be stressful), these mutations become visible. If such mutations have a direct benefit in survival, they might be selected and eventually become fixed by genetic assimilation. This was an intriguing theory, however, it was quite controversial. Most importantly, it was not clear what such a mechanism for phenotype canalization could be, so critics contemptuously called the mechanism "Waddington's widget" (Ruden et al., 2003). Also, the phenotypes Waddington observed in his original experiments were not clearly adaptive or advantageous. Waddington acknowledges this by saying: "There is, of course, no reason to believe that the phenocopy would in nature have any adaptive value, but the point at issue is whether it would be eventually genetically assimilated if it were favored by selection, as it can be under experimental conditions."

HSP90 AS A CAPACITOR OF EVOLUTION

It took almost 50 years from when Waddington's theory was first developed before Susan Lindquist and coworkers found the molecular nature of such a canalization mechanism at work in Waddington's original experiment. In a landmark paper, Rutherford and Lindquist (1998) showed that suppressing a protein called HSP90 reveals previously cryptic genetic variation. The experiments were again performed using flies, however, these findings have now been confirmed in other organisms as well (Jarosz et al., 2010). When levels of HSP90 were reduced genetically or pharmacologically (but not entirely abolished—full HSP90 knockouts are lethal to the organism), Rutherford and Lindquist (1998) noted substantial phenotypic variation in many adult traits (including some wing phenotypes similar to the ones originally described by Waddington). As in the heat shock studies, these phenotypes could be selected for in subsequent generations and become quickly independent of the levels of HSP90. This important finding was the first characterization of a molecular mechanism of canalization.

HSP90 still remains the preeminent example of such a genetic buffering mechanism, although other candidate mechanisms have been identified as well (Siegal, 2013). Insight into how HSP90 might act to canalize phenotypes comes from the fact that it has a well established function as a chaperone during a heat shock, a role that gave the protein its name. Heat shock proteins (there are many others besides HSP90, like HSP70 or HSP80, the number standing for the size of the protein in kilodaltons) were initially discovered in 1962 when an Italian scientist noticed a strong transcriptional response in *Drosophila* cells after being accidentally placed under warmer conditions (Ritossa, 1962). Since then, heat shock proteins have been widely studied for their role in protein folding, especially when coping with proteins denatured by heat and other environmental stresses. Because of these dual roles (the roles of canalization and as a chaperone during stress), HSP90 would seem to be a great candidate for playing the role in evolution that Waddington theorized half a century earlier; it is conceivable that scenarios exist in nature, where an environmental stress might temporarily decrease HSP90 levels through titration by stress-damaged proteins, thereby uncovering selectable cryptic genetic variation. Indeed, Lindquist's team was able to show the same phenotypes they observed in their HSP90-reduced strains could be elicited by a simple environmental stress, such as raising the flies under warmer or colder conditions. The publication received great attention, partly because of the elegant metaphor used to describe the model; the authors called HSP90 a "capacitor" of evolution (Rutherford and Lindquist, 1998; Queitsch et al., 2002), in reference to the electrical device that can store and release energy. Extending this analogy, such a molecular capacitor is able to store genetic variation in form of cryptic genetic variation and release it upon demand in a stressful environment. In the years following the study, however, there have been arguments favoring alternative explanations for the phenotypes generated via HSP90 inhibition, such as the role of HSP90 in preventing transposon mutagenesis (Specchia et al., 2010) or due to epigenetic effects (Sollars et al., 2003). Recent results, however, clearly indicate a positive role of HSP90 on kinase evolution, as HSP90 clients show increased evolutionary rates compared with nonclients (Lachowiec et al., 2015), and theoretical calculations provide strong evidence that revealing cryptic variation can cause the frequent evolution of complex adaptations (Trotter et al., 2014).

HSP90 IN CAVEFISH EVOLUTION

Remarkably, despite tremendous interest in the capacitor function of HSP90, examples in nature for such a mechanism are still rare, especially in vertebrate evolution. In a recent paper (Rohner et al., 2013), we combined efforts with the Lindquist lab and set out to find a real-world example of HSP90's evolutionary role. We tested whether canalization through HSP90 could indeed hide cryptic variation that might play a role in cavefish evolution. Cavefish seemed to be the perfect system to study this phenomenon, given the argument that if such a mechanism contributed to evolution, it would most likely occur in a setting where members of a species are suddenly confronted with a completely new environment and hence might encounter physiological stress. This is exactly the scenario in cavefish evolution (especially in *Astyanax mexicanus*), where surface forms are carried into the caves by flooding events, and as soon as the flood retreats, are stuck in the cave, a completely unfamiliar environment. We aimed to answer two questions. First, do modern surface fish populations carry genetic variation that can be revealed by blocking HSP90 or by an environmental stress, i.e., do they contain cryptic genetic variation? This point is particularly interesting for the cavefish system, as there are actually very few examples of the presence of cryptic genetic variation in natural populations (Paaby and Rockman, 2014). Second and maybe even more importantly, did such cryptic variation contribute to the evolution of morphological traits in cavefish? Similar to previous studies, we used a well defined chemical inhibitor of HSP90, radicicol (Yeyati et al., 2007). Radicicol binds the highly conserved ATP binding domain of HSP90, which has allowed this inhibitor to be applicable across the entire eukaryotic domain from yeast to plants to humans. This inhibitor is also known to be very specific to HSP90, minimizing off-target effects (Queitsch et al., 2002). When applied to developing embryos of natural surface fish populations (for the first week of development), we noticed an increase in the variation in eye size in larval and adult fish. Interestingly, the mean (the average of all tested individuals) of the fish's eye size was not changed; however, the variation was larger on both ends, leading to fish that had larger and smaller eyes compared with fish that were raised in a control group. After selecting and breeding adults with smaller eyes, we found (within a single generation) that offspring fish had smaller eyes than control fish. As this second generation of fish was raised in the absence of the inhibitor, we were able to exclude teratogenic effects of the inhibitor. Thus, the increase in phenotypic variation could only be explained by underlying cryptic genetic variation, which became visible only after taxing the HSP90 reservoir. This observation was intriguing, as it indicated that there is cryptic genetic variation for eye size present in surface fish populations that, when released, could be rapidly selected if adaptive (see the section below, "Is eye loss in cavefish an adaptive trait"); however, this finding begged the question, what could tax the HSP90 reservoir in the cave, what could be such an environmental stress that would deplete levels of HSP90 and unmask the genetic variants? To answer this question, we went to the caves in Mexico and measured multiple parameters in the cave environment and in the environment where the surface fish live, the rivers surrounding the caves. Although it feels pretty hot and humid, when spelunking in these caves, we did not detect any big difference in the temperature of the cave water and the surface rivers, eliminating an obvious candidate of the list from potential environmental stressors; however, we did observe a striking difference in the conductivity (which essentially reflects the dissolved salt content in the water) between the cave and surface waters. In some extreme cases, we found conductivity in the caves to be ten times lower than that of the surface rivers. Conductivity is an important environmental parameter for fish, given that their skin is water permeable, and it is known that low conductivity can elicit a stress response similar to a heat shock (Choi and An, 2008). Furthermore, sticklebacks reared in low-salinity conditions display an increase in cryptic genetic variation in body size (McGuigan et al., 2011), making the differential conductivity in the caves a prime candidate for an environmental stressor acting through HSP90. When we raised surface fish populations in the lab under similar low-conductivity conditions, we indeed detected a heat shock response. Raised to adulthood, these fish displayed an increase in the variation in eye size, the same as the fish raised in the HSP90 inhibitor. The results connect the natural, ecological conditions with our observations in the laboratory where we artificially tampered with HSP90 function. Lastly, we wanted to see if we could find evidence in the cavefish for a signature of such a selection event. We repeated the exact experiments of reducing HSP90 function using radicicol that we performed with the surface fish, but this time in a laboratory using a cavefish population raised in Tinaja. In this context, we couldn't measure eye size directly (because cavefish don't have eyes), but we used the orbit size as a surrogate. Cavefish still have an orbit at the position where the eye used to be, which can be easily measured using a bone stain (Figure 7.2). We previously showed that it is possible to use the orbit size as a substitute for eye size, since orbit size correlates with eye size in surface fish (Rohner et al., 2013). When we measured orbit size in the fish that were treated with radicicol, we did not observe an increase in the variation of eye size, contrasting with the surface fish experiments. This means that indeed there was less cryptic variation present in the cavefish populations than in the surface fish population. This may be not particularly unexpected given that cavefish population sizes are smaller and derived from just a few founder fish, generally leading to a lower genomic variation due to a genetic bottleneck (Bradic et al., 2012). But what we found was that the treated fish showed a further decrease in eye size relative to the untreated groups. This observation strongly suggests that the alleles that have been selected for in cavefish evolution (for smaller eye size) are the alleles that are interacting with HSP90, giving credence to the idea that HSP90-mediated cryptic variation contributed to eye loss in cavefish evolution. In summary, this study suggests that the surface fish originally had cryptic (standing) genetic variation for eye size, masked by HSP90. When the fish ended up in the caves, the environmental factors (low conductivity and maybe

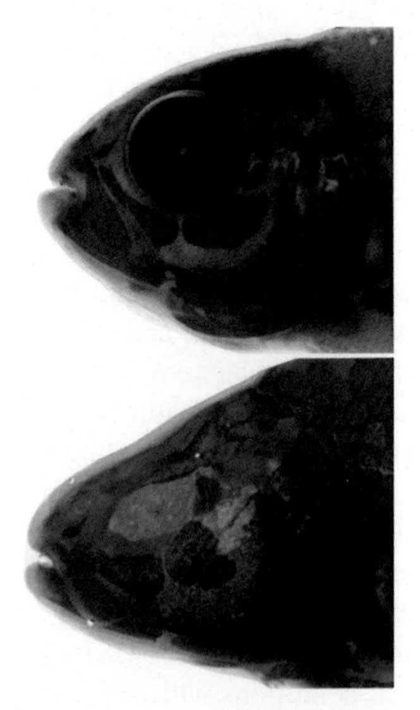

FIGURE 7.2 Bone stain of surface fish and cavefish heads. Though cavefish no longer have an eye, they still have an orbit, formed by the so-called orbital bones. Recent results indicate that the loss of eyes in cavefish populations was accompanied by changes in the genome derived from standing genetic variation present in surface fish populations. *Skull bones were stained with alizarin red. Credit Nicolas Rohner.*

others as well) physiologically stressed the fish, leading to a reduction of their HSP90 activity. The previously cryptic eye-size variation became unmasked, and nature selected for the small-eye phenotype. Such an environmental stress is of course believed to be only transient, because the cavefish would have eventually adapted to these lower conductivity conditions over subsequent generations. But our experiments showed that selection can have a measurable effect over a single generation. In this context, it might be of interest to note that cavefish have higher basal HSP90 levels than surface fish (Hooven et al., 2004), a likely adaptation to the stress conditions in the cave, rendering them potentially more stress-resistant. The lower conductivity is also most likely not the only stress factor in the cave that taxes the HSP90 system. For example, starvation also affects HSP90 expression (Antonopoulou et al., 2013). Given the low amount of food in most caves, starvation seems like another very good candidate to be studied in the future.

IS EYE LOSS IN CAVEFISH AN ADAPTIVE TRAIT

An important point of this study is that it provides further fodder for the question of whether losing eyes is an adaptive trait. The strongest argument for the hypothesis that eye loss is adaptive comes from genetic crossing experiments (see Borowsky, Chapter 6, this volume). When QTL analyses were performed for eye size in cavefish it was observed that eye size is a multigenic trait with at least twelve different loci contributing to the loss of eyes (Protas et al., 2007). When the effect size of these loci was analyzed individually, it was found that for all twelve individual QTL, the allele fixed within the cave population resulted in smaller eyes (Protas et al., 2007). This is a strong argument for selection based upon the theory of H. Allen Orr that the polarity of alleles fixed within a derived population can provide evidence for selection (Orr, 1998). In short, if there were no selection on a trait, one would expect that, just by chance, some of the random changes in the genome would lead to larger eyes as well. Given that this is not the case, and in all other known obligate cavefish in the world, there is a tendency to reduce the eyes (see Borowsky, Chapter 6, this volume) and not make them larger, it seems very likely that losing eyes in an environment in which there is no use for them is indeed advantageous. It is not hard to imagine what the reasons could be for such an advantage, as eyes are very metabolically costly organs to maintain; the retina alone is one of the most energy-consuming tissues in the body, and intriguingly the energy requirements in the dark are even higher than they are in the light (Okawa et al., 2008). In addition, if there is a selection to enhance other senses (as is the case for cavefish), it is likely helpful to free up some resources. For that matter, it is possible and even very likely that some eye QTL could have been selected indirectly through pleiotropy. For example, it has been shown that the expansion of superficial neuromasts is tightly negatively correlated with eye size (Yamamoto et al., 2009).

Finding the actual genetic loci underlying this cryptic genetic variation will be the next step. Although challenging, it will become possible using available whole-genome and eye QTL data, in conjunction with powerful tools developed for identifying HSP90 clients. Mapping such loci and understanding the genetic architecture of cryptic genetic variation is of broad interest (Gibson and Dworkin, 2004), not only for evolutionary genomics but also for human pathological conditions, as many diseases are threshold-dependent or modified by genetic factors, such as cystic fibrosis and others (Merlo and Boyle, 2003).

DETECTING STANDING GENETIC VARIATION

There are several ways to detect standing genetic variation on a whole-genome level, and now that the genome information of *Astyanax mexicanus* is available, such approaches will certainly be used in the future. It is, however, important to point out that none of these methods are flawless, and, ideally, multiple cases of evidence should be used to detect standing genetic variation.

(1) The signature of selection
Standing genetic variation leaves a different "signature of selection" in the genome than does *de novo* mutation. When a mutation is selected in a population, nearby polymorphisms (unrelated to the selective event) will hitchhike along until recombination in the following generations breaks this region apart. When *de novo* mutations are driven to fixation in the population, the region around the selected mutation displays only the haplotype of this particular founder (which are only a subset of the original haplotypes present in the entire population for that region), creating a region of low heterozygosity and low diversity around the selective site. Sweeps on *de novo* mutations are often termed "hard sweeps"; however, for alleles present as standing, segregating variation in the population, recombination has had sufficient time to move these alleles onto different genetic backgrounds. Thus, when alleles present as standing genetic variation are driven to fixation, the genetic diversity linked to such a variant is not as severely eroded, though the trough in diversity at the selected locus may be just as severe as in hard sweeps. Lastly, a "soft sweep" is a sweep on an allele that is initially on genetic backgrounds due to recurrent mutation or migration. In this case, the trough in diversity is not as severe as would be seen in a sweep from *de novo* mutation or from a sweep from standing genetic variation (Vitti et al., 2013).

(2) Standing genetic variation in the ancestral population
Another way of finding standing genetic variation is simply looking for the presence of a selected allele in the ancestral populations. Often, although the standing genetic variation persists at low frequencies, sampling and sequencing multiple individuals of the ancestral population can detect it. For example, in a study in the stickleback *Gasterosteus aculeatus* (Colosimo et al., 2005), the low-plate allele (the genetic locus responsible for the reduction of dermal plates in freshwater populations of sticklebacks compared with

the fully plated populations inhabiting ocean environments) was found to be present in at least one copy in roughly 4% of the sampled ocean populations. Further sampling of surface fish populations and whole-genome sequencing these sampled populations by using the cavefish genome as the reference (McGaugh et al., 2014) will clearly help in addressing this question.

(3) Parallel evolution
When available, parallel or repeated evolutionary events can be tremendously helpful in finding cases of standing genetic variation. If the same selected allele is present in multiple independent populations, it becomes likely that the allele stems from standing genetic variation, given that the chances for it occurring independently are rather small (though not entirely impossible). If such multiple events are available, it is even possible to estimate the age of the allele. If the estimated age of the allele (e.g., by examining the rate of neutral sequence evolution in the shared haplotype) is older than the separation event, it is the strongest evidence for evolution through standing genetic variation (Colosimo et al., 2005; Brawand et al., 2014). On the other hand, if we assume there are multiple ways to get the same phenotype, and the identified alleles differ among the parallel events (e.g., different mutations in the same gene or even mutations in different genes), it is more likely that these originated from *de novo* mutation.

EXAMPLES FROM *ASTYANAX MEXICANUS*

Taking advantage of the unique parallel population structure of *Astyanax mexicanus*, a recent study (Bradic et al., 2013) used SNP (single-nucleotide polymorphism) data from multiple cave and surface fish populations to address the question of repeatability of evolution in nature. The authors specifically focused on SNPs that were differentiated between the cave and surface fish populations. When they compared these SNPs among the different cave populations, however, they found that many of the SNPs were shared among these caves. Such a finding of such a genomic pattern (similar reduction in heterozygosity and shared haplotypes) among the different and independently derived cave populations strongly suggests that standing genetic variation has been an important contributor to cavefish evolution, because it is very unlikely to obtain such a repeated pattern just by chance through *de novo* mutations. Such an involvement of standing genetic variation indicates there must have been some role for selection, direct or indirect, in cavefish evolution. In this respect, it is interesting to note that many of the SNPs comapped with known QTL for the different cave traits (Bradic et al., 2013). In another peculiar observation, there is even evidence that standing genetic variation has contributed to the loss of pigmentation in cavefish evolution, a trait believed to have the strongest evidence of being derived by neutral evolution (Wilkens, 1988). But in a recent study (Gross and Wilkens, 2013) the same mutation in the gene *oca2* (ocular and cutaneous albinism-2) was identified in two geographically distinct and independently derived populations

(Pachón and Micos). The *oca2* gene has been shown to be widely implicated in pigmentation loss and has previously been identified as the causative gene underlying pigmentation loss in *Astyanax mexicanus* (Protas et al., 2006). Given that this mutation potentially also derived from standing genetic variation argues rather for selection than neutral evolution (though data from only a single locus is not enough for a definite statement, and the mutation was not recovered in the wild population in a recent sampling effort [Espinasa et al., 2014]). The increasing number of incidences where standing genetic variation has contributed to cavefish evolution makes its prevalence and importance in adaption indisputable.

GENE FLOW

An important contributor to standing genetic variation is gene flow. Gene flow is the transfer of genetic material from one population to another. Studies using microsatellites have, in general, found much stronger gene flow into the caves than out (Bradic et al., 2012). Conversely, however, in some caves (e.g., the Micos cave), the same study found evidence for equally strong gene flow in the opposite direction (Bradic et al., 2012). The ecological reason for this is unknown so far. In addition, migration rates among the caves are very low and limited to geographically close caves; it has been argued that some of the caves are connected to some extent (personal communication with Louis Espinasa). In general, gene flow among the caves separated by longer geographic distances can be discounted. Combining the findings, it is assumed that gene flow is happening mainly from surface populations into cave populations. That it is possible to have some recurring gene flow into the cave suggests that standing genetic variation may be even more important in cavefish evolution. Standing genetic variation could have contributed substantially to the cavefish genetic diversity, and the small population sizes in the caves may be less constrained in terms of their evolutionary potential than their census population size might suggest. It is known that in smaller populations, the relative importance of genetic drift (chance) is higher and any new allele, deleterious or beneficial, would be more likely to be lost from the population, making adaptation challenging (Kliman et al., 2008); therefore, it is likely that standing genetic variation from the surface fish population is an important contributor to cavefish evolution. Likewise and more hypothetical, it has been shown that hybridization among different species or among populations on the verge of speciation (like the cave and surface fish) elicits a genomic stress (Madlung and Comai, 2004) that could principally deplete reservoirs of HSP90 and reveal even more cryptic genetic variation. Although this is rather speculative, it is known in many species that mixed genetic backgrounds show different variations than purer backgrounds (Montagutelli, 2000).

It is worth mentioning that some caves provide unique settings that make the idea of gene flow/hybridization and standing genetic variation from the surface very conceivable. For example, in the Río Subterráneo cave found in the Micos area is a distinct series of pools at different distances into the cave. In the first pools during spring 2013, there were some surface fish surviving, in the intermediate pools many intermediate forms, possibly hybrids are found among the cavefish. In deeper pools further into the cave, exclusively cavefish exist. Although all these pools are equally exposed to a complete absence of light, it is conceivable that the selective pressure follows somewhat of a gradient into the cave. The deeper the pools in the cave are located, the more stressful the environment could be (pools closer to the outside may have, for example, more food due to bats passing these locations more frequently, or animals migrating into the caves). Correlating with this hypothesis, we found the values for conductivity decline the deeper the pools are located within the caves. Certainly, a more rigorous sampling will be needed to elucidate the selective pressure, the population structure, and the evolutionary age of these intermediate populations.

OPEN QUESTIONS

There are many aspects of cavefish biology where a further study and identification of standing genetic variation are tremendously important and will potentially help in solving critical open questions. One of the important questions in the field is why, among all the fish in the Sierra de El Abra region in Mexico, *Astyanax mexicanus* is the species that has consistently adapted to the caves. Speculations about this include the fact that *Astyanax mexicanus* do not rely exclusively on visual cues to propagate, and chemical cues are sufficient. Though their mating dance is quite elaborate and most likely involves visual cues as well (personal observation), it is possible to induce mating in the dark in the laboratory. A different, more intriguing (and not mutually exclusive) explanation might be that the unique genetic reservoir of standing genetic variation of these fish contributed to this phenomenon as well.

Another question that remains to be solved, is the length of time it takes for a surface population of *Astyanax* to become adapted to the cave environment. Though we can roughly estimate the divergent times from cave and surface fish populations, we don't know very much about how long the process of adaptation takes. Given the fact that the selection pressure is likely very strong (surface fish do not survive for long in the cave environment), and it is known that under strong selective pressure, evolution can be very fast (Stuart et al., 2014), it is reasonable to assume that cavefish evolution is a prime candidate for fast evolution, and that standing genetic variation is thus particularly important for cavefish evolution.

There are still more questions than answers, but the increasingly possible and affordable field of population genomics will help in addressing these questions in the future.

REFERENCES

Antonopoulou, E., Kentepozidou, E., Feidantsis, K., Roufidou, C., Despoti, S., Chatzifotis, S., 2013. Starvation and re-feeding affect Hsp expression, MAPK activation and antioxidant enzymes activity of European sea bass (Dicentrarchus labrax). Comp. Biochem. Physiol. A Mol. Integr. Physiol. 165 (1), 79–88.

Barrett, R.D., Schluter, D., 2008. Adaptation from standing genetic variation. Trends Ecol. Evol. 23 (1), 38–44.

Bersaglieri, T., Sabeti, P.C., Patterson, N., Vanderploeg, T., Schaffner, S.F., Drake, J.A., Rhodes, M., Reich, D.E., Hirschhorn, J.N., 2004. Genetic signatures of strong recent positive selection at the lactase gene. Am. J. Hum. Genet. 74 (6), 1111–1120.

Borowsky, R., Cohen, D., 2013. Genomic consequences of ecological speciation in astyanax cavefish. PLoS One 8 (11), e79903.

Bradic, M., Beerli, P., García-de León, F.J., Esquivel-Bobadilla, S., Borowsky, R.L., 2012. Gene flow and population structure in the Mexican blind cavefish complex (*Astyanax mexicanus*). BMC Evol. Biol. 12, 9.

Bradic, M., Teotónio, H., Borowsky, R.L., 2013. The population genomics of repeated evolution in the blind cavefish Astyanax mexicanus. Mol. Biol. Evol. 30 (11), 2383–2400.

Brawand, D., Wagner, C.E., Li, Y.I., Malinsky, M., Keller, I., Fan, S., Simakov, O., Ng, A.Y., Lim, Z.W., Bezault, E., et al., 2014. The genomic substrate for adaptive radiation in African cichlid fish. Nature 513 (7518), 375–381.

Chan, Y.F., Marks, M.E., Jones, F.C., Villarreal Jr., G., Shapiro, M.D., Brady, S.D., Southwick, A.M., Absher, D.M., Grimwood, J., Schmutz, J., Myers, R.M., Petrov, D., Jónsson, B., Schluter, D., Bell, M.A., Kingsley, D.M., 2010. Adaptive evolution of pelvic reduction in sticklebacks by recurrent deletion of a Pitx1 enhancer. Science 327 (5963), 302–305.

Choi, C.Y., An, K.W., 2008. Cloning and expression of Na+/K+ -ATPase and osmotic stress transcription factor 1 mRNA in black porgy, Acanthopagrus schlegeli during osmotic stress. Comp. Biochem. Physiol. B: Biochem. Mol. Biol. 149 (1), 91–100.

Colosimo, P.F., Hosemann, K.E., Balabhadra, S., Villarreal Jr., G., Dickson, M., Grimwood, J., Schmutz, J., Myers, R.M., Schluter, D., Kingsley, D.M., 2005. Widespread parallel evolution in sticklebacks by repeated fixation of Ectodysplasin alleles. Science 307 (5717), 1928–1933.

Domingues, V.S., Poh, Y.P., Peterson, B.K., Pennings, P.S., Jensen, J.D., Hoekstra, H.E., 2012. Evidence of adaptation from ancestral variation in young populations of beach mice. Evolution 66 (10), 3209–3223.

Espinasa, L., Centone, D.M., Gross, J.B., 2014. A contemporary analysis of a loss-of-function of the oculocutaneous albinism type II (Oca2) allele within the micos astyanax cave fish population. Speleobiol. Notes 6, 48–54.

Gibson, G., Dworkin, I., 2004. Uncovering cryptic genetic variation. Nat. Rev. Genet. 5 (9), 681–690.

Gross, J.B., Wilkens, H., 2013. Albinism in phylogenetically and geographically distinct populations of Astyanax cavefish arises through the same loss-of-function Oca2 allele. Heredity 111 (2), 122–130.

Haldane, J.B.S., 1927. A mathematical theory of natural and artificial selection, part V: selection and mutation. Proc. Camb. Philos. Soc. 23, 838–844.

Hedrick, P.W., 2013. Adaptive introgression in animals: examples and comparison to new mutation and standing variation as sources of adaptive variation. Mol. Ecol. 22 (18), 4606–4618.

Hooven, T.A., Yamamoto, Y., Jeffery, W.R., 2004. Blind cavefish and heat shock protein chaperones: a novel role for hsp90alpha in lens apoptosis. Int. J. Dev. Biol. 48 (8-9), 731–738.

Jarosz, D.F., Taipale, M., Lindquist, S., 2010. Protein homeostasis and the phenotypic manifestation of genetic diversity: principles and mechanisms. Annu. Rev. Genet. 44, 189–216.

Kliman, R., Sheehy, B., Schultz, J., 2008. Genetic drift and effective population size. Nat. Educ. 1 (3), 3.

Lachowiec, J., Lemus, T., Borenstein, E., Queitsch, C., 2015. Hsp90 promotes kinase evolution. Mol. Biol. Evol. 32 (1), 91–99.

Madlung, A., Comai, L., 2004. The effect of stress on genome regulation and structure. Ann. Bot. 94 (4), 481–495.

McGaugh, S.E., Gross, J.B., Aken, B., Blin, M., Borowsky, R., Chalopin, D., Hinaux, H., Jeffery, W.R., Keene, A., et al., 2014. The cavefish genome reveals candidate genes for eye loss. Nat. Commun. 5, 5307.

McGuigan, K., Nishimura, N., Currey, M., Hurwit, D., Cresko, W.A., 2011. Cryptic genetic variation and body size evolution in threespine stickleback. Evolution 65 (4), 1203–1211.

Merlo, C.A., Boyle, M.P., 2003. Modifier genes in cystic fibrosis lung disease. J. Lab. Clin. Med. 141 (4), 237–241.

Montagutelli, X., 2000. Effect of the genetic background on the phenotype of mouse mutations. J. Am. Soc. Nephrol. 11, S101–S105.

Okawa, H., Sampath, A.P., Laughlin, S.B., Fain, G.L., 2008. ATP consumption by mammalian rod photoreceptors in darkness and in light. Curr. Biol. 18 (24), 1917–1921.

Orr, H.A., 1998. Testing natural selection vs. genetic drift in phenotypic evolution using quantitative trait locus data. Genetics 149 (4), 2099–2104.

Orr, H.A., Betancourt, A.J., 2001. Haldane's sieve and adaptation from the standing genetic variation. Genetics 157, 875–884.

Paaby, A.B., Rockman, M.V., 2014. Cryptic genetic variation: evolution's hidden substrate. Nat. Rev. Genet. 15 (4), 247–258.

Protas, M.E., Hersey, C., Kochanek, D., Zhou, Y., Wilkens, H., Jeffery, W.R., Zon, L.I., Borowsky, R., Tabin, C.J., 2006. Genetic analysis of cavefish reveals molecular convergence in the evolution of albinism. Nat. Genet. 38 (1), 107–111.

Protas, M., Conrad, M., Gross, J.B., Tabin, C., Borowsky, R., 2007. Regressive evolution in the Mexican cave tetra, *Astyanax mexicanus*. Curr. Biol. 17 (5), 452–454.

Queitsch, C., Sangster, T.A., Lindquist, S., 2002. Hsp90 as a capacitor of phenotypic variation. Nature 417 (6889), 618–624.

Ritossa, F., 1962. A new puffing pattern induced by temperature shock and DNP in Drosophila. Experientia 18, 571–573.

Rohner, N., Jarosz, D.F., Kowalko, J.E., Yoshizawa, M., Jeffery, W.R., Borowsky, R.L., Lindquist, S., Tabin, C.J., 2013. Cryptic variation in morphological evolution: HSP90 as a capacitor for loss of eyes in cavefish. Science 342 (6164), 1372–1375.

Ruden, D.M., Garfinkel, M.D., Sollars, V.E., Lu, X., 2003. Waddington's widget: Hsp90 and the inheritance of acquired characters. Semin. Cell Dev. Biol. 14 (5), 301–310.

Rutherford, S.L., Lindquist, S., 1998. Hsp90 as a capacitor for morphological evolution. Nature 396 (6709), 336–342.

Siegal, M.L., 2013. Crouching variation revealed. Mol. Ecol. 22 (5), 1187–1189.

Sollars, V., Lu, X., Xiao, L., Wang, X., Garfinkel, M.D., Ruden, D.M., 2003. Evidence for an epigenetic mechanism by which Hsp90 acts as a capacitor for morphological evolution. Nat. Genet. 33 (1), 70–74.

Specchia, V., Piacentini, L., Tritto, P., Fanti, L., D'Alessandro, R., Palumbo, G., Pimpinelli, S., Bozzetti, M.P., 2010. Hsp90 prevents phenotypic variation by suppressing the mutagenic activity of transposons. Nature 463 (7281), 662–665.

Stuart, Y.E., Campbell, T.S., Hohenlohe, P.A., Reynolds, R.G., Revell, L.J., Losos, J.B., 2014. Rapid evolution of a native species following invasion by a congener. Science 346 (6208), 463–466.

Trotter, M.V., Weissman, D.B., Peterson, G.I., Peck, K.M., Masel, J., 2014. Cryptic genetic variation can make "Irreducible Complexity" a common mode of adaptation in sexual populations. Evolution 68 (12), 3357–3367.

Vitti, J.J., Grossman, S.R., Sabeti, P.C., 2013. Detecting natural selection in genomic data. Annu. Rev. Genet. 47, 97–120.

Waddington, C.H., 1942. Canalization of development and the inheritance of acquired characters. Nature 3081, 563–565.

Waddington, C.H., 1953. Genetic assimilation of an acquired character. Evolution 7 (2), 118–126.

Wilkens, H., 1988. Evolution and genetics of epigean and cave Astyanax fasciatus (Characidae, Pisces). Evol. Biol. 23, 271–367.

Yamamoto, Y., Byerly, M.S., Jackman, W.R., Jeffery, W.R., 2009. Pleiotropic functions of embryonic sonic hedgehog expression link jaw and taste bud amplification with eye loss during cavefish evolution. Dev. Biol. 330 (1), 200–211.

Yeyati, P.L., Bancewicz, R.M., Maule, J., van Heyningen, V., 2007. Hsp90 selectively modulates phenotype in vertebrate development. PLoS Genet. 3 (3), e43.

Part III

Morphology and Development

Chapter 8

Pigment Regression and Albinism in *Astyanax* Cavefish

William R. Jeffery, Li Ma, Amy Parkhurst and Helena Bilandžija
Department of Biology, University of Maryland, College Park, Maryland, USA

INTRODUCTION

Pigmentation has important functions in animals, including protection from solar ultraviolet radiation, regulation of body temperature, immunity, camouflage, mimicry, and provision of visible markings for species and sex recognition (Protas and Patel, 2008); however, the beneficial roles of pigmentation in lighted habitats may not be the same for life in perpetual darkness. Animals that spend their entire lives in dark environments, including the soil, deep sea, parasites in the bodies of their hosts, and caves, can reduce or even entirely lose pigmentation without any obvious effects on fitness. Consequently, the reduction or absence of pigmentation is a common phenotype in cave-dwelling animals, including flatworms, annelids, mollusks, crustaceans, insects, and vertebrates (Culver and Pipan, 2009; Protas and Jeffery, 2012). These cave-adapted animals show other regressive traits, most famously the loss of eyes, and constructive traits involving the enhancement of nonvisual sensory systems and behaviors (Porter and Crandall, 2003; Juan et al., 2010). The suite of constructive and regressive traits of cave-adapted animals is known as troglomorphy.

The mechanisms and causes of troglomorphic pigment loss are poorly understood in most cave animals. The Mexican cavefish, *Astyanax mexicanus*, however, is a notable exception. This species consists of two different morphs, an eyed and pigmented surface-dwelling form (surface fish) and an eyeless and pale cave-dwelling form (cavefish) (Jeffery, 2001). The regression of pigmentation has been studied in *Astyanax* cavefish since the 1940s (Rasquin, 1947), shortly after the description of the first cavefish population (Hubbs and Innes, 1936). Subsequently, 28 additional cavefish populations with various levels of pigment loss were discovered (Mitchell et al., 1977). The cavefish populations were derived from several different colonizations of surface fish, which are still present in the rivers and streams surrounding the caves and have, in some cases,

Biology and Evolution of the Mexican Cavefish. http://dx.doi.org/10.1016/B978-0-12-802148-4.00008-6
© 2016 Elsevier Inc. All rights reserved.

evolved depigmentation and other troglomorphic traits independently during the past few million years (Gross, 2012).

Astyanax is an excellent model system to study the biology of pigment regression due to the following special attributes. First, pigmentation is dramatically changed in the two *Astyanax* morphs: surface fish are strongly pigmented and most cavefish populations are pale (Figure 8.1(A)–(C)). Second, the differences in pigmentation have a known evolutionary polarity; pale cavefish are descended from pigmented ancestors likely to closely resemble present day surface fish. Third, changes in cavefish pigmentation are the evolutionary response to a known environmental cue, the absence of light. Finally, surface fish and cavefish are interfertile and can produce fully viable offspring, allowing the use of powerful genetic analysis to determine the mechanisms of depigmentation. Because of these attributes, studies of this system have revealed some of the mechanisms responsible for pigment degeneration, although many open questions remain to be addressed. In this article, we review the genetics, development, and evolution of pigment degeneration in *Astyanax* cavefish.

ASTYANAX PIGMENTATION AND DEPIGMENTATION

Astyanax has three different types of pigment cells (chromatophores): melanophores, iridophores, and xanthophores. Melanophores (see Box 1) synthesize black melanin pigment, which is deposited in membrane-bound organelles called melanosomes. Melanophores are located in the retina (retinal pigment epithelium, RPE), the iris, and the choroid layer of the eye, as well as in the hypodermal layer of the skin, which includes the scales and the fins (Figure 8.1(D) and (E)). They are also found to a lesser extent in internal tissues, including parts of the brain, along the nerve tracts and the linings of blood vessels, in the kidneys and the liver, and in the peritoneum (Rasquin, 1947). There are differences in the distributional pattern of melanophores and other chromatophores between larva and adults (Figure 8.1(A)–(C)), but the timing and mechanisms underlying this change have not been studied in *Astyanax*. In adults, skin melanophores are concentrated in the dorsal regions of the head, torso, and within a dark lateral stripe. The lateral stripe extends from the base of the head to the tail and ends in a definitive diamond-shaped spot near the beginning of the caudal fin (Figure 8.1(B) and (C)). Iridophores (also called guanophores) contain stacks of iridescent guanine platelets and are responsible for the glittery silver appearance of the body and eyeballs. They are concentrated in the same regions as melanophores, including the dorsal part of the head and torso, the choroid layer of the eyes, and the lateral stripe. Xanthophores synthesize yellow pteridine and/or orange carotenoid pigments. They are dispersed among the other types of pigment cells throughout the torso, with high concentrations in the dorsal and caudal fins (Figure 8.1(D)–(E)), but are absent from the eyes. In contrast to birds and mammals, in which pigment is transferred from the dermal chromatophores to adjacent epidermal cells that form plumage or fur, the

FIGURE 8.1 (A)–(C) Pigmentation in *Astyanax* surface fish and cavefish larvae (A), young adults (B), and mature adults (C). Surface fish are at the top and Pachón cavefish at the bottom of each frame. Photographs by (A). A.G. Strickler (A and B) and Y. Yamamoto (C). (D) and (E) Chromatophores in surface fish (D) and Pachón cavefish (E) caudal fins. Cavefish have lost melanophores, but not xanthophores. (D) and (E) from McCauley et al. (2004).

Box 1

Melanophore	A pigment cell (chromatophore) that contains melanin in a melanosome
Melanosome	An organelle where melanin is synthesized and stored in a melanophore
Melanoblast	Undifferentiated melanophore that is normally lacking melanin
Catecholamine	Monoamine derived from the amino acid tyrosine that can act as a hormone and/or a neurotransmitter; catecholamines include dopamine, norepinephrine, and epinephrine

pigment-containing organelles in *Astyanax* and other teleosts remain in chromatophores throughout life.

The pale color of *Astyanax* cavefish is due to large decreases in melanophores (Figure 8.1) and iridophores (Wilkens, 1988). Because of the reduction or loss of melanophores and iridophores, cavefish are opaque and not as glittery as surface fish; they also appear slightly pink, because red blood cells show through the skin. There is currently no further information available about iridophore regression; however, melanophores have been extensively studied and are known to show three modifications in cavefish relative to surface fish (Table 8.1). The first modification is melanophore reduction; the number of differentiated dark melanophores is decreased to varying degrees in different cavefish populations. Subterráneo and Chica cavefish contain the most melanophores and (non-albino) Pachón cavefish the least melanophores. Consequently, depigmented cavefish populations no longer show the lateral stripe or diamond-shaped spot of pigment typical of surface fish (Figure 8.1(C)). The second modification is brown; the extent of melanin deposition in the residual melanophores is reduced, resulting in eye and skin melanophores appearing brown rather than black. Brown melanophores occur in most of the cavefish populations. The third modification, found in Molino, Yerbaníz, Japonés, and most Pachón cavefish, is albinism, the absence of melanin pigment. Albino populations have no melanophores, although the existence of colorless melanoblasts has been demonstrated in some of these populations by L-3,4-dihydroxyphenylalanine (L-DOPA) administration (see later). Pachón cavefish are a special case, because the majority of individuals in the population are albino, but a few individuals have small numbers of residual brown melanophores (Table 8.1). In contrast to other chromatophores, there is no significant change in cavefish xanthophores (Rasquin, 1947; Wilkens, 1988) (Figure 8.1(D) and (E)). Because xanthophores are more clearly visible, cavefish have a yellowish hue, which is reinforced by the overproduction of yellow fat cells in the dermal layer of the skin (Rasquin, 1947).

TABLE 8.1 Melanin Pigmentation Phenotypes in *Astyanax* Surface Fish and Cavefish Populations

Fish Population	Phenotype (% of SF melanophores)	Mutation		Sum of Melanophores and Melanoblasts	References
		Brown/*mc1r*	Albinism/*oca2*		
Surface	(100)	No	No		1-8
Subterráneo	(40)	No	No		5, 8
Chica	(30/Variable)	Yes	No	Not reduced	1, 3, 7, 8
Sabinos	(30)	Yes	No	Not reduced	3-5, 7, 8
Curva	(20)	Yes	No	Not reduced	3, 6-8
Piedras	(20)	Yes	No		3, 6, 8
Pachón	(15)	Yes[a]	No	Reduced	5, 6-8
	(0)	Yes[a]	Yes		
Japonés/Yerbaníz[b]	(0)	Yes[a]	Yes		7, 8
Molino	(0)	No	Yes		7, 8

Blank spaces indicate no data or data not reported.
References are as follows: 1. Rasquin (1947); 2. Sadoglu (1957a,b); 3. Sadoglu and McKee (1969); 4. Wilkens (1988); 5. Wilkens and Strecker (2003); 6. McCauley et al. (2004); 7. Protas et al. (2006); 8. Gross et al. (2009).
[a]*Indicates that brown melanophores can be unmasked in appropriate crossings.*
[b]*Indicates likely interconnected caves.*

Teleosts exposed to different light conditions can change their coloration via either physiological or morphological color changes. Physiological color change is an immediate response to light intensity involving the translocation of pigment organelles within the chromatophore cytoplasm (Nery and Castrucci, 1997). In intense light, pigmented organelles move into dendrite-like cytoplasmic projections extending through the hypodermis, which makes the skin appear darker. In more shaded conditions, or in the absence of light, pigment organelles are packed tightly around the nucleus, which makes the skin appear lighter. Cyclic adenosine monophosphate (cAMP) signaling controls the dispersal of melanosomes, whereas the attenuation of melanosomes is regulated hormonally by the catecholamine (CAT) epinephrine. Physiological color change occurs in both surface fish and all studied cavefish populations (Burgers et al., 1962; Wilkens, 1970, 1988). Cavefish melanophores normally show attenuated melanosomes in their natural environment, but can undergo dispersal after exposure to light, making the fish appear darker. Reciprocally, epinephrine treatment of cavefish that have been adapted to light causes melanosome attenuation and even lighter skin. Thus, despite millions of years of evolution in the dark, cavefish melanophores have conserved the ability to respond to light.

Morphological color change is darkening of the skin in response to environmental light changes as well, but in contrast to physiological color change, it involves the differentiation of new chromatophores (Leclercq et al., 2010). Similar to other teleosts, *Astyanax* surface fish have a morphological color change, but this phenomenon appears to be absent in cavefish (Rasquin, 1947; Wilkens, 1988). The secretion of alpha-melanocyte stimulating hormone (αMSH) by the pituitary gland and its subsequent binding to the melanocortin 1 receptor (*mc1r*) triggers melanophore differentiation (Figure 8.2). The *mc1r* gene is mutated in cavefish (see later) possibly explaining the absence of morphological color change. Interestingly, despite mutations in *mc1r*, considerable levels of αMSH have been detected in cavefish (Burgers et al., 1962). The reduction in number of mature melanosomes and the lack of a morphological color change suggest that melanophore differentiation has been affected during cavefish evolution.

CONTROL OF MELANOGENESIS

Melanophore differentiation and maturation is a complex process involving the generation of melanosomes and the activation of melanin synthesis (melanogenesis). Although beyond the scope of this review, the stimuli for normal melanosome differentiation and melanin synthesis are relevant to depigmentation in cavefish. Melanogenesis is best understood in mammals (Slominski et al., 2003). It is less well known in teleosts, and there may be differences at the molecular level between the two kinds of vertebrates (Lane and Lister, 2012); however, for simplicity, a generic summary of melanogenic events is provided in Figure 8.2.

Micropthalmia-related transcription factor (*mitf*) plays a central role in melanogenesis (Widlund and Fisher, 2003) (Figure 8.2). The αMSH signaling

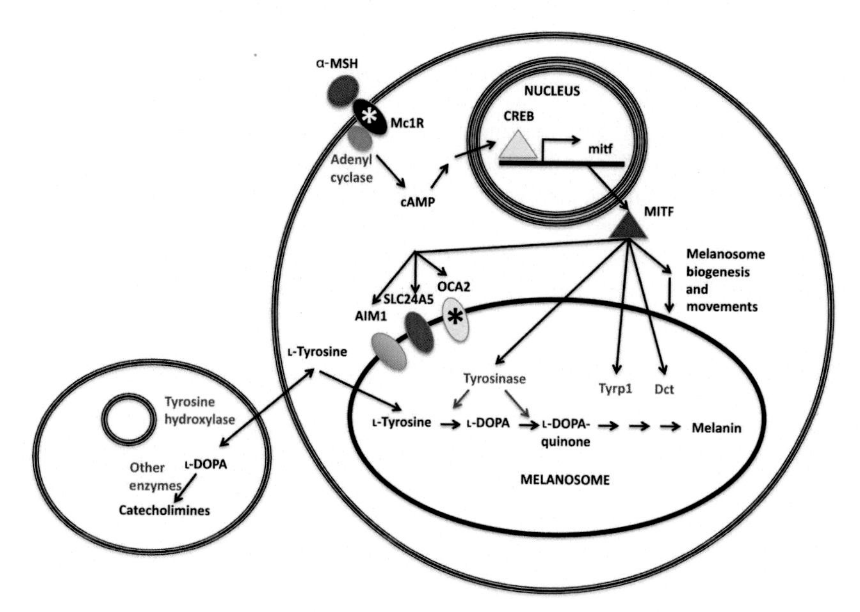

FIGURE 8.2 A simplified summary of the signaling cascade and reactions involved in melanophore differentiation, the melanin synthesis pathway, and the catecholamine synthesis pathway. Enzymes are shown in red lettering. All other factors, substrates, and products are shown in black lettering. Factors marked by an asterisk are mutated in cavefish. Single-headed arrows represent pathway directions. Double-headed arrows represent bidirectional movement of substrates.

pathway controls *mitf* transcription via sequential activation of *mc1r*, adenylyl cyclase, cAMP, and the cAMP response element-binding (CREB) protein, which binds to the *mitf* promoter. The *mitf* gene directly or indirectly controls more than 40 target genes (Chelli et al., 2009) involved in melanosome biogenesis and melanin synthesis, including *tyrosinase*, *dct*, *tryp1*, *oca2*, *aim1/mapt*, and *slc24a5* (Du and Fisher, 2002; Vachetenheim and Borovansky, 2009; Visser et al., 2012).

In differentiated melanophores, melanin synthesis involves the oxidation and polymerization of L-tyrosine (Figure 8.2). L-Tyrosine can be obtained from the diet or through oxidation of phenylalanine and is also the initial substrate in the CAT synthesis pathway. For melanin synthesis, L-tyrosine must be transported into the melanosome to become available as a substrate for tyrosinase, the rate-limiting enzyme acting at the first step of the pathway. Three melanosome membrane proteins are also involved in the first step in the pathway: AIM1/MATP, OCA2, and SLC24A5. The function of these proteins is not entirely understood; however, OCA2 has been proposed to control pH as a proton pump (Brilliant, 2001) or to serve as an L-tyrosine (Rosemblat et al., 1994) or tyrosinase (Toyofuku et al., 2002) transporter. Tyrosinase promotes the conversion of L-tyrosine to L-DOPA and L-DOPA to L-DOPAchrome. Melanin is produced from L-DOPAchrome via a sequence of downstream reactions promoted by tyrosinase-related protein 1 (*Tyrp1*), dopachrome tautomerase (DCT or *Tyrp2*), and tyrosinase. In humans, mutations affecting these proteins block melanin synthesis at different steps, leading to four types of oculocutaneous albinism (Okulicz et al., 2002). The first step is blocked by mutations in tyrosinase, *oca2* and *aim1/matp*, causing oculocutaneous albinism types 1, 2, and 4, respectively. A downstream step in the pathway is blocked by mutations in *Tyrp1*, resulting in oculocutaneous albinism type 3 albinism. The human oculocutaneous albinisms show that multiple steps and genes in the melanin synthesis pathway are potentially susceptible to mutations and could be responsible for cavefish albinism.

DEVELOPMENTAL BASIS OF CAVEFISH DEPIGMENTATION

The changes in adult pigment in cavefish arise during development. Eye and body chromatophores have different developmental origins in vertebrates. The melanophores of the RPE and iris are formed from the optic vesicle, a protrusion of the forebrain, and thus are derivatives of the neural ectoderm. The development of RPE pigmentation has not been studied extensively in *Astyanax*; however, RPE pigmentation, like body pigmentation (see below), is dependent upon normal expression of the *oca2* gene (Bilandžija et al., 2013). The *oca2* gene is expressed in the RPE of surface fish larvae, but not albino Pachón cavefish larvae. Furthermore, morpholino-based knockdown of *oca2* expression causes surface fish eyes to develop without pigmentation (Bilandžija et al., 2013). As described below, the defect responsible for the lack of eye and body pigmentation in albino cavefish is a mutation in the *oca2* gene (Protas et al., 2006).

All types of body chromatophores are formed from a specialized part of the ectoderm known as the neural crest (Le Douarin and Kalcheim, 1999). Neural crest cells delaminate from the surface epithelium during neural tube formation and migrate to different locations throughout the body. When they reach their target areas, neural crest cells differentiate into many different types of tissues and organs. They also produce stem cells that are set aside and later used to replenish neural crest-derived tissues and organs during growth. The neural crest cells that ultimately differentiate into chromatophores migrate following either a ventrolateral pathway into internal regions of the body or a dorsolateral pathway into the hypodermis of the skin. The early events of neural crest cell development and their migration pathways do not appear to be altered in Pachón cavefish (McCauley et al., 2004). Vital dye labeling studies have shown that neural crest cells migrate normally from the dorsal neural tube and invade distant locations in the body, including the skin. Furthermore, cavefish neural tube explants produce melanophore-like cells capable of synthesizing melanin in the presence of L-DOPA (see below) (Figure 8.3(A)). Thus, the effects on melanophore differentiation are not due to the absence of appropriate neural crest-derived progenitors or their ability to select the proper migration pathway. Instead, specific changes are likely to occur in the sequence of differentiation events after the divergence of pigment cell types from other neural crest precursor cells.

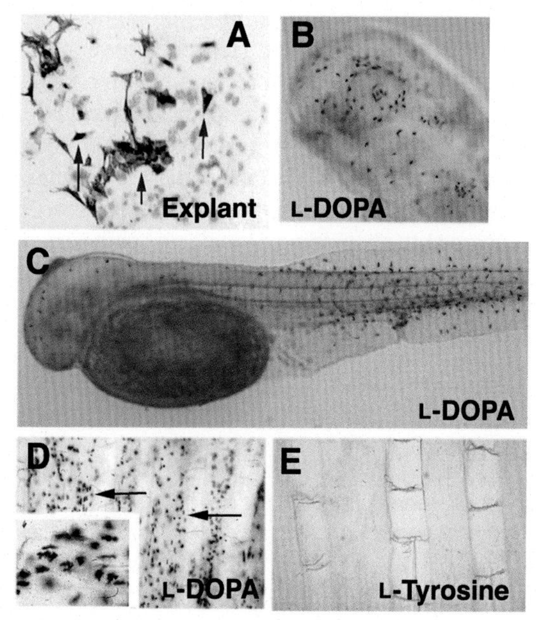

FIGURE 8.3 Melanogenic substrate assays in Pachón cavefish. (A) Melanophores capable of melanin synthesis after provision of L-DOPA substrate (arrows) migrate from neural tube explants in culture. (B) and (C) Melanoblasts produce melanin after treatment of albino Pachón larvae with L-DOPA. (D) and (E) Melanophores in adult caudal fins produce melanin after L-DOPA (arrows) (D) but not L-tyrosine (E) treatment. (A) and (C)–(E) from McCauley et al. (2004).

The block in chromatophore specification could either be controlled by events in the neural crest-derived progenitor cells themselves, by influences of the surrounding tissue environment on these cells, or by a combination of both internal and external stimuli. This issue has been addressed by examining *oca2* expression during body pigment cell development (Bilandžija et al., 2013). Migrating neural crest cells of surface fish, but not Pachón cavefish, express the *oca2* gene, and knockdown of surface fish *oca2* expression with morpholinos prevents melanophore differentiation in surface fish (Bilandžija et al., 2013). Thus, it seems that melanophore defects include changes in the precursor cells. Further research will be needed to determine if the tissue environment of cavefish melanophores also has effects on differentiation.

Developmental insights into the cavefish melanophore differentiation process have also been obtained using the melanogenic substrate assay (McCauley et al., 2004). In this assay, different substrates are provided to lightly fixed albino cavefish larvae or adults to determine which steps in the pathway are capable of rescue. The provision of exogenous L-DOPA to albino Pachón cavefish resulted in the deposition of melanin in melanoblasts of the eyes and body (Figure 8.3(A) and (B)), suggesting that tyrosinase and downstream enzymes are potentially functional in melanin synthesis. In contrast, the provision of

L-tyrosine as a substrate did not result in melanin production, suggesting that melanin synthesis is blocked at its first step, the conversion of L-tyrosine to L-DOPA. L-DOPA administration could restore the full complement of surface fish melanin-producing cells in Chica, Sabinos, Tinaja, and Curva cavefish (McCauley et al., 2004). In contrast, the rescue of melanin synthesis in Pachón cavefish by L-DOPA resulted in a smaller number of melanin-producing cells compared with surface fish. These results suggest that: (1) a defect in cavefish melanogenesis occurs at the beginning of melanin synthesis; (2) in some cavefish populations, the sum of melanophores and cells capable of melanin synthesis is about the same as in surface fish; and (3) the number of cells capable of producing melanin are reduced in Pachón cavefish.

GENETIC BASIS OF CAVEFISH DEPIGMENTATION

Cavefish pigmentation research benefits from the application of genetic analysis. In early studies, crossbreeding experiments among surface fish and different cavefish populations were used to determine the mode of inheritance of depigmentation phenotypes (Sadoglu, 1957a,b; Sadoglu and McKee, 1969; Wilkens, 1970, 1988; Wilkens and Strecker, 2003). Crosses between surface fish and Pachón or Sabinos cavefish yielded offspring with levels of pigmentation as intense as those of surface fish, showing that depigmentation is a recessive trait (see also Wilkens in Chapter 5). Intercrossing the F1 progeny resulted in F2 progeny showing a broad distribution of pigmentation levels ranging between the original parental types. Backcrossing of F1 progeny to the parental types also showed a wide range of pigmentation phenotypes. Together, these intermorph crosses implied that multiple genes control cavefish depigmentation. Crosses between different cavefish populations also gave insights into pigment regression. Depending upon the identity of the parental populations, the F1 progeny of intercavefish crosses showed either the same or greater levels of pigmentation than the parents (Wilkens and Strecker, 2003) (Table 8.2). For example, a cross between Molino cavefish, which exhibit albinism, and Curva or Piedras cavefish, which have retained about 30% of the surface fish melanophore complement (Table 8.1), yielded hybrids that were much more darkly pigmented than their parents. These results indicated that different genes control depigmentation in these cavefish populations. Later in the chapter, we describe two of the genes involved, *oca2* and *mc1r*, although classical genetic and quantitative trait loci (QTL) analysis (see below) suggests that many additional genes control melanophore changes (see also O'Quin and McGaugh in Chapter 7). Crossing of other cavefish populations, such as Piedras × Curva, yielded F1 hybrids with the same melanophore phenotypes as their parents, suggesting that pigment regression in these cavefish populations involves the same or similar genes (Wilkens and Strecker, 2003).

Genetic association analysis has been used to identify QTL that represent genomic regions controlling the cavefish depigmentation phenotype (Borowsky

TABLE 8.2 Pigmentation Phenotypes in *Astyanax* F1 Hybrids

Cross	Pigmentation Phenotype	References
Piedras × Curva	Brown melanophores/no albinism	4
Piedras × Yerbaníz	Brown melanophores/no albinism	4
Curva × Yerbaníz	Brown melanophores/no albinism	4
Curva × Pachón	Brown melanophores/no albinism	4
Molino × Curva	Dark melanophores/no albinism	4
Molino × Piedras	Dark melanophores/no albinism	4
Molino × Pachón	Albino	5
Japonés × Pachón	Albino	5
Pachón × Surface fish	Brown melanophores/no albinism	1-3
Sabinos × Surface fish	Brown melanophores/no albinism	1-3

References are as follows: 1. Sadoglu (1957a,b); 2. Sadoglu and McKee (1969); 3. Wilkens (1970); 4. Wilkens and Strecker (2003); 5. Protas et al. (2006).

and Wilkens, 2002; Protas et al., 2007, 2008). QTL analysis was done with the F2 progeny of genetic crosses between surface fish and the albino cavefish populations of Pachón, Molino, and Japonés. As expected from the classical genetic analysis described above, the results showed a large number of QTL (around 18 in the surface fish × Pachón cavefish cross) underpinning melanophore and pigmentation changes, each with a relatively small effect on the overall depigmentation phenotype. The genes underlying most of these QTL have not been identified; however, synteny with the zebrafish genome has allowed candidate genes to be recognized (Gross et al, 2008). For example, the *shroom2* gene, which regulates melanosome biogenesis in the RPE (Fairbank et al., 2006), is tightly linked to a prominent pigmentation QTL. In future studies, it should be possible to align QTL with the sequenced cavefish genome (McGaugh et al., 2014) and begin to identify and characterize more of the genes responsible for cavefish pigment degeneration.

Although pigmentation is an additive phenotype controlled by many distinct and independently segregating genes, there are two simple phenotypes, brown melanosomes and albinism, which are due to changes in single genes (Sadoglu, 1957b; Sadoglu and McKee, 1969). Brown and albinism are recessive traits that show classic Mendelian 3:1 ratios in the F2 generation of surface fish × cavefish crosses. The brown allele is fixed in most cavefish populations (Table 8.1). The albinism allele is homozygous in Molino, Yerbaníz, Japonés, and most Pachón cavefish; however, a small proportion of Pachón cavefish are heterozygous at the albinism locus and thus exhibit (brown) melanophores in relatively low numbers (Wilkens, 1988). Particularly insightful crosses have been done between albino Pachón and Molino or Japonés cavefish (Protas et al., 2006). The F1 generation was albino for both crosses, showing that the same gene is involved in the absence of melanin pigment in these cavefish populations (Protas et al, 2006). Genes and mutations underlying the brown and albinism traits have been identified and characterized.

The Brown Gene Encodes *mc1r*

Analysis of Pachón × surface fish F2 progeny revealed a single QTL related to the brown phenotype (Gross et al., 2009). The Pachón brown QTL was anchored to the zebrafish genome, and genes with known melanogenic activity were then surveyed. This analysis revealed that the *mc1r* gene is responsible for the brown phenotype. The *mc1r* gene activates the intracellular signaling pathway leading to transcription of *mitf*, which controls pigmentation (Figure 8.2).

The *mc1r* gene was sequenced in several different cavefish populations (Figure 8.4(A)). In Pachón cavefish, a 2 bp deletion was detected at the extreme 5′ end of the transcript that introduces a frame shift resulting in a downstream

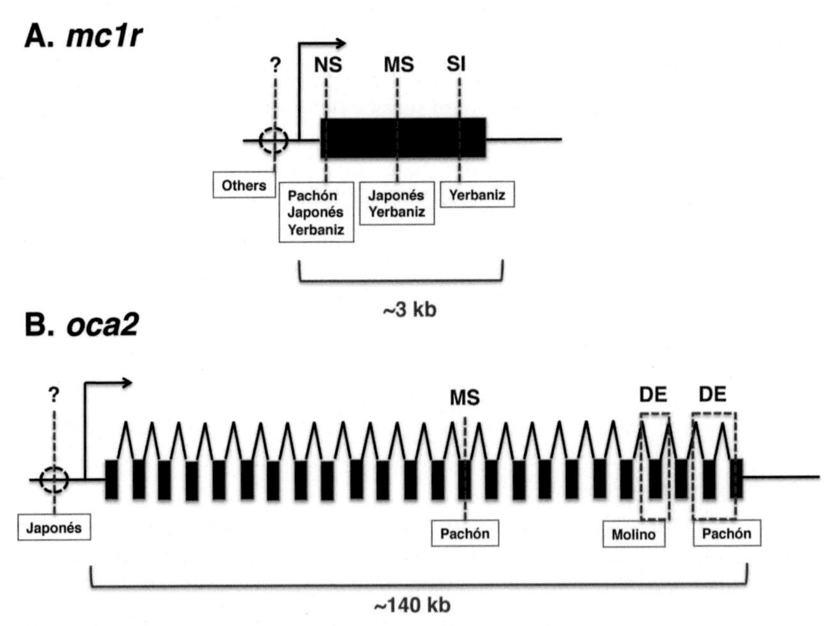

FIGURE 8.4 A diagram showing the types and locations of mutations in the *mc1r* (A) and *oca2* (B) genes in different cavefish populations. Dashed spheres: gene regulatory regions. Dashed lines and boxes: approximate locations of mutation. NS, nonsense mutation; MS, missense mutation; SI, silent mutation; DE, deletion. Line with right angle arrowhead: transcription start site. The two genes are not drawn at the same relative scales. The intron-exon sizes in *oca2* are standardized for clarity.

premature stop codon. The same mutation has been detected at this position in Japonés and Yerbaníz cavefish. Additionally, a missense mutation was discovered in the latter cavefish populations, and a silent mutation was uncovered in Yerbaníz cavefish (Gross et al., 2009). No sequence differences in the *mc1r*-coding region were found in Piedras or Curva cavefish, suggesting that gene regulatory regions, rather than coding regions, may be mutated. Functional analysis confirmed that the Pachón and Yerbaníz mutations have roles in melanogenesis (Gross et al., 2009). DNA constructs containing the wild type surface fish *mc1r* allele, the Pachón mutant allele, or the Yerbaníz mutant allele were introduced into zebrafish embryos whose endogenous *mc1r* was knocked down by a specific morpholino (Gross et al., 2009). The surface fish *mc1r* allele rescued pigmentation, but neither of the cavefish mutant alleles were able to promote this rescue (Gross et al., 2009). Thus, these mutations in *mc1r* are responsible for brown melanophores.

The Albinism Gene Encodes *oca2*

Genetic association analysis conducted on the progeny of surface fish × Pachón and surface fish × Molino cavefish crosses detected a single albinism QTL located on the same linkage group in both cavefish populations (Protas et al., 2006). A candidate gene approach then showed that the gene responsible for albinism is *oca2* (Figure 8.4(B)). More recently, *oca2* was confirmed as the gene responsible for albinism in Pachón cavefish by morpholino-based knockdown (Bilandžija et al., 2013) and transcription activator-like effector nuclease (TALEN)-based gene editing (Ma et al., 2015). The *oca2* gene encodes a putative 12-pass melanosome membrane protein. It is a large, complex gene consisting of 24 exons spanning about 140 kb in the Pachón cavefish genome (McGaugh et al., 2014). Protas et al. (2006) discovered three changes in Pachón relative to surface fish *oca2*: two point mutations resulting in conserved amino acid substitutions, and a large deletion extending from within intron 23 through most of exon 24. Only a single change was found in Molino cavefish, a large deletion encompassing exon 21. Thus, although *oca2* loss of function is responsible for albinism in both Molino and Pachón cavefish, the mutations are distinct. Deletions in both Molino and Pachón include regions predicted to be membrane-spanning domains. To conclusively demonstrate that these mutations are responsible for albinism, DNA constructs containing the wild-type surface fish *oca2* allele or the Pachón and Molino *oca2* alleles containing each of the point mutations and large deletions were introduced into a mouse *oca2*-deficient melanocyte cell line (Protas et al., 2006). The surface fish *oca2* DNA construct and Pachón cavefish *oca2* DNA constructs with amino acid polymorphisms were able to rescue melanogenesis, but the DNA constructs containing the large deletions were unable to promote melanin synthesis, implying that they are the mutations responsible for *oca2* loss of function. There is also evidence that albino Japonés cavefish have evolved albinism via changes in the *oca2* gene (Protas et al., 2006); however, no differences could be detected in the coding region of Japonés *oca2*, suggesting that a regulatory region may be mutated.

It has been reported that the progeny of pigmented Subterráneo cavefish allowed to interbreed in the laboratory developed albinism due to fixation of the same mutant allele as found naturally in Pachón cavefish (Gross and Wilkens, 2013). Assuming reliable laboratory husbandry, an interesting interpretation of these results could be that the mutant *oca2* allele was present as standing genetic variation in the original sampling of Subterráneo cavefish; however, subsequent analysis of the natural Subterráneo population was unable to confirm the presence of the mutant *oca2* allele (Espinasa et al., 2014). Thus the mutant *oca2* allele is either present at very low levels or has disappeared from the Subterráneo cavefish population since the original collections.

EVOLUTION OF CAVEFISH DEPIGMENTATION

The evolutionary mechanisms of cavefish pigment regression and albinism are poorly understood in *Astyanax* cavefish, as they are in other cave-adapted animals. The prevailing hypotheses for why cavefish have lost pigmentation are: (1) neutral mutation/genetic drift and (2) direct or indirect natural selection (see also Borowsky in Chapter 6). Interestingly, Darwin was the first to doubt that the evolution of regressive features in cave animals is caused by natural selection, specifically in reference to loss of eyes. Instead, he suggested that regressive traits gradually disappeared because of disuse (Darwin, 1859). In modern terms, this idea has been refined as the neutral mutation/genetic drift hypothesis, which states that during relaxed selection, genes controlling pigmentation could gradually accumulate mutations, eventually resulting in defective chromatophore development. The alternative hypothesis is that direct or indirect natural selection could account for depigmentation if pigmentation is costly to generate and maintain or if depigmentation is beneficial in the cave environment. It should be emphasized that current evidence does not allow a definitive conclusion about whether neutral mutation or natural selection is responsible for pigment degeneration in *Astyanax*. In fact, there is evidence that both processes may operate in different aspects of pigment regression.

The reduction in the number of melanophores is best explained by neutral mutation and genetic drift. Random mutations would be expected to result in an increase or decrease of traits over evolutionary time. The polarity of individual melanophore QTL, which can be either positive (increased numbers of melanophores) or negative (decreased numbers of melanophores) (Protas et al., 2007), supports the neutral mutation/genetic drift hypothesis (Orr, 1998). If neutral mutation/genetic drift is indeed the cause of reduced melanophores, then why have they not been completely lost in any of the cavefish populations? Perhaps this is only a matter of time, but different levels of residual melanophores in cavefish populations of roughly the same age, such as Sabinos and Pachón cavefish (Table 8.1) argue against this explanation. A second possible explanation may be developmental constraints; development of the melanophore lineage may be functionally linked to RPE and neural-crest development and impossible to eliminate completely without also negatively affecting one or both of these critical developmental centers. A third possibility may be that melanophore lineage has an unappreciated but essential role in caves, such as in immunity (Arciuli et al., 2012). The identification and characterization of some of the other genes involved in melanophore reduction might provide further information on the possible role of neutral mutation/genetic drift (or natural selection) in this process.

In contrast to the reduction in melanophores, several lines of evidence suggest that natural selection is involved in the evolution of albinism. First, in the 1960s, researchers reported that albinism (as well as complete eye loss) was rare in Pachón Cave (Kosswig 1963; Pfeiffer, 1966; Sadoglu and McKee, 1969) in which there is a cavefish population of relatively large size (Mitchell et al., 1977). Now, only about 50 years later, the frequency of albinism (and eye loss) is very high in Pachón Cave. It is unlikely that albinism could move toward fixation in such a short time unless it is influenced by strong selection. Second, in every albino cavefish population that has been studied (Molino, Pachón, and Japonés), *oca2* is the cause of albinism through different mutations (Protas et al., 2006) and melanin synthesis is blocked at its first step, the conversion of L-tyrosine to L-DOPA (McCauley, et al., 2004); however, it is clear from the situation in humans (where albinism is caused by mutations in at least four genes) that it is possible to interrupt melanin synthesis at multiple places in the pathway. The repeated mutation of *oca2* to block melanin synthesis at the same step seems unlikely to have evolved by chance (see also Borowsky in Chapter 6). Third, the fixation of mutant *oca2* alleles in the relatively "young" Molino cavefish population also supports the role of natural selection in the evolution of albinism. If natural selection is indeed involved in the evolution of albinism, however, what is the selective advantage of blocking melanin synthesis at its first step?

There are several possibilities. First, melanin synthesis could be blocked in order to conserve energy; however, it is not certain that melanin synthesis has a large energetic cost (Stoehr, 2006). Second, it is possible that blocking the first step prevents the accumulation of potentially toxic intermediates (Graham et al., 1978). This seems to be a reasonable possibility, but there is no evidence for or against it. Finally, it might be advantageous to block melanin formation at the first step, because this could make excess L-tyrosine substrate available for use in the alternative CAT synthesis pathway, which may be necessary for the evolution of constructive neurological and behavioral traits in cavefish (Bilandžija et al., 2013). Recent evidence indicates that CAT levels are enhanced in Pachón cavefish (Bilandžija et al., 2013; Elipot et al., 2014). Furthermore, when *oca2* expression is knocked down in surface fish, L-tyrosine and dopamine levels are significantly increased (Bilandžija et al., 2013), although it has not been determined whether CAT enhancement has phenotypic effects. Numerous changes in *Astyanax* cavefish have evolved compared with surface fish, and many have been previously shown to be CAT-related; there are changes in stress, feeding and foraging behaviors, and sleep (Gallo and Jeffery, 2012; Kowalko et al., 2013; Duboué et al., 2011, 2012). Thus, it is possible that a link exists between albinism and CAT-dependent processes that could be exploited by natural selection.

CONCLUSIONS AND FUTURE PROSPECTS

Astyanax offers important advantages for studying the loss of pigmentation and its relationship to cave adaptation. Most importantly, powerful genetic approaches are available and have been exploited to identify the *oca2* and *mc1r* genes underlying the albinism and brown phenotypes respectively (Protas et al., 2006; Gross et al., 2009). Similar approaches in combination with sequenced *Astyanax* genomes will be likely to identify additional genes involved in loss of pigmentation (see also, Tabor and Burgess in Chapter 20). Furthermore, the existence of 29 cave populations originating through multiple colonization events enables insights into the mechanisms of convergent and parallel loss of pigmentation. Accordingly, we know that albinism evolved due to convergent mutation of the same gene, *oca2*, in at least three different cavefish populations, which provides insight into the evolutionary mechanisms involved, probably natural selection.

The loss of pigmentation is a broadly convergent phenotype present in every phylum that has cave-dwelling representatives. There is a wealth of natural history observations on depigmentation of cave animals (Vandel, 1965), but only a handful of studies addressing the molecular and evolutionary mechanisms underlying this convergent phenotype. A recent study showed that albinism is under the control of either a single gene or at most two gene loci in the cave isopod *Asellus aquaticus* (Protas et al., 2011). Using the same technique as in *Astyanax* (McCauley et al., 2004), it has also been shown that cave planthoppers from Hawaii and Croatia have lost pigmentation by convergence involving a lesion in the first step of melanin synthesis (Bilandžija et al., 2012). Interestingly, the enzyme tyrosine hydroxylase, which acts at the first step in the insect melanin synthesis pathway (Gorman and Arakane, 2010), seems to be intact, implying that another factor operating at the first step may be involved, a situation similar to *oca2* in *Astyanax*. On the contrary, it was shown that both the first and downstream steps were disrupted in diverse cavefish species from Brazil (Felice et al., 2008). Ongoing research in our laboratory is focused upon determining the step of melanin synthesis that is disrupted in cave animals from various phyla and understanding the evolutionary forces driving this convergence. In *Astyanax*, many open questions also remain: what are the genes responsible for partial loss of pigmentation in diverse cave populations, are coding region or regulatory changes mutations involved, why are *oca2* and *mc1r* frequent targets of mutations, and if they are changed due to natural selection, what kind of selective advantages do they offer for life in dark caves?

REFERENCES

Arciuli, M., Fiocco, D., Cicero, R., Maida, I., Zanna, P.T., Guida, G., Horberg, T.E., Koppang, E.O., Gallone, A., 2012. Melanogenesis in the visceral tissues of *Salmo salar*. A link between immunity and pigment production. Biochem. Cell Biol. 90, 769–778.

Bilandžija, H., Ćetković, H., Jeffery, W.R., 2012. Evolution of albinism in cave planthoppers by a convergent defect in the first step of the melanin biosynthesis pathway. Evol. Dev. 14, 196–203.

Bilandžija, H., Ma, L., Parkhurst, A., Jeffery, W.R., 2013. A potential benefit of albinism in *Astyanax* cavefish: downregulation of the *oca2* gene increases L-tyrosine and catecholamine levels as an alternative to melanin synthesis. PLoS One 8 (11), e80823. http://dx.doi.org/10.1371/journal.pone.0080823.

Borowsky, R., Wilkens, H., 2002. Mapping a cavefish genome: polygenic systems and regressive evolution. J. Hered. 93, 19–21.

Brilliant, M., 2001. The mouse *p* (*pink-eyed dilution*) and human *P* genes, oculocutaneous albinism type 2 (OCA2), and melanosomal pH. Pigment Cell Res. 14, 86–93.

Burgers, A.C.J., Bennick, P.J.H., von Oordt, G.J., 1962. Investigations into the regulation of the pigmentary system of the blind Mexican cavefish, *Anoptichthys jordani*. Proc. K. Ned. Akad. Wet. 66, 189–195.

Chelli, Y., Ohanna, M., Balloti, R., Bertolotto, C., 2009. Fifteen-year quest for micropthalmia-associated transcription factor target genes. Pigment Cell Melamona Res. 23, 27–40.

Culver, D.C., Pipan, T., 2009. The Biology of Caves and Other Subterranean Habitats. Oxford University Press, Oxford, UK.

Darwin, C., 1859. On the Origin of Species by Means of Natural Selection, or, the Preservation of Favoured Races in the Struggle for Life. J. Murray, London.

Du, J., Fisher, D.E., 2002. Identification of Aim-1 as the underwhite mouse mutant and its transcriptional regulation by MITF. J. Biol. Chem. 277, 402–408.

Duboué, E.R., Keene, A.C., Borowsky, R.L., 2011. Evolutionary convergence on sleep loss in cavefish populations. Curr. Biol. 21, 671–676.

Duboué, E.R., Borowsky, R.L., Keene, A.C., 2012. β-Adrenergic signaling regulates evolutionarily derived sleep loss in the Mexican cavefish. Brain Behav. Evol. 80, 233–243.

Elipot, Y., Hinaux, H., Callebert, J., Launay, J.M., Blin, M., Rétaux, S., 2014. A mutation in the enzyme monoamine oxidase explains part of the *Astyanax* cavefish behavioural syndrone. Nat. Commun. 5 (3647). http://dx.doi.org/10.1038/ncomms4647.

Espinasa, L., Centone, D.M., Gross, J.B., 2014. A contemporary analysis of a loss-of-function oculocutaneous albinism type II (Oca2) allele within the Rio Subterráneo *Astyanax* cavefish population. Speleobiol. Notes 6, 48–54.

Fairbank, P.D., Lee, C., Ellis, A., Hildebrand, J.D., Gross, J.M., Wallingford, J.B., 2006. Shroom2 (APXL) regulates melanosome biogenesis and localization in the retinal epithelium. Development 133, 4109–4118.

Felice, V., Visconti, M.A., Trajano, E., 2008. Mechanisms of pigmentation loss in subterranean fishes. Neotrop. Ichthyol. 6, 657–662.

Gallo, N.D., Jeffery, W.R., 2012. Evolution of space dependent growth in the teleost *Astyanax mexicanus*. PLoS One 7 (8), e41443. http://dx.doi.org/10.1371/journal.pone.0041413.

Gorman, M.J., Arakane, Y., 2010. Tyrosine hydroxylase is required for cuticle sclerotization and pigmentation in *Tribolium casteneum*. Insect Biochem. Mol. Biol. 40, 267–273.

Graham, D.G., Tiffany, S.M., Vogel, F.S., 1978. The toxicity of melanin precursors. J. Invest. Dermatol. 70, 113–116.

Gross, J.B., 2012. The complex origin of *Astyanax* cavefish. BMC Evol. Biol. 12, 105.

Gross, J.B., Wilkens, H., 2013. Albinism in phylogenetically and geographically distinct *Astyanax* cavefish arises through the same loss-of-function *Oca2* allele. Heredity 111, 122–130.

Gross, J.B., Protas, M., Conrad, M., Scheid, P.E., Vidal, O., Jeffery, W.R., Borowsky, R., Tabin, C.J., 2008. Synteny and candidate gene prediction using an anchored linkage map of *Astyanax mexicanus*. Proc. Natl. Acad. Sci. U. S. A. 105, 20106–20111.

Gross, J.B., Borowsky, R., Tabin, C.J., 2009. A novel role for Mc1r in the parallel evolution of depigmentation in independent populations of the cavefish *Astyanax mexicanus*. PLoS Genet. 5 (1), e10000326.

Hubbs, C.L., Innes, W.T., 1936. The first known blind fish of the family Characidae: a new genus from Mexico. Occ. Pap. Mus. Zool. Univ. Mich. 342, 1–7.

Jeffery, W.R., 2001. Cavefish as a model system in evolutionary developmental biology. Dev. Biol. 231, 1–12.

Juan, C., Guzik, M.T., Jaume, D., Cooper, S.J.B., 2010. Evolution in caves: Darwin's "wrecks of ancient life" in the molecular era. Mol. Ecol. 19, 3865–3880.

Kosswig, C., 1963. Genetische analyse konstruktiver und degenerativer evolutionsprozesse. Z. Zool. Syst. Evol. 1, 290–309.

Kowalko, J.E., Rohner, N., Linden, T.A., Rompani, S.B., Warren, W.C., Borowsky, R., Tabin, C.J., Jeffery, W.R., Yoshizawa, M., 2013. Convergence in feeding posture occurs through different genetic loci in independently evolved cave populations of *Astyanax mexicanus*. Proc. Natl. Acad. Sci. U. S. A. 110, 16933–16938.

Lane, B.M., Lister, J.A., 2012. Otx but not Mitf transcription factors are required for zebrafish retinal pigment epithelium development. PLoS One 7 (11), e49357.

Le Douarin, N., Kalcheim, C., 1999. The Neural Crest. Cambridge University Press, Cambridge, UK.

Leclercq, E., Taylor, J.F., Migaud, H., 2010. Morphological skin color changes in teleosts. Fish Fish. 11, 159–193.

Ma, L., Jeffery, W.R., Essner, J.J., Kowalko, J.E., 2015. Genome editing using TALENs in blind Mexican cavefish. PLoS One 10 (3), e0119370. http://dx.doi.org/10.1371/journal.pone.0119370.

McCauley, D.W., Hixon, E., Jeffery, W.R., 2004. Evolution of pigment cell regression in the cavefish *Astyanax*: a late step in melanogenesis. Evol. Dev. 6, 209–218.

McGaugh, S.E., Gross, J.B., Aken, B., Blin, M., Borowsky, R., Chalopin, D., Hinaux, H., Jeffery, W.R., Keene, A., Ma, L., Minx, P., Murphy, D., O'Quin, K.E., Rétaux, S., Rohner, N., Searle, S.M.J., Stahl, B., Tabin, C., Volff, J.N., Yoshizawa, M., Warren, W., 2014. The cavefish genome reveals candidate genes for eye loss. Nat. Commun. 5 (5307). http://dx.doi.org/10.1038/ncomms6307.

Mitchell, R.W., Russell, W.H., Elliott, W.R., 1977. Mexican eyeless characin fishes, genus *Astyanax*: environment, distribution, and evolution. Spec. Publ. Mus. Texas Tech. Univ. 12, 1–89.

Nery, L.E., Castrucci, A.M., 1997. Pigment cell signaling for physiological color change. Comp. Biochem. Physiol. A Physiol. 118, 1135–1144.

Okulicz, J.F., Shah, R.S., Schwartz, R.A., Janniger, C.K., 2002. Oculocutaneous albinism. J. Eur. Acad. Dermatol. Venereol. 17, 251–260.

Orr, H.A., 1998. Testing natural selection vs. genetic drift in phenotypic evolution using quantitative trait locus data. Genetics 149, 2099–2104.

Pfeiffer, W., 1966. Uber die vererbung der Schreckreaktion bei *Astyanax* (Characidae, Pisces). Z. Vererbungsl. 98, 97–105.

Porter, M.L., Crandall, K., 2003. Lost along the way: the significance of evolution in reverse. Trends Ecol. Evol. 18, 541–547.

Protas, M., Jeffery, W.R., 2012. Evolution and development in cave animals: from fish to crustaceans. WIREs Dev. Biol. 1 (6), 823–845.

Protas, M.E., Patel, N.H., 2008. Evolution of color patterns. Annu. Rev. Cell Dev. Biol. 24, 425–446.

Protas, M., Hersey, E., Kochanek, C., Zhou, Y., Wilkens, H., Jeffery, W.R., Zon, L.T., Borowsky, R., Tabin, C.J., 2006. Genetic analysis of cavefish reveals molecular convergence in the evolution of albinism. Nat. Genet. 38, 107–111.

Protas, M., Conrad, M., Gross, J.B., Tabin, C., Borowsky, R., 2007. Regressive evolution in the Mexican tetra, *Astyanax mexicanus*. Curr. Biol. 17, 452–454.

Protas, M., Tabansky, I., Conrad, M., Gross, J.B., Vidal, O., Tabin, C.J., Borowsky, R., 2008. Multi-trait evolution in a cave fish, *Astyanax mexicanus*. Evol. Dev. 10, 196–209.

Protas, M.E., Trontelj, P., Patel, N.H., 2011. Genetic basis of eye and pigment loss in the cave crustacean, *Asellus aquaticus*. Proc. Natl. Acad. Sci. U. S. A. 108, 5702–5707.

Rasquin, P., 1947. Progressive pigmentary regression in fishes associated with cave environments. Zoologica 32, 35–44.

Rosemblat, S., Durham-Pierre, D., Gardner, J.M., Nakatsu, Y., Brilliant, M.H., Orlow, S.J., 1994. Identification of a melanosomal membrane protein encoded by pink eyed dilution (type II oculocutaneous albinism) gene. Proc. Natl. Acad. Sci. U. S. A. 91, 12071–12075.

Sadoglu, P., 1957a. Mendelian inheritance in hybrids between the Mexican blind fish and their overground ancestors. Verh. Dtsch. Zool. Ges. 1957, 432–439.

Sadoglu, P., 1957b. A Mendelian gene for albinism in natural cave fish. Experientia 13, 394.

Sadoglu, P., McKee, A., 1969. A second gene that affects eye and body color in Mexican blind cavefish. J. Hered. 60, 10–14.

Slominski, A., Tobin, D.J., Shibahara, S., Wortsman, J., 2003. Melanin pigmentation in mammalian skin and hormonal regulation. Physiol. Rev. 84, 1155–1228.

Stoehr, A.M., 2006. Costly melanin ornaments: the importance of taxon? Funct. Ecol. 20, 276–281.

Toyofuku, K., Valencia, J.C., Kushimoto, T., Costin, G.E., Virador, V.M., Ferrans, V.J., Hearing, V.J., 2002. The etiology of oculocutaneous albinism (OCA) type II: the pink protein modulates the processing and transport of tyrosinase. Pigment Cell Res. 15, 217–224.

Vachetenheim, J., Borovansky, J., 2009. "Transcription physiology" of pigment formation in melanocytes: central role of MITF. Exp. Dermatol. 19, 617–627.

Vandel, A., 1965. Biospeleology: The Biology of Cavernicolous Animals. Pergamon, Oxford, UK.

Visser, M., Kayser, M., Palstra, R.-J., 2012. *HERC2* RS12913832 modulates human pigmentation by attenuating chromatin-loop formation between a long-range enhancer and the *OCA2* promoter. Genome Res. 22, 446–455.

Widlund, H.R., Fisher, D.E., 2003. *Micropthalamia*-associated transcription factor; a critical regulator of pigment cell development and survival. Oncogene 22, 3035–3041.

Wilkens, H., 1970. Beitrage zur degeneration des melaninpigments bei cavernicolen Sippen des *Astyanax mexicanus* (Filippi). (Characidae, Pisces). Z. Zool. Syst. Evolutionforsch. 8, 173–199.

Wilkens, H., 1988. Evolution and genetics of epigean and cave *Astyanax fasciatus* (Characidae, Pisces). Evol. Biol. 23, 271–367.

Wilkens, H., Strecker, U., 2003. Convergent evolution of the cavefish *Astyanax* (Characidae, Teleostei): genetic evidence from reduced eye-size and pigmentation. Biol. J. Linn. Soc. 80, 545–554.

Chapter 9

Molecular Mechanisms of Eye Degeneration in Cavefish

Yoshiyuki Yamamoto
Department of Cell and Developmental Biology, University College London, London, UK

ADULT CAVEFISH EYE

Astyanax fasciatus mexicanus has at least 30 different cavefish populations in northern Mexico. Each population has a different degree of eye degeneration (Jeffery, 2009; Gross, 2012). Most populations have nonfunctional, small, degenerated eyes in the head. The rudimental eye is hidden in an orbit that is covered by a flap of skin. In the teleost (same as in other vertebrates), the eye has three major components: the anterior segment (cornea, ciliary body, and lens), the posterior segment (retia, optic nerve, choroid, and sclera), and extraocular tissue (extraocular muscle and its nerves) (Figure 9.1).

The small eyeball in cavefish is surrounded by a cartilaginous sclera, which has thin extraocular muscles attached. Cranial nerves (oculomotor, trochlear, and abducens) are innervated to the extraocular muscles, but the degenerated eyeball may not be able to move voluntarily in the cavefish head. A frail optic nerve can also be seen to be connected between the small eyeball and the brain in the mildly degenerated eye; however, in the severely degenerated eye, the optic nerve is not identifiable (Wilkens, 1988, 2010). The anterior eye structures are severely degenerated on the inside of the eyeball. In particularly, differentiated crystalline lenses do not exist in the eye. The posterior eye structure, like the neural retina, is also severely degenerated. In the mildly degenerated eye, the layers of the retina can still be identified, but the total cell number is significantly reduced (Wilkens, 1988). Interestingly, immunohistochemistry shows presence of *Pax6* (paired box 6) protein in some retinal cells. In zebrafish, *Pax6* protein is expressed in mature amacrine, ganglion, and retinal progenitor cells (Hitchcock et al., 1996). These retinal cells may still exist in the degenerated retina (Yamamoto and Jeffery, 2000). Photoreceptor cells (rod and cone cells) are severely impaired. Most of the degenerated eye loses the outer nuclear layer of the retina where photoreceptor cells are located (Wilkens, 2010; Langecker et al., 1993). Outer segments of the existing photoreceptor cells are shortened,

Biology and Evolution of the Mexican Cavefish. http://dx.doi.org/10.1016/B978-0-12-802148-4.00009-8
© 2016 Elsevier Inc. All rights reserved.

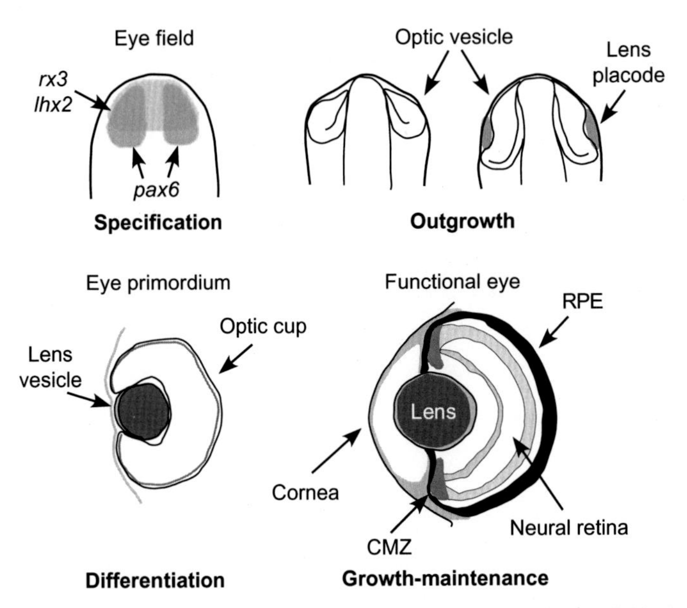

FIGURE 9.1 The four phases of eye formation. Specification (establishing eye field), Outgrowth (budding out of the optic vesicle from the neural tube and induction of lens placode), Differentiation (differentiating neural retina and lens cells), Growth-maintenance phases (increasing the size of fully functional eyes). RPE, retinal pigment epithelium; CMZ, ciliary marginal zone.

and cells are not well differentiated within the layer. Retinal pigment epithelium (RPE) also shows an abnormal flattened shape, and some cavefish populations also lose the pigmentation. In the most severely degenerated eye, there are no neuronal cells in the sclera, and only choroid tissue can be seen in the small eye-cyst. The size of the eye-cyst is variable within individuals, and variation is very high in adult cavefish. The size differences may reflect various degrees of eye degeneration (Wilkens, 2010).

EYE DEVELOPMENT

Eye formation can be divided into four phases: a *specification phase* with expression of transcription factors and signaling molecules to determine the eye field in anterior neural plate; an *outgrowth phase* when the specified cells in the eye field are budding out from the anterior neural tube and form two optic vesicles bilaterally; a *differentiation phase* with differentiation of retinal cells to form functional eyes; and a *growth-maintenance* phase to increase the size of fully functional eyes along with body growth (Figure 9.1).

The eye is part of the central nervous system, so eye development starts in the anterior neural plate. The transcription factors *pax6*, *rx3*, *lhx2/9*, and *six3* and signaling molecules *wnts*, *bmps*, and *shh* are expressed in the anterior neural plate and specified retinal field at the specification phase (Chuang and

Raymond, 2002) (Figure 9.1). Then, during the outgrowth phase, the optic vesicles are evaginated from the anterior neural tube bilaterally. Transcription factors *pax2* and *vax1* are expressed in the anterior part of the optic vesicles that becomes optic stalks at later stage, and *pax6* is expressed in the posterior part that becomes neural retinas in future development (Chuang and Raymond, 2002) (Figures 9.1 and 9.2). At a later stage, optic vesicles become optic cups and induce lens vesicles in both sides from the surrounding ectoderm. At this stage, the optic cup has two layers: the inner layer becomes the future neural retina, and the outer layer becomes the future RPE (Figure 9.1). During the differentiation phase, the lens vesicle starts to accumulate crystallin protein and becomes a crystalline lens. The inner layer of the optic cup constructs several layers and differentiates into ganglion, amacline, bipolar, Müller glial,

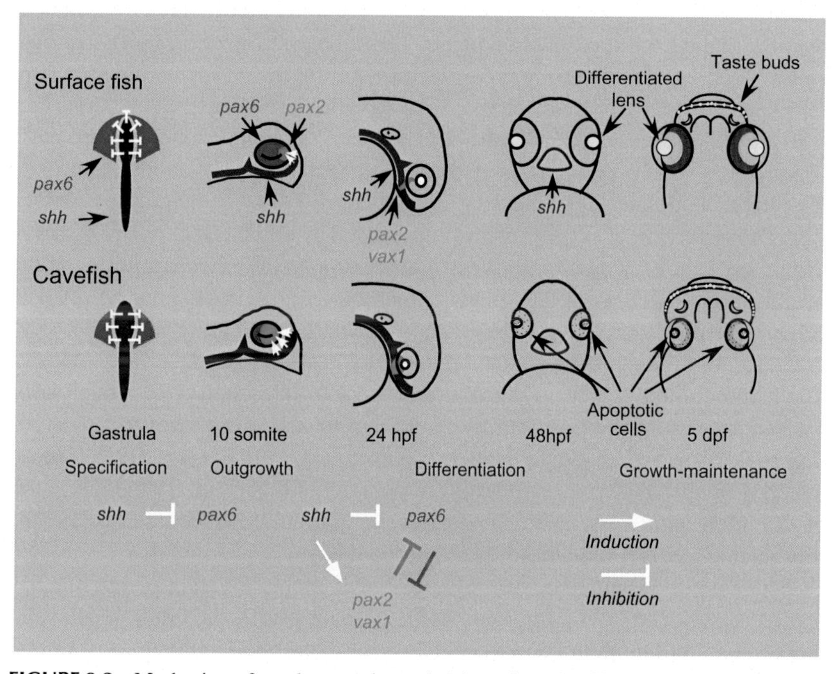

FIGURE 9.2 Mechanism of eye degeneration and pleiotropic trade-off in cavefishes. Midline signaling molecule *shh* suppresses expression of *pax6* to separate a single eye field into two at gastrula (Specification phase). In cavefishes, expanded expression of *shh* causes reduction of the eye field, which makes small optic vesicle during outgrowth phase; *shh* is continuously up-regulated in the anterior midline and induces enhanced expression of *pax2* and *vax1* at anterior optic vesicle. The enhancement inhibits *pax6* and reduces its expression at posterior optic vesicle, which is associated with formation of robust optic stalks and small lens vesicle at 24 hpf (Differentiation phase). Cells in the retina and lens start apoptosis at differentiation phase and degenerate during growth and maintenance phase; *shh* also expresses in the edge of oral epithelium and tastebud. The enhancement of *shh* at the anterior midline is also associated with a wider jaw and increased number of tastebuds in the cavefish.

horizontal, and photoreceptor cells (Fadool and Dowling, 2008). The ganglion cells send axons toward the optic tectum along with optic stalks and form synapses with cells in the tectum. The outermost layer of the optic cup becomes pigmented and differentiates into the RPE. At the end of the differentiation phase, the eye becomes functional and grows along with the body during the growth-maintenance phase. The ciliary marginal zone (CMZ) is located at the periphery of the retina where the retinal stem cells are located (Figure 9.1). These cells in the CMZ actively proliferate, and the newly added cells differentiate into retinal cells (Perron and Harris, 2000). The RPE maintains photoreceptor cells by involvement in the phagocytosis of the outer segment and also the vitamin A cycle (Strauss, 2005).

EYE DEGENERATION

The mechanism of eye degeneration of *A. fasciatus mexicanus* is well documented. During the specification phase, 7-10 h post-fertilization (hpf), eye field genes, such as *six3*, *pax6*, *lhx2/9*, and *rx3* are expressed in the anterior neural plate, but the expression domains in cavefish are smaller than in surface fish (Strickler et al., 2001; Pottin et al., 2011; Yamamoto, unpublished data) (Figure 9.2). The *pax6* gene is also expressed in the lens placode anterior to the retinal field, and the expression domain is also smaller than in the surface fish. The reduction of *pax6* expression is caused by enhanced expression of *shh* (Sonic hedgehog), which inhibits *pax6* expression at the anterior midline of the neural plate as well as in lens field (Yamamoto et al., 2004) (Figure 9.2). The enhancement of *shh* activates expression of another signaling molecule, *fgf8*, in the forebrain 2 h earlier, compared to the expression in surface fish embryos. The heterochronic (early) *fgf8* expression is also associated with reduction of *lhx2* expression in the posterior eye field that then becomes part of the hypothalamus instead of an eye in the cavefish embryos (Pottin et al., 2011). The reductions of *pax6* and *lhx2* expressions in the eye fields account for the formation of the small optic vesicles in cavefish during the outgrowth phase (10-24 hpf) (Pottin et al., 2011; Yamamoto, unpublished data).

The continuous enhancement of *shh* signals along the anterior midline upregulates *pax2* and *vax1* expression in the anterior part of the optic vesicle, and these transcription factors down-regulate *pax6* expression in the posterior optic vesicle during the outgrowth phase. This creates enhanced optic stalks and reduced optic cups at a later stage of development (Yamamoto et al., 2004).

Early in the differentiation phase (24 hpf), smaller lens vesicles containing lens epithelium and primary lens fiber cells are induced in cavefish embryos. The cells, however, cannot differentiate properly and cannot form secondary lens fiber, and subsequently fail to accumulate the crystallin protein in the lens (Figure 9.2). The lens starts to undergo programmed cell death (PCD) around 36 hpf and continues until the lens eventually disappears (Jeffery, 2009) (Figure 9.2). Interestingly, expression of an anti-apoptotic factor, the alpha-A-crystallin gene, and some other crystalline genes are down-regulated in the cavefish lens before PCD commences. These genes are likely to be responsible for the lens PCD in cavefish (Strickler et al., 2007a; Ma et al., 2014; Hinaux et al., 2014). After starting the lens PCD, cells in the retina also undergo PCD around 72 hpf (Strickler et al., 2007b; Alunni et al., 2007) (Figure 9.2). Although PCD is detected in all layers of the differentiating cavefish retina, most of retinal cell types (ganglion, amacline, horizontal, bipolar, and Müller glial cells) are still identifiable by molecular markers in the degenerating retina (Langecker et al., 1993; Jeffery et al., 2000; Strickler et al., 2002). Photoreceptor cells appear transiently and have abnormal short outer segments, and the cell numbers are subsequently reduced at a later stage (Langecker et al., 1993). These lost photoreceptor cells may be associated with a mutation in some opsin genes. Although a functional experiment has not been done yet, several fixed mutations are found in cavefish opsin genes and a blue opsin gene has a deletion of 12 consecutive nucleotides that may cause degeneration of photoreceptor cells (Yokoyama et al., 1995). Some cavefish populations have also lost pigmentation in the RPE where PCD has also been detected (Strickler et al., 2007b; Alunni et al., 2007) (Figure 9.2).

During the growth-maintenance phase (after 5 days post-fertilization [dpf]), the cavefish lens continues to have PCD and becomes smaller and eventually disappears. At later stages, anterior eye structures such as mature thin cornea and ciliary body fail to form. Although the eye is small and nonfunctional, cavefish develop almost complete set of extraocular structures, including the extraocular muscles and its nerves, sclera, and an optic nerve connected to the tectum (Wilkens, 1988; Soares et al., 2004; Yamamoto, unpublished data) (Figure 9.2). Surprisingly, cells in the CMZ are actively proliferating, although PCD has also been detected in the retina. BrdU cell-labeling experiments have shown that newly proliferated cells in the CMZ disappear in the retina at a later stage (Strickler et al., 2007b). In other words, the CMZ is actively proliferating and adding new cells to the cavefish retina, while simultaneously these cells are removed from the retina by PCD. As a result, the eye is kept small for 2-3 months, and then these small eyes sink into the head and degenerate (Strickler et al., 2007b) (Figure 9.2).

The mechanisms involved for the final eye degeneration step are still obscure. It is possible that the sinking of the eye into the orbit and covering by a flap of skin itself induces the final stages of retinal degeneration. Another possible mechanism is the disruption of the balance between number of proliferating cells and activity of PCD in the retina. Most adult cavefish keep degenerated eye-cysts and the sizes of the cysts are varied among individuals.

In sum, cavefish embryos develop small eye anlagen during the outgrowth phase, which are associated with a smaller eye field caused by enhancement of the midline signaling *shh* during the specification phase. The small eye develops all eye structures at the differentiation phase, but after lens vesicle starts PCD, other retinal cells also start PCD. Cells in CMZ, however, continue to actively proliferate, so the fry maintains small eyes for several months. The small eyes eventually sink into the head and degenerate.

The lens is an organizing center for eye development and degeneration.
PCD in the lens is the first sign of eye degeneration of *Astyanax* cavefish. When lens degeneration starts, cells in the retina also start to degenerate. This suggests that the lens may have an important role for eye degeneration. Indeed, a lens from eyed surface fish can rescue cavefish eye degeneration after transplantation into an optic cup at the beginning of the differentiation phase (24 hpf) (Yamamoto and Jeffery, 2000). The transplanted lens differentiates into a crystalline lens without undergoing PCD and induces formation of anterior eye structures such as a cornea and ciliary body in the host cavefish eye. The lens transplantation also rescues some of the posterior eye structures. The rescued eye has more antirhodopsin antibody-positive cells compared to a nontransplanted control, suggesting there are more rod photoreceptor cells in the transplanted eye. The eyeball in the transplanted side is much bigger than in the nontransplanted control side, and the majority of the rescued eyes do not sink into the orbit and remain outside of the head; however, the control eye is always found recessed in the orbit and covered by a flap of skin. The number of proliferating cells in the CMZ is similar between the lens-transplanted and nontransplanted controls, but the amount of PCD in the retina is much less in the transplanted side. Therefore, the transplanted surface fish lens does not affect proliferation in CMZ, but it is able to stop PCD in the cavefish retina directly or indirectly (Strickler et al., 2007b). The rescued eyeball also has both robust extraocular muscles and prominent optic nerve connected to the contralateral side of optic tectum. Although it is still not clear whether the rescued eye is fully functional (particularly photoreceptor cells), a behavior experiment suggests that lens-transplanted cavefish are in fact still blind (Romero et al., 2003).

Reversed transplantation (cavefish lens into surface fish) shows that transplanted cavefish lenses undergo PCD with similar timing as control cavefish lenses, and the lenses degenerate in the surface fish optic cup. The cavefish lens is not able to induce anterior eye structures, such as cornea and ciliary body in the host surface fish eye. The result is very similar as lens excision in surface fish. The transplanted cavefish lens does not have any effects for the host eye and may disappear quickly. Interestingly, PCD is not detected in the surface fish retina after the cavefish lens transplantation or the lens excision. Absence of a functional lens does not induce PCD in the surface fish retina. Strickler and his colleagues suggest that surface fish may possess a proper RPE that may prevent the retinal PCD; however, cavefish have neither proper RPE nor functional lenses (Wilkens, 1988; Strickler et al., 2007b). It is proposed that the absence of both factors may induce PCD in the cavefish retina (Strickler et al., 2007b).

Although the amount of PCD in the host surface fish retina is not affected by the transplanted cavefish lens, the manipulated eye becomes much smaller than the eye of the control side. The small eye possesses thin extraocular muscles and a frail optic nerve that is connected to the tectum (Soares et al., 2004). The small surface fish eye has sunken into the orbit, but without being covered by a flap of skin. Also, the retinal layers become thicker and distorted at a later stage, without complete degeneration like the severely degenerated cavefish eye (Yamamoto and Jeffery, 2000). This suggests that absence of a functional lens is not sufficient to induce severe retinal degeneration in the surface fish, and another mechanism is needed to explain the degeneration of the retina further, like in a cavefish. Interestingly, the cavefish lens transplanted into a surface fish eye has a very small vitreous chamber similar to that of the cavefish. It is possible that because of the lack of a ciliary body, which is the main producer of vitreous humor, these degenerated eyes may have failed to accumulate vitreous humor in the vitreous chamber (Coulombre and Coulombre, 1964). Furthermore, the presence of a thicker neural retina in the degenerated eyes is reminiscent of an eye with lower intraocular pressure (IOP) in chicken embryos (Coulombre, 1956; Neath et al., 1991). Coulombre suggests IOP creates an important force to grow the eyeball in the chick. While the source of the IOP in the embryonic eye is still not clear, Beebe suggested that the lens may have a role in producing IOP by inducing a ciliary body that also produces the aqueous humor, which has an important role for maintaining IOP in the adult eye (Beebe, 1986).

All together, the degenerating lens in cavefish, perhaps in combination with an impaired RPE, induces PCD in the retina to prevent an increase in the number of retinal cells. It also prevents the size of the eyeball from increasing in cavefish, possibly due to a lack of increased IOP and/or a failure to create the vitreous chamber. These results suggest that the lens has a crucial role for retinal cell degeneration as well as total size of the eye in cavefish.

MECHANISMS OF PCD IN THE CAVEFISH LENS

Because differentiated crystalline lenses are not detected in many *Astyanax* cavefish populations, researchers are focusing on the expression of crystalline genes. Particularly, *alphaA-crystallin* (*cryaa*) gene, an anti-apoptotic chaperone, which could have an important role for lens PCD in cavefish. Intriguingly, Behrens and his colleagues showed that the expression of *cryaa* is down-regulated in the embryonic lens of the *Piedras* cavefish population; however, the protein coding sequence and lens-specific promoter elements are highly conserved between surface fish and cavefish. Therefore, the authors concluded that *Cryaa* protein may have additional functions in non-ocular tissues and the conserved promoter elements may have a role for the non-ocular expression in the cavefish (Behrens et al., 1998). This was the first research on expression of *crystallin* gene in *Astyanax* cavefish and it had a big impact on other researchers; however, the promoter sequence is quite short (about 120 bp) and it still remains possible that other parts of the regulatory elements may have a role for the down-regulation of *cryaa* in the cavefish lens. In 2007, Strickler and his colleagues carefully compared expression patterns of *crystallin* genes between surface fish and the *Pachón* cavefish of *Astyanax*. They found that the *cryaa* gene is weakly expressed in the cavefish lens at 24 and 36 hpf, and the expression is lost just before the lens starts PCD. Although other structural genes such as *beta1*- and *gammaM2-crystallin* genes are also down-regulated in the lens, these genes continue to be expressed in the posterior side of the lens even after the lens has started PCD. The authors concluded that the down-regulation of *cryaa* might have a crucial role for PCD in the lens (Strickler et al., 2007a). This paper also showed *Cryaa* gene protein sequences are highly conserved among the *Astyanax* populations; however, the expression level in the lens is severely down-regulated. It was still not clear whether the gene regulatory elements of *cryaa* are responsible for the down-regulation (changing cis-acting elements) or if upstream-genes responsible for the expression of *cryaa* are also down-regulated (changing trans-acting factors). Quantitative trait loci (QTL) analysis suggested 12 loci in the *Astyanax* genome could be associated with eye degeneration in *Pachón* cavefish, and recent *Astyanax* genomic data suggested *cryaa* is located in the one of the eye size QTL (Protas et al., 2007; Gross et al., 2008; McGaugh et al., 2014). These genomic data suggested that altered cis-acting elements might have a crucial role for the down-regulation of *cryaa* expression and eye degeneration; however, two papers have suggested that multiple factors may associate with the lens PCD. Hinaux and her colleagues did extensive research on transcriptome data and identified a sequence of 14 *crystallin* genes. Although the coding sequences are very conserved between cavefish and surface fish, three *crystallin* genes were found with fixed mutations, and two of them could have radical mutations in the cavefish. Moreover, they showed that not only *cryaa* but other two *crystallin* genes, *crybb1c* and *crybgx*, are not expressed in *Pachón* cavefish before the lens starts PCD. This suggests that these defects found in *crystallin* genes (not only *cryaa*) may have an important role for the eye degeneration (Hinaux et al., 2014, 2015).

Ma and her colleagues suggested that the down-regulation of *cryaa* might be associated with the down-regulation of the upstream transcription factor gene, *sox2* rather than an altered gene regulatory element of *cryaa* in the genome. They compared about 10 kb of genomic sequences around *cryaa* among 5 teleost and identified 10 conserved sequences that are very similar between *Pachón* cavefish and surface fish. They also found a cavefish allele of *cryaa* expressed in lens of F1 hybrids (between *Pachón* cavefish and surface fish) that is equally expressed with the surface fish allele, suggesting that down-regulation of *cryaa* in cavefish is not associated with changing cis-acting elements because, if so, cavefish allele would not be expressed in the F1 hybrids. They also showed that the *sox2* gene is not expressed in the cavefish lens; moreover, the down-regulation of the *sox2* gene by antisense oligonucleotide abolished *cryaa* expression and triggered PCD in the lens of surface fish embryos (Ma et al., 2014). These two papers concluded that degeneration of cavefish lens could be associated with many different genes, including *crystallin* and its upstream transcription factors.

Interestingly, up-regulation of *shh* in surface fish embryos by *shh* messenger RNA (mRNA) injection induces PCD in the lens. It is still not clear how enhancement of *shh* induces PCD in the lens, but it would be interesting to see whether *sox2* is down-regulated in the lens after up-regulating *shh* in the surface fish.

EVOLUTIONARY FORCES WHY CAVEFISH FISH HAVE LOST THEIR EYE

There are two main hypotheses on how eyeless phenotype has been spread in cavefish populations, neutral mutation with genetic drift and natural selection. The neutral mutation theory suggests that, under a relaxed selective pressure for vision in a dark cave, mutations occur for eye-related genes and the mutations are spread in the population by random genetic drift. Natural selection theory suggests eye degeneration is due to a mutation that leads to make an advantageous effect for the cavefish. In this case, the mutations are spread in the population by selective pressure. Evolutionary constraints also contribute when a mutation that has deleterious effect for the cavefish is excluded from the population by negative selection (Wilkens, 2010; Jeffery, 2009). Comparing mechanisms of eye degeneration among different cavefish species could help us to elucidate the evolutionary trend for the eyeless phenotype. The trend, if any, could tell us what kind of evolutionary forces are involved in eye degeneration in cavefish.

THE TREND OF EYE DEGENERATION

In *Astyanax* cavefish, five degeneration mechanisms commonly occur among different populations:

1. *Forming a small embryonic lens and optic cup*: results in small eye field and optic vesicle during specification and the outgrowth phase
2. *Apoptosis and the absence of crystalline accumulation in lens*: down-regulation of gene expression in lens during differentiation phase
3. *The loss of photoreceptor cells*: missing lens and down-regulation of photoreceptor genes during differentiation and growth-maintenance phase
4. *Small eyeballs*: loss of the lens that fails to maintain retinal cells and increase IOP during the growth-maintenance phase
5. *Eye buried and covered by a flap of skin*: unknown mechanisms occurring during the growth-maintenance phase

Four of the degeneration mechanisms are associated with lens degeneration. It is still not clear what kind of mechanisms are involved with the burying of the eye. Interestingly, embryonic lens deletions lead to small eyes in surface fish, but these eyes are never buried and covered by skin within the orbit (Jeffery, 2009). Thus, there must be some other mechanisms involved.

EYE DEGENERATION IN OTHER CAVEFISH SPECIES

Eye degeneration of the cave catfish species, *Rhamdia*, is well described. Geological data indicates cave dwelling *Rhamdia* may have entered the cave about 1 million years ago (mya), which makes it younger than *Astyanax* cavefish, which were trapped in the caves about 2-3 mya (Wilkens, 2001; Bradic et al., 2012). Although the degree of eye degeneration differs between individuals, most adult cave catfish have small, degenerated eyes that are buried under the skin. The retinal layers are very thin, suggesting retinal cells are reduced in number. Outer nuclear and plexiform layers (photoreceptor cells) are missing, and RPE is flattened in the small eye, but all other retinal layers exist and may contain other retinal cell types. The optic nerve also exits properly and may connect with the brain. During the embryonic development, the cave-dwelling catfish develops slightly small eyes compared with its closely related surface catfish. This could be associated with a small eye field during the specification phase. Primary lens fibers develop, but secondary fibers do not elongate properly. Also in the lens, cell proliferation is reduced and necrosis occurs during the differentiation phase. Although most of the cavefish do not accumulate *Crystallin* proteins, some individuals develop almost normal lenses (Wilkens, 2001). This suggests that the lens degeneration allele might not be fixed in the species. During the growth-maintenance phase, there is a much greater diversity of eye size in the cave catfish individuals when compared to individuals of the surface dwelling *Rhamdia* at the same stage. This suggests that the degree of eye degeneration may be variable among individuals at this phase.

Recently, eye degeneration and expression of the associated genes were reported for another phylogenetically young cavefish, *Sinocyclocheilus* (Meng et al., 2013). Although this cavefish is blind, a small complete eye (differentiated lens and retinal cells) was found buried under the skin. All retinal layers exist, but cell density in outer and inner nuclear layers is significantly reduced. Immunohistochemistry labeling showed that this cavefish has a reduced number of rod, cone, and müller glial cells in the retina. Also, the photoreceptor cells are shorter and the outer segment is disrupted when compared to the closely related surface dwelling *Sinocyclocheilus* species. Cell proliferation in the retina is also reduced in the cavefish and the authors suggest that it may play a role for the formation of the small eye. Embryological data is not yet available, but a comparison of transcriptome data suggested that the transcription factor, cone-rod homeobox (*crx*) and *wnt* signaling molecules may be associated with the reduction of retinal cells in this cavefish (Meng et al., 2013).

The eye degeneration process of a fourth cavefish species, the Somalian cave-dwelling fish, *Phreatichthys andruzzii*, has also been described (Berti et al., 2001). This cavefish was isolated in caves about 1-2 mya, and is therefore slightly younger than *Astyanax* cavefish; however, its degree of eye degeneration is much more advanced than *Astyanax*, as adult cavefish have lost their eyes completely. During the second day of embryonic development (in the differentiation phase), the Somalian cavefish develops small lens vesicles and optic cups. Surprisingly, the small eyes start degenerating at this stage and pyknotic processes are seen in the lens and optic cup. These lenses never accumulate *Crystallin* proteins and completely degenerate within 4 days. The retinal cells have a reduced proliferation rate and continue to have pyknotic processes during the differentiation phase. The eye subsequently becomes smaller and the degenerated eye is buried under the skin. At this stage. the small eye-cyst has an optic nerve, sclera, and extraocular muscles, but the cyst disappears in a month, during growth-maintenance phase (Berti et al., 2001).

These four cavefish species shared several degeneration mechanisms. Starting with a small lens and optic cup, a degeneration and reduction of photoreceptor cells is followed by decreasing retinal cell numbers. The process of eye degeneration ends with the small eyeballs being buried under a skin flap.

EVOLUTIONARY MECHANISMS: NEUTRAL MUTATION WITH GENETIC DRIFT

Wilkens suggested that the increased diversity of eye size in cavefish is a sign of the relaxing selective pressure for eye loss (Wilkens, 2010). Because, if the degenerating eye has any functions, stabilizing selection may occur for the eye and the selective pressure would force a reduction in diversity of the eye size. This concept fits with *Astyanax* and *Rhamdia* cavefish during the growth-maintenance phase, but still fails to explain two phenomena:

- the diversity of eye size is not high at the early embryonic stage (outgrowth and differentiation phase)
- cavefish eyes are smaller than surface fish at all stages

Interestingly, just a single mutation is able to cause enlarged eyes in vertebrates. In zebrafish and humans, mutants of a transmembrane protein, *Lrp2*, showed enlarged eyes. *Lrp2* mRNA is expressed in the ciliary body, and RPE and can contribute to an increased IOP that may result in the presence of bigger eyes (Veth et al., 2011; Kantarci et al., 2007). Furthermore, dragoneye goldfish have an increased expression of transcription factor *six3* in photoreceptor cells during the growth-maintenance phase, which may be associated with increased retinal cell proliferation in this species (Ma et al., 2014). These mutants suggest that a single mutation is able to make bigger eyes, but these phenotypes have not been seen in any cavefish (Romero and Green, 2005). QTL analysis of *Astyanax* cavefish suggested that 12 QTL affect eye size and that all cavefish alleles are associated with decreased eye size rather than a size increase. For melanophore QTL, eight cavefish alleles control the decrease in the melanophore number, but five of them show an increase in the numbers. These polarities of QTL suggest that the reduction of the eye size may have evolved by natural selection, since all cavefish alleles result in reduced eye size, but the reduction of the pigmentation phenotype may be generated by neutral mutation with genetic drift, since there is no polarity to the cavefish alleles (Protas et al., 2007). Thus, evolution of a small eye phenotype in cavefish does not fit well within the neutral mutation with genetic drift concept.

Morphological analysis showed that lens and photoreceptor cells are targeted during eye degeneration more than any other tissues (such as ganglion and amacrine cells) (Wilkens, 1988, 2001; Meng et al., 2013). The failure in differentiation of secondary lens fiber cells and the missing outer segment of photoreceptor cells could be explained by the accumulation of mutations in developmental and/or structural genes powered by neutral mutation with genetic drift. So why do these cells degenerate more than other retinal cells in the cavefish species? One possibility is due to the amount of genes associated with the lens and photoreceptor cells; if lens and photoreceptor cells require more genes, these cells may have more chance to accumulate mutations, therefore these cells may degenerate more than other cells. Another possibility is that these cells are able to escape from negative selection. Lens and photoreceptor cells are highly specialized for sensing light. Many genes associated with these cells are expressed only in the photosensing organs, the eye and pineal, that may be relaxed from selective pressures in a cave. Genes expressed in other types of retinal cells, such as bipolar and ganglion cells, are also expressed in other nervous tissues such as brain. Mutations for those genes may have harmful effects for the eyes as well as these neural tissues, therefore these mutations would be eliminated in cavefish by negative selection, unable to spread through the cavefish population. Interestingly, only one deleterious mutation for an *opsin* gene has been reported in *Astyanax* cavefish (Yokoyama et al., 1995). Also fixed mutations in other *opsin* and *crystallin* genes may change protein functions radically (Yokoyama et al., 1995; Hinaux et al., 2014, 2015). Further experiments are needed to confirm whether the mutations are associated with the degeneration of the lens and photoreceptor cells. More intense sequence analyses, such as comparative genomics and/or transcriptome analysis and functional experiments, are required to confirm whether deleterious mutations have accumulated in the cavefish and whether neutral mutation with random drift may be involved with cavefish eye loss.

EVOLUTIONARY MECHANISMS: TRADE-OFF HYPOTHESIS

Trade-off is part of the positive selection concept; when several traits are linked to each other and are controlled by one pleiotropic gene, trade-off is the result of the enhancement of one of the linked traits and diminishing the other traits at same time. For *Astyanax* cavefish, half of eye size QTL (6 out of 12) colocalized with feeding ability QTL that affect jaw-bone size, number of teeth, and tastebuds. These data suggest that pleiotropic genes in the QTL may control both eye size and feeding ability, and selective pressure acts upon one gene to enhance feeding ability at the same time it sacrifices the eyes (Protas et al., 2007).

One of the candidates for the pleiotropic gene that controls both traits in *Astyanax* cavefish. During the enhancement of the midline signaling, *shh* leads to eye degeneration in the cavefish, but also controls the larger feeding apparatus comprised of a wider larval jaw with more tastebuds (Yamamoto et al., 2004, 2009). The *shh* gene may also be involved in inducing the development of a larger hypothalamus and olfactory pits and bulb that also improve feeding behavior (Pottin et al., 2011; Menuet et al., 2007; Bibliowicz et al., 2013);

however, the *shh* locus is not located near any of the eye size and feeding ability QTL. It will, therefore, be interesting to know whether genes that are associated with up-regulating *shh* are located in the pleiotropic QTL.

A recent paper also suggested another constructive trait linked to the eye size QTL. Vibration attraction behavior (VAB) is an important behavior to improve finding food in a cave. Superficial neuromasts near the eye orbits have important roles for feeding behavior in *Pachón* cavefish and one QTL was found to colocalize with both the number of the superficial neuromasts and VAB, as well as the eye size QTL (Yoshizawa et al., 2012). It is possible that these three traits may be controlled by one pleiotropic gene, the result being that the eye is sacrificed to enhance the advantageous traits. In this case, it would be interesting to compare genomic sequences of the pleiotropic QTL to identify important differences between the cavefish and surface fish. Identified genomic differences can be tested using genome-editing techniques, such as ZFN, CRISPR, or TALEN to confirm their role in the phenotypic differences (Gaj et al., 2013).

EVOLUTIONARY MECHANISMS: ENERGY CONSERVATION HYPOTHESIS

The energy conservation hypothesis forms part of a positive selection process in which losing eyes has advantageous effects for the cavefish. The eye is an energetically expensive organ similar to the brain; losing eyes may save energy to survive in nutrient-poor caves (Niven and Laughlin, 2008). This hypothesis fits well when considering mechanisms of eye degeneration in cavefish as discussed above. First, photoreceptor cells are energetically expensive to keep even when they are not in use. Oxygen consumption data suggests that photoreceptor cells consume a significant amount of oxygen in the mammalian retina (Wangsa-Wirawan and Linsenmeier, 2003). Also, adenosine triphosphate (ATP) consumption of both rod and cone photoreceptor cells are very high, even in darkness, due to maintaining membrane potential of the cells (Okawa et al., 2008). Therefore, the loss or reduction of these cells may help save energy.

Losing the lens also has two effects on energy conservation. First, degeneration of the lens leads to reduced retinal cell numbers, including photoreceptor cells. The degenerated eye also leads to thin extraocular muscles, small optic tectum, and a frail optic nerve (Yamamoto, unpublished data). Brain and muscular tissues also consume a lot of energy, so having small eyes has a direct advantage for the cavefish (Soares et al., 2004; Protas et al., 2007; Niven and Laughlin, 2008). Second, reducing the size of the eye allows the cavefish to use the space to store more fat. Cavefish increase their amount of fatty tissues during the food-rich season, including in orbital sockets, and the accumulation of fat will help survival during the nutrient-poor season (Hüppop, 2005); therefore, reducing eye size will help store more fat in the orbit and increase storage efficiency.

Recent data suggested that cavefish have evolved to improve food-finding ability, such as possessing a higher number of tastebuds and increased chemosensory sensitivity, and change behavior to reduce energy demand (Wilkens, 1988; Protas et al., 2008; Yamamoto et al., 2009; Bibliowicz et al., 2013; Elipot et al., 2013; Soares and Niemiller, 2013; Moran et al., 2014). Energy conservation is obviously important for surviving in a cave, but it is still not clear how much energy cavefish are able to save from degenerating their eyes and whether the energy saved is considered enough to push evolution toward eye degeneration. Further studies are necessary to analyze whether energy conservation is an important force for the eye degeneration in the cavefish. Interestingly, if losing eyes has an advantage for surviving in a cave, the enhancement of *shh* would have a dual selective pressure: energy conservation from inducing eye degeneration, and enhancement of both feeding apparatus and behaviors.

WHY "BUILD AND DESTROY"?

Three evolutionary processes—neutral mutation with random genetic drift, energy conservation, and pleiotropic trade-off hypothesis—could explain why these cavefish lost their eyes (Jeffery, 2009; Wilkens, 2010; Retaux and Casane, 2013); however, these three hypotheses cannot explain why blind organisms have eyes at embryonic stages and retain them for several months or even into adult life, instead of destroying the eye anlagen immediately or even not building them at all. For *Astyanax* and *Rhamdia* cavefish, diversity in eye size is increased during the growth-maintenance phase and could be explained by relaxed selective pressures (Wilkens, 2010); however, the eye size diversity is not high during the outgrowth and differentiation phases, which suggests that selective pressure may act upon the eye size, and stabilizing selection may occur during the early embryonic stages. Losing the eye anlagen during embryonic stages may not occur due to evolutionary constraints (Wilkens, 2010; Retaux and Casane, 2013). Stopping the initiation of eye development and losing the eye anlagen during early embryonic stages may have a deleterious effect on the development of other organs, and this may be associated with the fact that the eyeless phenotype may not be selected during evolution. Indeed, evidence can be taken from the *chokh* zebrafish mutant, which has lost expression of the eye retinal homeobox gene, *rx3*, in the presumptive eye field during the specification phase and fails to make optic vesicles in the embryo (Loosli et al., 2003). In this mutant, the expression of the telencephalon field genes, *tlc* and *foxg1*, expand posteriorly into the eye field where *rx3* is normally expressed. The mutant fish have completely lost their eyes and have abnormally enlarged the telencephalon (Stigloher et al., 2006). In the same *rx3* mutant, loss of the optic vesicle affects neural crest migration and induces craniofacial malformation (Langenberg et al., 2008). Thus, the loss of the eye anlagen at embryonic stages could have deleterious effects on the telencephalon and craniofacial formations during development. In *Astyanax*, the eye also has an important role for craniofacial development. Lens transplantation experiments have shown that the size of the eyeball is associated with the size of the orbit, and losing the eye causes severe malformation of the skull (Yamamoto et al., 2003; Dufton et al., 2012). Therefore, eye loss could also have a deleterious effect on craniofacial formation at early embryonic stages (such as the outgrowth and differentiation phases), and these phenotypes would not be selected during evolution; however, once this deleterious stage has passed and the cavefish has finished craniofacial development during the later growth-maintenance phase, the eye size will be free from selective pressure and thus results the high variation in eye size at this later stage. From the above discussion, it is clear that degenerating the eye from the outset would be unlikely to be selected, because the change would impair the development of other organs, such as brain and craniofacial components. "Build and Destroy" is a more effective mechanism, because once organogenesis has finished, the degenerating organ would be less closely associated with other organogenesis, meaning that its degeneration would be less deleterious.

Above all, the eye degeneration in cavefish provides significant insight of adaptation to the perpetual darkness though balancing craniofacial structure, gustatory organs, and cranial sensory systems. Comparative genomics studies using an available cavefish genome sequence (Ensembl: AstMex102) will reveal relaxed selection or selective pressures on the genes' loci involved in eye development. This will provide further understandings of the evolution of the eye degeneration, whether it is underlain by either or a combination of evolutionary processes: neutral mutation or natural selection. In addition, developmental biology with genome-editing technologies (TALEN and/or CRISPR) will reveal which combination of responsible loci is the most crucial for the fine "Build and Destroy" process. By investigating the multiple independently evolved cavefish populations, which could be using different gene sets for eye degenerations, we will have a general idea of how developmental refinement is conducted in evolution in a model system, eye development. Together with the analyses of energy consumption of the eye, the long-standing question, "why did cavefish lose the eye?" from the Charles Darwin era, will have a clearer answer with ecological relevance and fine molecular mechanisms.

REFERENCES

Alunni, A., Menuet, A., Candal, E., Penigault, J.B., Jeffery, W.R., Retaux, S., 2007. Developmental mechanisms for retinal degeneration in the blind cavefish *Astyanax mexicanus*. J. Comp. Neurol. 505, 221–233.

Beebe, D.C., 1986. Development of the ciliary body: a brief review. Trans. Ophthalmol. Soc. U. K. 105 (Pt 2), 123–130.

Behrens, M., Wilkens, H., Schmale, H., 1998. Cloning of the alphaA-crystallin genes of a blind cave form and the epigean form of *Astyanax fasciatus*: a comparative analysis of structure, expression and evolutionary conservation. Gene 216, 319–326.

Berti, R., Durand, J.P., Becchi, S., Brizzi, R., Keller, N., Ruffat, G., 2001. Eye degeneration in the blind cave-dwelling fish *Phreatichthys andruzzii*. Can. J. Zool. 79, 1278–1285.

Bibliowicz, J., Alie, A., Espinasa, L., Yoshizawa, M., Blin, M., Hinaux, H., Legendre, L., Pere, S., Retaux, S., 2013. Differences in chemosensory response between eyed and eyeless *Astyanax mexicanus* of the Rio Subterraneo cave. EvoDevo 4, 25.

Bradic, M., Beerli, P., Garcia-de Leon, F.J., Esquivel-Bobadilla, S., Borowsky, R.L., 2012. Gene flow and population structure in the Mexican blind cavefish complex (*Astyanax mexicanus*). BMC Evol. Biol. 12, 9.

Chuang, J.C., Raymond, P.A., 2002. Embryonic origin of the eyes in teleost fish. BioEssays 24, 519–529.

Coulombre, A.J., 1956. The role of intraocular pressure in the development of the chick eye. I. Control of eye size. J. Exp. Zool. 133, 211–255.

Coulombre, A.J., Coulombre, J.L., 1964. Lens development. I. Role of the lens in eye growth. J. Exp. Zool. 156, 39–47.

Dufton, M., Hall, B.K., Franz-Odendaal, T.A., 2012. Early lens ablation causes dramatic long-term effects on the shape of bones in the craniofacial skeleton of *Astyanax mexicanus*. PLoS One 7, e50308.

Elipot, Y., Hinaux, H., Callebert, J., Retaux, S., 2013. Evolutionary shift from fighting to foraging in blind cavefish through changes in the serotonin network. Curr. Biol. 23, 1–10.

Fadool, J.M., Dowling, J.E., 2008. Zebrafish: a model system for the study of eye genetics. Prog. Retin. Eye Res. 27, 89–110.

Gaj, T., Gersbach, C.A., Barbas 3rd., C.F., 2013. ZFN, TALEN, and CRISPR/Cas-based methods for genome engineering. Trends Biotechnol. 31, 397–405.

Gross, J.B., 2012. The complex origin of *Astyanax* cavefish. BMC Evol. Biol. 12, 105.

Gross, J.B., Protas, M., Conrad, M., Scheid, P.E., Vidal, O., Jeffery, W.R., Borowsky, R., Tabin, C.J., 2008. Synteny and candidate gene prediction using an anchored linkage map of *Astyanax mexicanus*. Proc. Natl. Acad. Sci. U. S. A. 105, 20106–20111.

Hinaux, H., Blin, M., Fumey, J., Legendre, L., Heuzé, A., Casane, D., Rétaux, S., 2014. Lens defects in *Astyanax mexicanus* Cavefish: evolution of crystallins and a role for alphaA-crystallin. Dev. Neurobiol. http://dx.doi.org/10.1002/dneu.22239.

Hinaux, H., Blin, M., Fumey, J., Legendre, L., Heuze, A., Casane, D., Retaux, S., 2015. Lens defects in *Astyanax mexicanus* Cavefish: evolution of crystallins and a role for alphaA-crystallin. Dev. Neurobiol. 75 (5), 505–521.

Hitchcock, P.F., Macdonald, R.E., VanDeRyt, J.T., Wilson, S.W., 1996. Antibodies against Pax6 immunostain amacrine and ganglion cells and neuronal progenitors, but not rod precursors, in the normal and regenerating retina of the goldfish. J. Neurobiol. 29, 399–413.

Hüppop, K., 2005. Adaptation to low food. In: Culver, D.C., White, W.B. (Eds.), Encyclopedia of Caves. Elsevier Academic Press, Amsterdam, pp. 4–10.

Jeffery, W.R., 2009. Regressive evolution in *Astyanax* cavefish. Annu. Rev. Genet. 43, 25–47.

Jeffery, W., Strickler, A., Guiney, S., Heyser, D., Tomarev, S., 2000. Prox 1 in eye degeneration and sensory organ compensation during development and evolution of the cavefish *Astyanax*. Dev. Genes Evol. 210, 223–230.

Kantarci, S., Al-Gazali, L., Hill, R.S., Donnai, D., Black, G.C., Bieth, E., Chassaing, N., Lacombe, D., Devriendt, K., Teebi, A., Loscertales, M., Robson, C., Liu, T., MacLaughlin, D.T., Noonan, K.M., Russell, M.K., Walsh, C.A., Donahoe, P.K., Pober, B.R., 2007. Mutations in LRP2, which encodes the multiligand receptor megalin, cause Donnai-Barrow and facio-oculo-acoustico-renal syndromes. Nat. Genet. 39, 957–959.

Langecker, T.G., Schmale, H., Wilkens, H., 1993. Transcription of the opsin gene in degenerate eyes of cave dwelling *Astyanax fasciatus* (Teleostei, Characidae) and its conspecific ancestor during early ontogeny. Cell Tissue Res. 273, 183–192.

Langenberg, T., Kahana, A., Wszalek, J.A., Halloran, M.C., 2008. The eye organizes neural crest cell migration. Dev. Dyn. 237, 1645–1652.

Loosli, F., Staub, W., Finger-Baier, K.C., Ober, E.A., Verkade, H., Wittbrodt, J., Baier, H., 2003. Loss of eyes in zebrafish caused by mutation of chokh/rx3. EMBO Rep. 4, 894–899.

Ma, L., Parkhurst, A., Jeffery, W.R., 2014. The role of a lens survival pathway including sox2 and alphaA-crystallin in the evolution of cavefish eye degeneration. EvoDevo 5, 28.

McGaugh, S.E., Gross, J.B., Aken, B., Blin, M., Borowsky, R., Chalopin, D., Hinaux, H., Jeffery, W.R., Keene, A., Ma, L., Minx, P., Murphy, D., O'Quin, K.E., Retaux, S., Rohner, N., Searle, S.M., Stahl, B.A., Tabin, C., Volff, J.N., Yoshizawa, M., Warren, W.C., 2014. The cavefish genome reveals candidate genes for eye loss. Nat. Commun. 5, 5307.

Meng, F., Zhao, Y., Postlethwait, J.H., Zhang, C., 2013. Differentially-expressed genes identified in cavefish endemic to China. Curr. Zool. 59, 170–174.

Menuet, A., Alunni, A., Joly, J.S., Jeffery, W.R., Retaux, S., 2007. Expanded expression of Sonic Hedgehog in *Astyanax* cavefish: multiple consequences on forebrain development and evolution. Development 134, 845–855.

Moran, D., Softley, R., Warrant, E.J., 2014. Eyeless Mexican cavefish save energy by eliminating the circadian rhythm in metabolism. PLoS One 9, e107877.

Neath, P., Roche, S.M., Bee, J.A., 1991. Intraocular pressure-dependent and —independent phases of growth of the embryonic chick eye and cornea. Invest. Ophthalmol. Vis. Sci. 32, 2483–2491.

Niemiller, M.L., Higgs, D.M., Soares, D., 2013. Evidence for hearing loss in amblyopsid cavefishes. Biol. Lett. 9, 20130104.

Niven, J.E., Laughlin, S.B., 2008. Energy limitation as a selective pressure on the evolution of sensory systems. J. Exp. Biol. 211, 1792–1804.

Okawa, H., Sampath, A.P., Laughlin, S.B., Fain, G.L., 2008. ATP consumption by mammalian rod photoreceptors in darkness and in light. Curr. Biol. 18, 1917–1921.

Perron, M., Harris, W.A., 2000. Retinal stem cells in vertebrates. BioEssays 22, 685–688.

Pottin, K., Hinaux, H., Retaux, S., 2011. Restoring eye size in *Astyanax mexicanus* blind cavefish embryos through modulation of the Shh and Fgf8 forebrain organising centres. Development 138, 2467–2476.

Protas, M., Conrad, M., Gross, J.B., Tabin, C., Borowsky, R., 2007. Regressive evolution in the Mexican cave tetra, *Astyanax mexicanus*. Curr. Biol. 17, 452–454.

Protas, M., Tabansky, I., Conrad, M., Gross, J.B., Vidal, O., Tabin, C.J., Borowsky, R., 2008. Multi-trait evolution in a cave fish, *Astyanax mexicanus*. Evol. Dev. 10, 196–209.

Retaux, S., Casane, D., 2013. Evolution of eye development in the darkness of caves: adaptation, drift, or both? EvoDevo 4, 26.

Romero, A., Green, S.M., 2005. The end of regressive evolution: examining and interpreting the evidence from cave fishes. J. Fish Biol. 67, 3–32.

Romero, A., Green, S.M., Romero, A., Lelonek, M.M., Stropnicky, K.C., 2003. One eye but no vision: cave fish with induced eyes do not respond to light. J. Exp. Zool. B Mol. Dev. Evol. 300, 72–79.

Soares, D., Yamamoto, Y., Strickler, A.G., Jeffery, W.R., 2004. The lens has a specific influence on optic nerve and tectum development in the blind cavefish *Astyanax*. Dev. Neurosci. 26, 308–317.

Soares, D., Niemiller, M.L., 2013. Sensory adaptations of fishes to subterranean environments. Bioscience 63, 274–283.

Stigloher, C., Ninkovic, J., Laplante, M., Geling, A., Tannhauser, B., Topp, S., Kikuta, H., Becker, T.S., Houart, C., Bally-Cuif, L., 2006. Segregation of telencephalic and eye-field identities inside the zebrafish forebrain territory is controlled by Rx3. Development 133, 2925–2935.

Strauss, O., 2005. The retinal pigment epithelium in visual function. Physiol. Rev. 85, 845–881.

Strickler, A.G., Byerly, M.S., Jeffery, W.R., 2007a. Lens gene expression analysis reveals down-regulation of the anti-apoptotic chaperone alphaA-crystallin during cavefish eye degeneration. Dev. Genes Evol. 217, 771–782.

Strickler, A.G., Famuditimi, K., Jeffery, W.R., 2002. Retinal homeobox genes and the role of cell proliferation in cavefish eye degeneration. Int. J. Dev. Biol. 46, 285–294.

Strickler, A.G., Yamamoto, Y., Jeffery, W.R., 2007b. The lens controls cell survival in the retina: evidence from the blind cavefish *Astyanax*. Dev. Biol. 311, 512–523.

Strickler, A.G., Yamamoto, Y., Jeffery, W.R., 2001. Early and late changes in Pax6 expression accompany eye degeneration during cavefish development. Dev. Genes Evol. 211, 138–144.

Veth, K.N., Willer, J.R., Collery, R.F., Gray, M.P., Willer, G.B., Wagner, D.S., Mullins, M.C., Udvadia, A.J., Smith, R.S., John, S.W., Gregg, R.G., Link, B.A., 2011. Mutations in zebrafish lrp2 result in adult-onset ocular pathogenesis that models myopia and other risk factors for glaucoma. PLoS Genet. 7, e1001310.

Wangsa-Wirawan, N.D., Linsenmeier, R.A., 2003. Retinal oxygen: fundamental and clinical aspects. Arch. Ophthalmol. 121, 547–557.

Wilkens, H., 1988. Evolution and genetics of epigean and cave *Astyanax fasciatus* (Characidae, Pisces). Evol. Biol. 23, 271–367.

Wilkens, H., 2001. Convergent adaptations to cave life in the *Rhamdia laticauda* catfish group (Pimelodidae, Teleostei). Environ. Biol. Fish 62, 252–261.

Wilkens, H., 2010. Genes, modules and the evolution of cave fish. Heredity 105, 413–422.

Yamamoto, Y., Byerly, M.S., Jackman, W.R., Jeffery, W.R., 2009. Pleiotropic functions of embryonic sonic hedgehog expression link jaw and taste bud amplification with eye loss during cavefish evolution. Dev. Biol. 330, 200–211.

Yamamoto, Y., Espinasa, L., Stock, D.W., Jeffery, W.R., 2003. Development and evolution of craniofacial patterning is mediated by eye-dependent and—independent processes in the cavefish *Astyanax*. Evol. Dev. 5, 435–446.

Yamamoto, Y., Jeffery, W.R., 2000. Central role for the lens in cave fish eye degeneration. Science 289, 631–633.

Yamamoto, Y., Stock, D.W., Jeffery, W.R., 2004. Hedgehog signalling controls eye degeneration in blind cavefish. Nature 431, 844–847.

Yokoyama, S., Meany, A., Wilkens, H., Yokoyama, R., 1995. Initial mutational steps toward loss of opsin gene function in cavefish. Mol. Biol. Evol. 12, 527–532.

Yoshizawa, M., Yamamoto, Y., O'Quin, K.E., Jeffery, W.R., 2012. Evolution of an adaptive behavior and its sensory receptors promotes eye regression in blind cavefish. BMC Biol. 10, 108.

Chapter 10

The Evolution of the Cavefish Craniofacial Complex

Joshua B. Gross and Amanda K. Powers
Department of Biological Sciences, University of Cincinnati, Cincinnati, Ohio, USA

DISCOVERY, CHARACTERIZATION, AND THE HISTORICAL RELEVANCE OF CRANIOFACIAL EVOLUTION IN CAVEFISH

Although the developmental, molecular, and genetic underpinnings mediating evolutionary changes to the cavefish skull are just now being elucidated, the craniofacial complex was among the very first morphological traits characterized in this remarkable system. Roughly 10 years after discovery of the first *Astyanax* cavefish at the Chica cave locality (formerly "*Anoptichthys hubbsi*," ca. 1936), Alvarez (1946) described unusual morphologies associated with the complex of dermal bones encircling the eye. This intramembranous set of bones, termed the "circumorbital" series, is comprised of a supraorbital bone (situated superior to the orbit of the eye) and six suborbital (synonymous with "infraorbital") bones (Figure 10.1(A)). In surface-dwelling populations inhabiting the rivers and streams surrounding the Sierra de El Abra, these bones are entirely distinct from one another, representing singular bony structures reminiscent of other teleost fish (Valdez-Moréno and Contreras-Balderas, 2003). In his observations, however, Alvarez (1946) discovered cave-dwelling fish harbored a spectrum of dramatic alterations in this series, including fusions between bones (e.g., the third and fourth suborbital bones) and "fragmentations" within particular bones (Figure 10.1(B)). For example, the third suborbital bone ("SO3"), situated inferior to the orbit of the eyes, is normally present as a singular ossified element; however, this bone has been described as fragmented into as many as ten distinct elements in cavefish (Alvarez, 1946).

During the era in which Alvarez performed his analyses, cave-dwelling forms were regarded as an entirely separate genus ("*Anoptichthys*") from their surface-dwelling relatives (Hubbs and Innes, 1936; Innes, 1937; Breder, 1943). Further, only three (of the 29) cave populations of the Sierra de El Abra had been discovered by the late-1940s (Mitchell et al., 1977). Interestingly, extreme alterations to the craniofacial complex were observed in geographically- and evolutionarily-distinct populations (Figure 10.1(C)). The precise patterns of

Biology and Evolution of the Mexican Cavefish. http://dx.doi.org/10.1016/B978-0-12-802148-4.00010-4
© 2016 Elsevier Inc. All rights reserved.

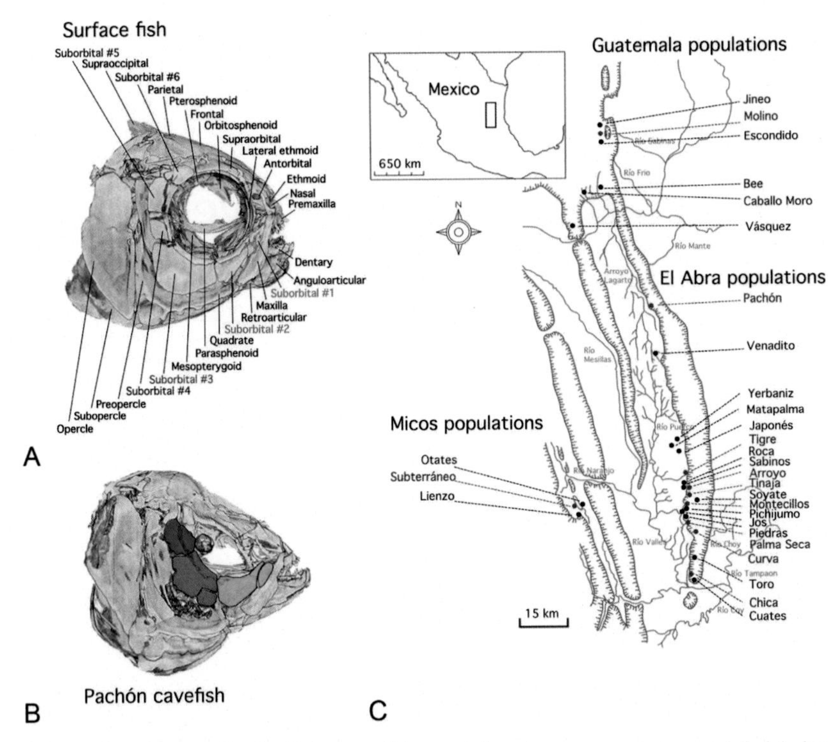

FIGURE 10.1 Geographically distinct cavefish populations converge on similar craniofacial aberrations. Representative surface (A) and Pachón cavefish (B) individuals were scanned using micro-computed tomography (micro-CT) to create 3D surface-rendered images. Aberrant craniofacial phenotypes are highlighted as follows: SO1+2 fusion/fragmentation in green, SO3 fragmentation in red, and SO4+5 fusion in blue (B). Of the 29 documented caves in the Guatemala, El Abra, and Micos regions in northeastern Mexico, 10 populations exhibit craniofacial anomalies (C; highlighted in red). The map of the Sierra de El Abra region of northeastern Mexico was adapted from Mitchell et al. (1977).

fragmentation and fusion, however, differed among populations. These morphological characters were used, in part, to justify classification of the three early known cavefish populations as entirely separate species (Chica cave, *Anoptichthys hubbsi*; Sabinos cave, *Anoptichthys jordani*; Pachón cave, *Anoptichthys antrobius*; Alvarez, 1947; reviewed in Jeffery, 2009b).

Alvarez (1947) was the first to describe the presence, and therefore recurrent evolution, of extreme patterns of circumorbital bone fusions and fragmentations in each of these three cave populations (Alvarez, 1947). The same morphological alterations had not been described in the surface-dwelling populations, and it is assumed that these changes arise as a consequence of the light- and nutrient-restricted subterranean microenvironment. More recently, Jeffery (2009a) has also characterized analogous craniofacial alterations in the Micos (Río Subterráneo) cave population. Therefore, although fusions/fragmentations of the circumorbital bone series can be regarded as convergent phenotypes associated with cave colonization, the stereotypical changes affecting these bones appear to be specific to each population (Table 10.1; Jeffery, 2009b).

TABLE 10.1 Summary of Craniofacial Alterations Identified in *Astyanax* Cavefish

Cave-associated Craniofacial Phenotype	Documented Cave Populations	References
SO1 fragmentation	Curva, Palma Seca, Montecillos, Tinaja, Arroyo, Tigre, and Pachón	Mitchell et al. (1977)
Fusions between SO1 and SO2	Curva, Palma Seca, Montecillos, Tinaja, Arroyo, Tigre, and Pachón[a]	Mitchell et al. (1977) and Gross et al. (2014)
Fusion between SO3 and SO4	Chica, Pachón, Sabinos, Guatemala (Molino), and Micos (Río Subterráneo)	Alvarez (1946, 1947) and Jeffery and Yamamoto (2000)
SO3 fragmentation	Chica, Sabinos, Guatemala (Molino), Micos (Río Subterráneo), Tinaja, and Pachón[a]	Alvarez (1946, 1947), Mitchell et al. (1977), Protas (2005) and Gross et al. (2014)
Decreased SO3 width	Pachón[a]	Protas et al. (2008)
Fusions between SO4 and SO5	Curva, Palma Seca, Montecillos, Tinaja, Arroyo, Tigre, and Pachón[a]	Mitchell et al. (1977) and Gross et al. (2014)
Fusions between SO5 and SO6	Curva, Palma Seca, Montecillos, Tinaja, Arroyo, Tigre, and Pachón[a]	Mitchell et al. (1977)
Fusions between SO4, SO5, and SO6	Curva, Palma Seca, Montecillos, Tinaja, Arroyo, Tigre, and Pachón[a]	Mitchell et al. (1977)
Increase in size of SO bones	Chica, Pachón, Sabinos, Guatemala (Molino), and Micos (Río Subterráneo)	Jeffery and Yamamoto (2000) and Gross et al. (2014)
Supraorbital shape	Pachón and Tinaja	Dufton et al., 2012
Convex opercular shape	Pachón	Yamamoto et al. (2003)
Expanded olfactory pit (nares)	Pachón	Yamamoto et al. (2003)
Increase in number of maxillary teeth	Curva, Los Sabinos, Tinaja, Molino, and Pachón[a]	Yamamoto et al. (2003) and Protas et al. (2008)

[a]*Includes Surface x Pachón F_2 hybrids.*

CRANIOFACIAL CHANGES ACROSS INDEPENDENTLY DERIVED CAVE POPULATIONS

Expeditions to the Sierra de El Abra region of NE Mexico from the 1940s through the 1970s led to the discovery of 26 caves populated by troglomorphic fish (reviewed in Mitchell et al., 1977), beyond the first three characterized populations (Hubbs and Innes, 1936; Alvarez, 1946, 1947). By this time, the genus *Anoptichthys* had been largely dismissed (however this nomenclature has persisted in the literature; see Hassan, 1986), and cave-dwelling forms were regarded as distinct morphotypes of the genus *Astyanax* (Schemmel, 1967). This shift in nomenclature followed the discovery that cave and surface populations (as well as different cave populations) were capable of hybridizing (Jordan, 1946) and producing viable offspring (Şadoğlu, 1957). Moreover, classic molecular analyses (based upon allozyme structure) supported a close relationship between surface- and cave-dwelling *Astyanax* fish (Avise and Selander, 1972). At present, a consensus does not exist regarding the taxonomic status of cave populations. For instance, some authors regard extant cave- and surface-dwelling forms as different morphotypes of the species *Astyanax mexicanus* (Protas et al., 2006). Others regard different cave populations as distinct species from one another (Ornelas-García et al., 2008), and recent evidence suggests that multiple species of *Astyanax* surface-dwelling fish seeded the caves of the Sierra de El Abra (Bradic et al., 2012; Coghill et al., 2014). Irrespective of the precise phylogenetic relationships among cave and surface populations, each of these cave forms is capable of hybridizing with other cave and surface forms to produce viable offspring. Thus, cave and surface fish demonstrate proximate phylogenetic relationships that diverged variably over the past several million years (Ornelas-García et al., 2008).

Mitchell et al. (1977) provided the first comprehensive description of the geology and demography of these cave populations. Included in their landmark publication were the first morphological comparisons of the skull among *Astyanax* cave forms (Mitchell et al., 1977). These studies sought to determine the level of independence (or "discreteness") of different cave populations based upon their degree(s) of alteration in craniofacial morphology. Accordingly, the authors evaluated eight geologically and geographically diverse cave populations (Curva, Palma Seca, Jos, Montecillos, Tinaja, Arroyo, Tigre, and Pachón caves) alongside three separate surface-dwelling populations (Mitchell et al., 1977). Among the craniofacial traits evaluated were head depth, head width, gape, mandibular length, maxillary length, gill raker number on the upper and lower limbs of the first gill arch, and suborbital fragmentation and fusion.

Although fusion of the suborbital bones was observed with some significant frequency, it was only found to impact particular bones (Mitchell et al., 1977). For instance, the authors noted common fusions between SO1 and SO2 (Figure 10.2(E) and (G), green); SO4 and SO5 (Figure 10.2(E)–(H), blue); and SO5 and SO6. They further noted that the only fusion event involving three bones involved the SO4, SO5, and SO6 bones. Their work also supported prior studies (Alvarez, 1946), indicating these anomalies "rarely" (if ever) occur naturally in surface-dwelling fish populations. At the level of independent cave populations, the most informative (i.e., discriminating) craniofacial characters were maxillary length, the total number of suborbital bones, gill raker number on the lower limb of the first gill arch, right SO1 fragment number, left SO1 fragment number, right SO3 fragment number, and head width (Mitchell et al., 1977). Interestingly, the left-sided SO3 fragment number was not found to be informative in their study, suggestive of an "asymmetric" bias in SO3 fragmentation (Figure 10.2(E)–(H), red). Interestingly, Gross et al. (2014) recently

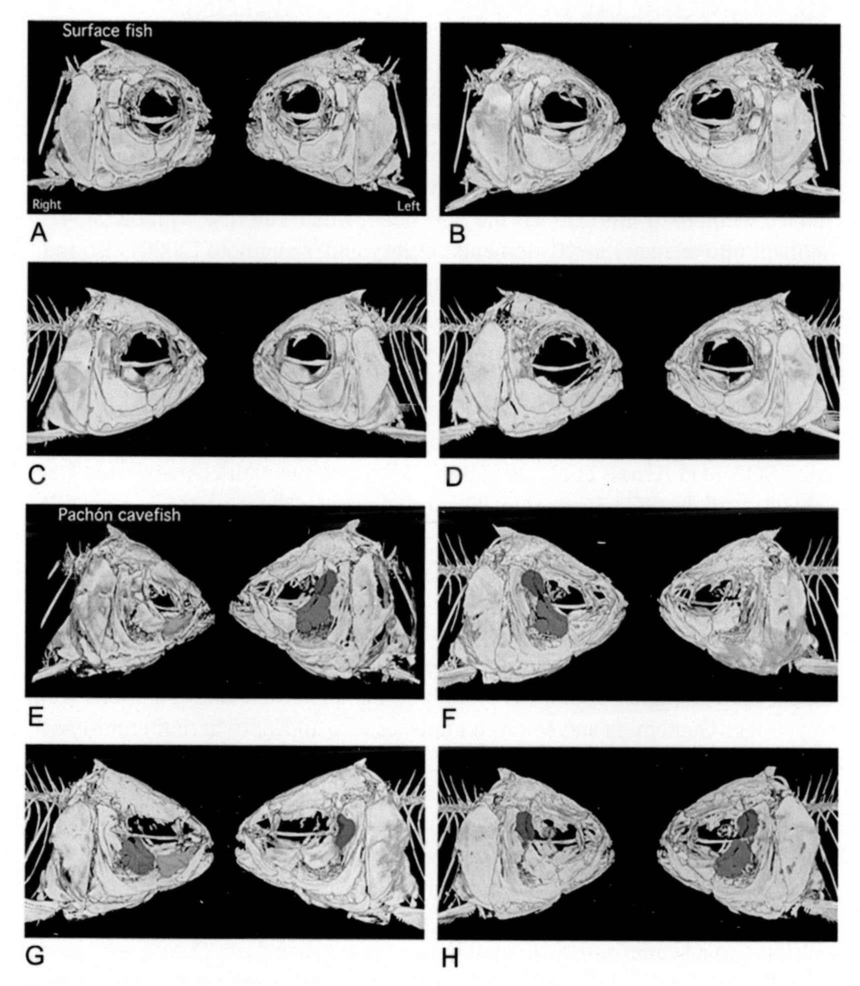

FIGURE 10.2 Cavefish demonstrate frequent left-right asymmetries in the craniofacial complex. Micro-CT surface-rendered images of four surface fish (A–D) and four Pachón cavefish (E–H) illustrate common left-right asymmetries for suborbital bones in cavefish. Four surface fish specimens demonstrate lateral symmetry in the circumorbital series (A–D). All Pachón specimens (E–H) demonstrate multiple left-right asymmetric aberrations.

described an asymmetric genetic signal detected based upon the SO3 fragmentation number for the right (but not left) cranial side. Thus, this asymmetric bias in craniofacial fragmentation appears to be quite stable within cave-dwelling populations (Gross et al., 2014), having persisted in the wild for several decades. Although surface fish demonstrated far less variability associated with the craniofacial complex, head length was an informative discriminating character for surface-dwelling populations (Mitchell et al., 1977).

THE NATURE OF MORPHOLOGICAL CHANGES TO THE CRANIOFACIAL COMPLEX IN CAVE-DWELLING POPULATIONS

One of the most significant and well characterized alterations associated with cavefish craniofacial evolution is the structure of the suborbital bony series, which demonstrates changes to the size, shape, and number in cave-dwelling populations. Jeffery and Yamamoto (2000) reported that the most drastically affected member of this series is the SO3 bone, which can fuse with the SO4, or fragment into as many as 10 elements. Jeffery and Yamamoto (2000) also indicated that suborbital bones are increased in size in cave-dwelling forms (however this phenotype depends upon which cave is being evaluated). Additionally, the supraorbital and suborbital bones are laterally displaced into the empty orbital space of the eye in cave compared to surface forms (Yamamoto et al., 2003). Consistent with this, a recent genetic analysis of Pachón x surface fish hybrids identified genetic signals associated with the bony area of suborbital bones SO2-SO5 (Gross et al., 2014). In addition to the well characterized fragmentation of the SO3 bone into discrete elements, these elements expanded into the neuromast canal (with smaller elements located on the ventral side of the largest elements; Yamamoto et al., 2003). Owing to fusions of the suborbital bones, the total number of bones in the SO4-6 complex are reduced in cavefish (Yamamoto et al., 2003). Further, although many morphological alterations were first identified in members of the older El Abra cave complex (Chica, Sabinos, and Pachón caves), additional subterranean populations from the younger Guatemala and Micos complexes also appear to harbor craniofacial modifications (Jeffery, 2009a).

The constellation of changes affecting the craniofacial complex in other cave populations includes differences in the nasal aperture size, the operculum (i.e., opercle bone), and the number of maxillary teeth. Specifically, the distance between the nasal and antorbital bones may have become larger to create an expanded olfactory pit ("nares"). This morphological alteration may contribute to a keener sense of smell in the cave environment (Yamamoto et al., 2003; Bibliowicz et al., 2013). The opercle bone is larger and convex (rather than concave) in cavefish compared to surface fish (Yamamoto et al., 2003). Further, while surface-dwelling fish have one maxillary tooth, cavefish from the Pachón, Sabinos, Tinaja, Curva, and Molino caves typically possess two or more maxillary teeth (Yamamoto et al., 2003). Jeffery (2001) described additional changes to the feeding apparatus in cavefish. Specifically, compared to surface-dwelling forms, cavefish demonstrate lower jaw protrusion in addition to premature ossification of the maxilla. The maxilla may have evolved a larger size in order to provide more space for teeth (Jeffery, 2001).

Collectively, cave-associated changes to the craniofacial complex, alongside population genetic and ecological evidence, support the notion that multiple ancestral stocks of epigean forms colonized the El Abra, Micos, Guerrero, and (possibly) Guatemala caves over the past several million years (Jeffery, 2009a). Across different cave populations, the Pachón cave is often appreciated as harboring the most extreme troglomorphic characteristics based upon morphology (reviewed in Mitchell et al., 1977; Wilkens, 1988) and genetic analyses (Bradic et al., 2012). These extreme phenotypes include severe forms of suborbital bone fragmentation (Alvarez, 1946), likely owing to their isolated geographic position and nutrient-poor microenvironment (Mitchell et al., 1977). Although the cave environment is generally nutrient-poor, cavefish reared in the lab under normal lighting and feeding conditions also converge on cave-associated cranial defects, which bolsters support for genetic control of these traits (Gross et al., 2014).

MECHANISMS OF CRANIOFACIAL EVOLUTION IN CAVE-DWELLING POPULATIONS

Although it remains unclear precisely what selective pressure could be driving cave-associated craniofacial alterations, they clearly arise convergently across the cave population landscape of northeastern Mexico. In the context of cave evolution, phenotypic losses (such as eye and pigmentation) are hypothesized to regress as a consequence of one of three evolutionary mechanisms. The first explanation is that traits without an adaptive or phenotypic value are lost through accumulation of neutral genetic mutations (Wilkens, 1988). A second explanation is that unnecessary traits pose an energetic cost that is selected against in the cave environment (Protas et al., 2007). A third explanation is that certain regressive and constructive features are united through the same genetic regulation, and influence one another through antagonistic pleiotropy (Jeffery, 2010).

Jeffery (2008) described a number of troglomorphic features that can be categorized as regressive (e.g., eye and pigmentation loss) and other features that are constructive (e.g., increased tooth number and cranial neuromast size). The craniofacial characters that have been identified, however, are not easily categorized as regressive or constructive. Regarding the suborbital bones, Mitchell et al. (1977) first noted that they have "begun to regress accompanying regression of the eyes." Furthermore, they reasoned that if eyeless populations were discrete (i.e., arose from independent invasions), then patterns of suborbital bone regression should similarly be "discrete" among populations (Mitchell et al., 1977).

Jeffery (2008) characterized craniofacial features evolving after colonization of the subterranean environment as "other changes." This is perhaps the most appropriate category. The craniofacial complex in cavefish is *changed* (e.g., from a single bone to multiple bone fragments); however, it is not obviously a "regressed" character. Further, with respect to size, the third suborbital bone (SO3) becomes smaller in certain cave populations (e.g., Pachón cavefish; Jeffery, 2008), but appears to become somewhat larger in other cave populations (e.g., Río Subterráneo cavefish; see Jeffery, 2008). Therefore, in the absence of a clear and uniform polarity of change (e.g., a smaller eye, or more numerous tastebuds), categorical labels for suborbital alterations to the skull are less relevant. Furthermore, alterations to the craniofacial skeleton do not have any obvious adaptive significance in the context of a cave habitat (Jeffery, 2005).

At present, it is unclear what role development plays in the generation of craniofacial changes within cavefish. The suborbital series are dermal bones that develop directly from condensations of the craniofacial mesenchyme (Jeffery, 2001). These bones begin to form at ~22 mm body length, which is generally between ~3 and 4 months of age under normal laboratory rearing conditions. By this point, the fish have reached juvenilehood. This bony series starts to ossify in the following order: SO2 first, followed by SO1 and SO3, and then SO4-SO6 (Yamamoto et al., 2003). The SO3 bone begins to form from several primary ossification centers, which eventually fuse. Following the fusion of this element, when the fish is ~33 mm, the failure of one or more secondary ossification centers ventral to this element results in the observed SO3 fragmentation (Yamamoto et al., 2003).

This bony series is induced late in development through interactions between the cranial neural crest and underlying neural tissues. The neural crest gives rise to numerous tissue types: pigmentation, components of the peripheral nervous system, bone and cartilage of the facial skeleton, portions of the heart and endocrine system (Jeffery, 2009a), and several visual system structures (e.g., cornea, iris, sclera; Jeffery, 2009b). Although this is the source population for both body pigmentation and the facial skeleton (Jeffery, 2009a), some evidence suggests the neural crest does not differ significantly among cave and surface populations. For instance, migratory pathways of the cranial neural crest do not appear to be altered in Pachón cavefish based upon vital dye labeling (Jeffery, 2009a). Second, xanthophore pigment cells (which are derived from melanophores) do not appear to be affected in cavefish. Third, *in vitro* experiments demonstrate that pigment-like cells appear to migrate away from cultured neural tube explants (Jeffery, 2009a). Thus, at least with respect to pigmentation cells, Jeffery (2009a) suggested that precursors and their migratory pathways are largely unaffected in cavefish. Rather, differentiation events are perhaps affected after the emigration of pigmentation precursors from the cranial neural crest. Future studies should discern if specification of cranial neural crest cells destined to give rise to mesenchymal cell types are similarly unaffected. If this crest-derived tissue is similarly unaffected in cavefish, it may indicate that craniofacial fusions/fragmentations are rooted in later developmental events (e.g., fusion failures during ossification) rather than through misspecification of neural crest cell types.

Jeffery (2001) suggested that changes to the morphology of the circumorbital bones could be governed by expansion of dermal bone ossification into empty orbital space and the formation of additional ossification centers during development. Thus, supernumerary ossification centers could be related to architectural changes associated with the cranial lateral line elements (neuromasts), which may "seed" the ossification centers of the suborbital series (Jeffery, 2001). The vertebrate skull is a highly complex, integrated structure that draws from multiple developmental tissues of origin. Therefore, improving our understanding of the origin of cave-associated alterations to the craniofacial complex will require integrative approaches evaluating the developmental, genetic, and molecular underpinnings of this complicated structure.

COORDINATED CHANGES BETWEEN THE CRANIOFACIAL COMPLEX AND OTHER CAVE-ASSOCIATED TRAITS

Early studies in cavefish assumed suborbital bone alterations arose strictly as indirect consequences of eye loss (Mitchell et al., 1977). Wilkens (1988) referenced suborbital bone fragmentation as a contributor, along with the loss of a large eye, to the changes to the suborbital lateral line canal between cave and surface morphs. Specifically, the suborbital canal is shorter, interrupted, and lacks a deep ventral flexure that is normally observed in surface fish (Wilkens, 1988). Thus, the pattern of suborbital bone fragmentation observed in cavefish may contribute (directly or indirectly) to changes in the lateral line canal. Neuromasts may affect bone ossification (Jeffery, 2001), and it has been suggested that cranial neuromast expansion in cavefish may influence the size of certain craniofacial elements, such as the mandible (see Franz-Odendaal and Hall, 2006).

More recent studies have documented an experimental link between visual system regression and changes to the craniofacial complex. Given the important structural position of the eyes within the head, disruption of visual system growth could lead to serious craniofacial defects (Mathers and Jamrich, 2000). The first experimental approach reported in the literature utilized lens transplantation experiments (Yamamoto and Jeffery, 2000). Cavefish embryos receiving a surface fish lens sustained the growth of a structurally intact eye, and demonstrated partial recapitulation of the ancestral surface fish skeleton (including the circumorbital bone series). Conversely, surface fish experimentally manipulated to undergo regression of the visual system harbored an orbital skeleton resembling cavefish (Jeffery and Yamamoto, 2000; Yamamoto et al., 2003). Additional craniofacial changes occurring beyond the orbital skeleton were not evaluated; however, these results demonstrated that the lens dictates certain craniofacial changes evolving in cavefish, indirectly, through sustained or lost eye growth (Jeffery and Yamamoto, 2000; Jeffery, 2005, 2009b).

Although the structural presence of an eye has a clear effect on the craniofacial skeleton, some evolutionary changes affecting the cavefish craniofacial skeleton are unrelated to eye loss (Jeffery, 2008). Lens transplantation and rescue experiments in cavefish only partially restored suborbital and nasal bones to the surface fish phenotype. Certain changes not associated with eye loss included opercle bone shape and morphology of the tooth-bearing maxillary bones (Jeffery, 2008). With respect to the underlying causes, Jeffery (2009b) suggested that the majority of the craniofacial changes in cavefish are caused by loss of an eye. These phenotypic alterations demonstrate how seemingly small changes in the early developmental program can have substantial consequences on the adult morphology (Jeffery, 2009b).

Also using a lens ablation strategy, Dufton et al. (2012) characterized the effects of eye removal on the adult craniofacial complex. Surface fish individuals across multiple critical developmental stages (24 hours post-fertilization (hpf)-4 days post-fertilization (dpf)) were subjected to lens removal, grown through adulthood (~1 year), and evaluated for changes to the craniofacial skeleton (Dufton et al., 2012). Earlier removal of the lens during development was associated with more dramatic effects on the adult cranial skeleton. These impacts included changes in supraorbital bone position and shape, expansion of the SO4 and SO5 bones in the orbital and SO6 areas, and an increased number of mandibular teeth (Dufton et al., 2012). Thus, altered developmental and tissue-level patterning, mediated through the absence of a structural visual system, influences many of the characterized changes to the cavefish craniofacial skeleton.

GENETIC ANALYSES OF CRANIOFACIAL EVOLUTION

The genetic basis for craniofacial evolution has been investigated in other natural fish systems. For instance, freshwater three-spine stickleback fish in Alaska have undergone reduced size and shape changes to their opercle facial bone compared with marine populations. Genetic analyses revealed a major effect locus strongly contributes to this phenotypic change (Kimmel et al., 2005). African cichlid fish have similarly undergone adaptive changes to their skull and jaws through diverse genetic changes mediated by *Wnt* and *hedgehog* signaling (Roberts et al., 2011; Parsons et al., 2014). Antarctic notothenioid fishes have evolved changes to their craniofacial complex through heterochronic shifts in key cranial gene expression patterns that result in delayed bone formation as juveniles (Albertson et al., 2010). Many of these characterized changes, however, are likely adaptive in nature. By examining blind Mexican cavefish, we have the opportunity to study changes to the craniofacial skeleton that may evolve in the absence of obvious selective pressures. Furthermore, certain extreme craniofacial alterations, including asymmetric fusions and fragmentations, are largely uncharacterized in the literature. Therefore, *Astyanax* offers the opportunity to better understand the genetic basis for these unique aberrations, the molecular basis for left-right craniofacial asymmetry, and the evolution of craniofacial aberrations in the absence of obvious selective forces.

The first genetic analysis of a cave-associated craniofacial phenotype in *Astyanax* was performed for the third suborbital bone (SO3). Protas et al. (2008) developed a metric for this bone by dividing SO3 width by the distance from the preopercular canal to the orbit. This bone size metric was smaller for cavefish compared to surface fish (Protas et al., 2008). Subsequent studies revealed several quantifiable craniofacial phenotypes that demonstrated significant genetic associations. Gross et al. (2014) investigated the genetic underpinnings of 33 craniofacial phenotypes in the context of an experimental F_2 pedigree generated from a surface x Pachón cavefish cross (Gross et al., 2014). Interestingly, many of the hybrid individuals harbored cave-associated craniofacial anomalies despite having fully formed and functional eyes (see Figure 10.3(A) and (B)). This finding is consistent with the work of Yamamoto et al. (2003) who found that craniofacial alterations are mediated by a mixture of both eye-dependent and eye-independent processes.

This quantitative trait locus (QTL) study evaluated 237 individuals using a linkage map populated by 175 polymorphic markers. A genetic basis was discovered for three classes of traits: suborbital bone fragmentation, fusion, and area. Interestingly, the genetic signals associated with the bony area (for SO2-SO5) were symmetric, i.e., present whether the trait was scored on the right or left side of the head; however, the fusion and fragmentation phenotypes, were asymmetric, i.e., only evident when scoring one (but not the other) side of the head (Gross et al., 2014). The authors concluded that the craniofacial complex in cavefish is mediated by genetically symmetric loci in the case of the bony area, but through asymmetric loci for aberrant phenotypes, such as SO3 fragmentation, SO1 + 2 fusion and SO4 + 5 fusion.

Some craniofacial phenotypes and other regressive/constructive traits are associated with the same genomic regions, based upon overlap or colocalization of QTL (Gross et al., 2014); however, recent studies did not find an association between feeding posture (as a behavior) and other morphological phenotypes, including craniofacial morphology, body depth, and distribution/number of taste sensory organs (Kowalko et al., 2013). Further exploration may reveal a relationship, however, between documented changes in feeding angle (Schemmel, 1980; see also Kowalko et al., 2013) and other craniofacial morphologies—perhaps those involving changes to the lower jaw. In sum, the spectrum of diverse craniofacial phenotypes is likely mediated by a large number of genes (Jeffery, 2009b), and these genetic signals are both symmetric (i.e., bony area) and asymmetric (i.e., suborbital bone fragmentation/fusion; Gross et al., 2014). Future advances will be aided by the sequenced genome, identification of candidate genes, deep RNA sequencing, and implementation of transgenic approaches (CRISPRs, TALENS) to functionally validate how genetic changes mediate craniofacial aberrations in cavefish. These studies will be necessary to determine the precise genetic basis for these features, and whether craniofacial alterations are a direct or indirect consequence of other regressive and constructive traits evolving in cavefish.

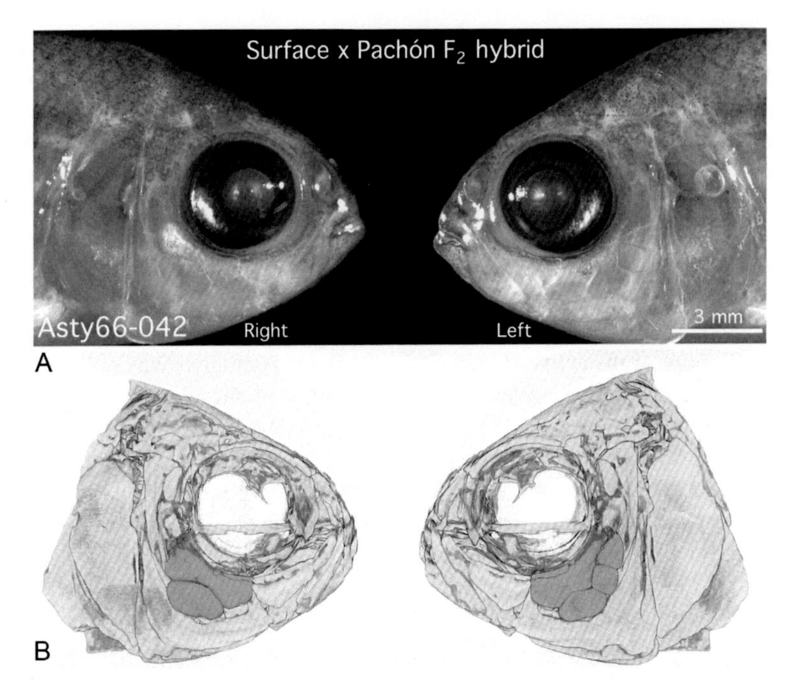

FIGURE 10.3 Asymmetric bone fragmentation persists in hybrid individuals despite the presence of a structural eye. Light microscopic images of specimen Asty66-042 (a surface x Pachón cavefish F_2 hybrid) illustrate a well formed eye on the right and left sides of the head (A). Micro-CT surface-rendered images of the same specimen show two distinct, fragmented SO3 elements on the right side and four elements on the left side of the face (shaded; B). Magnification in A = 7.81×.

CONCLUSIONS

Cavefish populations that are both geographically and phylogenetically distinct have converged on several changes to the craniofacial complex relative to surface-dwelling fish. Many of these alterations, such as the positioning and shape of bones, are related to the development of an eye. In the blind river eel, a primordial eye is retained, which is believed to play a key nonvisual role in craniofacial patterning (Kish et al., 2011). Why would the eye primordia form, if these eyes are eventually lost anyway? One reason may be that these primordial tissues are essential for early developmental (pleiotropic) signaling, perhaps through *shh*, which is necessary for increased jaw and nasal pit sizes (Kish et al., 2011). Alternatively, the primitive eye may also be crucial for generating and/or maintaining a circadian rhythm critical during the early phases of the life history. In this scenario, the developing eye is likely maintained through selective processes acting upon this early developmental program. Given that facial morphogenesis is clearly a critical developmental process, the early presence of visual system tissue may play an indispensable role for maintaining migratory pathways of the cranial neural crest (Kish et al., 2011). The evolutionary mechanisms by which some craniofacial changes emerge remain unknown. Adaptive changes to the skull have been well characterized in a number of natural model systems, such as Darwin's finches (Abzhanov et al., 2004) and cichlid fishes (Albertson et al., 2003). Craniofacial alterations in cavefish, however, cannot be easily categorized as regressive or constructive changes evolving under selective pressures. Indirect selection may play a role in the evolution of craniofacial alterations. Jeffery (2005) suggested that eye loss may be an antagonistic pleiotropic consequence associated with constructive (or modified) changes affecting other tissue or organ systems. Since midline *shh* signaling plays highly conserved roles in visual system development (Stock et al., 2006), perhaps other systems affected by *shh* signaling (craniofacial structures, tastebuds, teeth) could also indirectly affect, or be affected by, eye loss (Jeffery, 2005). Understanding the underlying causes of these changes may help identify novel genetic targets for analogous human disorders. For example, nonsyndromic cleft palate in humans is a failure-to-fuse phenotype that can be modeled by natural systems like the *Astyanax*. Working with natural systems, such as cavefish, will enable us to capture genetic diversity as well as identify mutations affecting both coding and regulatory regions of candidate genes. For instance, certain genes such as *ptc1* (a component of the *shh* signaling pathway) have been shown to play a critical role in craniofacial morphogenesis in a variety of traditional and natural model systems (Schartl, 2014). With the recent availability of a cavefish genome sequence to complement ongoing QTL studies, our emerging field is well poised to provide exciting new insight into many of these longstanding problems.

ACKNOWLEDGMENTS

The authors are grateful to the editors, Alex Keene, Masato Yoshizawa, and Suzanne McGaugh, for the invitation to contribute to this volume. The authors also acknowledge members of the Gross lab, especially Brian Carlson and Bethany Stahl, for helpful conversations and discussions during the production of this chapter. The Gross lab gratefully acknowledges funding from the National Institutes of Health (NIDCR; grant no. DE022403), which has supported our research into craniofacial asymmetry in the *Astyanax* system.

REFERENCES

Abzhanov, A., Protas, M., Grant, B.R., Grant, P.R., Tabin, C.J., 2004. *Bmp4* and morphological variation of beaks in Darwin's finches. Science 305, 1462–1465.

Albertson, R.C., Streelman, J.T., Kocher, T.D., 2003. Genetic basis of adaptive shape differences in the cichlid head. J. Hered. 94, 291–301.

Albertson, R.C., Yan, Y.-L., Titus, T.A., Pisano, E., Vacchi, M., Yelick, P.C., Detrich, H.W., Postlethwait, J.H., 2010. Molecular pedomorphism underlies craniofacial skeletal evolution in Antarctic notothenioid fishes. BMC Evol. Biol. 10, 4.

Alvarez, J., 1946. Revisión del género *Anoptichthys* con descipción de una especie nueva (Pisces, Characidae). An. Esc. Nac. Cien. Biol., Mexico 4, 263–282.

Alvarez, J., 1947. Descripcion de *Anoptichthys hubbsi* caracindo ciego de La Cueva de Los Sabinos, S. L. P. Rev. Soc. Mexicana Hist. Nat. 8, 215–219.

Avise, J.C., Selander, R.K., 1972. Evolutionary genetics of cave-dwelling fishes of the genus *Astyanax*. Evolution 26, 1–19.

Bibliowicz, J., Alié, A., Espinasa, L., Yoshizawa, M., Blin, M., Hinaux, H., Legendre, L., Père, S., Rétaux, S., 2013. Differences in chemosensory response between eyed and eyeless *Astyanax mexicanus* of the Río Subterráneo cave. Evodevo 4, 25.

Bradic, M., Beerli, P., García-de León, F.J., Esquivel-Bobadilla, S., Borowsky, R.L., 2012. Gene flow and population structure in the Mexican blind cavefish complex (*Astyanax mexicanus*). BMC Evol. Biol. 12, 9.

Breder Jr., C.M., 1943. Apparent changes in phenotypic ratios of the characins at the type locality of *Anoptichthys jordani* Hubbs and Innes. Copeia 1943, 26–30.

Coghill, L.M., Darrin Hulsey, C., Chaves-Campos, J., García-de León, F.J., Johnson, S.G., 2014. Next generation phylogeography of cave and surface *Astyanax mexicanus*. Mol. Phylogenet. Evol. 79, 368–374.

Dufton, M., Hall, B.K., Franz-Odendaal, T.A., 2012. Early lens ablation causes dramatic long-term effects on the shape of bones in the craniofacial skeleton of *Astyanax mexicanus*. PLoS ONE 7. e50308.

Franz-Odendaal, T.A., Hall, B.K., 2006. Modularity and sense organs in the blind cavefish, *Astyanax mexicanus*. Evol. Dev. 8, 94–100.

Gross, J.B., Krutzler, A.J., Carlson, B.M., 2014. Complex craniofacial changes in blind cave-dwelling fish are mediated by genetically symmetric and asymmetric loci. Genetics 196, 1303–1319.

Hassan, E.S., 1986. On the discrimination of spatial intervals by the blind cave fish (*Anoptichthys jordani*). J. Comp. Physiol. A 159, 701–710.

Hubbs, C.L., Innes, W.T., 1936. The first known blind fish of the family Characidae: a new genus from Mexico. Occas. Papers Mus. Zool., Univ. Michigan 342, 1–7.

Innes, W.T., 1937. A cavern characin *Anoptichthys jordani*, Hubbs and Innes. Aquarium 5, 200–202.

Jeffery, W.R., 2001. Cavefish as a model system in evolutionary developmental biology. Dev. Biol. 231, 1–12.

Jeffery, W.R., 2005. Adaptive evolution of eye degeneration in the Mexican blind cavefish. J. Hered. 96, 185–196.

Jeffery, W.R., 2008. Emerging model systems in evo-devo: cavefish and microevolution of development. Evol. Dev. 10, 265–272.

Jeffery, W.R., 2009a. Evolution and development in the cavefish *Astyanax*. Curr. Top. Dev. Biol. 86, 191–221.

Jeffery, W.R., 2009b. Regressive evolution in *Astyanax* cavefish. Annu. Rev. Genet. 43, 25–47.

Jeffery, W.R., 2010. Pleiotropy and eye degeneration in cavefish. Heredity 105, 495–496.

Jeffery, W.R., Yamamoto, Y., 2000. The lens is a regulator of craniofacial development and evolution in the teleost *Astyanax*. Dev. Biol. 222, 239.

Jordan, C.B., 1946. *Anoptichthys* x *Astyanax* hybrids. Aquarium 15, 198.

Kimmel, C.B., Ullmann, B., Walker, C., Wilson, C., Currey, M., Phillips, P.C., Bell, M.A., Postlethwait, J.H., Cresko, W.A., 2005. Evolution and development of facial bone morphology in threespine sticklebacks. Proc. Natl. Acad. Sci. U. S. A. 102, 5791–5796.

Kish, P.E., Bohnsack, B.L., Gallina, D., Kasprick, D.S., Kahana, A., 2011. The eye as an organizer of craniofacial development. Genesis 49, 222–230.

Kowalko, J.E., Rohner, N., Linden, T.A., Rompani, S.B., Warren, W.C., Borowsky, R., Tabin, C.J., Jeffery, W.R., Yoshizawa, M., 2013. Convergence in feeding posture occurs through different genetic loci in independently evolved cave populations of *Astyanax mexicanus*. Proc. Natl. Acad. Sci. U. S. A. 110, 16933–16938.

Mathers, P.H., Jamrich, M., 2000. Regulation of eye formation by the *rx* and *pax6* homeobox genes. Cell. Mol. Life Sci. 57, 186–194.

Mitchell, R.W., Russell, W.H., Elliott, W.R., 1977. Mexican Eyeless Characin Fishes, Genus *Astyanax*: Environment, Distribution, and Evolution. Texas Tech Press, Lubbock, TX.

Ornelas-García, C.P., Domínguez-Domínguez, O., Doadrio, I., 2008. Evolutionary history of the fish genus *Astyanax* Baird & Girard (1854) (Actinopterygii, Characidae) in Mesoamerica reveals multiple morphological homoplasies. BMC Evol. Biol. 8, 340.

Parsons, K.J., Taylor, A.T., Powder, K.E., Albertson, R.C., 2014. *Wnt* signalling underlies the evolution of new phenotypes and craniofacial variability in Lake Malawi cichlids. Nat. Commun. 5, 3629.

Protas, M.E., 2005. The genetic basis of morphological evolution in the Mexican cave tetra, *Astyanax mexicanus*. Doctoral dissertation, Harvard University.

Protas, M.E., Hersey, C., Kochanek, D., Zhou, Y., Wilkens, H., Jeffery, W.R., Zon, L.I., Borowsky, R., Tabin, C.J., 2006. Genetic analysis of cavefish reveals molecular convergence in the evolution of albinism. Nat. Genet. 38, 107–111.

Protas, M.E., Conrad, M., Gross, J.B., Tabin, C., Borowsky, R., 2007. Regressive evolution in the Mexican cave tetra, *Astyanax mexicanus*. Curr. Biol. 17, 452–454.

Protas, M.E., Tabansky, I., Conrad, M., Gross, J.B., Vidal, O., Tabin, C.J., Borowsky, R., 2008. Multi-trait evolution in a cave fish, *Astyanax mexicanus*. Evol. Dev. 10, 196–209.

Roberts, R.B., Hu, Y., Albertson, R.C., Kocher, T.D., 2011. Craniofacial divergence and ongoing adaptation via the hedgehog pathway. Proc. Natl. Acad. Sci. U. S. A. 108, 13194–13199.

Schartl, M., 2014. Beyond the zebrafish: diverse fish species for modeling human disease. Dis. Model Mech. 7, 181–192.

Şadoğlu, P., 1957. Mendelian inheritance in the hybrids between the Mexican blind cave fishes and their overground ancestor. Verh. Dtsch. Zool. Ges., Graz. 1957, 432–439.

Schemmel, C., 1967. Comparative studies of the cutaneous sense organs in epigean and hypogean forms of *Astyanax* with regard to the evolution of cavernicoles. Z. Morphol. Oekol. Tiere 61, 255–316.

Schemmel, C., 1980. Studies on the genetics of feeding behaviour in the cave fish *Astyanax mexicanus f. Anoptichthys*. Z. Tierpsychol. 53, 9–22.

Stock, D.W., Jackman, W.R., Trapani, J., 2006. Developmental genetic mechanisms of evolutionary tooth loss in cypriniform fishes. Development 133, 3127–3137.

Valdez-Moréno, M.E., Contreras-Balderas, S., 2003. Skull osteology of the characid fish *Astyanax mexicanus* (Teleostei: Characidae). Proc. Biol. Soc. Wash. 116, 341–355.

Wilkens, H., 1988. Evolutionary biology. In: Hecht, M.K., Wallace, B. (Eds.), Evolution and Genetics of Epigean and Cave *Astyanax fasciatus* (Characidae, Pisces): Support for the Neutral Mutation Theory. Plenum Publishing, New York, pp. 271–367.

Yamamoto, Y., Jeffery, W.R., 2000. Central role for the lens in cave fish eye degeneration. Science 289, 631–633.

Yamamoto, Y., Espinasa, L., Stock, D.W., Jeffery, W.R., 2003. Development and evolution of craniofacial patterning is mediated by eye-dependent and -independent processes in the cavefish *Astyanax*. Evol. Dev. 5, 435–446.

Chapter 11

Evolution and Development of the Cavefish Oral Jaws: Adaptations for Feeding

A.D.S. Atukorala and Tamara A. Franz-Odendaal
Department of Biology, Mount Saint Vincent University, Halifax, Nova Scotia, Canada

INTRODUCTION

Most cavefish live their entire lives in dark caves, completely dependent upon the limited food resources that can be found in the cave pools (Hüppop, 1987; Schemmel, 1980; Soares and Niemiller, 2013). The cavefish, *Astyanax mexicanus*, is an omnivorous feeder that eats flush-out debris and leftovers from other cave-dwelling animals, such as cave-dwelling bats and some small organisms (Culver and Pipan, 2009). The sighted surface morph, which is also omnivorous, feeds on a variety of insects and occasionally small fish. Although both are omnivorous, the blind cavefish lacks the major sensory organ required to find food, namely eyes. As such, these cavefish have, over the last 1 million years, evolved mechanisms to search, identify, and consume food in these resource-limiting environments. These mechanisms include the development of constructive traits, such as large jaws to accommodate more teeth and an increased number of tastebuds around the mouth and ventral surface of the head (Jeffery, 2005, 2009). Coincident with the evolution of these traits are changes in feeding behavior. Adult cavefish tend to scurry around bottom surfaces looking for food, and as such feed with their body at a 45° angle with respect to the food surface, compared to surface fish that feed in a more upright position (average angle 74°) (Kowalko et al., 2013). Cavefish eggs have more yolk compared to the surface morphs, enabling them to delay active feeding in these sparse food habitats (Hinaux et al., 2011). These adaptations have enabled cavefish to be the top predators in their subterranean ecosystems. In this chapter, we discuss these morphological adaptations with respect to feeding, with particular focus on the jaws and teeth.

Biology and Evolution of the Mexican Cavefish. http://dx.doi.org/10.1016/B978-0-12-802148-4.00011-6
© 2016 Elsevier Inc. All rights reserved.

Adult Dentition and Jaw bones

The dentition of the adult tetra are multicuspid in morphology (Valdez-Moreno and Contreras-Balderas, 2003; Figure 11.1). In adult cavefish, teeth can be found in both the lower and upper oral jaws. There are more teeth in the oral jaws of cavefish than in the sighted surface morph. All teeth are attached to the bone with one root and teeth have between one and eight cusps (Atukorala et al., 2013).

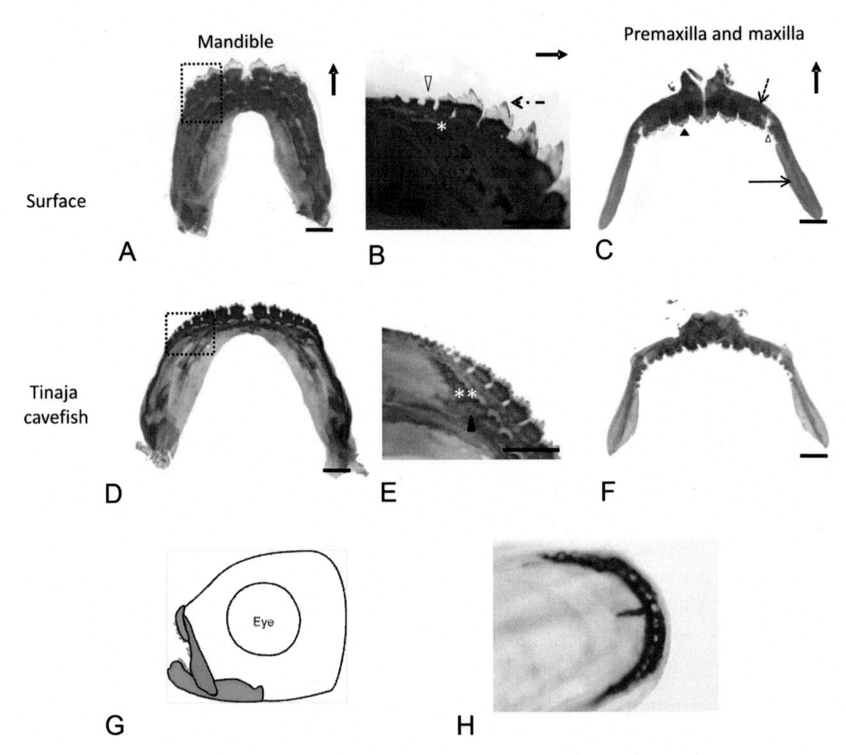

FIGURE 11.1 The adult dentition of *Astyanax mexicanus*. (A)–(F) Alizarin red-stained adult surface fish (A–C), Tinaja cavefish (D–F). (A), (D), show the mandibles, dorsal view. The dotted boxes in (A) and (D) are enlarged in (B) and (E), respectively. In (A), (D), teeth are arranged in one row in the rostral rim of the mandible. Rostral teeth are multicuspid and large. The black-dashed arrow in (B) indicates the cusp tip of a multicuspid tooth, which is yellow due to iron deposition. The white arrowheads in (B) and black arrow heads in (H) indicate the caudal teeth, which are smaller compared with the rostral teeth. The white asterisk in (B) indicates the tooth-free zone that can be found between the rostral and caudal teeth of the surface fish, but which cannot be seen in cavefish (double white asterisk in (E)). The black arrowhead in (E) indicates the successional tooth row. (C) and (F) show the right and left premaxillae and maxillae, rostral view. In (C), premaxilla is indicated by a black dotted arrow and the premaxillary tooth is indicated by a black arrowhead. The maxilla is indicated by a black arrow, and maxillary teeth are indicated by a black arrowhead in (C). The black arrow in the upper right corners of (A) and (B) indicate the rostral direction. The black arrow in (C) indicates dorsal direction. All scale bars are 100 μm. (G) shows a schematic of the adult dentition. Mandible, green; premaxilla, purple; maxilla, orange. (H) shows the tooth sockets in a 7.7 mm SL sighted tetra stained with alkaline phosphatase.

Tooth cusps are conical and pointed. They are positioned mediolaterally, and the enameloid cap shows iron deposition (Figure 11.1(B)). Interestingly, most F1 hybrids made by crossing Tanija cavefish with surface fish have intermediate numbers of teeth and tooth cusps. The lower jaw (i.e., the mandible) consists of the left and right dentary bone, and the upper jaw consists of the premaxilla and maxillary bone (Figure 11.1(G); Valdez-Moreno and Contreras-Balderas, 2003).

The mandible is robust; each dentary bone is attached in the midline by the mandibular symphysis, a hinge joint that enables the mandibular halves to spread out when swallowing and also to function as a strong unit (William and Conrad, 1936). Each dentary articulates dorsoposteriorly with the anguloarticular and the retroarticular and posteromedially with coromeckelian bone (Valdez-Moreno and Contreras-Balderas, 2003). The dentary itself bears a single row of teeth that extends from the mandibular symphysis to the caudal region of the mandible. In the rostral region, there are four large, tightly packed, multicuspid teeth pointing rostrodorsally, whereas in the caudal region, there are several short, small teeth that point medioventrally toward the oral cavity (Figure 11.1(A) and (D)). These caudal teeth are haphazardly arranged. The multicuspid rostral teeth have four to eight cusps, and teeth are firmly attached to the bone by one root in a well developed tooth socket (Atukorala et al., 2013). The caudal teeth are also attached to the bone, but the most caudally positioned tooth sockets are not as deep as those of the rostral teeth. Interestingly, some caudal teeth are large and unicuspid, while others are small and multicuspid. Compared with the cavefish, the sighted surface fish have in general fewer teeth in total in the lower jaw (8 ± 1 teeth vs. 12 ± 2 teeth in Tinaja cavefish; Atukorala et al., 2013). Their multicuspid teeth also have fewer cusps (five cusps vs. seven cusps in cavefish), and these teeth are also separated from the caudal teeth by a small tooth-free zone (Atukorala et al., 2013). Interestingly, most F1 hybrids (Tinaja cavefish x surface fish) also lack this tooth-free zone separating rostral and caudal teeth. The mandible in the cavefish is larger than in the surface morph. In particular, it has a larger rostral width (more square-shaped), and a larger anterior-posterior dimension, whereas caudal width appears to be constrained. In surface fish, the mandible is long and narrower rostrally. Thus there are significant differences in the lower jaw between the cavefish and its sighted ancestor.

The premaxillae are triangular bones (left and right) that are attached dorsally with the ethmoid and nasal bones, and laterally with the maxillary bone (Figure 11.1(B), 11.1(C), 11.1(E), and 11.1(F)). Each premaxillary bone has two rows of teeth that run rostrocaudally. Typically, the anterior row has four to five teeth, while the posterior row has five to seven teeth (Figure 11.1(B), 11.1(C), 11.1(E), and 11.1(F)). The teeth are large, multicuspid, and short. Tooth sockets are similar to those in the mandible and are firmly attached to the bone by one root. Tinaja cavefish have on average 9 ± 0.9 teeth, while F1 hybrids have 8 ± 0.8 teeth, and surface fish have an average of 8 ± 0.9 teeth (Atukorala et al., 2013).

The maxillary bone is long and thin, and articulates dorsally with the premaxilla as in other derived teleosts. It is also typical that the maxilla of derived teleosts bears teeth unlike in ancestral fishes. In ancestral fishes, the toothless maxilla is located behind the premaxilla (Nelson, 2006). The posterior ramus of the maxilla overlaps the dentary (Figure 11.1). The maxillary teeth have four and six cusps per tooth, and these teeth are always positioned 180° to the occluding jaws. These teeth always develop intraosseously and attach to the bone by one strong root (Atukorala and Franz-Odendaal, 2014). Tooth numbers in the maxillary bone vary across the different *A. mexicanus* morphs. Tinaja cavefish have on average of 3.54 ± 0.5 teeth on each maxillary bone. F1 hybrids have 2 ± 0.6 teeth, while surface fish have an average of 1 ± 0.9 teeth.

Similar to the mandible, the width of the premaxillae and the length and width of the maxillary bone is different when comparing cavefish, sighted fish, and F1 hybrids (Atukorala et al., 2013). These morphological differences in jaw size are likely correlated with the differences in tooth and cusp number. This is similar to cichlids in which adaptive variation in jaw size (and changes in dentition) in response to differences in food types has been attributed their success (Albertson et al., 2003; Albertson and Kocher, 2006). The underlying developmental mechanisms for these differences will be discussed later.

Development of the Jaws Supporting the Dentition

Vertebrate oral jaw bones are derived from the first pharyngeal arch (Cobourne and Sharpe, 2003; Fraser et al., 2009). In teleosts, the first evidence of the jaws is the presence of Meckel's cartilage (Cubbage and Mabee, 1996). Meckel's cartilage is present during embryonic development at 55 h post-fertilization (hpf) in *A. mexicanus*. Meckel's cartilage articulates posteriorly with the more dorsally located palatoquadrate cartilage (Cubbage and Mabee, 1996). This early jaw development is similar in both morphs of *A. mexicanus*.

No detailed study of early jaw formation in the blind cavefish has been conducted; however, the surface morph has been described (Milligan et al., 2012). An overview image of the adult *A. mexicanus* mandible is shown in Figure 11.2. In the surface morph, the lower jaw begins to ossify along the buccal and lingual curves of Meckel's cartilage by 4.5 mm standard length (SL), while the majority of the mandible remains unossified. Ossification progresses toward the coronoid and retroarticular processes. At this stage, there does not appear to be a gap at the mandibular symphysis between the two halves of Meckel's cartilage. Ossification progresses on the lingual surface of Meckel's cartilage at the coronoid and retroarticular processes. Osteoblast activity (as evidenced by alkaline phosphatase staining) is observed in the perichondrium (outer layer) of Meckel's cartilage and is particularly intense in the labial area and at the articulation sites of the palatoquadrate-mandible, as well as within these elements. Teeth first appear at 4 days post-fertilization (dpf) (development of dentition is discussed later). Between 6.0 and 6.5 mm SL, osteoblast activity around the

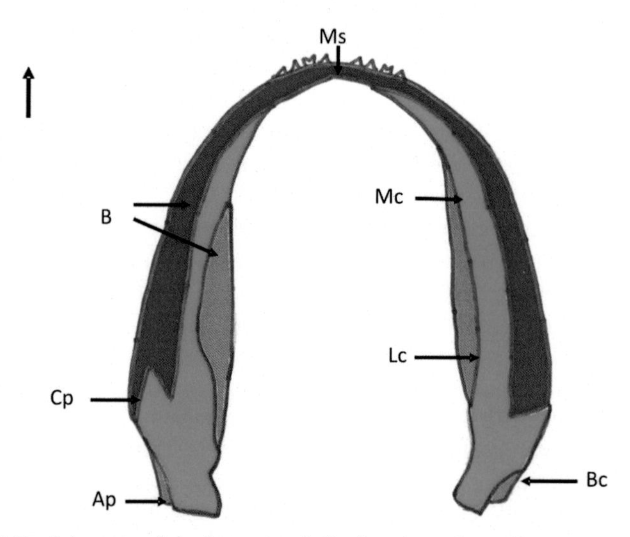

FIGURE 11.2 Schematic of the lower jaw indicating the early ossification. Meckel's cartilage, blue; ossified areas of the mandible, pink and red. Mc, Meckel's cartilage; B, new bone; Ap, Articular process; Cp, Coronoid process; Ms, Mandibular symphasis; Lc, Lingual curve; Bc, Buccal curve.

tooth sockets and at the bone surface near these structures is observed with alkaline phosphatase staining (Milligan et al., 2012; Figure 11.1(H)).

The larval premaxilla is derived from the ethmoid cartilage, which is a part of the neurocranium (Cubbage and Mabee, 1996). Similar to the larval mandible, the premaxillae bear teeth at 4 dpf (Atukorala and Franz-Odendaal, 2014). Ossification of the premaxillae starts at 4.5 mm SL, and the length and width of the premaxillae increases with age. The increase in length of the tooth-bearing rostral edge corresponds to the growth of the occluding mandibular bone.

A pair of maxillary bones forms part of the upper jaw. In the Mexican tetra, as in other teleosts, the palatoquadrate cartilage acts as the larval maxilla. The palatoquadrate cartilage is roughly triangular with slightly concave dorsal edges and convex ventral edges. Although the larval Mexican tetra does not have teeth in the maxilla, juveniles and adult fish do; this may relate to different feeding ecologies over their lifespan. The ossification of the maxillary bone starts very early (at 10 dpf; around 4.7 mm SL) as in the mandible and the premaxilla; however, teeth only appear much later in juveniles at 50 dpf (Atukorala and Franz-Odendaal, 2014; Trapani et al., 2005). Ossification is slow and the maxilla is well ossified at 40 dpf (9.0 mm SL).

Development of the Dentitions

As described earlier, adult cavefish and surface morphs of *A. mexicanus* have teeth in all their oral jaws. The teeth of this species start developing as simple unicuspid teeth that, after several replacement cycles, are replaced by strong multicuspid teeth (Atukorala and Franz-Odendaal, 2014; Trapani et al., 2005). The row of successional teeth is always positioned labially to its functional teeth. Some caudal teeth can be found toward the labial side of the mandible, while in some groups, these teeth are positioned laterally to the rostral teeth (Atukorala et al., 2013).

The inner core of the adult teeth is made of dentine and pulp, while the outer surface is enameloid, a unique well mineralized tissue that has the same functions as mammalian tooth enamel (Atukorala et al., 2013). Tooth development is generally well conserved through vertebrate evolution, and the majority of the genes regulating tooth development (at least in mammals) has been identified in the past two decades (a comprehensive database can be found at http://bite-it.helsinki.fi).

Histologically, the first sign of tooth development is the thickening of the epithelium or the formation of the epithelial tooth placode. This is followed by the bud stage, during which the epithelium invaginates into the underlying epithelium, and then the cap and bell stage, during which differentiation of the final tooth structure occurs; however, in less derived vertebrates, the teeth do not have a root structure and are directly attached to the underlying bone. In the Mexican tetra, rostral oral teeth have well developed sockets to hold the teeth as described earlier.

Early tooth development events have not been described in the cavefish; however, in the surface form, the earliest signs of jaw teeth can be found at the age of 4 dpf (Figure 11.3). These first-generation teeth form on Meckel's cartilage and the ethmoid cartilage of the oral jaws and they erupt into the oral cavity at the age of 4 dpf (Atukorala and Franz-Odendaal, 2014). At this age, the mouth opens and larval fish start eating live food, such as brine shrimp in the laboratory. Tooth development may be delayed in cavefish, since their eggs are larger and the larval fish hatch with a larger yolk sac compared with surface fish (Hinaux et al., 2011). This delay may enable larval cavefish to survive longer in their food-limited environments.

As growth proceeds, there are several changes to the dentition (Figure 11.3). These include changes in tooth number, changes in cusp number per tooth,

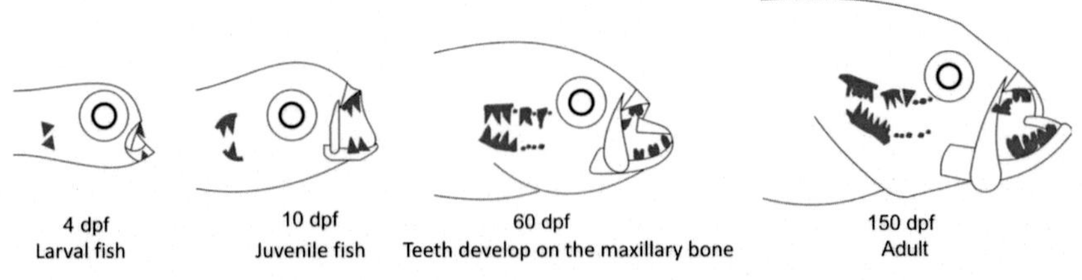

FIGURE 11.3 (A) Spatial and temporal pattern of tooth eruption in *Astyanax mexicanus*. Schematic showing the spatial distribution of teeth in oral and pharyngeal jaw bones over ontogeny. At 4 dpf, teeth are present in the mandible, premaxilla, and upper and lower pharyngeal bones. At 30 dpf, teeth are present in the gill rakers and on the third suspensory pharyngeal bone. At 60 dpf, teeth are present in the maxilla and second suspensory pharyngeal bone.

changes in tooth type and size, and changes to the location of the tooth sockets (Trapani et al., 2005; Atukorala et al., 2013; Atukorala and Franz-Odendaal, 2014). The first-generation unicuspid dentition is progressively replaced by bi, and then tricuspid dentitions (Atukorala and Franz-Odendaal, 2014; Trapani et al., 2005). With the growth of the jaw bones, development of these teeth also changes from an epithelial location to the intramedullary region of the jaw bone. In adult cavefish, the multicuspid rostral teeth develop inside the bone, a process known as intraosseus tooth development (Trapani et al., 2005). The maxillary dentition, which appears much later, in juvenile life, has the same developmental and morphological sequence of tooth development as the mandibular teeth (Atukorala and Franz-Odendaal, 2014). That is, over the lifespan of the fish, the number of unicuspid teeth decreases, while the number of teeth with five or more cusps increases (Trapani et al., 2005). Furthermore, the first-generation rostral and caudal teeth are uniform in height, unlike in the adult dentition, when these teeth differ greatly in size (height and width). It is after the first tooth replacement cycle that the height of the rostral teeth drastically increases compared with the caudal teeth (Atukorala and Franz-Odendaal, 2014).

The transition of unicuspid first-generation teeth to subsequent multicuspid dentition in oral jaws in tetra may be associated with the differing feeding demands at different life history stages (e.g., soft vs. hard prey). A study conducted in two cichlids species from Lake Malawi demonstrated that the timing of turnover from first-generation to replacement teeth differs among species and is related to their feeding ecology (Muschick et al., 2011). The hardness of the food may influence the forces applied to the jaw when eating, resulting in changes in jaw shape and hence also tooth morphology over ontogeny.

Development of Tastebuds on the Oral Jaws

The larger jaws of the cavefish result in larger mouths to maximize the opportunities to capture food. With their larger mouths, there is more space on the lips for expansion of the gustatory system, namely tastebuds. Schemmel (1967) showed that the numbers of tastebuds in cavefish were five- to sevenfold greater compared to the sighted morph. Furthermore, in the sighted morph, tastebuds are mainly found in the labial epithelium, whereas in the blind morph, they are also found in the skin of the maxilla, lower jaw, and ventral aspect of the head (Boudriot and Reutter, 2001).

Tastebuds in teleosts are typically pear- or onion-shaped intraepithelial sensory structures that consist of multiple cell types (light and dark receptor cells, basal cells, Merkel cells; Figure 11.4(A)). Surrounding the tastebuds but not part of the organ *sensu strictu* are marginal cells; these cells are situated between the sensory cells of the tastebud and the squamous epithelium of the skin. The light and dark sensory receptor cells make up the bulk of the tastebud. The apical ends of these receptor cells protrude into the oral cavity or the external environment, and provide feedback to the chemoreception area

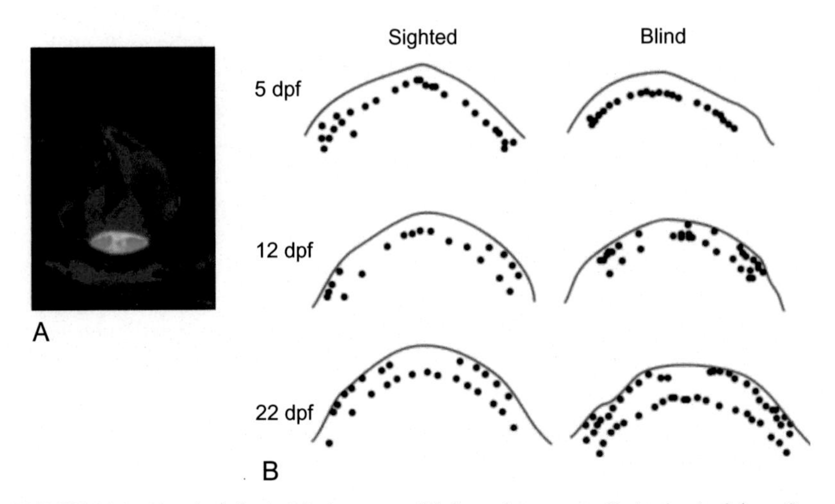

FIGURE 11.4 Tastebuds in the Mexican tetra. (A) Several receptor cells (anti-calretinin, red) are present within each tastebud and are colocalized with a single serotonergic basal cell (antiserotonin, green). (B) Schematic diagrams showing the distribution of tastebuds within lower jaws of the surface tetra and cavefish with increasing age. Each diagram is based upon a single representative specimen. *Modified from Varatharasan et al. (2009).*

of the telencephalon. Although Schemmel (1980) showed that the morphology of the tastebuds do not differ between the two morphs, more recent evidence has shown using immunohistochemistry that the tastebuds in cavefish larvae contain more receptor cells and are innervated by more axon profiles (Varatharasan et al., 2009).

Tastebuds begin their development in the same manner as teeth do in the form of epithelial primordia. Soon after the induced epithelium has evaginated above the level of the rest of the epithelial tissue, the tastebud will enter the differentiation stage of development (Northcutt, 2004). The fully developed tastebud is buried within the epithelium, and only small, chemosensory microvilli are exposed to the superficial environment in the oropharyngeal cavity (Boudriot and Reutter, 2001). At the base of the tastebud is the nerve fiber plexus, where the tastebud is innervated (Torrey, 1940). At 5 dpf in both cavefish and surface morphs of *A. mexicanus*, fish have one row of tastebuds on the lips (Varatharasan et al., 2009; Figure 11.4(B)). A second (inner) row starts developing at 12 dpf from the central midpoint of the jaw, in the area of the mandibular symphysis. Varatharasan et al. (2009) reported that this second row of tastebuds appears quicker in the cavefish lower jaw than at similarly aged surface fish (Figure 11.4), possibly suggesting that there is a more rapid induction of tastebuds in the cavefish. By 22 dpf, there are significantly more tastebuds in and around the cavefish mouth (Varatharasan et al., 2009; Figure 11.4(B)). Thus there is an expansion of the gustatory system in both numbers of tastebuds and receptor cells, and in their innervation in cavefish.

RELATIONSHIP BETWEEN THE CONSTRUCTIVE TRAITS OF TEETH AND TASTEBUD EXPANSION AND EYE LOSS

Gene Networks Underlying These Traits

Although much research has been conducted on the loss of eyes and the effect of lens degeneration on associated or surrounding eye structures (Yamamoto and Jeffery, 2000; Jeffery, 2005; Yamamoto et al., 2003; Protas et al., 2007; Dufton et al., 2012), few studies have focused on multiple orofacial structures, such as teeth and tastebuds. Indeed, these sensory systems are likely to be linked via common gene networks (Franz-Odendaal and Hall, 2006) and some (e.g., taste and eyes) have shown to be linked via sonic hedgehog (*shh*) signaling pathway (Yamamoto et al., 2009). In this section, we discuss how these regressive and constructive changes may be interconnected via genetic pathways.

Teeth

Vertebrate teeth develop from the reciprocal interaction between the ectoderm and the neural crest-derived ectomesenchyme. Numerous key developmental genes are essential to regulate tooth development (e.g., *shh*, *wnt*, *bmp4*, and *dlx2)* (Vainio et al., 1993; Weiss et al., 1995; Sarkar and Sharpe, 1999; Cobourne et al., 2001; Cobourne and Sharpe, 2003). The presence, absence, or overlapping expression pattern of these genes along the tooth-forming odontogenic band can produce a wide range of subtle differences in tooth number and shapes (Graber, 1978; Cobourne and Sharpe, 2003; Fleischmannova et al., 2008). Several studies have shown that across different vertebrate groups, the most posterior tooth of each tooth type is the most vulnerable to variation (Cobourne and Sharpe, 2003). For example, in humans, the incisor-canine region and the most posterior molar exhibit the highest percentage of developmental defects (Graber, 1978). A similar mechanism might explain the natural variation in tooth number and tooth shape of the caudal teeth in *Astyanax*, while the rostral tooth number and cuspal morphology is relatively constant. The transitional zone from rostral to caudal teeth is more variable, and the teeth in this zone do indeed vary in number, size, and cusp morphology (Cobourne and Sharpe, 2003; Trapani et al., 2005).

The *shh* gene shows highly localized expression in early dental placodes of the first branchial arch (Cobourne et al., 2001; Cobourne and Sharpe, 2003). Overexpression of *shh* in mutant mouse models was found to increase tooth number and induce tooth formation in the diastema region (Graber, 1978; Sarkar and Sharpe, 1999; Cobourne et al., 2001). In addition, there are several signaling families that interact with the hedgehog (*Hh*) pathway to induce normal tooth development, such as the ectodysplasin pathway. Other molecular interactions serve to increase numbers of teeth (Sarkar and Sharpe, 1999; Tucker et al., 2004; Harjunmaa et al., 2012). For example, bone morphogenic protein (*Bmp*) and *Wnt* signaling pathways are closely associated with the *Hh* pathway in supernumerary tooth formation (Vainio et al., 1993; Sarkar and Sharpe, 1999).

The expression pattern of *Bmp2* and *Bmp4* in tooth development of three teleost fish species (i.e., zebrafish, tetra, and medaka) has been investigated and shown to be conserved (Wise and Stock, 2006); however, the same authors, in a later study, concluded that *Bmp2* and *Bmp4* are not essential for pharyngeal tooth formation in zebrafish, and suggest that there may be functional redundancy of these genes with other genes such as *Bmp7* (Wise and Stock, 2010). *Bmp* signaling is modulated by several proteins at different levels. In the extracellular compartment, *Bmp* signaling is modulated by secreted *Bmp* antagonists, such as Noggin, which binds preferentially to *Bmp2*, *Bmp4*, and *Bmp7* to prevent the *Bmp* pathway. *Bmp4* can also act as an inhibitor of *shh* and *fgf4* in the development of other epithelial appendages (Jung et al., 1998), and the localized inhibitory action of *Bmp4* has been shown to control tooth number in cichlid dentitions (Streelman et al., 2003).

Furthermore, the application of Noggin to mandibular explants of mouse embryos is sufficient to transform the normally unicuspid incisors into multicuspid teeth (Tucker et al., 1998). Similarly, in the pharyngeal teeth of zebrafish and Mexican tetra, overexpression of Noggin can change the tooth morphology from unicuspid to bicuspid (Jackman et al., 2013). In mice and voles, molar cusp shape is determined by interactions between *shh* and *fgf4*, *lef1* and *p21* genes (Harjunmaa et al., 2012). Interestingly, it was determined that overexpression of *shh*, which is seen in cavefish midline at the early development (Yamamoto et al., 2004), could inhibit enamel knot formation, thereby reducing the cusp number in the tooth crown (Harjunmaa et al., 2012); however, in cavefish, we found a higher number of cusps compared to the surface morphs (Atukorala et al., 2013), suggesting that several possible genes are responsible for the highly developed dentition in cavefish.

The distinct expression pattern of the *Bmp* antagonist, Noggin, suggests that Noggin plays an inhibitory role by preventing early tooth development in the maxillary bone in *A. mexicanus* (Atukorala and Franz-Odendaal, 2014). Interestingly, Noggin-mutant mice ($Nog^{-/-}$) have a slight delay in early mandibular tooth and maxillary molar development, which recovers with advancing age, but the maxillary incisors arrest in the early stages of development, and the residual tooth bud regresses thereafter (Hu et al., 2012). These mutants have a single maxillary incisor tooth that is malformed. This further supports our hypothesis that Noggin might act as a regulator of maxillary tooth development in *A. mexicanus* (Atukorala and Franz-Odendaal, 2014). The expression of Noggin in the external epithelium of the developing multicuspid tooth germs of maxillary and mandibular teeth at 50 dpf is suggestive of a role for Noggin in determining tooth morphology in teleost fish, as it is already documented in murine tooth development (Tucker et al., 1998). Moreover, Noggin is known to play a role in cytodifferentiation (ameloblast and odontoblast cell formation) in mice (Ohazama et al., 2005). In 50 dpf tooth germs, we detect the expression of Noggin in the inner enamel epithelium, suggesting a conserved cytodifferentiation role regulated by Noggin in Mexican tetra tooth development (Atukorala and Franz-Odendaal, 2014). Future comparative and/or gene knockdown studies in surface fish and cavefish will reveale the specific role of Noggin in multicuspid tooth development.

TOOTH-TASTEBUD LINKAGES

Fraser et al. (2010) proposed that tooth and tastebud development are coupled, since they are colocalized, and since these two structures share a common gene regulatory network. Teeth and tastebuds do indeed both develop from a thickened epithelial placode and from a common epithelium (Fraser et al., 2010). Teeth generally form before tastebuds during development in cichlids (MacDougall, 2009) and this colocalization is, however, decoupled in some vertebrates, such as mammals (Sarkar and Sharpe, 1999). Decoupled colocalization is also observed in some teleost fish. For example, while zebrafish have lost oral teeth, they have retained the capacity to develop oral tastebuds and have also gained further extra-oral sites (Fraser et al., 2010; Hansen et al., 2002). Interestingly, teeth and tastebuds are colocalized in the zebrafish posterior pharynx. Similarly, in the largemouth bass, *Micropterus salmoides*, teeth and tastebuds are colocalized on the pharyngeal jaws (Linser et al., 1998). Further research is required to fully understand the developmental and genetic relationship between teeth and tastebud development in teleost jaws.

Ectodermal and endodermal epithelia give rise to tastebuds; however, it is unclear which signals are necessary from the underlying mesenchyme for their development (Barlow and Northcutt, 1997; Fraser et al., 2010). Interestingly, innervation is not required for the initiation of teeth nor for tastebuds. Indeed, the first-generation teeth lack innervation in teleosts (Sire et al., 2002), yet innervation is necessary for subsequent morphogenesis and continued tooth replacement. In the Mexican tetra, both teeth and tastebuds get increasingly more complex as development proceeds (Figure 11.4). More evidence is needed to explore the decoupled colocalization of teeth and tastebud development in Characiformes. The Mexican tetra has teeth in the oral jaws, and these teeth become structurally more complex over ontogeny; this is unlike zebrafish, who only have pharyngeal teeth. Hence how patterning of the dentition and tastebuds potentially interact during development can best be studied using the Mexican tetra model. Furthermore, since the Mexican tetra has a complex dentition, it is also an excellent model organism in which to study the formation of complex dentitions and tooth replacement cycles.

TEETH-EYE LINKAGES

Two studies have investigated the relationship between tooth and eye development by experimentally manipulating the larval Mexican tetra. In one study, the embryonic lens was transplanted between the surface and Pachón cavefish, and maxillary tooth number was examined; however, no differences were noted

(Yamamoto et al., 2003). This latter study concluded that maxillary teeth are not influenced by a transplanted cavefish lens. In another study with lens ablation experiments of surface fish larvae at 1-4 dpf, Dufton et al. (2012) showed that mandibular tooth number is greater on the surgery side compared to the nonsurgery side, despite eye loss. This increase in tooth number is in the caudal region only; rostral teeth are unaffected. The first unicuspid rostral teeth start to develop at the time of surgery, while caudal teeth only start developing at 12 dpf. These results show that either rostral teeth patterning is tightly regulated or that early initiation of tooth development in the rostral jaw region is unaffected by the lens ablation compared to caudal teeth in the mandible. In addition, these results show that the teeth in the other jaw bones are not affected by the lens, thus hinting that the smaller, mostly unicuspid teeth of the oral jaws might be linked to eye development via common gene regulatory networks.

Although there is currently no clear explanation for the increase in caudal tooth number, the spatial and temporal expression of shared genes in eye and jaw development may play a role. Both eye and lower jaw development are highly coordinated events in craniofacial development. Apart from the common molecular markers (e.g., *dlx, pitx2, shh, wnt*), lens removal and subsequent eye regression may also cause changes in mechanical forces around the eye and perhaps the oral region (Dufton et al., 2012). These changes in forces may consequently affect the patterning of the teeth. Interestingly, in the small-eyed mouse mutant *Pax6*^{Sey}/*Pax6*^{Sey}, in which the lens fails to form, 80% of the mutants show an increase in anterior upper tooth number by one to two teeth (Kaufman et al., 1995). Indeed, changes in *Pax6* has been shown to have pleiotropic effects (Strickler et al., 2001). The genetics underlying these adaptive morphological features that enhance feeding are discussed in other chapters in this book.

MODULARITY AND ADAPTIVE EVOLUTION

In cavefish, while regressive changes such as eye and pigment loss evolved, so did constructive changes, such as enlarged jaws, increased tastebuds and increased tooth number. One could consider each of the traits as independent, yet interacting, modules (Franz-Odendaal and Hall, 2006; Wilkens, 2010). For example, as the gustatory module expanded, the eye module was reduced or *vice versa*. In support of this hypothesis, each of these changes in sensory structures also has a coincident change in the respective area of the brain that processes this information. The optic tectum and optic nerves are reduced, while the telenchephalon (processing chemoreception) is expanded by 40% relative to that of the sighted morph (Peters et al., 1993). Natural selection could have acted upon a particular module (e.g., jaw size) that enabled more teeth and tastebuds to develop; this module may be connected by shared gene networks with other modules (e.g., eye module) that therefore also changed. Alternatively, natural selection could also have acted upon genes directly. Modularity is discussed later in this chapter.

A change in feeding behavior or food types could also have been the key evolutionary adaptation in cavefish (Schemmel, 1980). Indeed, Kowalko et al. (2013) have shown that cavefish from different caves evolved independently and have converged on similar feeding postures (e.g., Pachón and Tinaja). This altered feeding posture enables more tactile contact with the substrate and likely evolved as a result of selection for improved feeding. Evidence for this comes from studies that have shown (i) that cavefish are better at finding food in the dark than their sighted ancestors (Hüppop, 1987); (ii) that sighted fish trapped in caves show signs of starvation (Mitchell et al., 1977); and (iii) that different feeding postures have been shown to be important for different feeding strategies in other fish species (e.g., Barel, 1983; Otten, 1983). Other authors have shown that jaw anatomy and diet of fish with divergent dentitions are correlated (Albertson et al., 2003; Parsons et al., 2011, 2012). The hardness of different types of food may influence the forces applied to the jaw when eating, which could have been the driving force behind the changes in jaw shape and hence also tooth morphology over ontogeny. Together, these highlight the complexity of the evolutionary response of *A. mexicanus* to the extreme cave environment.

The mandible houses the teeth and supports associated sensory structures (tastebuds and some neuromasts). Thus the mandible can be considered a module itself. With the regression of the eye, this entire mandibular module seems to have expanded within cavefish. The gene regulatory network underlying these modules is likely closely interlinked. Indeed, *shh* was found to play a central role in cavefish eye degeneration (Yamamoto and Jeffery, 2000; Retaux et al., 2008; Yamamoto et al., 2009), and *shh* is involved in tetra oral jaw and tooth development (Yamamoto et al., 2009). A study overexpressing *shh* in surface fish during early development determined that there was an increase in tastebud number and jaw size coincident with eye reduction (Yamamoto et al., 2009). The pleiotropic effect of this gene pathway might underlie the increase in tooth number in cavefish (Yamamoto et al., 2009). The increase in jaw size likely facilitated the expansion of tastebuds in the oral jaw region. Thus this increase in jaw size observed in cavefish (and after lens ablation in the sighted fish) is likely caused by a change in the spatial and temporal expression profile for *Bmps*. Since altering *Bmp4* expression in zebrafish induces variation in jaw shape, and cichlid jaw shape variation is associated with *Bmp4* expression (Albertson and Kocher, 2006), the larger jaw likely enables more teeth and more tastebuds to develop in and around the larger mouth. Thus it is possible that natural selection acted to favor large jaw sizes in *A. mexicanus* as it became isolated within caves.

CONCLUSION

Blind cavefish have a remarkable ability to search for, detect, and find food, despite lacking eyes. This chapter discusses some of the recent work that has been conducted, focusing on jaw, tooth, and tastebud development, and how these traits might be linked with one another and with eye development. Adult dentition consists of robust multicuspid teeth in the mandible and premaxillae and smaller teeth more caudally positioned. These smaller teeth vary in shape and size in cavefish, F1 hybrids, lens-ablated fish, and surface tetra, with cavefish having the highest number of teeth and largest jaws. In addition, the Mexican tetra exhibits a complex tooth replacement cycle during its lifespan. Slightly after jaw and tooth development commences, tastebuds appear around the oral jaws. Again, these are more numerous, have more receptor cells, and more innervation in the cavefish compared with the sighted surface fish. Together, these increases in jaw size, tooth number and complexity, and tastebuds have greatly contributed to their success in the resource-limiting environments in which they live.

The recent genome analysis of cavefish will enable us to further explore the molecular genetics behind the constructive and regressive traits of cavefish. Together with its surface counterpart, this model organism offers a number of advantages over the current animal models in understanding complex phenotypes from an evolutionary-developmental biology perspective. Moreover, future molecular and genomic analysis of tooth, tastebud, and jaw development of this powerful natural model organism will uncover unforeseen relationships between eye and tooth development in vertebrates.

ACKNOWLEDGMENTS

We thank the Natural Sciences and Engineering Research Council of Canada (NSERC) and Mount Saint Vincent University for funding this research. We also thank Christine Hammer for the schematic in Figure 1(G) and all the Franz-Odendaal Bone Development Lab members for their encouragement and support.

REFERENCES

Albertson, R.C., Kocher, T.D., 2006. Genetic and developmental basis of adaptive variation in the cichlid feeding apparatus. Heredity 97, 211–222.

Albertson, R.C., Streelman, J.T., Kocher, T.D., 2003. Genetic basis of adaptive shape differences in the cichlid head. J. Hered. 94, 291–301.

Atukorala, A.D.S., Franz-Odendaal, T.A., 2014. Spatial and temporal events in tooth development of *Astyanax mexicanus*. Mech. Dev. 134, 42–54.

Atukorala, A.D.S., Hammer, C., Dufton, M., Franz-Odendaal, T.A., 2013. Adaptive evolution of the lower jaw dentition in Mexican tetra (*Astyanax mexicanus*). Evodevo 4, 28.

Barel, C.D.N., 1983. Towards a constructional morphology of cichlid fishes (Teleostei, Perciformes). Neth. J. Zool. 33, 357–424.

Barlow, L.A., Northcutt, R.G., 1997. Taste buds develop autonomously from endoderm without induction by cephalic neural crest or paraxial mesoderm. Development 124, 949–957.

Boudriot, F., Reutter, K., 2001. Ultrastructure of the taste buds in the blind cave fish *Astyanax jordani* ("Anoptichthys") and the sighted river fish *Astyanax mexicanus* (Teleostei, Characidae). J. Comp. Neurol. 434, 428–444.

Cobourne, M.T., Sharpe, P.T., 2003. Tooth and jaw: molecular mechanisms of patterning in the first branchial arch. Arch. Oral Biol. 48, 1–14.

Cobourne, M.T., Hardcastle, Z., Sharpe, P.T., 2001. Sonic hedgehog regulates epithelial proliferation and cell survival in the developing tooth germ. J. Dent. Res. 80, 1974–1979.

Cubbage, C.C., Mabee, P.M., 1996. Development of the cranium and paired fins in the zebrafish (Danio rerio) (Ostariophysi, Cyprinidae). J. Morphol. 229, 121–160.

Culver, D.C., Pipan, T., 2009. The Biology of Caves and Other Subterranean Habitats. Oxford University Press, Oxford.

Dufton, M., Hall, B.K., Franz-Odendaal, T.A., 2012. Early lens ablation causes dramatic long-term effects on the shape of bones in the craniofacial skeleton of *Astyanax mexicanus*. PLoS ONE 7, e50308.

Fleischmannova, J., Matalova, E., Tucker, A.S., Sharpe, P.T., 2008. Mouse models of tooth abnormalities. Eur. J. Oral Sci. 116, 1–10.

Franz-Odendaal, T.A., Hall, B.K., 2006. Modularity and sense organs in the blind cavefish, *Astyanax mexicanus*. Evol. Dev. 8, 94–100.

Fraser, G.J., Hulsey, C.D., Bloomquist, R.F., Uyesugi, K., Manley, N.R., Streelman, J.T., 2009. An ancient gene network is co-opted for teeth on old and new jaws. PLoS Biol. 7, e31.

Fraser, G., Robert, C., Soukup, V., Marianne Bronner-Fraser, M.B., Streelman, J.T., 2010. The odontode explosion: the origin of tooth-like structures in vertebrates. Bioessays 32, 808–817.

Graber, L.W., 1978. Congenital absence of teeth: a review with emphasis on inheritance patterns. J. Am. Dent. Assoc. 96, 266–275.

Hansen, A., Reutter, K., Zeiske, E., 2002. Taste bud development in the zebrafish, Danio rerio. Dev. Dyn. 223, 483–496.

Harjunmaa, E., Kallonen, A., Voutilainen, M., Hamalainen, K., Mikkola, M.L., Jernvall, J., 2012. On the difficulty of increasing dental complexity. Nature 483, 324–327.

Hinaux, H., Pottin, K., Chalhoub, H., Père, S., Elipot, Legendre L., Retaux, S., 2011. A developmental staging table for *Astyanax mexicanus* surface fish and Pachón cavefish. Zebrafish 8, 155–165.

Hu, X., Wang, Y., He, F., Li, L., Zheng, Y., Zhang, Y., Yp, C., 2012. Noggin is required for early development of murine upper incisors. J. Dent. Res. 91, 394–400.

Hüppop, K., 1987. Food-finding ability in cave fish (*Astyanax fasciatus*). Int. J. Speleol. 16, 59–66.

Jackman, W.R., Davies, S.H., Lyons, D.B., Stauder, C.K., Denton-Schneider, B.R., Jowdry, A., Aigler, S.R., Vogel, S.A., Stock, D.W., 2013. Manipulation of Fgf and Bmp signaling in teleost fishes suggests potential pathways for the evolutionary origin of multicuspid teeth. Evol. Dev. 15, 107–118.

Jeffery, W.R., 2001. Cavefish as a model system in evolutionary developmental biology. Dev. Biol. 231, 1–12.

Jeffery, W.R., 2005. Adaptive evolution of eye degeneration in the Mexican blind cavefish. J. Hered. 96, 185–196.

Jeffery, W.R., 2008. Emerging model systems in evo-devo: cavefish and microevolution of development. Evol. Dev. 10, 265–272.

Jeffery, W.R., 2009. Regressive evolution in *Astyanax* cavefish. Annu. Rev. Genet. 43, 25–47.

Jung, H.S., Francis-West, P.H., Widelitz, R.B., Jiang, T.X., Ting-Berreth, S., Tickle, C., Wolpert, L.A., Chuong, C.M., 1998. Local inhibitory action of BMPs and their relationships with activators in feather formation: implications for periodic patterning. Dev. Biol. 196, 11–23.

Kaufman, M.H., Chang, H.H., Shaw, J.P., 1995. Craniofacial abnormalities in homozygous Small eye (Sey/Sey) embryos and newborn mice. J. Anat. 186 (Pt 3), 607–617.

Kowalko, J., Rohner, N., Linden, T.A., Rompani, S.B., Warren, W.C., Borowsky, R., Tabin, C.J., Jeffery, W.R., Yoshizawa, M., 2013. Convergence in feeding posture occurs through different genetic loci in independently evolved cave populations of *Astyanax mexicanus*. Proc. Natl. Acad. Sci. U. S. A. 110, 16933–16938.

Linser, P.J., Carr, W., Cate, H.S., Derby, C.D., Netherton, J.C., 1998. Functional significance of the co-localization of taste buds and teeth in the pharyngeal jaws of the largemouth bass, *Micropterus salmoides*. Biol. Bull. 195, 273–281.

MacDougall, A.I., 2009. Coupling the developmental programs of teeth and tastebuds in Malawi Cichlids. In: School of Biology. Georgia Institute of Technology, Georgia, p. 23.

Milligan, B., Harris, N., Franz-Odendaal, T.A., 2012. Understanding morphology: a comparative study on the lower jaw in two teleost species. J. Appl. Ichthyol. 28, 346–352.

Mitchell, R., Russell, W.H., Elliott, W.R., 1977. Mexican Eyeless Characin Fishes, Genus Astyanax: Environment, Distribution, and Evolution. Texas Tech University Press, Lubbock, TX.

Muschick, M., Barluenga, M., Salzburger, W., Meyer, A., 2011. Adaptive phenotypic plasticity in the Midas cichlid fish pharyngeal jaw and its relevance in adaptive radiation. BMC Evol. Biol. 11, 116.

Nelson, J.S., 2006. Fishes of the World. Wiley, New York.

Niemiller, M.L., Near, T.J., Fitzpatrick, B.M., 2012. Delimiting species using multilocus data: diagnosing cryptic diversity in the southern cavefish, *Typhlichthys subterraneus* (Teleostei: Amblyopsidae). Evol. Dev. 66, 846–866.

Northcutt, R., 2004. Taste buds: development and evolution. Brain Behav. Evol. 64, 198–206.

Ohazama, A., Tucker, A., Sharpe, P.T., 2005. Organized tooth-specific cellular differentiation stimulated by BMP4. J. Dent. Res. 84, 603–606.

Otten, E., 1983. The jaw mechanism during growth of a generalized Haplochromis species: *H. elegans* Trewaves 1933 (Pisces, Cichlidae). Neth. J. Zool. 33, 55–98.

Parsons, K.J., Cooper, W.J., Albertson, R.C., 2011. Modularity of the oral jaws is linked to repeated changes in the craniofacial shape of african cichlids. Int. J. Evol. Biol. 2011, 641501.

Parsons, K.J., Marquez, E., Albertson, R.C., 2012. Constraint and opportunity: the genetic basis and evolution of modularity in the cichlid mandible. Am. Nat. 179, 64–78.

Peters, N., Schacht, V., Schmidt, W., Wilkens, H., 1993. Gehirnproportionen und Ausprägungsgrad der Sinnesorgane von *Astyanax mexicanus* (Pisces, Characinidae): ein Vergleich zwischen dem Flußfish und sienen Hoehlenderivaten "Anoptichthys". Z. Zool. Syst. Evolutionsforsch. 31, 144–159.

Protas, M., Conrad, M., Gross, J.B., Tabin, C., Borowsky, R., 2007. Regressive evolution in the Mexican cave tetra, *Astyanax mexicanus*. Curr. Biol. 17, 452–454.

Retaux, S., Pottin, K., Alunni, A., 2008. Shh and forebrain evolution in the blind cavefish *Astyanax mexicanus*. Biol. Cell. 100, 139–147.

Sarkar, L., Sharpe, P.T., 1999. Expression of Wnt signalling pathway genes during tooth development. Mech. Dev. 85, 197–200.

Schemmel, C., 1967. Vergleichende Untersuchungen an den Hautsinnesorganen ober-and unterirdisch lebender Astyanax-Foramen. Z. Morph. Tiere. 61, 255–316.

Schemmel, C., 1980. Studies on the genetics of feeding behaviour in the cave fish *Astyanax mexicanus f. Anoptichthys*. An example of apparent monofactorial inheritance by polygenes. Z. Tierpsychol. 53, 9–22.

Sire, J.Y., Davit-Beal, T., Delgado, S., Van Der Heyden, C., Huysseune, A., 2002. First-generation teeth in nonmammalian lineages: evidence for a conserved ancestral character? Microsc. Res. Tech. 59, 408–434.

Soares, D., Niemiller, M.L., 2013. Sensory adaptations of fishes to subterranean environments. Bioscience 63, 274–283.

Streelman, J.T., Webb, J.F., Albertson, R.C., Kocher, T.D., 2003. The cusp of evolution and development: a model of cichlid tooth shape diversity. Evol. Dev. 5, 600–608.

Strickler, A.G., Yamamoto, Y., Jeffery, W.R., 2001. Early and late changes in Pax6 expression accompany eye degeneration during cavefish development. Dev. Genes Evol. 211, 138–144.

Torrey, T., 1940. The influence of nerve fibers upon taste buds during embryonic development. Proc. Natl. Acad. Sci. U. S. A. 26, 627–634.

Trapani, J., Yamamoto, Y., Stock, D.W., 2005. Ontogenetic transition from unicuspid to multicuspid oral dentition in a teleost fish: *Astyanax mexicanus*, the Mexican tetra (Ostariophysi: Characidae). Zool. J. Linn. Soc. 145, 523–538.

Tucker, A.S., Matthews, K.L., Sharpe, P.T., 1998. Transformation of tooth type induced by inhibition of BMP signaling. Science 282, 1136–1138.

Tucker, A.S., Headon, D.J., Courtney, J.M., Overbeek, P., Sharpe, P.T., 2004. The activation level of the TNF family receptor, Edar, determines cusp number and tooth number during tooth development. Dev. Biol. 268, 185–194.

Vainio, S., Karavanova, I., Jowett, A., Thesleff, I., 1993. Identification of BMP-4 as a signal mediating secondary induction between epithelial and mesenchymal tissues during early tooth development. Cell 75, 45–58.

Valdez-Moreno, M.E., Contreras-Balderas, S., 2003. Skull osteology of the characid fish *Astyanax mexicanus* (Teleostei: Characidae). Proc. Biol. Soc. Wash. 116, 341–355.

Varatharasan, N., Croll, P.C., Franz-Odendaal, T.A., 2009. Taste bud development and patterning in sighted and blind morphs of *Astyanax mexicanus*. Dev. Dyn. 238, 3056–3064.

Weiss, K.M., Ruddle, F.H., Bollekens, J., 1995. Dlx and other homeobox genes in the morphological development of the dentition. Connect. Tissue Res. 32, 35–40.

Wilkens, H., 2010. Genes, modules and the evolution of cave fish. Heredity (Edinb) 105, 413–422.

William, K.G., Conrad, M., 1936. The structure and development of the complex symphysial hinge-joint in the mandible of hydrocyon lineatus bleeker, a characin fish. J. Zool. 106, 975–984.

Wise, S.B., Stock, D.W., 2006. Conservation and divergence of Bmp2a, Bmp2b, and Bmp4 expression patterns within and between dentitions of teleost fishes. Evol. Dev. 8, 511–523.

Wise, S.B., Stock, D.W., 2010. bmp2b and bmp4 are dispensable for zebrafish tooth development. Dev. Dyn. 239, 2534–2546.

Yamamoto, Y., Jeffery, W.R., 2000. Central role for the lens in cave fish eye degeneration. Science 289, 631–633.

Yamamoto, Y., Espinasa, L., Stock, D.W., Jeffery, W.R., 2003. Development and evolution of craniofacial patterning is mediated by eye-dependent and -independent processes in the cavefish *Astyanax*. Evol. Dev. 5, 435–446.

Yamamoto, Y., Stock, D.W., Jeffery, W.R., 2004. Hedgehog signalling controls eye degeneration in blind cavefish. Nature 431, 844–847.

Yamamoto, Y., Byerly, M.S., Jackman, W.R., Jeffery, W.R., 2009. Pleiotropic functions of embryonic sonic hedgehog expression link jaw and taste bud amplification with eye loss during cavefish evolution. Dev. Biol. 330, 200–211.

Chapter 12

Neural Development and Evolution in *Astyanax mexicanus*: Comparing Cavefish and Surface Fish Brains

Sylvie Rétaux, Alexandre Alié, Maryline Blin, Lucie Devos, Yannick Elipot and Hélène Hinaux
Development and Evolution of the Forebrain, DECA Group, Neuroscience-Paris Saclay Institute, CNRS avenue de la terrasse, Paris, France

INTRODUCTION

Anybody who visits an *Astyanax* facility hosting surface fish and cavefish in adjacent tanks and who has a good sense of observation will at first have a hard time believing that they truly belong to the same species. Indeed, the two morphs look so different. Usually, the visitor first sees that one is depigmented and albino while the other is nicely colored. Then he rapidly gets a feeling that there is "something wrong" with the head of the cave morph, and he finds that the eyes are missing. These are the two main, obvious morphological differences.

The attentive observer will further compare the two types of fish and he will find much more. He will see that surface morphs swim in the water column and school, while cave morphs have a tendency to occupy the bottom half of the tank, and swim constantly on their own. He will notice that in the surface fish groups, one or two individuals constantly strike at some others, behaving as dominant in the school, and that this does not apply to cave morphs. On the other hand, he will be surprised by how well blind morphs navigate in their tank, almost never bumping on the aquarium walls or into their congeners. If he has the chance to visit the fish facility at feeding time, he will be struck by the fast and furious way the surface fish swim toward food; and he will appreciate the special feeding posture taken by cave morphs, which allows them to clean the food from the bottom of the tanks efficiently within a few minutes. These differences, together with some others that are not obvious at first sight (e.g., reduced sleep/increased wakefulness, attraction to vibrations or olfactory capabilities),

Biology and Evolution of the Mexican Cavefish. http://dx.doi.org/10.1016/B978-0-12-802148-4.00012-8
© 2016 Elsevier Inc. All rights reserved.

correspond to major behavioral differences between the two *Astyanax* morphs. They correspond to behaviors that can be classified as various types: (1) sensory; (2) motor; and (3) other, more complex and motivated behaviors, which all are governed by various parts of the nervous system.

What are the developmental and evolutionary mechanisms underlying the above-listed changes in the cavefish nervous system and its associated behaviors? Research in the field has mainly explored two directions that will be reviewed here.

First, comparative neurodevelopment and comparative neuroanatomy studies have revealed quantitative variations in the size of specific regions of the brain or in the number or size of specific sensory organs between cave and surface morphs. This type of variation can be coined as "neural specialization," supposedly in adaptation to environmental changes. For example, in the dark, it is probably advantageous to be "olfactory-oriented" to find food and mates, while in a lighted environment, it is important to maintain visual function. Classical cases of such brain evolutionary specialization come, for example, from nocturnal rodents in which the visual cortex is reduced, but the auditory and somato-sensory cortex is expanded (Campi and Krubitzer, 2010; Krubitzer et al., 2011). Among fishes, similar processes are described in cichlid fishes. In African lakes, very closely related cichlid species with distinct ecological specializations have significantly different brains: rock-dwellers (Mbuna) live in complex environments, engage in complex social interactions, and have a large telencephalon; while sand-dwellers (non-Mbuna) live in a simple environment, essentially use visually driven behaviors, and have a large optic tectum and thalamus. Interestingly, it has been shown that differences between Mbuna and non-Mbuna arise early in development, and that boundaries between brain regions, hence the respective sizes of these brain regions, are set up through antagonisms among signaling systems (Sylvester et al., 2010, 2013). In cave *Astyanax* as well, we will see that natural variations in nervous system patterning occur through early signaling modulations.

Second, some recent evidence suggests evolution of "brain neurochemistry" between cave and surface *Astyanax*. Indeed, even subtle changes in neuromodulatory systems are prone to generate significant variations in complex behaviors, such as motivated or social behaviors. This can be achieved if the number of neurons using a given neurotransmitter is changed (a situation that resembles the possibility discussed above, in which the size of a brain region or a neuronal group varies), or if the intensity or amount of neurotransmission is affected at the level of the synthesis, release, reception, modulation, or transduction of the signal. From a neurophysiological point of view, the behavioral syndrome—which is a correlated suite of behavioral phenotypes across multiple situations (Sih et al, 2004), such as those described above—exhibited by cave *Astyanax* clearly evokes the possibility of such disequilibrium in neuromodulatory transmitters.

ADULT BRAIN ANATOMY AND BRAIN NETWORKS

Figure 12.1 presents the comparative anatomy of adult Pachón cavefish and surface fish brains at the macroscopic level. As described by Riedel (1997), the cavefish brain is "slender and elongated." This impression is mainly due to the difference of shape of the telencephalon (trapezoidal in cavefish, ovoid in surface fish) and to the severe reduction in the width and global size of the optic tectum in cavefish (Figure 12.1(A) and (B)).

Concerning the telencephalon, volumetric studies indicate it is enlarged in the Pachón population, but not in other populations, such as Micos or Chica (Peters et al., 1993). The authors of this study hypothesized that telencephalic enlargement was due to the enhancement of the sense of taste, but this should be confirmed by connectivity studies. Qualitative and quantitative observations done in our laboratory in adults and juveniles also suggest that the olfactory bulbs are larger in cavefish (Figure 12.1(A) and (B); Rétaux and Bibliowicz, unpublished). Concerning olfactory connectivity, Riedel and Krug have documented

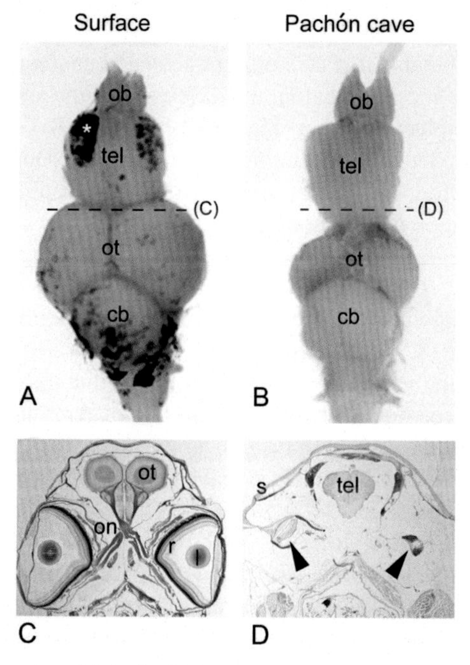

FIGURE 12.1 Comparing adult brains in *Astyanax* surface fish and Pachón cavefish. (A and B) show dorsal views of adult brains after dissection (anterior is up). The two individuals were of identical size (4 cm standard length). The dotted lines indicate the approximate section levels shown in (C) and (D). ob, olfactory bulbs; tel, telencephalon; ot, optic tectum; cb, cerebellum. (C and D) show frontal sections through the head of adult fish, after Klüver and Barrera coloration. The arrowheads on the Pachón picture show degenerated and cystic eye, partially calcified (dark/purple) and covered by skin. r, retina; l, lens; on, optic nerve; ot, optic tectum; s, skin.

that the projections of the olfactory bulb onto the cavefish telencephalon resemble a "simple Bauplan," and they concluded that the telencephalon is not dominated by olfactory inputs (Riedel and Krug, 1997); however, as they only analyzed the cavefish olfactory projections, no comparative surface fish data exists to determine olfactory specialization, or lack thereof, at this level.

At the diencephalic level, the major difference between the two morphs is the absence of eyes in cavefish, and is accompanied by a very severe reduction of the optic nerves (Riedel, 1997; Figure 12.1(C) and (D)). Despite evolution in complete darkness, cavefish have nevertheless conserved their "pineal eye"; the dorsal diencephalic pineal gland (or epiphysis) is structurally intact (Grunewald-Lowenstein, 1956; Herwig, 1976; Langecker et al., 1993; Omura, 1975) and has conserved the ability in larvae to detect light, probably thanks to correct rhodopsin expression (Yoshizawa and Jeffery, 2008). A specific and progressive regression of the regular outer-segment organization of pineal sensory cells nevertheless occurs in 3-, 9-, and 18-month-old cavefish, without affecting other parts of the pineal gland (Herwig, 1976). Interestingly, this regression begins earlier and is more obvious when cavefish are reared in constant darkness than when they are reared in light/dark conditions (Herwig, 1976), and it also occurs in constant light (Omura, 1975). This suggests that part of the degenerative process is attributable to a lack of light-activated neuronal activity.

Caudally, in the mesencephalon, the difference in the size of the optic tectum of the two morphs is striking (Figure 12.1(A) and (B)). This applies to all *Astyanax* cavefish populations examined, including Micos and Chica for which no difference was found with the surface fishes' telencephalon (Peters et al., 1993), or Los Sabinos (our personal observations), and to various extents that seem related to the degree of eye reduction. Logically, tectal hypomorphy has been linked to eye rudimentation (see below). Regarding the connection, as stated above, the optic nerve is greatly reduced, and projections from the retinal cyst are very sparse (Voneida and Sligar, 1976; Figure 12.1(C)). Some residual fibers can be seen in the superficial layers of the medial third of the tectum (as well as in the nucleus opticus hypothalamicus and lateral geniculate nucleus); however, this remnant visual connection is unresponsive to visual cues, and no electrophysiologically detectable signal can be recorded from the optic cyst onto the tectum (Voneida and Fish, 1984). This poses the question of the function of the cavefish "optic" tectum, which does indeed contain efferent pyramidal cells. The only evoked activity that is recordable in the cavefish tectum is generated after somatosensory stimulation (but not lateral line or auditory stimulation, which are invariably evoked in the torus semicircularis), and in a topographical manner (Voneida and Fish, 1984). As it is not unusual to find extravisual modality in the vertebrate tectum, the authors interpreted this finding as a decrease in visual inputs paralleled by an increase in somatic inputs, a situation that is comparable to experimental models following enucleation (e.g., Benedetti, 1992; Chabot et al., 2007; Champoux et al., 2008; Mundinano and Martinez-Millan, 2010). An interesting question would be to know whether the visual-to-somatic rewiring in cavefish occurs during development as a plasticity phenomenon, in parallel to the progressive degeneration of the eye and the loss of visual innervation, or whether this rewiring is genetically programmed and has already been fixed during evolution in the dark. Comparison with visually deprived surface fish would start answering this question. More generally, cavefish are useful models when studying vision-related and vision-dependent neural plasticity phenomena.

A SPECIAL CASE: DEVELOPMENT AND DEGENERATION OF THE CAVEFISH VISUAL SYSTEM

The genetic mechanisms underlying eye loss in cavefish are reviewed in Chapter 11 of this book (Yamamoto et al.), and the evolutionary forces leading to eye loss have been discussed recently (Rétaux and Casane, 2013). Here, we will only briefly describe the progressive remodeling of the visual system that occurs in cavefish between embryonic and adult stages.

During cavefish early embryogenesis and larval development, an eye is formed from the diencephalic neuroepithelium and its adjacent lens placode. This eye starts forming retinal layers (Alunni et al., 2007) and the proliferative zones of both the retina and the lens are active, although they are smaller than in surface fish larvae (Alunni et al., 2007; Hinaux et al., 2015; Strickler et al., 2002; Figure 12.2(C′) and (D′)). In fact, retinal cells are constantly born and are incorporated into the retina, while concomitantly many retinal cells die by apoptosis. At the end, cell death will win the battle against neurogenesis, and the eye will disappear (Figure 12.1(D)). The initial trigger for eye degeneration in cavefish is thought to be lens apoptosis; transplantation of a surface fish lens into a cavefish optic cup is able to rescue the eye of the cavefish while the reciprocal experiment induces the degeneration of the surface fish eye (Yamamoto and Jeffery, 2000).

The optic tectum, a brain region that derives from the alar plate of the mesencephalon and that constitutes the major retinorecipient structure in the brains of fishes and amphibians, is also patterned and regionalized properly in cavefish. The presumptive optic tectum expresses *Pax6*, *Pax2*, *Engrailed2* in domains of equivalent sizes in the two morphs (Soares et al., 2004; see *Lhx9* in Figure 12.2(A) and (B)). During the first days of development, proliferation is also equivalent in the dorsal mesencephalon of cave and surface larvae (Menuet et al., 2007; Blin and Rétaux, unpublished observations) (see proliferating cell nuclear antigen (PCNA) in Figure 12.2(C) and (D)). When the DiI tracing technique is applied on the initial cavefish retinotectal projection at 36 or 72 h post-fertilization (hpf), it is found that the optic nerve develops from the axons of the first generated retinal ganglion cells, reaches the tectum, and even arborizes on its target (Soares et al., 2004). This may explain why cavefish may be able to see, or at least to have some visual abilities, very transiently, during their first days of life, as suggested by positive electroretinograms and

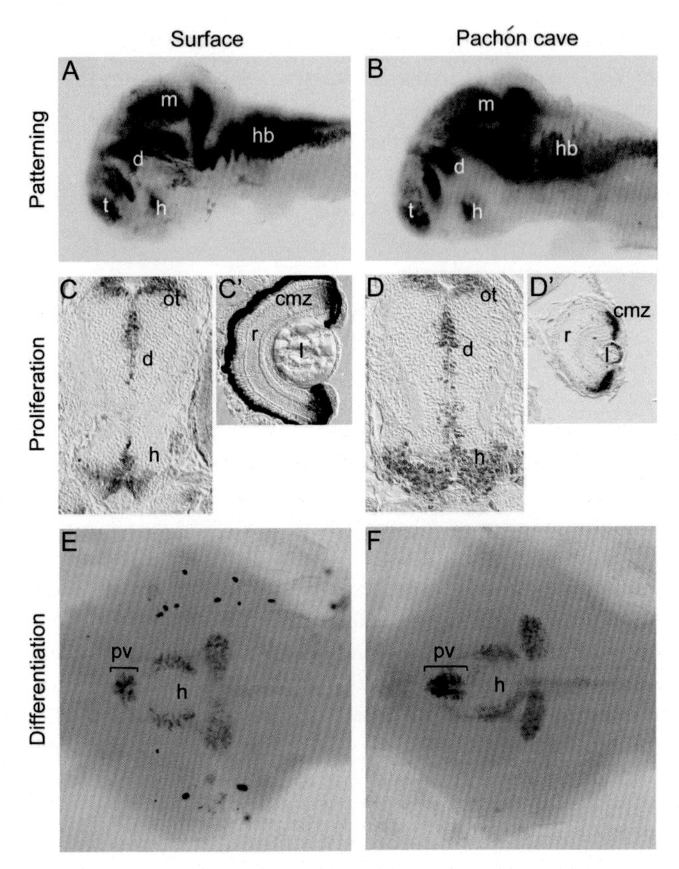

FIGURE 12.2 Comparing larval brain development in *Astyanax* surface fish and Pachón cavefish. (A and B) Patterning of the 36 hpf brain, as observed through the expression of the LIM-homeodomain transcription factor *Lhx9*, *in toto*, on a lateral view. Anterior is left and dorsal is up. t, telencephalon; d, diencephalon; h, hypothalamus; m, midbrain; hb, hindbrain. (CC′ and DD′), proliferation in the 60 hpf larval brain (C and D) and in the 7 dpf eye (C′ and D′), as viewed through PCNA immunohistochemistry on frontal sections. d, diencephalon; h, hypothalamus; ot, optic tectum; r, retina; l, lens; cmz, ciliary marginal zone. (E and F) Differentiated neurons in the hypothalamus (h) and the paraventricular nucleus (pv) at 7 dpf, illustrated here for the serotonergic system on a ventral view of a dissected brain after serotonin immunostaining, *in toto*. Anterior is left.

prey-catching behavior (Daphne Soares, communication at AIM2009); however, as rhodopsin is apparently not expressed in the cavefish retina around these stages (Yoshizawa and Jeffery, 2008), the underlying visual circuit and visual transduction mechanism is therefore unclear and deserves further studies.

As mentioned above, though, the adult cavefish tectum is much reduced in volume (−50%) and contains less neurons (−20%) than in surface fish (Soares et al., 2004). To our knowledge, cell death (by apoptosis, necrosis, or autophagy) has not been investigated in the cavefish optic tectum during the period of tectal shrinkage. To investigate directly whether tectal regression was a secondary consequence of eye degeneration, the lens transplantation model was used. In lens-transplanted cavefish with a restored eye (on one side only), the size of the corresponding optic nerve and the extant of contralateral tectal innervation are increased; however, the procedure results in only a slight increase in tectal volume (+13%) and tectal neuron number (+8%) (Soares et al., 2004), which hardly compares to the surface fish situation. Importantly, it seems that lens transplantation in cavefish restores the eye as an organ, but does not restore vision-based response, tested in a phototaxis assay (Romero et al., 2003). Thus, in cavefish that received embryonically a surface fish lens, an eye and a retinotectal projection is present (Soares et al., 2004), but this visual system is probably not active or functional. This data strongly suggests that the reduced size of the optic tectum in cavefish is indeed a secondary consequence of eye degeneration, and indicates activity-dependent mechanisms that are probably lacking at the tectal level to maintain the integrity of the structure.

EARLY EMBRYONIC DEVELOPMENT: THE ORIGIN OF CAVEFISH DIFFERENCES?

Fortunately for EvoDevo studies, surface fish and cavefish embryos develop at the same pace, allowing rigorous comparisons of early embryogenesis and larval stages (Hinaux et al., 2011). At the end of gastrulation, at 9.5-10 hpf, Pachón cavefish embryos have a slightly ovoid shape resembling a rugby ball, whereas surface fish embryos exhibit a rounder shape (Hinaux et al., 2011). According to the literature, such a phenotype suggests a slight change in dorsoventral patterning during early embryogenesis (e.g., Barth et al., 1999; Kishimoto et al., 1997; Neave et al., 1997). *Bmp* and *Wnt* signaling molecules and activities have not yet been investigated significantly and in a comparative manner between cavefish and surface fish embryos, but it is well established that Hedgehog expression (Sonic and Tiggy-Winckle) is expanded at the anterior ventral midline during gastrulation (Pottin et al., 2011; Yamamoto et al., 2004; Figure 12.3(A)). Such Hedgehog hypersignaling from the mesoderm in the cavefish gastrula is indirectly responsible, through unknown mechanisms, for lens apoptosis and subsequent eye loss (Yamamoto et al., 2004). As demonstrated by pharmacological manipulations, Hedgehog hypersignaling also affects the precise onset of expression of other signaling molecules such as *Fgf8*, although it does not apparently change *Fgf3* expression (Pottin et al., 2011; Figure 12.3(A)). In fact, Hedgehog heterotopy (expanded expression) and *Fgf8* heterochrony (earlier expression) at the anterior margin of the cavefish neural plate affects the patterning and morphogenesis of its future forebrain. Both the expression patterns of several transcription factors that prefigure the presumptive territories of the retina and other forebrain regions (*Pax6*, *Lhx2*/*Lhx9*) and the neural plate fate map are slightly modified in cavefish, precisely at midline level where Hedgehog and *Fgf8* signaling are

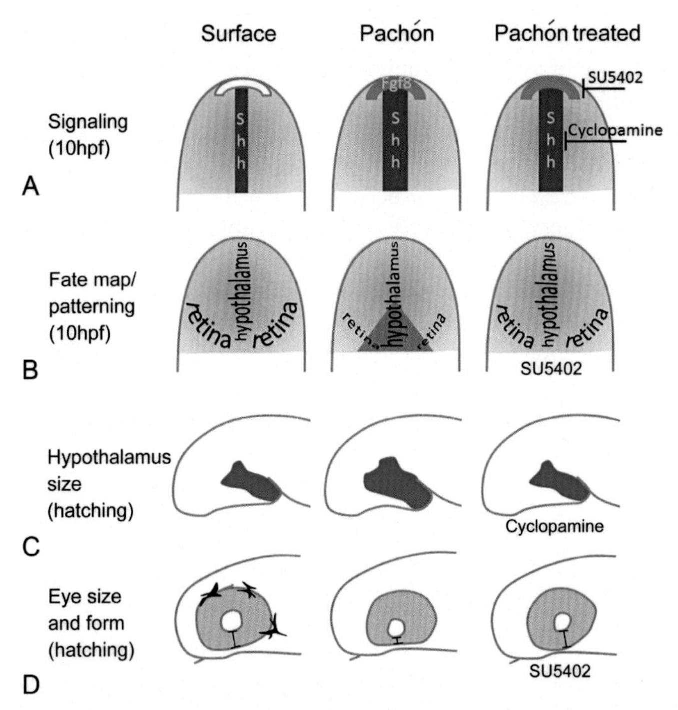

FIGURE 12.3 Comparing the neural plate and embryonic brain morphogenesis in *Astyanax* surface fish and Pachón cavefish. The first and second columns show surface fish and cavefish, respectively. In the third column, the patterning and morphogenetic effects of Hedgehog signaling inhibition by cyclopamine or *FgfR* signaling inhibition by SU5402 on Pachón embryos are depicted and resemble the surface fish phenotype. Schemas are drawn according to experimental evidences from several articles (Menuet et al., 2007; Pottin et al., 2011; Rétaux et al., 2008; Yamamoto et al., 2004). (A) Signaling systems at neural plate stage, schematized on a dorsal view (anterior is up). *Shh* expression (dark/purple at the midline) is larger and *Fgf8* expression (gray/orange at the anterior neural border) is earlier in cavefish. Note: *Fgf8* is not expressed at this early stage in surface fish. (B) Neural fate map and neural plate patterning. The regions of the neural plate fated to become the hypothalamus and the retina are indicated. Several anterior neural plate genes (*Lhx2*, *Lhx9*, *Zic1*, and *Pax6*) show a lack of expression at the cavefish midline (gray). In surface fish embryos, these midline cells are fated to give rise to the ventral quadrant of the retina, which is absent in cavefish. In cavefish, cells located in the equivalent zone of the neural plate (gray triangle) contribute to the hypothalamus or the dorsal retina. (C) Size of the hypothalamus, as labeled by *Nkx2.1a*, on a lateral view of a schematic brain at 24 hpf (anterior is left and dorsal is up). (D) Morphology of the eye. Black brackets indicate the ventral quadrant of the retina, strongly reduced in cavefish and restored after SU5402 treatment. Pigment cells on the surface fish eye are drawn.

changed (Pottin et al., 2011; Strickler et al., 2001; Figure 12.3(B)). We have proposed that medial neural plate cells that are normally fated to become the ventral part of the retina instead contribute to the hypothalamus in cavefish (Pottin et al., 2011). At the end of neurulation, the resulting morphology is an eye with a missing ventral quadrant and a forebrain with an enlarged presumptive hypothalamic territory (Figure 12.3(C) and (D)).

Importantly, *Shh* expression is also expanded in Chica and Los Sabinos embryos, and their *Pax6* medial neural plate pattern is modified the same way as in Pachón embryos (Jeffery, 2009; Strickler et al., 2001; Yamamoto et al., 2004). This tells us that similar developmental processes are modulated in independently evolved cavefish populations, and give rise to the same phenotypes. In fact and more generally, in cavefish embryos from all populations examined so far, the eyes first develop and then regress. This observation can even be extended to other cave vertebrates, including mammals, fishes, and amphibians (reviewed in Rétaux and Casane, 2013). Although this could be viewed as a waste of energy for a developing embryo, we have proposed that optic cup morphogenesis corresponds to a developmental constraint and probably cannot be circumvented (Pottin et al., 2011; Rétaux and Casane, 2013). Indeed, from a morphogenetic point of view, the vertebrate forebrain cannot develop properly without undergoing coordinated cell movements that include the initial formation of the visual organ (e.g., England et al., 2006; Rembold et al., 2006).

LARVAL BRAIN DEVELOPMENT: ESTABLISHING SUBTLE DIFFERENCES

During neurulation and after hatching, cavefish continuously display expanded *Shh* expression in all anterior basal forebrain domains (Menuet et al., 2007). In conjunction with the above described consequences of midline-dependent early morphogenetic events, sustained *Shh* hypersignaling affects neuronal patterning in a subtle manner—and to an extent that is developmentally tolerable, viable, and possibly even adaptive. Indeed, the global regionalization of the cavefish brain remains correct and unaffected, as shown by standard expression patterns of all "developmental" genes investigated (see *Lhx9*, e.g., in Figure 12.2(A)).

In the cavefish ventral telencephalon or subpallium, the expression domains of *Shh*, *Nkx2.1b* (a marker of the medial part of the subpallium and the preoptic region) and the *Nkx2.1*-dependent LIM-homeodomain factors *Lhx6* and *Lhx7* (Grigoriou et al., 1998; Sussel et al., 1999) are enlarged (Menuet et al., 2007). This enlargement appears specific to ventral telencephalic neural components under the control of this particular developmental "cascade," as the *Dlx2* or *Nkx2.2* expression domains are unchanged when compared to surface fish. Interestingly, *Nkx2.1b* and *Lhx6* happen to label a population of GABAergic interneurons that migrate tangentially to populate the olfactory bulbs (Menuet et al., 2007), and that we hypothesized to be the equivalent of the mammalian rostral migratory stream (e.g., Lois and Alvarez-Buylla, 1994). As a positive correlation between the abundance of olfactory bulb interneurons and olfactory performance is reported, it is tempting to speculate that the *Shh*-dependent increase in GABA/*Lhx6*/*Nkx2.1b*-positive migratory stream in cavefish could be advantageous for their life in perpetual darkness.

The cavefish hypothalamus, as labeled with *Shh*, with *Nkx2.1a* and *Nkx2.1b* regional markers, with *Lhx6* subterritory marker, or else as assayed for proliferation (Figure 12.2(B)), appears larger and actively proliferating compared with surface fish (Menuet et al., 2007; Rétaux et al., 2008). Increased proliferation is observed specifically in the hypothalamic and preoptic territories, but not in the more dorsally located diencephalic or mesencephalic regions (Figure 12.2(B)), and a treatment of cavefish embryos with the *Shh*-signaling inhibitor cyclopamine diminishes hypothalamic proliferation and size. This suggests a region-specific, *Shh*-dependent control of proliferation and possibly neurogenesis in the hypothalamus, opening the interesting possibility that this neuroendocrine brain region, which contains many neuronal groups expressing neuromodulatory transmitters, such as monoamines and neuropeptides, has evolved in cavefish.

Is the entire hypothalamus/preoptic region, then, enlarged in cavefish? Or are specific neuronal groups affected, while others are unchanged? Some recent insights came from the comparative analysis of the serotonin neurotransmitter system (Elipot et al., 2013). First, a 4-h heterochrony exists between the appearance of the first serotonin-expressing neurons in the anterior hypothalamus of cavefish (at 18 hpf) and surface fish (at 22 hpf). This may be related to the differences in proliferation/neurogenesis control discussed above. Second, the resulting serotonergic group is larger and contains more cells in cavefish; however, the size of other, more posterior hypothalamic serotonin neuronal groups is identical in cave and surface larvae (Figure 12.2(C)), showing a finely regulated and group-specific regulation of neuron numbers. Third, the size difference in the anterior group is *Shh*-dependent. And finally, this anatomical variation in serotonin circuits seems to translate into behavioral differences, namely an increase in foraging behavior (Elipot et al., 2013).

Much remains to be investigated in a comparative manner on neural patterning, differentiation, and wiring in the larval cavefish brain. But the few aspects that have been investigated so far indicate that we will probably discover discrete, specific, and multiple variations in neuronal patterning in cavefish that result from early embryonic events that change subtle aspects of behavior, and that illustrate the morphogenetic and functional outcomes of developmental evolution and variations.

SENSORY SYSTEMS

The idea of a sensory compensation for absence of vision in animals living in the dark is "classical" and was proposed by early authors, and has been regularly reviewed since (Barr, 1968; Niemiller and Poulson, 2010; Soares and Niemiller, 2013; Wilkens, 1988). Longer appendages in insects and lateral line modifications in fish were often cited. More recently, the idea of sensory modules that would either be developmentally and genetically independent, or that would interact together, and upon which natural selection could act, has been put forward (Franz-Odendaal and Hall, 2006; Wilkens, 2010). Below, we briefly review available data on the developmental evolution of chemosensation (gustation, olfaction) and mechanosensation (lateral line) systems in *Astyanax*. Of note, although hearing is an important sense for aquatic organisms, differences in auditory capacities have not been reported for cavefishes, including for *Astyanax* (Popper, 1970).

Chemosensory System

A better chemical sense has long been suggested for cave *Astyanax* (Breder and Rasquin, 1943; Humbach, 1960). Strikingly, according to Humbach's observation, blind cavefish would have a sense of taste 300 times more acute for bitter and 2000-4000 times more acute for salty, acid, and sweet substances than *Phoxinus* (a minnow, cyprinid); however, these early studies did not strictly discriminate between olfaction and gustation.

Olfaction is surprisingly poorly studied. A "classical" development of the olfactory organs and their lamellae from ectodermal placodes was described in *Astyanax* (Schemmel, 1967). Quantification of the continuously increasing number of olfactory lamellae in surface fish and the Pachón and Los Sabinos cavefishes throughout their lives shows no significant difference. Schemmel concluded that the olfactory modality cannot be considered as specialized in cavefish. He noted, however, that nasal capsules are more opened and flattened in cavefish, so that lamellae are more exposed. More recently, using *in situ* recordings in the Subterráneo cave, which hosts a hybrid population of mixed troglomorphic and epigean characters, we have found that troglomorphic fish present significantly larger naris size, and this was associated with a strong behavioral response elicited by food extracts (Bibliowicz et al., 2013), opening the possibility that olfactory abilities might have evolved in cave-dwelling *Astyanax*.

Concerning gustation, Schemmel was also the first to describe an increased number of tastebuds in Los Sabinos, and even more in Pachón cavefish (Schemmel, 1967). He reported a several-fold increase in tastebud numbers in adults (Schemmel, 1974). More precisely, three different types of tastebuds are distributed on the lips and oral cavity of both *Astyanax* morphs, but only cavefish harbor some on their lower jaws (Boudriot and Reutter, 2001). Moreover, the nerve fiber plexuses of type II and III tastebuds contain more axons in cavefish (Boudriot and Reutter, 2001), and there are more sensory receptor cells per tastebud in cavefish (Varatharasan et al., 2009). Such an enlarged and predominantly ventrally spread gustatory area on the skin of the head was interpreted as functionally relevant to localize food situated on the bottom, and considered as a compensatory improvement of the sense of taste. Interestingly, Substance P is found in tastebuds of cavefish, but not surface fish or other teleosts (Bensouilah and Denizot, 1991). These authors have proposed that the presence of this neurotransmitter could modulate the threshold of excitability of the taste cells.

In fact, tastebud number amplification in cavefish is already present in the first days post-fertilization (dpf) (Varatharasan et al., 2009; Yamamoto et al.,

2009), and the rate of tastebud development is accelerated in cavefish larvae; the difference with surface fish larvae is small at 5 dpf, but threefold at 22 dpf (Varatharasan et al., 2009). That these differences are detectable only after the onset of eye degeneration and that they increase during the degeneration process suggests a link between gustatory and visual development, a notion that is supported by functional experiments (Yamamoto et al., 2009): (1) *Shh* hypersignaling in the oropharyngeal region of cavefish embryos is responsible for tastebud number amplification; (2) early conditional overexpression of *Shh* in surface fish induces positive effects on later tastebud development and negative effects on eye development in the same embryos; and (3) there is an inverse relationship between eye size and tastebud number in the progeny of crosses between surface and cave *Astyanax*. This constitutes the only example to date of a direct link between the development of two sensory organs involving a pleiotropic effect of *Shh* and suggesting indirect selection as an evolutionary driving force for eye loss in cavefish.

Lateral Line

It has been long known that cavefish (Pachón, Los Sabinos, Chica) possess more free (superficial) neuromasts in the suborbital region of their face, and more fragmentation of infraorbital canal neuromasts than surface fish (Bensouilah and Denizot, 1991; Jeffery et al., 2000; Schemmel, 1967). More recently, Yoshizawa and colleagues (2010, 2012) have reported that neuromasts found at a high density in the suborbital and eye orbit region of cavefish mediate the vibration attraction behavior (VAB). Indeed, cavefish are specifically attracted by vibrations at about 35 Hz at the surface of the water, a behavior that clearly seems advantageous to find food in the dark (Yoshizawa et al., 2010). Of note, the morphology of cavefish sensory receptors also differ; the cupula (hair stereo-cilia covered by gelatinous case) of their free head neuromasts is up to 300 μm in length, compared to about 42 μm in surface fish (Teyke, 1990; see also Varatharasan et al., 2009), and is also larger, as are the neuromasts themselves (Yoshizawa et al., 2010). As the height and diameter of the cupula regulate sensitivity, these large, free neuromasts are twice as sensitive in young adults than smaller ones (Yoshizawa et al., 2014). This could explain why cavefish neuromasts can detect low frequency stimuli (below 50 Hz) in otherwise calm cave pools.

The cavefish VAB-mediating neuromasts develop late, after 2 months of age, when the eye is completely gone (Yoshizawa et al., 2010), which may explain why they invade the cavefish eye orbit region. Moreover, and contrarily to the case of tastebuds discussed above, experimental induction of eye regression in surface fish via Hedgehog overexpression is insufficient to increase the number of orbital neuromasts or to promote the appearance of the VAB (Yoshizawa et al., 2012). It will, therefore, be crucial to understand the developmental mechanisms controlling the timing of head neuromast organogenesis and the size of individual sensory organs, as well as to determine the neuronal circuits underpinning the VAB. Considering that some cavefish populations (Pachón, Los Sabinos, or Piedras), but not others (Molino), exhibit a strong form of VAB, this system is ideal to investigate the origin of neural and behavioral novelty during evolution.

CAVEFISH BRAIN NEUROCHEMISTRY

The ensemble of cavefish behavioral modifications described in the introduction is sometimes referred to as a "behavioral syndrome," which would appear quite pathological to a clinician, to whom the cavefish condition would probably evoke disorders involving neuromodulatory and aminergic transmission. Actually, we currently know two genes that are important players in these neurotransmitter systems, and which carry mutations in their coding sequence in cavefish: *Oca2* and monoamine oxidase (MAO) (Elipot et al., 2014a; Protas et al., 2006). Moreover, several relevant genes (such as the 14-3-3 protein YWHAE, the glutamate receptor AMPA2 or the cannabinoid receptor CB1) whose expression is altered in cavefish were recently identified through a microarray study (Strickler and Soares, 2011). These genes play roles in neural networks controlling learning, feeding, or addiction, and could therefore underlie some of the cavefish behavioral phenotypes.

Oca2 (ocular and cutaneous albinism-2) is a transmembrane protein involved in the transport of L-tyrosine, the precursor of melanine, into melanosomes. Its mutation in cavefish (Protas et al., 2006) explains the depigmented phenotype that is reviewed in Chapter 8 of this book (Jeffery et al.). But L-tyrosine also happens to be the precursor of dopamine and noradrenalin, two central monoamines, therefore opening the possibility that an expanded L-tyrosine pool is available as a precursor for dopamine in cavefish. In line with this idea, *Oca2* morpholino knockdown in surface fish embryos increases both L-tyrosine and dopamine levels (Bilandžija et al., 2013b), and dopamine and noradrenalin levels are very high in the brains of young adult cavefish when compared with surface fish (Elipot et al., 2014a). Importantly, *Oca2* carries different loss of function mutations in various *Astyanax* cavefish populations (Protas et al., 2006) and more generally, the first step in melanin synthesis is also affected in other cave-dwelling animals, including insects (Bilandžija et al., 2013a), showing striking convergence on a defect in this particular pathway. Of note, the direct link between a large pool of available L-tyrosine due to the *Oca2* deficiency and the high levels of dopamine in the brain of cavefish remains unclear, because the activity of tyrosine hydroxylase, the rate-limiting enzyme of dopamine synthesis, is identical in the brains of surface and cave morphs (Elipot et al., 2014a). An interesting research venue may be offered by the fact that the dopamine synthesis activator gene *YWHAE* is up-regulated in cavefish (Strickler and Soares, 2011). Finally and behaviorally, high noradrenalin levels in cavefish probably play a role in reduced sleep/increased wakefulness, as suggested by sleep rescue after beta-noradrenergic receptor blockade (Duboue et al., 2012). And high dopamine levels may underlie feeding drive and reward-associated responses (Singh, 2014).

MAO is the serotonin-degrading enzyme. It carries a partial loss-of-function point mutation in Pachón cavefish, leading to very high serotonin levels in the brain (Elipot et al., 2014a). Note that there is only one form of MAO in teleosts, whereas mammals have two. Combined with the larger anterior hypothalamic serotonergic group, this mutation could contribute to the cavefish's persistent foraging behavior (Elipot et al., 2013). The MAO mutation is also likely to explain other cavefish behavioral phenotypes; in surface fish treated with deprenyl, a specific MAO inhibitor, serotonin levels (but not dopamine or noradrenalin) are increased, therefore mimicking the cavefish condition (Elipot et al., 2014a), and both schooling behavior (Kowalko et al., 2013) and hierarchical aggressiveness (Elipot et al., 2013) are lost. The serotonergic raphe nucleus in the hindbrain is probably involved in the loss of aggressiveness. Indeed, low raphe serotonin levels are associated with dominant individuals in surface fish groups, and it is possible to elicit some aggressiveness in cavefish by embryonic manipulations that reduce the size of their raphe serotonergic nucleus (Elipot et al., 2013). On the other hand, the serotonin-dependent brain circuits involved in loss of schooling are unknown, but we have proposed that the lack of collective behaviors in cavefish is probably tightly related to their loss of hierarchical and aggressive behavior (Rétaux and Elipot, 2013 and see also this book, Chapter 17).

In sum, we are at the beginnings of our understanding of cavefish brain neurochemistry, but all the data accumulated so far suggest that the subtle equilibrium and neurotransmission homeostasis present in vertebrate brains are unbalanced in more than a few ways in cavefish. Along the same lines, some "general" neurotransmission genes, such as neuroligin or the neurofilament protein M seem up- or down-regulated in cavefish, respectively, according to cross-species (zebrafish) microarray experiments (Strickler and Jeffery, 2009). Neuroligin is a postsynaptic adhesion molecule thought to control the balance between excitatory and inhibitory synapses (Mackowiak et al., 2014) and NF-M is a cytoskeleton component that regulates axonal growth and homeostasis (Yuan et al., 2012). Therefore, such dysregulations might also have general effects on cavefish neural functions, but their origin and their exact impacts have not yet been investigated.

CONCLUSIONS AND PERSPECTIVES

The *Astyanax* model system is now entering the genome era, with the available Pachón genome (McGaugh et al., 2014) and the exciting possibilities offered by transgenesis and genome-editing techniques (Elipot et al., 2014b; see also Chapter 19 by Burgess et al. in this book). Such progress will render possible many new lines of investigations. Comparative genomics will allow the investigation of *cis*-regulatory aspects in the evolution of gene regulation. Loss and gain of function experiments in cave and surface morphs will decipher the exact roles and effects of mutations identified in cavefish. With the generation of transgenic fluorescent reporter cavefish and surface fish lines, researchers will be able to compare early cavefish brain morphogenesis and growth by 3D live imaging, or to analyze and manipulate neuronal activity *in vivo*. In other words, the cavefish has become a top model for neuroscience research for investigators interested in brain evolution and morphological, functional, and behavioral adaptation.

ACKNOWLEDGMENTS

Work in the group was supported by ANR grant (Astyco) and (Blindtest) and CNRS. Many thanks to Franck Bourrat for sectioning and coloration of adult *Astyanax* heads, to Stéphane Père, Magalie Bouvet and Diane Denis for taking care of our *Astyanax* colony, and to Laurent Legendre for collaborative help on husbandry methods.

REFERENCES

Alunni, A., et al., 2007. Developmental mechanisms for retinal degeneration in the blind cavefish *Astyanax mexicanus*. J. Comp. Neurol. 505, 221–233.

Barr, T.C., 1968. Cave ecology and the evolution of troglobites. In: Dobzhansky, Th., Hecht, M.K., Steere, W.C. (Eds.), Evolutionary Biology, vol. 2. Plenum Press, New York, pp. 35–101.

Barth, K.A., et al., 1999. Bmp activity establishes a gradient of positional information throughout the entire neural plate. Development 126, 4977–4987.

Benedetti, F., 1992. The development of the somatosensory representation in the superior colliculus of visually deprived mice. Brain Res. Dev. Brain Res. 65, 173–178.

Bensouilah, M., Denizot, J.P., 1991. Taste buds and neuromasts of *Astyanax jordani*: distribution and immunochemical demonstration of co-localized substance P and enkephalins. Eur. J. Neurosci. 3, 407–414.

Bibliowicz, J., et al., 2013. Differences in chemosensory response between eyed and eyeless *Astyanax mexicanus* of the Rio Subterráneo cave. EvoDevo 4, 25.

Bilandžija, H., et al., 2013a. Evolution of albinism in cave planthoppers by a convergent defect in the first step of melanin biosynthesis. Evol. Dev. 14, 196–203.

Bilandžija, H., et al., 2013b. A potential benefit of albinism in *Astyanax* cavefish: downregulation of the *oca2* gene increases tyrosine and catecholamine levels as an alternative to melanin synthesis. PLoS ONE 8, e80823.

Boudriot, F., Reutter, K., 2001. Ultrastructure of the taste buds in the blind cave fish *Astyanax jordani* ("Anoptichthys") and the sighted river fish *Astyanax mexicanus* (Teleostei, Characidae). J. Comp. Neurol. 434, 428–444.

Breder, C.M., Rasquin, P., 1943. Chemical sensory reactions in the Mexican blind characins. Zoologica 28, 169–200.

Campi, K.L., Krubitzer, L., 2010. Comparative studies of diurnal and nocturnal rodents: differences in lifestyle result in alterations in cortical field size and number. J. Comp. Neurol. 518, 4491–4512.

Chabot, N., et al., 2007. Audition differently activates the visual system in neonatally enucleated mice compared with anophthalmic mutants. Eur. J. Neurosci. 26, 2334–2348.

Champoux, F., et al., 2008. Effects of early binocular enucleation on auditory and somatosensory coding in the superior colliculus of the rat. Brain Res. 1191, 84–95.

Duboue, E.R., et al., 2012. β-Adrenergic signaling regulates evolutionarily derived sleep loss in the Mexican cavefish. Brain Behav. Evol. 80 (4), 233–243.

Elipot, Y., et al., 2013. Evolutionary shift from fighting to foraging in blind cavefish through changes in the serotonin network. Curr. Biol. 23 (1), 1–10.

Elipot, Y., et al., 2014a. A mutation in the enzyme monoamine oxidase explains part of the Astyanax cavefish behavioral syndrome. Nat. Commun. 5, 3647.

Elipot, Y., et al., 2014b. *Astyanax* transgenesis and husbandry: how cavefish enters the lab. Zebrafish 11 (4), 291–299.

England, S.J., et al., 2006. A dynamic fate map of the forebrain shows how vertebrate eyes form and explains two causes of cyclopia. Development 133, 4613–4617.

Franz-Odendaal, T.A., Hall, B.K., 2006. Modularity and sense organs in the blind cavefish, *Astyanax mexicanus*. Evol. Dev. 8, 94–100.

Grigoriou, M., et al., 1998. Expression and regulation of *Lhx6* and *Lhx7*, a novel subfamily of LIM homeodomain encoding genes, suggests a role in mammalian head development. Development 125, 2063–2074.

Grunewald-Lowenstein, M., 1956. Influence of light and darkness on the pineal body in *Astyanax mexicanus*. Zoologica 41, 119–128.

Herwig, H.J., 1976. Comparative ultrastructural investigations of the pineal organ of the blind cave fish, *Anoptichthys jordani*, and its ancestor, the eyed river fish, *Astyanax mexicanus*. Cell Tissue Res. 167, 297–324.

Hinaux, H., et al., 2011. A developmental staging table for *Astyanax mexicanus* surface fish and Pachón cavefish. Zebrafish 8, 155–165.

Hinaux, H., et al., 2015. Lens defects in *Astyanax mexicanus* cavefish: focus on crystallin evolution and function. Dev. Neurobiol 75 (5), 505–521.

Humbach, I., 1960. Geruch und Gesmack bei den augenlosen Höhlenfischen *Anoptichthys joedani*, Hubbs und Innes und *Anoptichtys hubbsi*, Alvarez. Naturwissenschaften 47, 551.

Jeffery, W.R., 2009. Regressive evolution in *Astyanax* cavefish. Annu. Rev. Genet. 43, 25–47.

Jeffery, W., et al., 2000. *Prox1* in eye degeneration and sensory organ compensation during development and evolution of the cavefish *Astyanax*. Dev. Genes Evol. 210, 223–230.

Kishimoto, Y., et al., 1997. The molecular nature of zebrafish swirl: BMP2 function is essential during early dorsoventral patterning. Development 124, 4457–4466.

Kowalko, J.E., et al., 2013. Loss of schooling behavior in cavefish through sight-dependent and sight-independent mechanisms. Curr. Biol. 23, 1874–1883.

Krubitzer, L., et al., 2011. All rodents are not the same: a modern synthesis of cortical organization. Brain Behav. Evol. 78, 51–93.

Langecker, T.G., et al., 1993. Transcription of the opsin gene in degenerate eyes of cave-dwelling *Astyanax fasciatus* and its conspecific epigean ancestor during early ontogeny. Cell Tissue Res. 273, 183–192.

Lois, C., Alvarez-Buylla, A., 1994. Long-distance neuronal migration in the adult mammalian brain. Science 264, 1145–1148.

Mackowiak, M., et al., 2014. Neuroligins, synapse balance and neuropsychiatric disorders. Pharmacol. Rep. 66, 830–835.

McGaugh, S.E., et al., 2014. The cavefish genome reveals candidate genes for eye loss. Nat. Commun. 5, 5307.

Menuet, A., et al., 2007. Expanded expression of Sonic Hedgehog in *Astyanax* cavefish: multiple consequences on forebrain development and evolution. Development 134, 845–855.

Mundinano, I.C., Martinez-Millan, L., 2010. Somatosensory cross-modal plasticity in the superior colliculus of visually deafferented rats. Neuroscience 165, 1457–1470.

Neave, B., et al., 1997. A graded response to BMP-4 spatially coordinates patterning of the mesoderm and ectoderm in the zebrafish. Mech. Dev. 62, 183–195.

Niemiller, M.L., Poulson, T.L., 2010. Chapter 7: subterranean fishes of North America: amblyopsidae. In: Trajano, E., Bichuette, M.E., Kapoor, B.G. (Eds.), The Biology of Subterranean Fishes. Science Publishers, Enfield, NH, pp. 1–112.

Omura, Y., 1975. Influence of light and darkness on the ultrastructure of the pineal organ in the blind cave fish, *Astyanax mexicanus*. Cell Tissue Res. 160, 99–112.

Peters, V.N., et al., 1993. Gehirnproportionen und Ausprägungsgrad des Sinnesorgane von *Astyanax mexicanus* (Pisces, Characinidae). Zool. Syst. Evolut. Forsch. 31, 144–159.

Popper, A.N., 1970. Auditory capacities of the Mexican blind cavefish (*Astyanax jordani*) and its eyed ancestor (*Astyanax mexicanus*). Anim. Behav. 18, 552–562.

Pottin, K., et al., 2011. Restoring eye size in *Astyanax mexicanus* blind cavefish embryos through modulation of the Shh and Fgf8 forebrain organising centres. Development 138, 2467–2476.

Protas, M.E., et al., 2006. Genetic analysis of cavefish reveals molecular convergence in the evolution of albinism. Nat. Genet. 38, 107–111.

Rembold, M., et al., 2006. Individual cell migration serves as the driving force for optic vesicle evagination. Science 313, 1130–1134.

Rétaux, S., Casane, D., 2013. Evolution of eye development in the darkness of caves: adaptation, drift, or both? EvoDevo 4, 26.

Rétaux, S., Elipot, Y., 2013. Feed or fight: a behavioral shift in blind cavefish. Commun. Integr. Biol. 6, e23166.

Rétaux, S., et al., 2008. Shh and forebrain evolution in the blind cavefish *Astyanax mexicanus*. Biol. Cell. 100, 139–147.

Riedel, G., 1997. The forebrain of the blind cave fish *Astyanax hubbsi* (Characidae). I. General anatomy of the telencephalon. Brain Behav. Evol. 49, 20–38.

Riedel, G., Krug, L., 1997. The forebrain of the blind cave fish *Astyanax hubbsi* (Characidae). II. Projections of the olfactory bulb. Brain Behav. Evol. 49, 39–52.

Romero, A., et al., 2003. One eye but no vision: cave fish with induced eyes do not respond to light. J. Exp. Zool. B Mol. Dev. Evol. 300, 72–79.

Schemmel, C., 1967. Vergleichende Untersuchungen an den Hautsinnesorganen ober-and unterirdisch lebender *Astyanax*-Foramen. Z. Morph. Tiere. 61, 255–316.

Schemmel, C., 1974. Genetische Untersuchungen zur Evolution des Geschmacksapparates bei cavernicolen Fischen. Z. Zool. Syst. Evolutionforsch. 12, 196–205.

Sih, A., et al., 2004. Behavioral syndromes: an ecological and evolutionary overview. Trends Ecol. Evol. 19, 372–378.

Singh, M., 2014. Mood, food, and obesity. Front. Psychol. 5, 925.

Soares, D., Niemiller, M.L., 2013. Sensory adaptations of fishes to subterranean environments. Bioscience 63, 274–283.

Soares, D., et al., 2004. The lens has a specific influence on optic nerve and tectum development in the blind cavefish *Astyanax*. Dev. Neurosci. 26, 308–317.

Strickler, A.G., Jeffery, W.R., 2009. Differentially expressed genes identified by cross-species microarray in the blind cavefish *Astyanax*. Integr. Zool. 4, 99–109.

Strickler, A.G., Soares, D., 2011. Comparative genetics of the central nervous system in epigean and hypogean *Astyanax mexicanus*. Genetica 139, 383–391.

Strickler, A.G., et al., 2001. Early and late changes in *Pax6* expression accompany eye degeneration during cavefish development. Dev. Genes Evol. 211, 138–144.

Strickler, A.G., et al., 2002. Retinal homeobox genes and the role of cell proliferation in cavefish eye degeneration. Int. J. Dev. Biol. 46, 285–294.

Sussel, L., et al., 1999. Loss *of Nkx2.1* homeobox gene function results in a ventral to dorsal molecular respecification within the basal telencephalon: evidence for a transformation of the pallidum into the striatum. Development 126, 3359–3370.

Sylvester, J.B., et al., 2010. Brain diversity evolves via differences in patterning. Proc. Natl. Acad. Sci. U. S. A. 107, 9718–9723.

Sylvester, J.B., et al., 2013. Competing signals drive telencephalon diversity. Nat. Commun. 4, 1745.

Teyke, T., 1990. Morphological differences in neuromasts of the blind cave fish *Astyanax hubbsi* and the sighted river fish *Astyanax mexicanus*. Brain Behav. Evol. 35, 23–30.

Varatharasan, N., et al., 2009. Taste bud development and patterning in sighted and blind morphs of *Astyanax mexicanus*. Dev. Dyn. 238, 3056–3064.

Voneida, T.J., Fish, S.E., 1984. Central nervous system changes related to the reduction of visual input in a naturally blind fish (*Astyanax hubsi*). Am. Zool. 24, 775–782.

Voneida, T.J., Sligar, C.M., 1976. A comparative neuroanatomic study of retinal projections in two fishes: *Astyanax hubbsi* (the blind cave fish), and *Astyanax mexicanus*. J. Comp. Neurol. 165, 89–105.

Wilkens, H., 1988. Evolution and genetics of epigean and cave *Astyanax fasciatus* (Characidae, Pisces). Support for the neutral mutation theory. In: Hecht, M.K., Wallace, B. (Eds.), Evolutionary Biology, vol. 23. Plenum, New York/London, pp. 271–367.

Wilkens, H., 2010. Genes, modules and the evolution of cave fish. Heredity (Edinb) 105, 413–422.

Yamamoto, Y., Jeffery, W.R., 2000. Central role for the lens in cave fish eye degeneration. Science 289, 631–633.

Yamamoto, Y., et al., 2004. Hedgehog signalling controls eye degeneration in blind cavefish. Nature 431, 844–847.

Yamamoto, Y., et al., 2009. Pleiotropic functions of embryonic sonic hedgehog expression link jaw and taste bud amplification with eye loss during cavefish evolution. Dev. Biol. 330, 200–211.

Yoshizawa, M., Jeffery, W.R., 2008. Shadow response in the blind cavefish *Astyanax* reveals conservation of a functional pineal eye. J. Exp. Biol. 211, 292–299.

Yoshizawa, M., et al., 2010. Evolution of a behavioral shift mediated by superficial neuromasts helps cavefish find food in darkness. Curr. Biol. 20, 1631–1636.

Yoshizawa, M., et al., 2012. Evolution of an adaptive behavior and its sensory receptors promotes eye regression in blind cavefish. BMC Biol. 10, 108.

Yoshizawa, M., et al., 2014. The sensitivity of lateral line receptors and their role in the behavior of Mexican blind cavefish (*Astyanax mexicanus*). J. Exp. Biol. 217, 886–895.

Yuan, A., et al., 2012. Neurofilaments at a glance. J. Cell Sci. 125, 3257–3263.

PART IV

Behavior

Chapter 13

The Evolution of Sensory Adaptation in *Astyanax mexicanus*

M. Yoshizawa
Department of Biology, University of Hawai'i Manoa, Honolulu, Hawaii, USA

INTRODUCTION

Animals have various morphological and physiological traits, many of which could have rapidly evolved through adaptation processes, especially when animals face novel or extreme environmental challenges (Eldredge and Gould, 1972). To understand in detail how these traits evolve, we first must identify selection pressures and then study their association with trait changes (reviewed in Irschick et al., 2013). This is difficult to do for most organisms, because the environmental conditions that originally led to shaping morphological and physiological traits—and afforded specific selection advantages—are ambiguous.

The cavefish represents an example of an animal that has experienced robust selection pressures. Upon the transition from inhabiting streams to the cave environment, cavefish ancestors likely experienced major challenges in finding mates and food under conditions of complete darkness. Thus, they adjusted their sensory modality to nonvisual sensing to improve performance in a dramatically different environment (Culver, 1982; Culver and Pipan, 2009; Poulson and White, 1969; Soares and Niemiller, 2013). With respect to the Mexican cavefish populations, the cave systems seem to have formed long before the first cavefish ancestors arrived, so as soon as they became trapped, the cavefish ancestors rapidly adapted to the novel cave environment (Gross, 2012; see Chapters 3 and 4); however, some caves might have had transitional periods in which there was some access to the surface, allowing hybrid populations to exist. For example, in the Micos cave, surface fish have recently been observed living in the front pools, while hybrids live in the cave's main pools farther back (Mitchell et al., 1977; and personal observation, 2013). Another cave containing a hybrid population, the Chica cave, may have reconnected with the surface environment after a prolonged separation (Wilkens and Hüppop, 1986; Bradic et al., 2012,

Biology and Evolution of the Mexican Cavefish. http://dx.doi.org/10.1016/B978-0-12-802148-4.00013-X
© 2016 Elsevier Inc. All rights reserved.

2013; Coghill et al., 2014). Given these conditions, how did cavefish ancestors adapt to this radical change of habitat? What mechanisms likely facilitated the survival and fecundity of this animal species? Because of the clear selection pressures in caves—no light and sparse food—this simple ecosystem provides insight into the selection pressures on sensory systems by testing the advantage of its associated behavior. Therefore, this chapter highlights certain sensory and neural systems of cavefish that are related to their foraging behavior. First, I discuss sensory systems that are enhanced, including the mechanosensory lateral line and chemosensory systems, and their associated behaviors. Then, I discuss sensory systems that are lost, namely the visual system, and the potential advantage of this loss. Also briefly mentioned are the auditory and tactile systems and the light-sensing pineal organ, which either show no obvious change in sensitivity or have not yet been studied. To conclude, I discuss the substantial potential of the *Astyanax mexicanus* system for revealing the selection and evolutionary mechanisms of sensory systems.

ENHANCED SENSORY SYSTEMS

The Mechanosensory Lateral Line System and Associated Vibration Attraction Behavior

Vibration Attraction Behavior: An Adaptive Cave-Associated Foraging Behavior

Vibration attraction behavior (VAB) describes the fish behavior of swimming toward a vibrating object in darkness, either in a natural cave pool or in a laboratory setting (Figure 13.1(A)). This behavior is assumed to be advantageous for capturing prey under food-sparse and no-predator environments, such as in caves. Indeed, compared to surface fish, cavefish are better at capturing vibrating food in darkness (see below). This idea is also supported by the fact that a similar behavior exists in Amblyopsid cavefish, a different genera than Characin, *A. mexicanus* (Eigenmann, 1909; Hill, 1969), and that at least three *Astyanax* cavefish populations have evolved VAB (Figure 13.1(B)–(D)) (Parzefall, 1983; Abdel-Latif et al., 1990; Yoshizawa et al., 2010). Considering that the Pachón, Piedras, and Toro cavefish populations, as well as the American Amblyopsid cavefish, have evolved separately under similar ecological conditions (Borowsky, 2008; Bradic et al., 2012; Gross, 2012; Hill, 1969; Mitchell et al., 1977; Ornelas-García et al., 2008; Strecker et al., 2012; Wilkens and Strecker, 2003), the convergence of VAB suggests that it has a distinct advantage in the cave environment. The extent of VAB is variable, and some individuals of one of the oldest cavefish populations, Pachón, even lack this behavior (Figure 13.1(E)); however, although the majority of surface fish lack VAB, a few individuals exhibit low levels of this behavior (Figure 13.1(E)). This variation of surface fish VAB, called a standing variation of VAB, may be controlled by different allelic sets among surface fish, and may have contributed to the survival of the initial colonization in the cave (see below). Taking advantage of

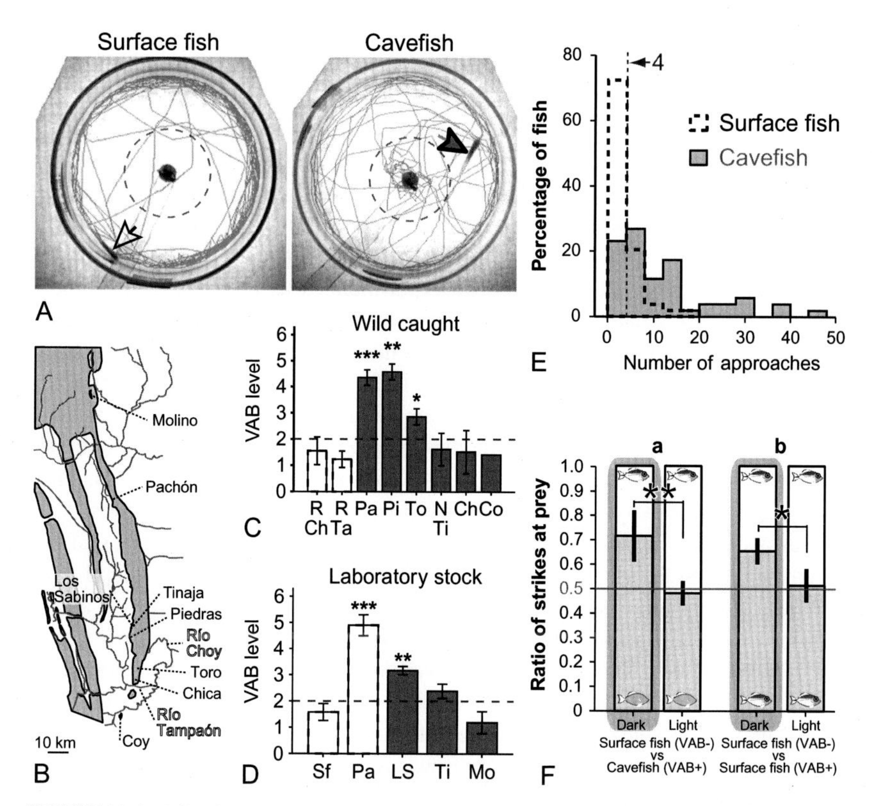

FIGURE 13.1 Vibration attraction behavior (VAB) and its repeated evolution. (A) Path (purple lines) of swimming in surface fish (left) and Pachón cavefish (right) during the 3-min assay period. Dotted lines represent the 4-cm diameter quantification area surrounding the glass rod (dark spot in the center of the chamber). Arrows indicate the starting positions of the fish. (B) Geographic location of caves in the Sierra de El Abra region of Northeast Mexico. Gray shading indicates limestone mountain ranges, and lines indicate primary river systems. (C) VAB levels measured as the square-root of the number of approaches into a 4-cm circle (see A) among wild-caught *Astyanax mexicanus* populations. Three separate cave populations showed VAB (above the threshold level of 2; Yoshizawa et al., 2010), but no VAB was apparent in either surface fish population. Kruskal-Wallis test ($\chi^2=28.3$, $df=7$, $P<0.001$) followed by post hoc Mann-Whitney tests with Bonferroni correction were performed between each cavefish population and grouped surface fish (Río Choy and Río Tampaón). $*P<0.05$; $**P<0.01$; $***P<0.001$. Río Choy surface fish (R Ch), $N=4$, Río Tampaón surface fish (R Ta), $N=13$; Pachón cavefish (Pa), $N=8$; Piedras cavefish (Pi), $N=4$; Toro cavefish (To), $N=6$; Nacimento Tinaja cavefish (N Ti), $N=2$; Chica cavefish (Ch), $N=3$; Coy cavefish (Co), $N=1$ were used. (D) VAB levels measured in the same way as in (C) among laboratory stock *A. mexicanus* populations. The Pachón (Pa) and Los Sabinos (LS) populations of cavefish display greater VAB than surface fish (Sf) and the Tinaja (Ti) and Molino (Mo) cave populations (Kruskal-Wallis $\chi^2=43.1$, $df=4$, $P<0.001$; Post hoc test with Bonferroni adjustment comparing surface fish with: Pachón, $P<0.001$; Los Sabinos, $P<0.01$; Tinaja, $P>0.05$). $N=19$, 19, 20, 20, and 10 for surface fish and the Pachón, Los Sabinos, Tinaja, and Molino cavefish, respectively. (E) VAB levels measured as number of approaches. Surface fish: $N=54$, area surrounded by dotted line; cavefish: $N=52$, gray area. Vertical dashed line represents the NOA cutoff value of 4 for classifying fish with (>4 approaches) and without (<4 approaches) VAB with a stimulus of 50 Hz (Yoshizawa et al., 2010). (Fa) and (Fb) Prey capture competition assays. Bars show the proportion of strikes at prey between pairs of surface fish (eyed fish cartoons) and cavefish (eye-less fish cartoons) with or without VAB during a 1-min assay period in darkness ((Fa) and (Fb), left bars) and in light ((Fa) and (Fb), right bars). A total of eight pairs of cavefish versus surface fish (Fa), and five pairs of surface fish with and without VAB (Fb) in the dark and light are shown. Values are mean ratio of strikes ± 95% confidence intervals of the mean. $*P<0.05$; $**P<0.01$. For details about the method, please see Yoshizawa et al. (2010).

this variation, the adaptive significance of VAB was tested in competitive prey-capture experiments using pairs of fish with and without VAB that were fed small amounts of brine shrimp (Figure 13.1(F)). Compared to surface fish, cavefish were better at capturing this vibrating food in darkness (Figure 13.1(Fa)), which is consistent with the observation of food-finding ability (Hüppop, 1987).

However, the key finding of these experiments was that surface fish individuals with VAB had significantly more brine shrimp captures in the dark than surface fish without VAB, a difference that disappeared in the light (Figure 13.1(Fb)). Thus, VAB plays a role in foraging and is likely adaptive for survival in dark caves. In wild populations of surface fish, VAB is presumably deleterious, since fish with VAB may swim toward predators at night, such as the nocturnal prawn (Wilson et al., 2004; Yoshizawa and Jeffery, 2011). This may be the reason why many surface fish do not show VAB. But, for those that do, surface fish can actually reduce their VAB level at night (data not shown, tested by Lauren Sohn). In contrast to VAB in surface fish, VAB is adaptive in cavefish, since it increases foraging in an environment with limited food availability, light, and macroscopic predators (Yoshizawa et al., 2010; Yoshizawa and Jeffery, 2011). Thus, the standing variation of VAB, similar to that seen in modern surface fish, could have been a trait of cavefish ancestors that was subject to positive selection during the initial cave colonization of *A. mexicanus*.

Perhaps after the colonization, VAB was further shifted beyond the standing variation, for example, in terms of its tuning. The ability of fish to "tune in" to certain vibration frequencies is absent in surface fish, including those with VAB (they have a broad response range from 5 to 50 Hz; Figure 13.2(A) and (B)). In contrast, cavefish show VAB with a significant peak at 35 Hz (Figure 13.2(A); Yoshizawa et al., 2010, 2014). Considering that many crustaceans in the water column produce 30-40 Hz water fluctuations while swimming (Lang, 1980; Montgomery and Macdonald, 1987), the tuning of VAB at 35 Hz is itself a novel trait that is likely adaptive in the cave ecosystem; however, because the exact frequencies of prey have not yet been tested in caves, future studies that sample the underwater vibrations (using an anemometer and/or multiple hydrophones) and characterize the food sources in each cave would be of particular interest. Such ecological information would reveal major relationships between ecological demands and the evolution of adaptive foraging behavior.

The Mechanosensory Lateral Line as a Sensory Organ for Detecting Vibrations

How has VAB evolved in *A. mexicanus*? Because cavefish have a detection peak at 35 Hz, the lateral line system (which fish use to detect low-frequency vibrations) is a good candidate for conferring the enhanced VAB observed in cavefish. Furthermore, the lateral line system has been implicated in the American Amblyopsid cavefish's ability to detect prey (Münz, 1989; Poulson, 1963; Schemmel, 1967). Accordingly, when the lateral line system is inhibited by two

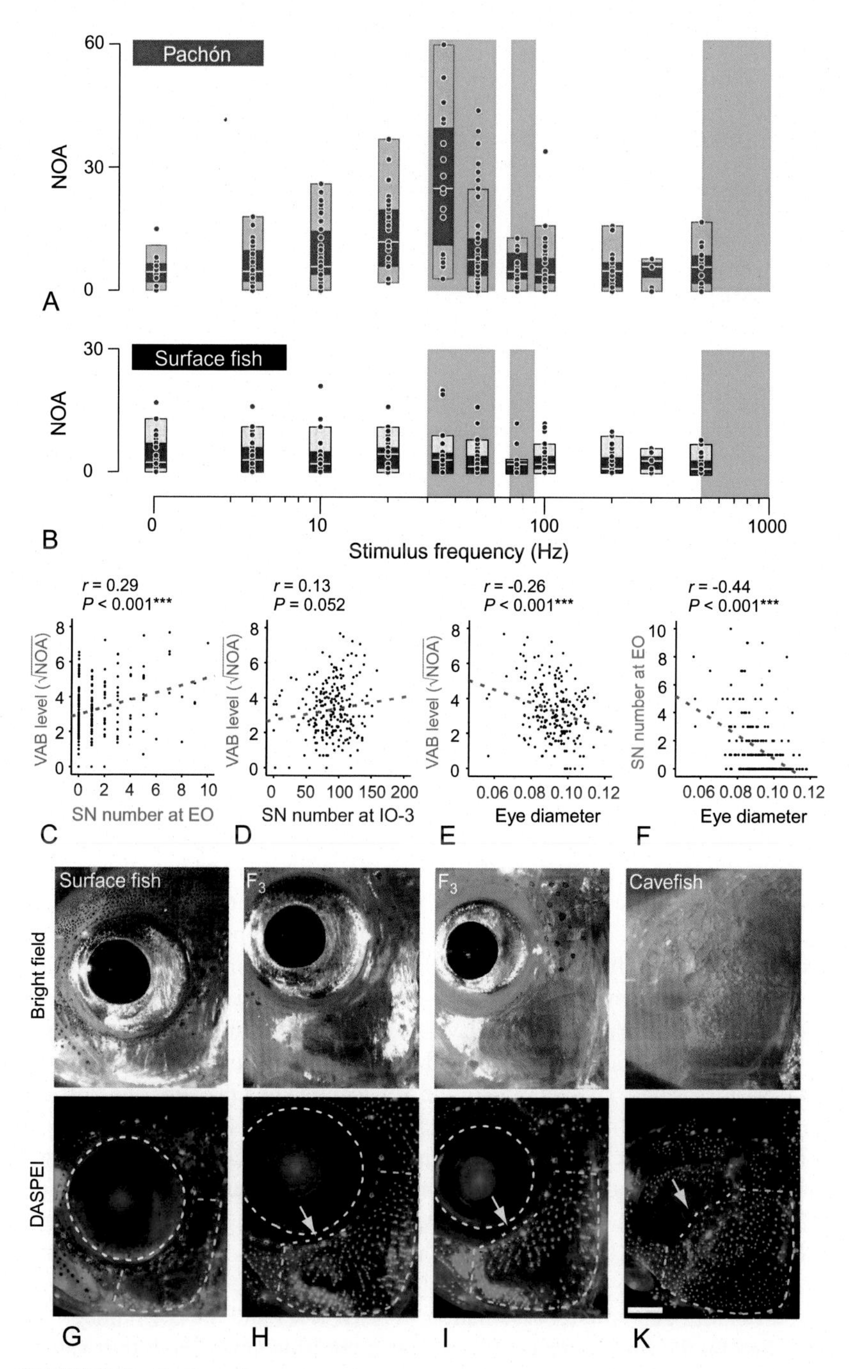

FIGURE 13.2 See legend next page

FIGURE 13.2 (A) and (B) Frequency-dependent response of vibration attraction behavior. The frequency at which individuals approached a vibration rod at the surface of the water varied with the stimulus frequency in blind cavefish from the (A) Pachón (red) population, but not in the (B) surface fish (gray) population. Circles represent individual measurements, the light rectangles indicate the range of values, the dark rectangles represent the bounds of the first (25%) and third (75%) quartiles, and the white line represents the median. Outliers were identified as exceeding 50% of the range between first and third quartiles. Variation in behavior is well apparent in the maximum and third quartile values. Green shading indicates the best sensing range of superficial neuromasts (SN, left), canal neuromasts (middle), and inner ear (right) in *A. mexicanus* (Münz, 1989; Popper, 1970). NOA: number of approaches. (C–K) Genetic analysis of VAB, SN, and eye regression. (C–F) Regression analysis in F_2 and F_3 generations showing the relationships between VAB and (C) SN number at the eye orbit (EO), (D) SN number at the third infraorbital bone area (IO-3), and (E) eye diameter. (F) The relationship of SN number at the eye orbit and eye diameter. SN number at the eye orbit was positively correlated with VAB level, which was negatively correlated with eye size. SN number at the eye orbit was also negatively correlated with eye size. Linear regression lines are shown in red. (G–K) Bright field images (upper) and DASPEI (2-(4-(dimethylamino)styryl)-N-ethylpyridinium iodide)-stained neuromasts (lower) compared among (G) surface fish, (H) and (I) two examples of F_3 hybrids, and (K) cavefish. Scale bar in (K) is equal to 1.0 mm. In (G–I), circles outlined by white dashed lines indicate the edges of the eye, red dashed lines indicate the lines of infraorbital CN in the head lateral line, the areas enclosed by the blue dotted lines indicate the approximate outline of the IO-3 region, and the areas shown by the yellow dotted lines and indicated by yellow arrows show the eye orbit regions (Yoshizawa et al., 2012).

different inhibitors (cobalt and gentamicin) VAB vanishes, indicating that the lateral line is required for VAB in Mexican cavefish (Yoshizawa et al., 2010).

The lateral line system consists of canal neuromasts (CN) and superficial neuromasts (SN) (Coombs et al., 1989). Adult cavefish and surface fish have similar numbers of CN, but cavefish have several-fold more SN than surface fish (Yoshizawa et al., 2010). These SN are a class of lateral-line sensory organs composed of ciliated hair cells and a gelatinous cupular matrix (Figure 13.2(G)–(K), see below), which are increased in both size and number in *A. mexicanus* cavefish (Schemmel, 1967). By SN ablation and genetic analysis, it has been demonstrated that SN—not CN—are the sensory receptors responsible for VAB (Yoshizawa et al., 2010). In these SN ablation experiments, a Vetbond superglue adhesive is applied to the cheek epidermis where only SN (not CN) are located, and ablation of the SN occurs when the adhesive falls off while fish are swimming. These adhesive-treated cavefish exhibit decreased VAB. To test whether SN and VAB are regulated by the same gene, Yoshizawa et al. tested their genetic association through a crossing experiment. A single surface fish was crossed to a Pachón cavefish to generate F_1 hybrids, then F_2 hybrids were generated by intercrossing a pair of F_1 hybrids. F_2 individuals show different phenotypes according to their unique allelic combination derived from a random genomic recombination. If VAB level and the SN number are controlled by the same gene or tightly linked multiple genes, both trait values are correlated in F_2 and further intercrossed F_3 hybrids. Although SN are found throughout the body and are particularly abundant in the head of cavefish, the phenotypic correlation in F_2 and F_3 hybrids and the ablation studies revealed that those SN within the orbit of the degenerated eye (SN at the eye orbit) have a particularly important role in VAB (Figure 13.2(C), see below; Yoshizawa et al., 2012). This result is noteworthy, because it suggests that the extra cranial space created by the loss of eyes in cavefish is an important factor in promoting this novel behavior. Furthermore, differences in the number of SN at the eye orbit are not seen among surface fish (in fact, no eye orbit-SN were observed in surface fish), suggesting that their appearance in cavefish could not arise from selection on a standing phenotypic variation in the surface-living ancestor (Yoshizawa et al., 2012).

The differences in function between SN subtypes raise the possibility that the sensitivity of eye orbit-SN may be specifically enhanced compared to the SNs in the cheek region. The morphology of a neuromast is a key parameter that defines its frequency-dependent sensitivity (van Netten and McHenry, 2014), and therefore morphology has the potential to cause the VAB response. Neuromasts contain a transparent extracellular mucus structure, the cupula, which surrounds the hair bundles that project from the apical surfaces of hair cells within the sensory epithelium (Coombs et al., 1989). In SN, the cupula is elongated and bends in the direction of flow over the surface of the body. These longer and less-stiff neuromasts can sense lower frequencies, while shorter and stiffer neuromasts can be tuned at higher frequencies (stiffness can be estimated from the numbers of hair bundles and width of neuromast; van Netten and McHenry, 2014; McHenry and Van Netten, 2007; McHenry et al., 2008; Teyke, 1988, 1990). A series of comparative studies of frequency-dependent sensitivity revealed that sensitivities are not significantly different between SN at the eye orbit and the ones at infraorbital area in either surface fish or Pachón cavefish (Yoshizawa et al., 2014), but they did show differences between the populations. The Pachón cavefish population showed significantly higher sensitivities of SN at both the eye orbit and the infraorbital area than the surface fish population, indicating that the cave population's neuromast units afford better sensing ability of water vibration/flow (Yoshizawa et al., 2014). Another important finding was that large, mature adult cavefish (>6 cm long) showed reduced sensitivities of both eye orbit-SN and infraorbital SN compared with smaller, younger adults (<4 cm long). Correspondingly, VAB is significantly attenuated in large cavefish and large surface fish (Yoshizawa et al., 2014), indicating that the sensitivity of a single neuromast unit provides a good estimate of VAB; however, in group tanks (~10 cavefish per tank) and in the field (Los Sabinos cave, personal observation by the author), large cavefish are still attracted to a 35-Hz vibration stimulus. Thus, the lack of VAB in large adults could be attributed to a change in the nervous system. Perhaps older fish's matured nervous systems pinpoint the more complex vortices generated while other cavefish are eating (Hüppop, 1987). Nonetheless, in natural situations among other individuals, mature cavefish can respond to a 35-Hz vibration stimulus, but with unknown mechanism(s), perhaps based upon the auditory or tactile systems (see section "Sensory Systems that Potentially Contributing to Cave Adaptation").

Future neural tracing and neurophysiological studies will reveal relationships between the gain/loss of VAB and the elaboration of the neural network in the octavolatelaris nucleus (the first projection area of neuromasts), torus semicircularis (frequency-dependent response), thalamus, tectum (sensory integration), and/or telencephalon (sensory integration and memory; Wullimann and Grothe, 2013). This exciting new research will help reveal how animals acquire novel behavioral tuning.

Genetic Basis for the Evolution of VAB and its Sensory System

To investigate the genetic relationship between VAB and SN, a series of quantitative genetic studies using F_2 and F_3 hybrids derived from a single cross of surface fish and cavefish were performed (Yoshizawa et al., 2012). Yoshizawa et al. also investigated a possible relationship of VAB and SN at the eye orbit with the most conspicuous cave-related trait, eye reduction (Figure 13.2(C)–(K)). Through these analyses, they found that VAB and SN at the eye orbit are significantly positively correlated (Figure 13.2(C)), and VAB and SN at the eye orbit both appeared to be strongly negatively correlated with eye size (Figure 13.2(E) and (F)). Note that "negative correlation" means one trait decreases in magnitude while the other trait increases; that is, both VAB and SN at the eye orbit increased while eye size decreased (Figure 13.2(E) and (F)). VAB was also negatively correlated with pigmentation (correlation between albinism and VAB: Kendall's tau $= -0.13$, $P = 0.020$). Thus, because F_2 and F_3 hybrids have a mosaic genomic structure of surface fish and cavefish (due to genomic recombination events), these findings reveal that VAB, SN at the eye orbit, eye size, and pigmentation can be regulated by shared genetic factors.

To identify genomic regions regulating VAB, the authors also performed a genome scan for quantitative trait loci (QTL) underlying these traits (see also Chapter 6) and found multiple genomic loci for VAB, SN at the eye orbit, and eye size (Figure 13.3). Importantly, the QTL for all three traits overlap each other in two regions of the *Astyanax* genome (linkage groups (LG) 2 and 17; Figure 13.3), and this clustering was significantly greater than expected by chance. Since the genetic clustering of multiple traits can provide a mechanism for their correlated evolution, adaptive evolution of VAB and SN at the eye orbit has likely contributed to the correlated loss of eyes in cavefish. That is to say, evolution may have given rise to a novel population of mechanosensory neuromasts within the lateral line that have a specialized function for detecting prey. Also, the loss of eyes yields new space for these neuromasts, providing another adaptive reasoning for eye loss, in addition to saving energy and the pleiotropy of sonic hedgehog (*shh*) gene (Chapter 10; Jeffery, 2005; Yamamoto et al., 2009; Yoshizawa et al., 2012, 2013).

It is possible that a single gene regulates multiple behavioral or morphological traits and, therefore, regulates both VAB and changes in eye size. Alternatively, multi-trait QTL clustering (i.e., where genes that regulate each different trait form a cluster in a confined area within the genome) may link suites of traits that are necessary for adaptation. The molecular pathways leading to VAB and

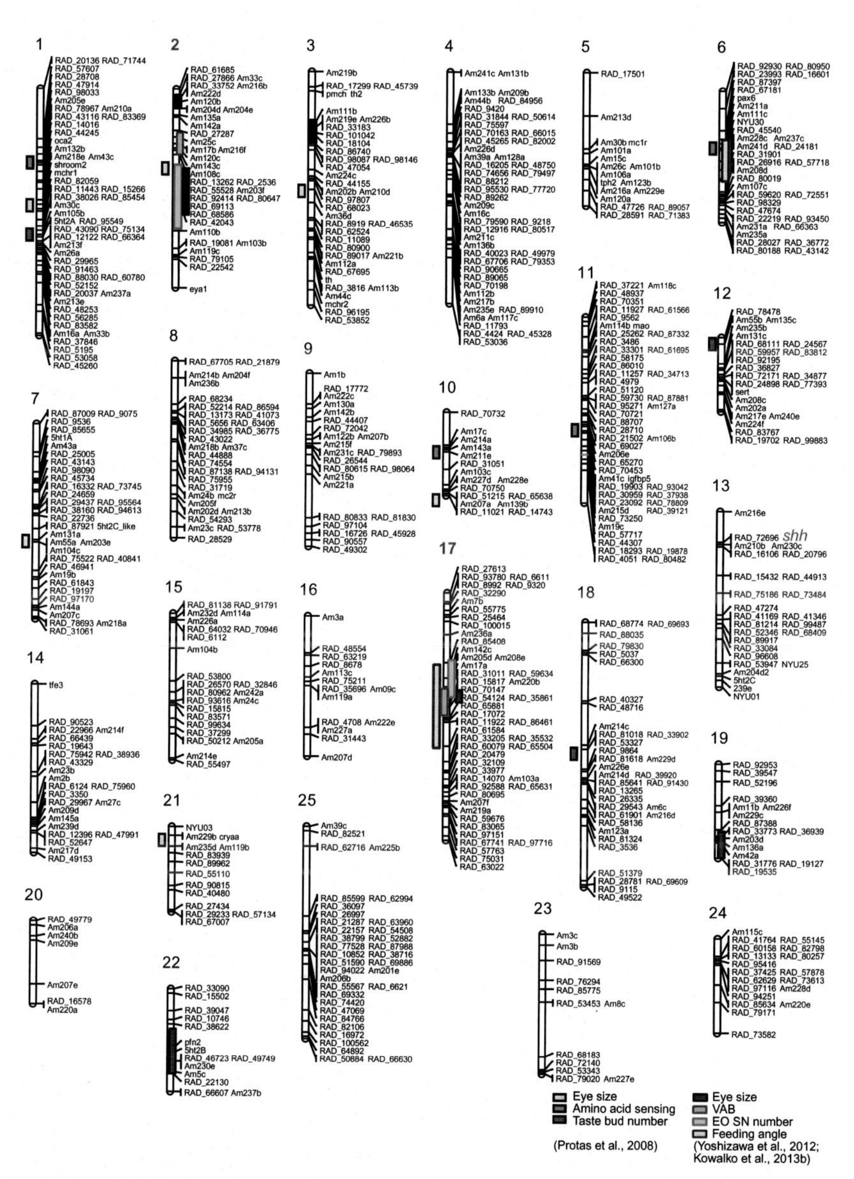

FIGURE 13.3 See legend next page

FIGURE 13.3 Linkage map and QTL clustering. (A) Linkage map constructed from 115 F_2 hybrid progeny of a cross between a single surface fish female (Texas) and a single cavefish male (Pachón). The map includes 699 markers assembled into 25 linkage groups that collectively span 1835.5 cM (O'Quin et al., 2013). Colored bars represent approximate position of QTL for eye size, chemical (amino acid) sensing ability, tastebud number, VAB level, EO SN number, and feeding angle (Kowalko et al., 2013b; Protas et al., 2008; Yoshizawa et al., 2012). Five overlapping Bayesian credible intervals of QTL for VAB, EO SN number, and eye size were found on LGs 2 and 17. Colored bars denote Bayesian credible intervals with probability coverage as 0.95 for each significant or suggestive QTL.

associated changes in SN number and eye size may be distinct (Yoshizawa et al., 2012), but the correlations among these traits suggest that the reduction of eye size provides more space for SN at the eye orbit and therefore promotes VAB. Thus, the clustering of QTL for eye size, VAB, and SN at the eye orbit provides a genetic mechanism for the co-inheritance and correlated evolution of these beneficial cave-related traits, for example, the preservation of co-adapted gene complexes (Figure 13.3; Mayr, 1963). Since multi-trait QTL clusters facilitate the co-inheritance of traits, either as a result of pleiotropy or tight physical linkage of their underlying mutations, these clusters provide a mechanism for integrating traits necessary for adaptation. In support of this theory, other rapidly evolving species, including domesticated lines, such as honeybee, fowl, and rice (Mougel et al., 2012; Onishi et al., 2007; Wright et al., 2010), as well as spectacular examples of adaption radiation, such as stickleback and African cichlid fishes (Albert et al., 2008, Albertson et al., 2003), reveal multi-trait QTL clustering. For example, QTL mapping using F_2 hybrids generated from crossing a "suction" feeder and a "biting" feeder from a single species of cichlid revealed QTL clusters that concurrently coordinate the depth and length parameters of the jaw apparatus. This QTL cluster may facilitate a rapid shift of feeding style from the one to the other (Albertson et al., 2003). Also, in a cross between a Pacific marine form and a lake benthic form, body shape evolution was also found to be coordinated by multiple QTL clusters (Albert et al., 2008). Therefore, QTL clustering is potentially a general mechanism by which multiple traits are shifted in a favored direction at once, allowing the species to express the best-fit combination of traits within the next generation.

These studies on the genetic basis and sensory evolution of VAB help deepen our understanding of the relationship between evolution and adaptive behaviors. Genetic studies reveal that a slightly simpler but efficient QTL cluster facilitates the rapid evolution and fixation of multiple adaptive traits.

While a greater variety of genetic tools are available in zebrafish, cavefish possess a number of advantages for genetically mapping behavioral and morphological traits. Many zebrafish mutants harbor defects in physiology and/or morphology. These defects serve well as disease models, but they do not provide a clear model for evolution, which must show advanced functions in sensory systems and behavioral traits without having deleterious effect in their habitat.

Above all, there is convincing evidence for why studying adaptive behavior is a useful window through which we can understand evolution of sensory systems. Additionally, future studies involving genomic sequencing will identify changes in the genes that allow for VAB tuning and increase SN at the novel cranial area. These studies will further advance the field of evolutionary adaptation through sensory enhancements.

Chemical Sensory System

Fish primarily detect chemical senses through the olfactory and tastebud systems. Both systems are significantly enhanced in cavefish (Schemmel, 1967; Wilkens, 1988). While the sensitivity of both systems is enhanced in cavefish, it is yet unknown whether they evolved specific sensory tuning. This section will discuss the olfactory system, while the next will discuss the tastebud system.

Enhancement in the Olfactory System

The nostril opening in Mexican cavefish, at least in Pachón and Los Sabinos cavefish, and the precursor area of the nasal epithelium in Pachón embryo, is significantly larger than surface fish (Menuet et al., 2007; Schemmel, 1967; Chapter 12), suggesting that cavefish are better able to sense the chemical stimulus from a food source. By using the natural Micos hybrid population, Bibliowicz et al. tested the ability of two groups of fish—eye-reduced fish resembling blind cavefish and eyed fish resembling surface fish (not taking into account pigmentation)—to respond to food smells (Bibliowicz et al., 2013). The cavefish-like fish responded more and were more attracted to the food smell than surface-like fish. Notably, these cavefish-like fish had a significantly larger nostril opening, suggesting that this behavioral difference occurs because of the sensing advantage of the olfactory system (Bibliowicz et al., 2013; Chapter 13). Furthermore, earlier studies have shown that the size of the nostril opening is negatively correlated with eye size (Yamamoto et al., 2003), and that regressed eyes are associated with higher levels of *shh* expression. Elevated *shh* expression also increases the interneurons of olfactory bulb (Menuet et al., 2007). Thus, pleiotropic hedgehog (*hh*) signaling may be the master controller of olfactory sensing capacity.

Enhancement in the Tastebud System

In addition to potentially controlling olfactory sensing, pleiotropic *hh* signaling also increases the number of the tastebuds and the jaw size, which is common in cavefish (Schemmel, 1967; Varatharasan et al., 2009; Wilkens, 1988; Yamamoto et al., 2009). Because cavefish show enhanced tastebud numbers mostly at the epidermis of the lower jaw facing the bottom of the pool, it was once suggested that these gustatory structures were linked to bottom-feeding posture (Schemmel, 1980), as cavefish feed at the bottom with their anteroposterior body axis oriented at a ~45° angle to the substrate. In contrast, when surface fish feed at the bottom, they orient their anteroposterior body axis perpendicular to the bottom and cannot stabilize their body posture in darkness; however, a series of genetic analyses using F_2 hybrids showed no correlation between bottom-feeding posture and these gustatory traits, as well as other cranial morphologies, suggesting that this bottom-feeding posture is instead based upon changes to the central nervous system (Kowalko et al., 2013b). Furthermore, QTL mapping revealed that the QTL for this bottom-feeding posture is not associated with the *shh* locus, revealing that these traits evolved without *shh* mutation; conversely, tastebud number is controlled by an *shh* overexpression (Figure 13.3). Above all, Kowalko et al. concluded that the evolution of the cavefish's feeding posture might be based upon the central, perhaps the motor control, rather than the sensory system.

In summary, both known chemical sensors—the olfactory system and the tastebuds—are likely controlled by *hh* signaling. Interestingly though, the chemical (amino acid) sensing QTL and the feeding posture QTL are colocalized on LG 6, but do not overlap with the *shh* locus at LG 13 (Figure 13.3; Kowalko et al., 2013b; Protas et al., 2008). Since the QTL for the number of tastebuds is not colocalized with the QTL for chemical sensing, chemical sensing might be regulated by the olfactory/nasal epithelium function, whose QTL has not been investigated, except for *shh* (Figure 13.3). It would be interesting to perform an ablation study on the nasal epithelium to determine if the shared mechanism between chemical sensing and feeding posture depends upon olfactory function. Such an ablation study would also evaluate the distinct role tastebuds may play in navigating fish toward chemical (food) sources.

REGRESSED SENSORY SYSTEMS

Visual System: Binocular Eyes

Eye regression is observed in cave-dwelling animals across phyla (Culver, 1982; Culver and Pipan, 2009), and thus, it is one of the most intensely investigated morphological traits in *A. mexicanus*. The three major potential advantages of eye loss are argued elsewhere, such as here in Chapters 5, 6, 8, 10, and 13 as well as in many articles. These advantages are: (1) energy conservation; (2) indirect advantages through pleiotropy for other beneficial traits; and (3) opening up space for nonvisual sensors (Jeffery, 2005; Rétaux and Casane, 2013; Rohner et al., 2013; Wilkens, 1988; Yoshizawa et al., 2012). In the previous section, we discussed three sensory systems—SN at the eye orbit, nostril opening, and tastebud number—that are regulated concurrently with eye size, and mentioned that their benefits are associated with eye regression. In this section, we highlight a cavefish behavior, loss of schooling, that was revealed as the first example of a behavioral change caused by the loss of the eye's visual function, rather than by reduced eye size.

Loss of Schooling: A Potential New Foraging Strategy

Schooling has multiple benefits, including to help fish avoid predators and to forage, but these benefits may not be valid in an environment where food is sparse and there are no predators, such as in caves (Krause, 1993; Partridge, 1982; Plath and Schlupp, 2008). As such, many *A. mexicanus* cavefish populations do not show schooling (Parzefall and Senkel, 1986). Their loss of schooling behavior has been shown to depend upon the fish's concurrent loss of visual sensing ability, indicating that schooling is primarily governed through visual inputs (Kowalko et al., 2013a). Other studies have indicated that the lateral line also plays an important role in schooling (Faucher et al., 2010; Partridge and Pitcher, 1980; Pitcher et al., 1976). Through various experiments—a correlation analysis with F_2 hybrids (derived from a surface fish and Tinaja cavefish cross) and pharmacological ablation—Kowalko et al. further investigated the role of the lateral line in schooling behavior, showing that the loss of schooling in cavefish parallels their increase in the lateral line (Kowalko et al., 2013a). Then, by mapping the QTL, they identified one QTL for a visual-related genetic factor (dark preference) that also controls schooling. This finding suggests that cavefish's loss of schooling is mainly due to eye loss.

Kowalko et al. argued that cavefish's loss of schooling is mainly a result of relaxed selection, meaning that the loss of schooling is due to mutations that became neutral for their survival and fecundity after the colonization in cave; however, it is possible that loss of schooling may help cavefish find food in a sparse environment. For example, because cavefish are mainly solitary and spread out over an entire cave pool, the chance that one of them finds a chunk of food is likely higher than if they schooled. Additionally, cavefish, both in the wild and the laboratory, have been shown to gather at a spot where another cavefish is eating, perhaps using VAB (Hüppop, 1987; Yoshizawa et al., 2010; author's personal observation in Los Sabinos cave; see below). In this way, other cavefish can follow the water noises produced by the first fish that finds food, which is perhaps an efficient way to forage. Loss of schooling might also be beneficial when food is scattered over the entire cave pool, as occurs in the Los Sabinos cave pool, where bat guano is continuously dropped from a large bat colony (Mitchell et al. 1977; author's personal observation). This relationship between ecology and foraging strategy is fascinating for understanding evolutionary processes of sensory systems and its associated behaviors. Future in-depth investigations of different cave ecosystems as well as food-finding behavioral tests performed under the cave-like conditions of a food-sparse environment will uncover advantages of regressed sensory systems (e.g., the eye) and their associated behaviors (e.g., schooling).

SENSORY SYSTEMS THAT POTENTIALLY CONTRIBUTE TO CAVE ADAPTATION

Inner Ear Sensory System: Sound Localization

Although the lateral line system is involved in identifying the location of oscillating objects, it is a relatively near-field sensory system and responds at low frequencies (<200 Hz among fish species; Coombs et al., 1989, 2013). Instead, the inner ear plays a major role in long-range sound localization and responds to high-frequency stimuli. *A. mexicanus* is especially equipped for sound perception; *A. mexicanus* and many other fish in the order Ostariophysi have a Weberian apparatus, which is a bony connection between the swim bladder and the inner ear that enhances the sensation of sound (Popper, 1971). To compare the hearing ability of surface fish and (Chica) cavefish, Popper first trained fish through operant conditioning; various high-frequency sounds were played at a volume fish could hear, and then an electrical shock was administered. After this training, fish learned to escape when they heard the sound. Then, the author tested the fish's responses when the sound volume was continuously lowered. Through this experiment, the lowest-hearing threshold was determined (Popper, 1970). Surprisingly, Popper found that the auditory capacity of cavefish and surface fish is not significantly different; however, the Weberian apparatus in cavefish has a slight but significant morphological difference from those of surface fish (Popper, 1971). In a recent study, Coffin et al. showed that the swim bladder plays a central role in sound localization, suggesting that the Weberian apparatus in cavefish may allow them to localize sound better, which may explain why cavefish can respond and be attracted to sounds from remote locations (Coffin et al., 2014; Parzefall, 1983); however, no comparative study has yet examined the sound-localization abilities between surface fish and cavefish. Thus, future studies should explore if Mexican cavefish have superior abilities to localize sound compared to surface fish, and if so, how this occurs. Such a study could potentially reintroduce the cavefish inner ear system as another sensory system that regulates foraging behavior.

Other Unsolved Sensory Systems: Tactile Sensing and the Pineal Light-Sensing Organs

In addition to the lateral line system, the tactile sensory system, whose somatosensory map localizes to the tectum (Voneida and Fish, 1984) provides mechanosensory input. In one study, Voneida and Fish found that surface fish showed a much weaker response in the tectum to tactile stimuli (Voneida and Fish, 1984). Although no reports have described a comparative histological study for this system, the recent studies started showing the involvement of tactile sensor in a startle response against the water jet (e.g., zebrafish) and rheotaxis (e.g., *A. mexicanus* cavefish) (Chapter 19; Bak-Coleman and Coombs, 2014; Kohashi and Oda, 2008; Van Trump and McHenry, 2013). It would indeed be interesting to resolve the changes in the somatosensory tactile system of cavefish and identify how they influenced the evolution of startle and/or rheotaxis behaviors in cavefish. Their influence on rheotaxis behaviors would be particularly interesting, because cavefish seem to mainly rely on the tactile system instead of the lateral line system (Bak-Coleman and Coombs, 2014; Van Trump and McHenry, 2013), whereas surface fish could express rheotaxis based upon the combination of tactile, lateral line, and visual systems as other fish species (Baker and Montgomery, 1999; Lyon, 1904; Montgomery et al., 1997).

Another sensory system of interest is the pineal eye, which is an extra light-sensing organ that fish have, separate from the eyes. Despite degeneration and the loss of visual capacity in their bilateral eyes, cavefish show little or no morphological differences in their pineal eye compared to surface fish (Grunewald-Lowenstein, 1956; Herwig, 1976; Langecker, 1992; Omura, 1975). For example, the photoreceptor segments of pineal sensory cells are still present in cavefish. Electrophysiological studies have also suggested that the pineal photosensory function still persists in adult blind cavefish (Tabata, 1982). In addition, surface fish and cavefish show the same level of the pineal-based shadow response in the larval stage; fish swim toward a water surface when a shadow is cast (Yoshizawa and Jeffery, 2008). Yoshizawa et al. assumed that the pineal organ is retained in cavefish because of its thermo-sensing ability, and its light-sensing ability can be just residual. For example, cavefish and surface fish start spawning according to changes in temperature, and importantly, the pineal organ responds to this temperature change and elicits a melatonin hormone that regulates both the circadian rhythm and the reproductive behavior in other fish species (Falcón et al., 2009; Zachmann et al., 1992). In the dark cave environment, the circadian rhythm may not important. Indeed, cavefish show largely reduced sleep and attenuated circadian rhythm (see Chapters 16 and 17; Beale et al., 2013; Duboué et al., 2011, 2012; Yoshizawa et al., 2015); however, reproduction is critical. In future studies, it will be exciting to determine if the cavefish pineal organ produces the melatonin hormone in response to temperature change, and also to determine what role melatonin plays in mating. Such studies will provide an example of how the physiological functions of two light-sensing organs—the binocular eyes and pineal organ—directed them toward different evolutionary trajectories.

CONCLUDING REMARKS

To summarize how the evolution of neural regulation affects foraging behavior in cavefish, we have seen that many cave-associated behaviors are under the control of eye regression and/or *hh* signaling (Figure 13.4). This is because (1) the eye is a large sensing organ in the cranial area, so a smaller or regressed eye opens up space for other sensing organs to dominate; and (2) pleiotropic *hh* signaling affects many developmental processes concurrently; however, we now know that these are not the only pathways that contribute to sensory and behavioral adaptation. Gaining a full picture of the adaptation mechanism will help us resolve how animals change in evolutionary processes.

A. mexicanus cavefish have many features that make them an excellent model vertebrate to investigate the adaptive evolution of sensory systems. These cavefish have evolved under simple and clear selection pressures, they are easy to rear in the laboratory, they require a relatively short generation time (5-6 months), and they display outstanding adaptive sensory systems and behaviors. The next frontier in cavefish research is, therefore, to identify more

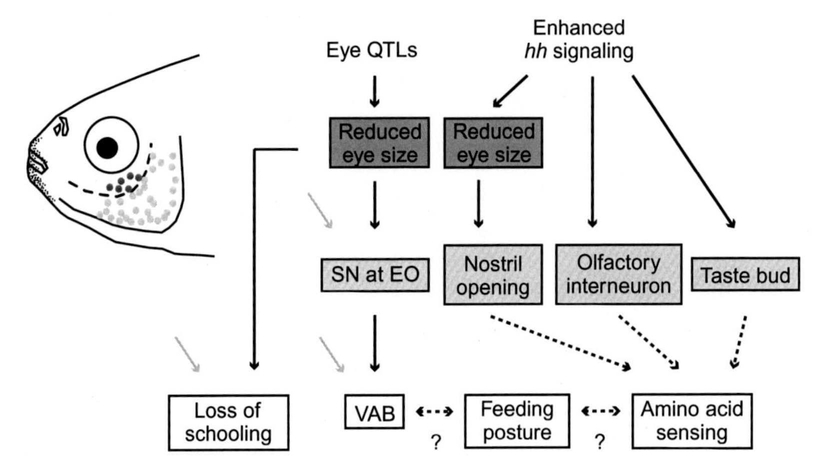

FIGURE 13.4 Schematic diagram of the relationship between molecular factors, sensory systems, and behaviors. Eye QTLs and enhanced *hh* signaling contribute to eye reduction (*hh* contribution was tested in Yamamoto et al., 2004), and reduced eye size promotes the increase of SN at the eye orbit (EO) (*shh*-independent, Yoshizawa et al., 2012) or the nasal epithelium area (Yamamoto et al., 2003). Enhanced *hh* signaling promotes the increase in the number of tastebuds (Yamamoto et al., 2009) and GABAergic interneurons in the olfactory bulb (Menuet et al., 2007). The nostril opening is positively correlated with amino acid-sensing ability (Bibliowicz et al., 2013), and it is likely that the increased tastebuds and olfactory interneurons also have a positive effect on amino acid-sensing. VAB level depends upon the number of SN at the eye orbit, but other factors (i.e., albinism) may also contribute to the VAB level (see also Chapter 9). The QTL for feeding posture is colocalized with the QTL for VAB and amino acid-sensing, suggesting a shared mechanism such as internal state (i.e., motivation). The increase of EO SN and loss of schooling are promoted by eye regression; other genetic factors may be involved (Kowalko et al., 2013a; Yoshizawa et al., 2012). Solid black arrows indicate defined relationships, and dotted arrows are presumptive relationships. Gray arrows indicate proposed involvement of other genetic factors.

of the genes and mutations involved in adaptation. This will open a new field where genetics, ontogeny, neuroscience, phylogeny, and ecology are integrated.

The recent development of powerful tools has led to a wealth of important information we can use to unravel evolutionary mysteries, including available genome sequences (AstMex102 from the Ensembl genome browser at www.ensembl.org), available embryology techniques that up- and down-regulate genes (Gross et al., 2009; Hinaux et al., 2014; Yamamoto and Jeffery, 2000; Yamamoto et al., 2004, 2009), defined embryonic and larval stages (Hinaux et al., 2011), transcriptome datasets (Gross et al., 2013; Hinaux et al., 2013), and defined evolutionary relationships among populations (Bradic et al., 2012, 2013; Coghill et al., 2014; Gross, 2012; Ornelas-García et al., 2008). Furthermore, advances in transgenic capability (Elipot et al., 2014) and genomic engineering methods (TALEN and CRISPR/Cas9 technology; Gaj et al., 2013; Hwang et al., 2013; Ma et al., 2015) now allow us to perform more pinpointed studies. These available techniques are filling the technical gap between the *A. mexicanus* system and other model animal systems, including zebrafish and medaka, and they highlight the powerful advantages of this system: clear selection pressures and many adapted traits.

Overall, with new information and better experimental techniques, we can explore the *A. mexicanus* system as an evolutionary vertebrate model, ultimately allowing us to comprehensively understand the evolutionary processes through which genomic and developmental shifts produce enhanced or co-opted adaptive sensory system.

REFERENCES

Abdel-Latif, H., Hassan, E.S., von Campenhausen, C., 1990. Sensory performance of blind Mexican cave fish after destruction of the canal neuromasts. Naturwissenschaften 77, 237–239.

Albert, A.Y.K., Sawaya, S., Vines, T.H., Knecht, A.K., Miller, C.T., Summers, B.R., Balabhadra, S., Kingsley, D.M., Schluter, D., 2008. The genetics of adaptive shape shift in stickleback: pleiotropy and effect size. Evolution 62, 76–85.

Albertson, R.C., Streelman, J.T., Kocher, T.D., 2003. Directional selection has shaped the oral jaws of Lake Malawi cichlid fishes. Proc. Natl. Acad. Sci. U. S. A. 100, 5252–5257.

Bak-Coleman, J., Coombs, S., 2014. Sedentary behavior as a factor in determining lateral line contributions to rheotaxis. J. Exp. Biol. 217, 2338–2347.

Baker, C.F., Montgomery, J.C., 1999. Lateral line mediated rheotaxis in the Antarctic fish *Pagothenia borchgrevinki*. Polar Biol. 21, 305–309.

Beale, A., Guibal, C., Tamai, T.K., Klotz, L., Cowen, S., Peyric, E., Reynoso, V.H., Yamamoto, Y., Whitmore, D., 2013. Circadian rhythms in Mexican blind cavefish *Astyanax mexicanus* in the lab and in the field. Nat. Commun. 4, 2769.

Bibliowicz, J., Alié, A., Espinasa, L., Yoshizawa, M., Blin, M., Hinaux, H., Legendre, L., Père, S., Rétaux, S., 2013. Differences in chemosensory response between eyed and eyeless *Astyanax mexicanus* of the Rio Subterráneo cave. EvoDevo 4, 25.

Borowsky, R., 2008. Restoring sight in blind cavefish. Curr. Biol. 18, R23–R24.

Bradic, M., Beerli, P., Garcia-de Leon, F.J., Esquivel-Bobadilla, S., Borowsky, R.L., 2012. Gene flow and population structure in the Mexican blind cavefish complex (*Astyanax mexicanus*). BMC Evol. Biol. 12, 9. http://dx.doi.org/10.1186/1471-2148-12-9.

Bradic, M., Teotónio, H., Borowsky, R.L., 2013. The population genomics of repeated evolution in the blind cavefish *Astyanax mexicanus*. Mol. Biol. Evol. 30, 2383–2400.

Coffin, A.B., Zeddies, D.G., Fay, R.R., Brown, A.D., Alderks, P.W., Bhandiwad, A.A., Mohr, R.A., Gray, M.D., Rogers, P.H., Sisneros, J.A., 2014. Use of the swim bladder and lateral line in near-field sound source localization by fish. J. Exp. Biol. 217, 2078–2088.

Coghill, L.M., Darrin Hulsey, C., Chaves-Campos, J., García de Leon, F.J., Johnson, S.G., 2014. Next generation phylogeography of cave and surface *Astyanax mexicanus*. Mol. Phylogenet. Evol. 79, 368–374.

Coombs, S., Görner, P., Münz, H., 1989. The Mechanosensory Lateral Line. Springer-Verlag, New York.

Coombs, S., Bleckmann, H., Fay, R.R., Popper, A.N., 2013. The lateral line system. In: Springer Handbook of Auditory Research, vol. 48. Springer, New York.

Culver, D.C., 1982. Cave Life, Evolution and Ecology. Harvard University Press, Cambridge.

Culver, D.C., Pipan, T., 2009. The Biology of Caves and Other Subterranean Habitats. Oxford University Press, Oxford.

Duboué, E.R., Keene, A.C., Borowsky, R.L., 2011. Evolutionary convergence on sleep loss in cavefish populations. Curr. Biol. 21, 671–676.

Duboué, E.R., Borowsky, R.L., Keene, A.C., 2012. β-Adrenergic signaling regulates evolutionarily derived sleep loss in the Mexican cavefish. Brain Behav. Evol. 80, 233–243.

Eigenmann, C.H., 1909. Cave Vertebrates of America: A Study in Degenerative Evolution. The Carnegie Institution of Washington, Washington, DC, p. 241.

Eldredge, N., Gould, S.J., 1972. Punctuated equilibria: an alternative to phyletic gradualism. In: Schopf, T.J.M. (Ed.), Models in Paleobiology. Freeman, Cooper and Company, San Francisco, CA, pp. 82–115.

Elipot, Y., Legendre, L., Père, S., Sohm, F., Rétaux, S., 2014. *Astyanax* transgenesis and husbandry: how cavefish enters the laboratory. Zebrafish 11, 291–299.

Falcón, J., Besseau, L., Fuentès, M., Sauzet, S., Magnanou, E., Boeuf, G., 2009. Structural and functional evolution of the pineal melatonin system in vertebrates. Ann. N. Y. Acad. Sci. 1163, 101–111.

Faucher, K., Parmentier, E., Becco, C., Vandewalle, N., Vandewalle, P., 2010. Fish lateral system is required for accurate control of shoaling behaviour. Anim. Behav. 79, 679–687.

Gaj, T., Gersbach, C.A., Barbas, C.F., 2013. ZFN, TALEN, and CRISPR/Cas-based methods for genome engineering. Trends Biotechnol. 31, 397–405.

Gross, J.B., 2012. The complex origin of *Astyanax* cavefish. BMC Evol. Biol. 12, 105. http://dx.doi.org/10.1186/1471-2148-12-105.

Gross, J.B., Borowsky, R., Tabin, C.J., 2009. A novel role for Mc1r in the parallel evolution of depigmentation in independent populations of the cavefish *Astyanax mexicanus*. PLoS Genet. 5 (1), e1000326. http://dx.doi.org/10.1371/journal.pgen.1000326.

Gross, J.B., Furterer, A., Carlson, B.M., Stahl, B.A., 2013. An integrated transcriptome-wide analysis of cave and surface dwelling *Astyanax mexicanus*. PLoS ONE 8, e55659.

Grunewald-Lowenstein, M., 1956. Influence of light and darkness on the pineal body in *Astyanax mexicanus* (Filippi). Zoologica 41, 119–128.

Herwig, H.J., 1976. Comparative ultrastructural investigations of pineal organ of blind cave fish, *Anoptichthys jordani*, and its ancestor, eyed river fish, *Astyanax mexicanus*. Cell Tissue Res. 167, 297–324.

Hill, L.G., 1969. Feeding and food habits of the spring cavefish, *Chologaster agassizi*. Am. Midl. Nat. 82, 110–116.

Hinaux, H., Pottin, K., Chalhoub, H., Pere, S., Elipot, Y., Legendre, L., Rétaux, S., 2011. A developmental staging table for *Astyanax mexicanus* surface fish and Pachon cavefish. Zebrafish 8, 155–165.

Hinaux, H., Poulain, J., da Silva, C., Noirot, C., Jeffery, W.R., Casane, D., Rétaux, S., 2013. De novo sequencing of *Astyanax mexicanus* surface fish and Pachón cavefish transcriptomes reveals enrichment of mutations in cavefish putative eye genes. PLoS ONE 8, e53553.

Hinaux, H., Blin, M., Fumey, J., Legendre, L., Heuzé, A., Casane, D., Rétaux, S., 2014. Lens defects in *Astyanax mexicanus* cavefish: evolution of crystallins and a role for *alphaA-crystallin*. Dev. Neurobiol. 75 (5), 505–521.

Hüppop, K., 1987. Food-finding ability in cave fish. Int. J. Speleol. 16, 59–66.

Hwang, W.Y., Fu, Y., Reyon, D., Maeder, M.L., Tsai, S.Q., Sander, J.D., Peterson, R.T., Yeh, J.-R.J., Joung, J.K., 2013. Efficient genome editing in zebrafish using a CRISPR-Cas system. Nat. Biotechnol. 31, 227–229.

Irschick, D.J., Albertson, R.C., Brennan, P., Podos, J., Johnson, N., Patek, S., Dumont, E., 2013. Evo-devo beyond morphology: from genes to resource use. Trends Ecol. Evol. 28, 267–273.

Jeffery, W.R., 2005. Adaptive evolution of eye degeneration in the Mexican blind cavefish. J. Hered. 96, 185–196.

Kohashi, T., Oda, Y., 2008. Initiation of Mauthner- or non-Mauthner-mediated fast escape evoked by different modes of sensory input. J. Neurosci. 28, 10641–10653.

Kowalko, J.E., Rohner, N., Rompani, S.B., Peterson, B.K., Linden, T.A., Yoshizawa, M., Kay, E.H., Weber, J., Hoekstra, H.E., Jeffery, W.R., et al., 2013a. Loss of schooling behavior in cavefish through sight-dependent and sight-independent mechanisms. Curr. Biol. 23, 1874–1883.

Kowalko, J.E., Rohner, N., Linden, T.A., Rompani, S.B., Warren, W.C., Borowsky, R., Tabin, C.J., Jeffery, W.R., Yoshizawa, M., 2013b. Convergence in feeding posture occurs through different genetic loci in independently evolved cave populations of *Astyanax mexicanus*. Proc. Natl. Acad. Sci. U. S. A. 110, 16933–16938.

Krause, J., 1993. The influence of hunger on shoal size choice by three-spined sticklebacks, *Gasterosteus aculeatus*. J. Fish Biol. 43, 775–780.

Lang, H.H., 1980. Surface wave discrimination between prey and nonprey by the backswimmer *Notonecta glauca L.* (Hemiptera, Heteroptera). Behav. Ecol. Sociobiol. 6, 233–246.

Langecker, T.G., 1992. Persistence of ultrastructurally well-developed photoreceptor cells in the pineal organ of a phylogenetically old cave-dwelling population of *Astyanax fasciatus* Cuvier, 1819 (Teleostei, Characidae). Zeitschrift Fur Zool. Syst. Und Evol. 30, 287–296.

Lyon, E.P., 1904. On rheotropism in fishes. Am. J. Physiol. 12, 149–161.

Ma, L., Jeffery, W.R., Essner, J.J., Kowalko, J.E., 2015. Genome editing using TALENs in blind Mexican cavefish, *Astyanax mexicanus*. PLoS ONE 10, e0119370.

Mayr, E., 1963. Animal Species and Evolution, first ed. Belknap Press, Cambridge.

McHenry, M.J., Van Netten, S.M., 2007. The flexural stiffness of superficial neuromasts in the zebrafish (*Danio rerio*) lateral line. J. Exp. Biol. 210, 4244–4253.

McHenry, M.J., Strother, J.A., van Netten, S.M., 2008. Mechanical filtering by the boundary layer and fluid–structure interaction in the superficial neuromast of the fish lateral line system. J. Comp. Physiol. A Neuroethol. Sens. Neural Behav. Physiol. 194, 795–810.

Menuet, A., Alunni, A., Joly, J.S., Jeffery, W.R., Rétaux, S., 2007. Expanded expression of sonic hedgehog in *Astyanax* cavefish: multiple consequences on forebrain development and evolution. Development 134, 845–855.

Mitchell, R.W., Russell, W.H., Elliott, W.R., 1977. Mexican eyeless characin fishes, genus *Astyanax*: environment, distribution, and evolution. In: Mackey, C., Barnett, G.E. (Eds.), Special Publications of the Museum of Texas Technological University, vol. 12. Texas Tech Press, TX.

Montgomery, J.C., Macdonald, J.A., 1987. Sensory tuning of lateral line receptors in Antarctic fish to the movements of planktonic prey. Science 235, 195–196.

Montgomery, J.C., Baker, C.F., Carton, A.G., 1997. The lateral line can mediate rheotaxis in fish. Nature 389, 960–963.

Mougel, F., Solignac, M., Vautrin, D., Baudry, E., Ogden, J., Tchapla, A., Schweitz, H., Gilbert, H., 2012. Quantitative traits loci (QTL) involved in body colour, wing morphometry, cuticular hydrocarbons and venom components in honeybee. Apidologie 43, 162–181.

Münz, H., 1989. Functional organization of the lateral line periphery. In: Coombs, S., Görner, P., Münz, H. (Eds.), The Mechanosensory Lateral Line. Springer-Verlag, New York.

O'Quin, K.E., Yoshizawa, M., Doshi, P., Jeffery, W.R., 2013. Quantitative genetic analysis of retinal degeneration in the blind cavefish *Astyanax mexicanus*. PLoS ONE 8 (2), e57281. http://dx.doi.org/10.1371/journal.pone.0057281.

Omura, Y., 1975. Influence of light and darkness on ultrastructure of pineal organ in blind cave fish, *Astyanax mexicanus*. Cell Tissue Res. 160, 99–112.

Onishi, K., Horiuchi, Y., Ishigoh-Oka, N., Takagi, K., Ichikawa, N., Maruoka, M., Sano, Y., 2007. A QTL cluster for plant architecture and its ecological significance in asian wild rice. Breed. Sci. 57, 7–16.

Ornelas-García, C.P., Domínguez-Domínguez, O., Doadrio, I., 2008. Evolutionary history of the fish genus *Astyanax* Baird & Girard (1854) (Actinopterygii, Characidae) in Mesoamerica reveals multiple morphological homoplasies. BMC Evol. Biol. 8, 340. http://dx.doi.org/10.1186/1471-2148-8-340.

Partridge, B.L., 1982. The structure and function of fish schools. Sci. Am. 246, 114–123.

Partridge, B.L., Pitcher, T.J., 1980. The sensory basis of fish schools: relative roles of lateral line and vision. J. Comp. Physiol. A Neuroethol. Sens. Neural Behav. Physiol. 135, 315–325.

Parzefall, J., 1983. Field observation in epigean and cave populations of Mexican characid *Astyanax mexicanus* (Pisces, Characidae). Mém. Biospéléol. 10, 171–176.

Parzefall, J., Senkel, S., 1986. Schooling behavior in cavernicolous fish and their epigean conspecifics. In: 9 Congr. Int. Espeleol. Barcelona, pp. 107–109.

Pitcher, T.J., Partridge, B.L., Wardle, C.S., 1976. A blind fish can school. Science 194, 963–965.

Plath, M., Schlupp, I., 2008. Parallel evolution leads to reduced shoaling behavior in two cave dwelling populations of Atlantic mollies (Poecilia mexicana, Poeciliidae, Teleostei). Environ. Biol. Fishes 82, 289–297.

Popper, A.N., 1970. Auditory capacities of the Mexican blind cave fish (*Astyanax jordani*) and its eyed ancestor (*Astyanax mexicanus*). Anim. Behav. 18, 552–562.

Popper, A.N., 1971. Morphology of Weberian ossicles of two species of genus *Astyanax* (Ostariophysi – Characidae). J. Morphol. 133, 179–188.

Poulson, T.L., 1963. Cave adaptation in amblyopsid fishes. Am. Midl. Nat. 70, 257–290.

Poulson, T.L., White, W.B., 1969. The cave environment. Science 165, 971–981.

Protas, M., Tabansky, I., Conrad, M., Gross, J.B., Vidal, O., Tabin, C.J., Borowsky, R., 2008. Multi-trait evolution in a cave fish, *Astyanax mexicanus*. Evol. Dev. 10, 196–209.

Rétaux, S., Casane, D., 2013. Evolution of eye development in the darkness of caves: adaptation, drift, or both? EvoDevo 4, 26.

Rohner, N., Jarosz, D.F., Kowalko, J.E., Yoshizawa, M., Jeffery, W.R., Borowsky, R.L., Lindquist, S., Tabin, C.J., 2013. Cryptic variation in morphological evolution: HSP90 as a capacitor for loss of eyes in cavefish. Science 342, 1372–1375.

Schemmel, C., 1967. Vergleichende Untersuchungen an den Hautsinnesorganen ober– und unterirdisch lebender *Astyanax*-Formen. Z. Morph. Tiere 61, 255–316.

Schemmel, C., 1980. Studies on the genetics of feeding behavior in the cave fish *Astyanax mexicanus f. anoptichthys*. An example of apparent monofactorial inheritance by polygenes. Z. Tierpsychol. 53, 9–22.

Soares, D., Niemiller, M.L., 2013. Sensory adaptations of fishes to subterranean environments. Bioscience 63, 274–283.

Strecker, U., Hausdorf, B., Wilkens, H., 2012. Parallel speciation in *Astyanax* cave fish (Teleostei) in Northern Mexico. Mol. Phylogenet. Evol. 62, 62–70.

Tabata, M., 1982. Persistence of pineal photosensory function in blind cave fish, *Astyanax mexicanus*. Comp. Biochem. Physiol.: Physiology 73, 125–127.

Teyke, T., 1988. Flow-field, swimming velocity and boundary layer; parameters which affect the stimulus for the lateral line organ in fish. J. Comp. Physiol. A 163, 53–61.

Teyke, T., 1990. Morphological differences in neuromasts of the blind cave fish *Astyanax hubbsi* and the sighted river fish *Astyanax mexicanus*. Brain Behav. Evol. 35, 23–30.

Van Netten, S.M., McHenry, M.J., 2014. The biophysics of the fish lateral line. In: The Lateral Line System. In: Coombs, S., Bleckmann, H., Fay, R.R., Popper, A.N. (Eds.), Springer Handbook of Auditory Research, vol. 48. Springer, New York, pp. 99–119.

Van Trump, W.J., McHenry, M.J., 2013. The lateral line system is not necessary for rheotaxis in the Mexican blind cavefish (*Astyanax fasciatus*). Integr. Comp. Biol. 53, 799–809.

Varatharasan, N., Croll, R.P., Franz-Odendaal, T., 2009. Taste bud development and patterning in sighted and blind morphs of *Astyanax mexicanus*. Dev. Dyn. 238, 3056–3064.

Voneida, T.J., Fish, S.E., 1984. Central nervous-system changes related to the reduction of visual input in a naturally blind fish (*Astyanax hubbsi*). Am. Zool. 24, 775–782.

Wilkens, H., 1988. Evolution and genetics of epigean and cave *Astyanax fasciatus* (Characidae, Pisces)—support for the neutral mutation theory. In: Hecht, M.K., Wallace, B. (Eds.), Evolutionary Biology. Plenum Publishing Corporation, New York, pp. 271–367.

Wilkens, H., Hüppop, K., 1986. Sympatric speciation in cave fishes? Studies on a mixed population of epigean and hypogean Astyanax (Characidae, Pisces). Zeitschrift Fur Zool. Syst. Und Evol. 24, 223–230.

Wilkens, H., Strecker, U., 2003. Convergent evolution of the cavefish *Astyanax* (Characidae, Teleostei): genetic evidence from reduced eye-size and pigmentation. Biol. J. Linn. Soc. Lond. 80, 545–554.

Wilson, E.V., Thomas, R.Q., Evans, L.M., 2004. Macrobrachium as a possible determinant of *Astyanax fasciatus* distribution in a neotropical lowland stream. Dartmouth Stud. Trop. Ecol. 2004, 99–102.

Wright, D., Rubin, C.-J., Martinez Barrio, A., Schütz, K., Kerje, S., Brändström, H., Kindmark, A., Jensen, P., Andersson, L., 2010. The genetic architecture of domestication in the chicken: effects of pleiotropy and linkage. Mol. Ecol. 19, 5140–5156.

Wullimann, M.F., Grothe, B., 2013. The central nervous organization of the lateral line system. In: The Lateral Line System. In: Coombs, S., Bleckmann, H., Fay, R.R., Popper, A.N. (Eds.), Springer Handbook of Auditory Research, vol. 48. Springer, New York, pp. 195–251.

Yamamoto, Y., Jeffery, W.R., 2000. Central role for the lens in cave fish eye degeneration. Science 289, 631–633.

Yamamoto, Y., Espinasa, L., Stock, D.W., Jeffery, W.R., 2003. Development and evolution of craniofacial patterning is mediated by eye-dependent and -independent processes in the cavefish *Astyanax*. Evol. Dev. 5, 435–446.

Yamamoto, Y., Stock, D.W., Jeffery, W.R., 2004. Hedgehog signalling controls eye degeneration in blind cavefish. Nature 431, 844–847.

Yamamoto, Y., Byerly, M.S., Jackman, W.R., Jeffery, W.R., 2009. Pleiotropic functions of embryonic sonic hedgehog expression link jaw and taste bud amplification with eye loss during cavefish evolution. Dev. Biol. 330, 200–211.

Yoshizawa, M., Jeffery, W.R., 2008. Shadow response in the blind cavefish *Astyanax* reveals conservation of a functional pineal eye. J. Exp. Biol. 211, 292–299.

Yoshizawa, M., Jeffery, W.R., 2011. Evolutionary tuning of an adaptive behavior requires enhancement of the neuromast sensory system. Commun. Integr. Biol. 4, 89–91.

Yoshizawa, M., Gorički, Š., Soares, D., Jeffery, W.R., 2010. Evolution of a behavioral shift mediated by superficial neuromasts helps cavefish find food in darkness. Curr. Biol. 20, 1631–1636.

Yoshizawa, M., Yamamoto, Y., O'Quin, K.E., Jeffery, W.R., 2012. Evolution of an adaptive behavior and its sensory receptors promotes eye regression in blind cavefish. BMC Biol. 10, 108.

Yoshizawa, M., O'Quin, K.E., Jeffery, W.R., 2013. QTL clustering as a mechanism for rapid multitrait evolution. Commun. Integr. Biol. 6, e24548.

Yoshizawa, M., Jeffery, W.R., Van Netten, S., McHenry, M., 2014. The sensitivity of lateral line receptors and their role in the behavior of Mexican blind cavefish (*Astyanax mexicanus*). J. Exp. Biol. 217, 886–895.

Yoshizawa, M., Robinson, B.G., Duboué, E.R., Masek, P., Jaggard, J.B., O'Quin, K.E., Borowsky, R.L., Jeffery, W.R., Keene, A.C., 2015. Distinct genetic architecture underlies the emergence of sleep loss and prey-seeking behavior in the Mexican cavefish. BMC Biol. 13, 15.

Zachmann, A., Falcon, J., Knijff, S.C.M., Bolliet, V., Ali, M.A., 1992. Effects of photoperiod and temperature on rhythmic melatonin secretion from the pineal organ of the white sucker (*Catostomus commersoni*) in vitro. Gen. Comp. Endocrinol. 86, 26–33.

Chapter 14

Feeding Behavior, Starvation Response, and Endocrine Regulation of Feeding in Mexican Blind Cavefish (*Astyanax fasciatus mexicanus*)

Hélène Volkoff
Departments of Biology and Biochemistry, Memorial University of Newfoundland, St. John's, Newfoundland and Labrador, Canada

INTRODUCTION

The Mexican tetra (*Astyanax fasciatus*) belongs to the family Characidae (Javonillo et al., 2010), which includes economically important species fish with regard to fisheries or aquaculture (e.g., piranhas, pacus, tetras; Hill and Yanong, 2010) and consists of eyed surface and eyeless cave forms (Culver and Pipan, 2009). The latter inhabit caves and are characterized by the absence of eyes and melanin pigmentation (Culver and Pipan, 2009; Protas and Jeffery, 2012). Cavefish also display behaviors that differ from those of surface forms: they show increased swimming/exploratory and feeding behaviors (Rétaux and Elipot, 2013); have lost sleep—perhaps to increase wakefulness and chances to find food (Duboué et al., 2011); do not present clear endogenous rhythmic activities (Boujard and Leatherland, 1992); aggressive (Espinasa et al., 2005) or schooling behaviors (Gregson and Burt de Perera, 2007); and have a reduced alarm behavior (Fricke, 1987).

FEEDING BEHAVIOR OF BLIND ASTYANAX

Cavefish have adapted to become very skilled at finding nourishment in an environment where food is often scarce (Rétaux and Elipot, 2013; Soares and Niemiller, 2013). Overall, surface fish placed in the dark are four times less efficient at finding food than cavefish (Hüppop, 1987) and rapidly show signs of starvation (Mitchell et al., 1977). The feeding success of cavefish can be attributed to

Biology and Evolution of the Mexican Cavefish. http://dx.doi.org/10.1016/B978-0-12-802148-4.00014-1
© 2016 Elsevier Inc. All rights reserved.

several characteristics. First, cavefish display increased food-searching behavior in part due to a constant locomotor activity, which is 1.5-fold higher than that of eyed fish (de Perera, 2004; Duboué et al., 2011; Salin et al., 2010). Cavefish also use a characteristic feeding posture, swimming at a 45° angle to the bottom, efficiently skimming and sampling the substrate, and increasing their food-finding ability (Kowalko et al., 2013; Schemmel, 1980), whereas surface fish stand vertically while feeding. Specialized anatomical features, such as increased number and size of tastebuds (Parzefall and Trajano, 2010), naris (Bibliowicz et al., 2013), teeth and lateral-line neuromasts (Yoshizawa et al., 2010) and larger jaws (Yamamoto et al., 2009), allow the cavefish to maximize detection of food (Elipot et al., 2013; Soares and Niemiller, 2013; see Chapter 13). Cavefish use vibration attraction behavior (swimming toward the source of a water disturbance; Yoshizawa et al., 2014) and generate suction flows that stimulate the lateral line by opening and closing their mouths when swimming (Holzman et al., 2014), which further increase their ability to find food. The mechanisms behind these specific cavefish behaviors might involve modified brain/hypothalamic areas and/or changes in neurotransmitter pathways. Indeed, cave and eyed *Astyanax* have different brain structures, with cavefish having larger hypothalami and olfactory bulbs (Langecker et al., 1995; Menuet et al., 2007; Peters, 1993), larger hypothalamic serotonergic cells (Elipot et al., 2013), and different brain transcription levels of receptors for neurotransmitters, such as glutamate or cannabinoids (Strickler and Soares, 2011).

METABOLISM AND RESPONSES TO FASTING OF BLIND *ASTYANAX*

Differences in energy homeostasis are seen between cave and surface fish. At the fed state, cavefish exhibit lower standard metabolic rates (Hervant and Malard, 2012), have different triglyceride and glycogen body contents (Hüppop, 1986; Salin et al., 2010; Strickler and Soares, 2011), and modified protein and lipid metabolism in the pineal gland (Omura, 1975), compared with surface fish. Low metabolic rates and high-energy reserves might enable cavefish to have higher survival rates than the surface form during fasting periods (Hervant and Malard, 2012; Salin et al., 2010). During fasting, both locomotor activity and oxygen consumption/metabolic rate decrease in the cave form, whereas they are little or not affected in the surface form (Salin et al., 2010). A number of metabolism-related genes are strongly expressed in cavefish, but not detected in surface fish, suggesting that cavefish possess a genome that is "enriched for metabolism-related processes," providing them with a more efficient metabolism (Gross et al., 2013). Paradoxically, if cavefish seem well adapted to long periods of fasting, in contrast with surface fish, they do not recover from a 2-month food deprivation after 12 days of re-feeding, suggesting that they might not be able to sustain multiple fasting events (Salin et al., 2010).

In addition, eyed-surface fish raised in the dark display endocrine and metabolic disturbances, such as reduced growth rates and gonad size, atrophy of the thyroid gland and adrenal cortex, large accumulations of adipose tissue and wasted musculature, suggesting that cavefish might have a modified endocrine system, enabling them to better function in darkness (Rasquin, 1949; Rasquin and Rosenbloom, 1954).

PEPTIDE SYSTEMS INVOLVED IN FEEDING AND FASTING IN *ASTYANAX*

In vertebrates, including mammals (Crespo et al., 2014; Sohn et al., 2013) and fish (Cerdá-Reverter, 2009; Shahjahan et al., 2014; Volkoff et al., 2005), feeding is regulated by several peptides produced by the brain—in particular the hypothalamus—and in the periphery. The structural differences seen between cave and surface fish brains, including the hypothalamus (Ito et al., 2007; Peters, 1993) might account for differences in foraging behavior and in mechanisms controlling hunger and satiety (Rétaux and Elipot, 2013; Sylvester et al., 2011). It is thus probable that brain peptides are involved in these behavioral changes. Recent studies show that hormones, metabolic enzymes and neurotransmitters might be involved in the regulation of the feeding of *Astyanax* (Table 14.1, Figure 14.1).

Orexins

Orexins (OXs) (or hypocretins), neuropeptides originally isolated in rats (Sakurai et al., 1998), have been identified in several fish species, including zebrafish, *Danio rerio* (Alvarez and Sutcliffe, 2002), goldfish, *Carassius auratus* (Hoskins et al., 2008), and red-bellied piranha, *Pygocentrus nattereri* (Volkoff, 2014a). In both mammals (de Lecea and Sutcliffe, 2013) and fish, OXs stimulate feeding and promote arousal (i.e., stimulate activity and reduce rest). In cavefish, peripheral injections of OX increase food consumption and locomotor activity (Penney and Volkoff, 2014), which is in line with previous observations in other fish species following peripheral injections [e.g., goldfish (Facciolo et al., 2012) and ornate wrasse *Thalassoma pavo* (Facciolo et al., 2009)] or central injections [e.g., goldfish (Volkoff et al., 1999; Nakamachi et al., 2006; Nisembaum et al., 2014) and zebrafish (Yokobori et al., 2011)].

It has been suggested that the primary action of OXs is to promote arousal, and that increases in feeding are a consequence of increased locomotor/searching behavior. In mammals, OX treatment induces arousal and locomotor activity (Hagan et al., 1999), and loss of OX neurons or mutations in the *ox* gene are both associated with excessive daytime sleepiness (narcolepsy) (Nishino and Mignot, 2011). Similarly, in zebrafish, larvae overexpressing OX display abnormal sleeping patterns (Prober et al., 2006) and increases in locomotor activity (Woods et al., 2014), and fish lacking the OX receptor show short and

TABLE 14.1 List of Appetite-Related Factors Examined in Cave *Astyanax* to Date, Showing the Effects of *In Vivo* Injections, and Fasting- and Meal-Induced Changes in Expression in *Astyanax* and in Other Fish Species, as well as the Known Interactions between Peptides and the Putative Role of Each Factor in Blind *Astyanax*

	Effects of *In Vivo* Injections on Feeding		Fasting-Induced Changes in Expression		Periprandial Changes in Expression		Interactions between Factors in *Astyanax*	Possible Role in Feeding in *Astyanax*
	Blind *Astyanax*	Other Fish Species (majority)	Blind *Astyanax*	Other Fish Species (Majority)	Blind *Astyanax*	Other Fish Species (Majority)		
Neuropeptides								
Orexin	↑	↑	↑	↑	✓	✓	Brain expression increases after apelin or ghrelin injection	Promotes feeding and locomotion
CART	?	↓	?	↓	?	✓	Brain expression not affected by apelin, ghrelin, orexin, or CCK injections	?

PYY	?	↓	×	↓	✓	✓		Inhibits feeding
CCK	↓	↓	×	↓	×	✓		Inhibits feeding
Ghrelin	↑	↑	?	↑	?	✓		Promotes feeding
Apelin	↑	↑	?	↑	?	?	Brain expression decreases after CCK injection	Promotes feeding
Enzymes								
TH	?	?	↑	?	?	?	Brain expression increases after apelin or orexin injection	Promotes feeding and locomotion?
TOR	?	?	?	?	?	?	Brain expression increases after apelin or ghrelin injection	Promotes feeding and locomotion?

↑: Increase; ↓: decrease; ×: no changes; ✓: changes; ?: not known/not examined. See text for other species names and for references.

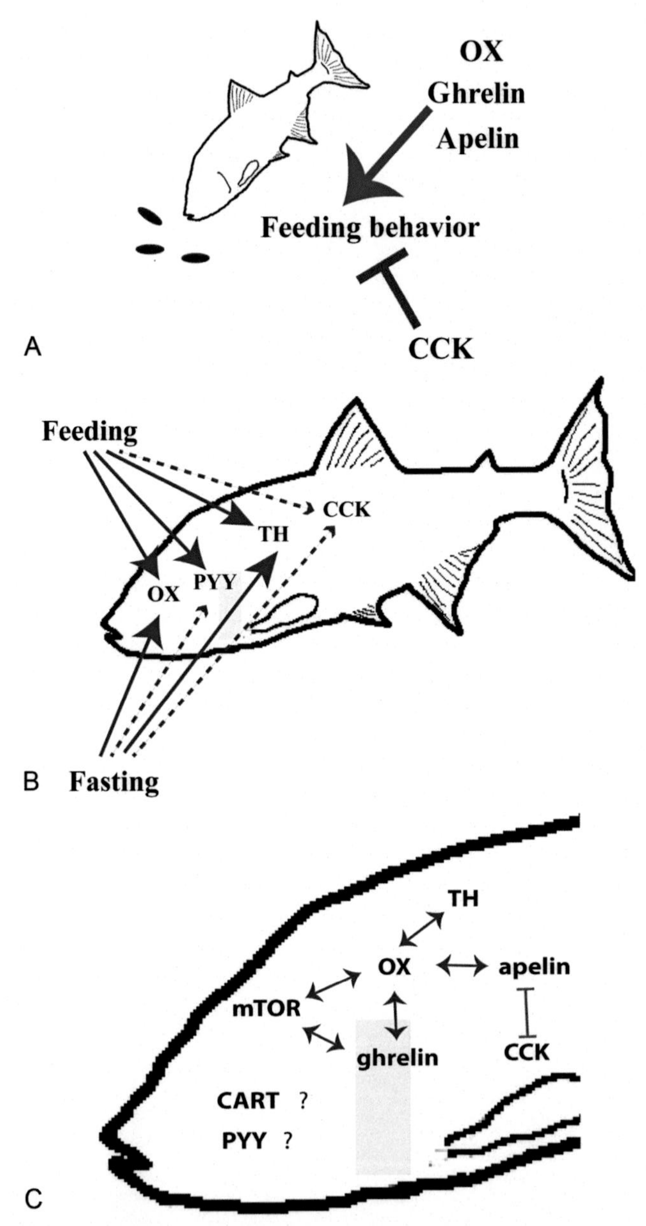

FIGURE 14.1 Summary of known appetite-regulating peptides in blind *Astyanax*. (A) Effects of injected peptides on feeding behavior in blind *Astyanax*. The arrows with arrowheads and bars indicate stimulation and inhibition, respectively. (B) Effects of feeding (meal) and fasting on the expression of brain peptides. Solid arrows indicate an effect, whereas dashed arrows indicate no effect. (C) Known interactions between brain peptides. Double-ended arrows with arrowheads and bars indicate stimulatory and inhibitory interactions, respectively. A "?" indicates that no interactions have been detected. CART: cocaine- and amphetamine-regulated transcript; CCK: cholecystokinin; TOR: target of rapamycin; OX: orexin; PYY: peptide YY; TH: tyrosine hydroxylase.

fragmented sleep (Appelbaum et al., 2009; Panula, 2010). This hypothesis is in line with the fact that periods of increased locomotor activity are associated with increased activity of hypothalamic OX neurons in zebrafish (Naumann et al., 2010) and that in both Atlantic cod, *Gadus morhua*, and goldfish, brain OX messenger RNA (mRNA) expression levels are higher during the day, when fish are active, than during the night (Hoskins and Volkoff, 2012a,b).

In cavefish, brain OX mRNA expression levels increase 1 h before and decrease 1 h after a scheduled mealtime (Wall and Volkoff, 2013), suggesting that OX acts as a short-term hunger signal, as previously shown in Atlantic cod (Hoskins and Volkoff, 2012a; Xu and Volkoff, 2007), orange-spotted grouper, *Epinephelus coioides* (Yan et al., 2011), and goldfish (Hoskins and Volkoff, 2012b). The peak in OX expression before mealtime might be linked to food anticipatory activity (FAA), an increase in appetite and locomotor activity seen prior to mealtime in other fish such as goldfish (Aranda et al., 2001), zebrafish (Sanchez and Sanchez-Vazquez, 2009; Yu et al., 2006), European catfish, *Silurus glanis* (Bolliet et al., 2001), Somalian cavefish, *Phreatichthys andruzzii* (Cavallari et al., 2011), and cave loach *Nemacheilus evezardi* (Biswas and Ramteke, 2008), for which a daily scheduled mealtime induces FAA, regardless of photoperiod or even the actual presence of food.

Also consistent with previous studies [e.g., goldfish (Abbott and Volkoff, 2011; Nakamachi et al., 2006); winter flounder, *Pseudopleuronectes americanus* (Buckley et al., 2010); *Verasper moseri* (Amiya et al., 2012); Atlantic cod (Tuziak et al., 2014; Xu and Volkoff, 2007), red-bellied piranha (Volkoff, 2014a) and zebrafish (Novak et al., 2005)], food restriction increases OX brain mRNA transcription levels in cavefish (Wall and Volkoff, 2013), indicating a role in the long-term regulation of feeding in cavefish.

Although there is no report comparing mRNA OX levels between surface and cave Astyanax forms, OX brain mRNA levels in cave *Astyanax* are higher than in Buenos Aires tetra, *Hyphessobrycon anisitsi*, a characid surface species closely related to *Astyanax* (Javonillo et al., 2010), suggesting that the higher overall locomotor/feeding activity in cave *Astyanax* compared to surface forms might be mediated by an increase in OX levels (Wall and Volkoff, 2013).

Cocaine- and Amphetamine-Regulated Transcript

Cocaine- and amphetamine-regulated transcript (CART) is a brain peptide that acts as an anorexigenic factor in mammals (Dockray and de Lartigue, 2013), birds (Tachibana et al., 2003), and fish, including goldfish (Volkoff and Peter, 2000, 2001b), carp, *Cyprinus carpio* (Wan et al., 2012), zebrafish (Nishio et al., 2012), and channel catfish, *Ictalurus punctatus* (Peterson et al., 2012). There is no direct evidence—for example, *in vivo* injection studies—of CART playing a role in the regulation of feeding in blind *Astyanax*. In this species, the response to fasting on CART brain expression levels has not been examined to date, but CART brain expression levels are not affected by peripheral injections of other appetite regulators, such as cholecystokinin (CCK), apelin, ghrelin, or OX (Penney and Volkoff, 2014), suggesting that central CART might not be a major regulator of feeding in cave *Astyanax*; however, multiple CART forms can be present in some fish species, and one form might be more affected by fasting. For example, in both goldfish (Volkoff and Peter, 2001a) and carp (Wan et al., 2012), two forms of CART (CART 1 and CART 2) have been identified, the brain expressions of both forms being affected by fasting. In medaka (*Oryzias latipes*), six CART cDNAs have been identified, with only one form (form 3) being affected by starvation (Murashita and Kurokawa, 2011). Four CART forms have been reported in zebrafish (CART 1, 2, 3, and 4) (Nishio et al., 2012), and a decrease in brain expression of all four CART forms have been reported by using quantitative polymerase chain reaction (qPCR) following starvation (Nishio et al., 2012), although another study, using gene-specific *in situ* hybridization show fasting-induced reductions in the population of brain CART 2 and CART 4-expressing cells, but no changes in CART 1 and CART 3-expressing cells (Akash et al., 2014). Analyses of the *Astyanax* genome (Ensembl, www.ensembl.org/Astyanax_mexicanus/) show that two CART forms (CART 2, corresponding to the form examined by (Penney and Volkoff, 2014), and CART4) are present in this species, and it is thus possible that another CART form might be more involved in the regulation of the feeding of cave *Astyanax*.

Peptide YY

Peptide YY (PYY) is a member of the neuropeptide Y (NPY) family, which includes NPY, two forms of peptide YY (PYYa and PYYb, previously called PY), and the tetrapod pancreatic polypeptide (Cerda-Reverter and Larhammar, 2000; Sundstrom et al., 2013; Wahlestedt and Reis, 1993).

NPY, produced by both the brain and gut, acts as an orexigenic factor in both mammals (Perboni et al., 2013) and fish [e.g., goldfish (Lopez-Patino et al., 1999), zebrafish (Yokobori et al., 2012), and Mozambique tilapia, *Oreochromis mossambicus* (Kiris et al., 2007)]. Although the NPY gene is present in cave *Astyanax* (based upon the Emsembl database), to date, this peptide has never been examined for its physiological actions in *Astyanax*.

In mammals, PYY is primarily released from the gastrointestinal tract (GIT) in response to food ingestion, thus inducing satiety (Blevins et al., 2008; Troke et al., 2013; Zhang et al., 2012). PYY inhibits feeding in both mammals (Karra et al., 2009) and fish, as injections of PYY in goldfish (Gonzalez and Unniappan, 2010) and grass carp, *Ctenopharyngodon idellus* (Chen et al., 2013) reduce food intake.

In cavefish, PYY brain mRNA levels increase after feeding (Wall and Volkoff, 2013), similar to what is seen in the brain of goldfish (Gonzalez and Unniappan, 2010) and in the gut of Atlantic salmon, *Salmo salar* (Valen et al., 2011), and grass carp (Chen et al., 2014), further suggesting a role for PYY as a short-term satiety factor.

In cavefish, a 10-day fasting period does not affect PYY brain mRNA expression, suggesting that central PYY is not involved in long-term fasting in this species (Wall and Volkoff, 2013). Consistent with these observations in cavefish, a 6-day fasting does not affect PYY mRNA expression in either the gut or brain of Atlantic salmon (Murashita et al., 2009); however, fasting decreases PYY mRNA expression levels in the brains of goldfish (Gonzalez and Unniappan, 2010), the guts of yellowtail, *Seriola quinqueradata* (Murashita et al., 2006), and the guts of the sea lamprey, *Petromyzon marinus* (Montpetit et al., 2005). In red-bellied piranha, a 7-day fasting has no effect on brain PYY expression, but decreases gut PYY expression (Volkoff, 2014a). These data suggest that the response of PYY to feeding and fasting might be species-specific and depend upon the tissue examined (i.e., brain versus gut) and the duration of fasting.

Cholecystokinin

In both mammals (Boguszewski et al., 2010; de La Serre and Moran, 2013; Dockray, 2012) and fish (Nelson and Sheridan, 2006; Takei and Loretz, 2010), CCK is a brain-gut peptide that functions primarily as a satiety signal, as it inhibits food intake and induces the release of digestive enzymes from the intestine/pancreas and gallbladder.

In cavefish, peripheral injections of CCK reduce food intake (Penney and Volkoff, 2014), which is consistent with results in other fish species, for which injections [goldfish (Himick and Peter, 1994)] or oral administration [sea bass, *Dicentrarchus labrax* (Rubio et al., 2008)] of CCK decreases feeding, and oral administration of CCK antagonists increases food consumption [trout, *Oncorhynchus mykiss*, and sea bass (Gélineau and Boujard, 2001; Rubio et al., 2008)]. CCK injections in cavefish also induce a decrease in brain apelin, an orexigenic factor (see the succeeding text), mRNA expression (Penney and Volkoff, 2014), further suggesting an anorexigenic role for CCK. Interestingly, in larval zebrafish, overexpression of CCK induces increases in waking activity, similar to OX (Woods et al., 2014), but this behavior was not observed in CCK-injected blind *Astyanax* (Penney and Volkoff, 2014), or reported for any other fish species.

CCK brain expression does not display periprandial variations in cavefish (Wall and Volkoff, 2013), which contrasts with the postprandial increases in CCK brain [e.g., goldfish (Peyon et al., 1999), and channel catfish (Peterson et al., 2012)] and intestine [e.g., yellowtail (Murashita et al., 2006)] mRNA levels seen in other fish. Also, in cavefish, a 10-day fasting has no effect on CCK brain mRNA expression (Wall and Volkoff, 2013), in contrast to the fasting-induced decrease in CCK mRNA expression seen in the brain and/or intestine of several fish, including zebrafish (Koven and Schulte, 2012), winter flounder (MacDonald and Volkoff, 2009), cunner, *Tauogolabrus adspersus* (Babichuk and Volkoff, 2013), and seabreams, *Diplodus puntazzo* and *Diplodus sargus* (Micale et al., 2014). Similar to CART, multiple forms of CCK are present in fish, with usually one isoform being more involved in feeding regulation than others [e.g., Atlantic salmon (Murashita et al., 2009), catfish (Peterson et al., 2012), seabreams (Micale et al., 2014)]; however, analysis of the available *Astyanax* genome (emsembl) only reveals one CCK gene, suggesting that the lack of effect of fasting and feeding on cavefish CCK brain expression might not be due to presence of multiple forms of CCK. As CCK is generally more expressed in the intestine than the brain, it is possible that changes in brain expression might not have been high enough to be detected. Further studies examining CCK gut expression are needed to determine if CCK expression levels in the gut respond to fasting and feeding.

Ghrelin

Ghrelin, a peptide produced mainly by the stomach, stimulates appetite in both mammals (Verhulst and Depoortere, 2012) and fish (Jonsson, 2013). In cavefish, peripheral injections of ghrelin increase food intake (Penney and Volkoff, 2014), which is in line with previous ghrelin injection studies in fish [goldfish (Matsuda et al., 2006; Unniappan et al., 2004), brown trout, *Salmo trutta* (Jonsson, 2013; Tinoco et al., 2014), Mozambique tilapia (Riley et al., 2005) and rainbow trout (Shepherd et al., 2007)]. In cavefish, similarly to what is seen in goldfish (Miura et al., 2007), ghrelin injections induce an increase in OX brain expression (Penney and Volkoff, 2014), suggesting that the orexigenic actions of ghrelin might be mediated by OX in cavefish.

There is no published information about the effects of feeding status on ghrelin expression in cavefish. Although the nature of the effects of feeding status on ghrelin in fish are sometimes inconsistent between studies (see Jonsson, 2013, for review), several studies have shown that fasting and refeeding affects ghrelin mRNA expression and/or ghrelin plasma levels in a number of species [e.g., zebrafish and goldfish (Amole and Unniappan, 2009; Unniappan et al., 2004), tilapia (Fox et al., 2009), Ya-fish, *Schizothorax davidi* (Zhou et al., 2014), and Atlantic cod (Xu and Volkoff, 2009)]. The variability in the results might indicate that the regulatory role of ghrelin on food intake in fish is dependent upon the species considered, the type of ghrelin injected, the route of administration, as well as the observation time (short- versus long-term studies).

Apelin

In mammals, apelin and its receptor are expressed in a number of tissues, including brain, and peripheral tissues, such as GIT, and cardiovascular and reproductive systems, and apelin regulates several physiological functions, including cardiovascular, fluid homeostasis, metabolic, and digestive processes (Knauf et al., 2013), but its role on the regulation of feeding is unclear (reviewed in Castan-Laurell et al., 2011).

In cavefish, apelin injections induce increases in food intake and in OX brain expression (Penney and Volkoff, 2014), and injections of CCK induce a decrease in brain apelin expression, suggesting an orexigenic role for apelin in cavefish (Penney and Volkoff, 2014). The orexigenic actions of apelin in fish have been shown in previous studies; apelin injections increase food intake in goldfish (Volkoff and Wyatt, 2009) and Ya-fish (Lin et al., 2014), while fasting induces increases in apelin mRNA expression in the brain of goldfish (Volkoff and Wyatt, 2009), red-bellied piranha (Volkoff, 2014a), and Ya-fish (Lin et al., 2014). In addition, in goldfish, *in vitro* treatment of brain fragments with apelin induces increases in the expression of orexigenic peptides, for example, OX, and a decrease of CART (Volkoff, 2014b), and brain injections of spexin, a peptide that inhibits both basal and OX-induced stimulation of feeding, reducing brain mRNA expression of apelin (Wong et al., 2013).

Monoamines and Metabolic Enzymes

Tyrosine Hydroxylase and Monoamine Neurotransmitters

Neuropeptides are large transmitter molecules composed of 3-36 amino acids that are ribosomally synthesized as large precursor molecules in soma/dendrites, and subsequently cut out by enzymes (convertases). They are released extrasynaptically, and can reach distant receptors by diffusion (Hokfelt et al., 2013; Purves et al., 2001). By contrast, classic (or "small-molecule") neurotransmitters are much smaller than neuropeptides, synthesized locally within the axon terminal, and mainly stored in synaptic vesicles. They are released in the synaptic cleft and act over distances less than a micrometer (Hokfelt et al., 2013; Purves et al., 2001). Monoamine neurotransmitters (also known as "biogenic amines") include catecholamine (dopamine [DA], norepinephrine, and epinephrine), histamine, and serotonin. They are derived from aromatic amino acids like phenylalanine, tyrosine, and tryptophan via a cascade of enzymatic reactions (Hokfelt et al., 2013; Purves et al., 2001).

Tyrosine hydroxylase (TH), an enzyme that catalyzes the conversion of L-tyrosine to L-*DOPA* (L-dihydroxyphenylalanine), is rate-limiting in catecholamine synthesis and is commonly used as a marker in the identification of catecholaminergic neurons (Daubner et al., 2011). In mammals, several studies have implicated dopaminergic pathways, TH and DA in the regulation of food intake and locomotor activity. For example, DA-deficient mice are hypophagic and rapidly die of starvation (Hnasko et al., 2004), TH knockout mice, which lack DA, exhibit low feeding and locomotor behaviors (Zhou and Palmiter, 1995), and in rats, the number of midbrain cells containing TH is higher in animals with high spontaneous locomotor activity (Jerzemowska et al., 2012). TH appears to interact with appetite regulators, as, for example, in both rodents (Puskas et al., 2010) and lungfish (Lopez et al., 2009), OX-positive cells are in close proximity to TH-positive neurons, and in rodents, treatment with PYY activates TH neurons in the medulla (Blevins et al., 2008), while CCK increases the activity of midbrain TH-containing neurons (Skirboll et al., 1981).

In cavefish, fasting induces increases in TH brain mRNA levels (Wall and Volkoff, 2013), and brain TH expression levels increase 1 h after a scheduled feeding time in unfed fish (Wall and Volkoff, 2013), suggesting a role of TH in regulating feeding. An overall increase in TH mRNA expression in the brain of fasted *Astyanax* might be due, in part, to an increase in locomotor activity and food searching, as, in vertebrates, the dopaminergic system has been shown to be involved in the sleep-wake cycle (Oganesyan et al., 2009). Consistent with this hypothesis, in the zebrafish brain, the mRNA levels of both TH and OX of dominant/aggressive males—which are characterized by more pronounced feeding and locomotor activities—are higher than in subordinate fish (Filby et al., 2010; Pavlidis et al., 2011). Also similar to OX, brain mRNA levels of TH in blind *Astyanax* are higher than in Buenos Aires tetra, suggesting that the higher overall locomotor activity in cavefish might be mediated by increases in TH levels. This is in line with higher levels of catecholamines, L-tyrosine (Bilandžija et al., 2013; Elipot et al., 2014) and YWHAE (or 14-3-3 protein epsilon, a DA synthesis activator that also activates TH) (Strickler and Soares, 2011) reported in the brains of cave *Astyanax* compared to the surface form. Although TH brain enzymatic activities between the cave and surface forms are similar (Elipot et al., 2014), the TH inhibitor 3-iodo-L-tyrosine, which disrupts synthesis of DA and norepinephrine, has no effect on sleep in surface fish, but induces a near significant increase in sleep in cavefish (Duboué et al., 2012).

It is noteworthy that the increase in monoamine and TH levels is also linked to the lack of pigmentation seen in the cave form, as the catecholamine and melanin synthesis pathways both use L-tyrosine as a substrate, and inhibition of melanin synthesis (via mutations in the *oca2* [ocular and cutaneous albinism-2] gene) provide more L-tyrosine substrate for catecholamine synthesis (Bilandžija et al., 2013; Jeffery, 2012; see Chapter 8.

Peripheral injections of both apelin and OX—but not ghrelin and CCK—induce an increase in TH mRNA expression in the blind *Astyanax* brain as compared to saline-injected fish (Penney and Volkoff, 2014), suggesting that TH/catecholamines mediate in part the orexigenic actions of apelin and OX—but not ghrelin or CCK—in cavefish. An interaction between OX and TH has been shown in voles, where feeding and food-hoarding behaviors induce an increase in the number of both OX- and TH-positive cells in the hypothalamus (Zhang et al., 2011).

Differences in the profiles of other neurotransmitters have also been reported between cave and surface *Astyanax*, suggesting that several neurotransmitters might be involved in the development of differences in behaviors in the two forms. For example, higher levels of serotonin and a higher number of hypothalamic serotonin neurons are seen in the cave *Astyanax* compared to the surface form (Elipot et al., 2013), and both the *N*-methyl-D-aspartate receptor antagonist MK-801 and the β-adrenergic antagonist propranolol increase sleep in cavefish, but have little effect on surface fish (Duboué et al., 2012).

Target of Rapamycin

Target of rapamycin (TOR) is a highly conserved serine-threonine protein kinase that plays a central role in the regulation of cell growth, proliferation, and metabolism in all eukaryotic species (Hall, 2008; Wullschleger et al., 2006). In mammals, TOR plays a key role in energy homeostasis; TOR mRNA is detected in areas of the brain linked to the regulation of energy balance—i.e., the hypothalamus—and TOR activity decreases in fasted animals, and increases after feeding (Stefater and Seeley, 2010). Recent evidence also shows interactions between the TOR pathway and feeding-related hormones. For example, in rats, TOR is colocalized with CART in the hypothalamus (Inhoff et al., 2010) and with ghrelin in the stomach (Xu et al., 2010), and CCK (Lembke et al., 2011) and ghrelin (Martins et al., 2012) both activate the hypothalamic TOR signaling pathway. In fish, TOR has also been linked to feeding, metabolism, and growth. In zebrafish, fasting decreases liver TOR expression (Craig and Moon, 2011) and hepatic lipid accumulation requires the TOR pathway (Sapp et al., 2014). In carp, TOR gut expression increases as the fish grows and the gut develops (Jiang et al., 2013), and is regulated by arginine, which improves fish growth, and digestive and absorptive abilities (Chen et al., 2012). In fine flounder, *Paralichthys adspersus*, TOR is inactivated during fasting (Fuentes et al., 2013) and in salmon, several major kinases involved in the TOR pathway are regulated by nutritional status (Seiliez et al., 2008).

In blind *Astyanax*, TOR brain expression is increased by both apelin and ghrelin injections, but is not affected by either OX or CCK injections (Penney and Volkoff, 2014), suggesting that the actions of apelin and ghrelin might involve the intracellular PI3K/AKT/TOR signaling pathway.

CONCLUDING REMARKS

In summary, blind *Astyanax* have evolved to maximize their energy intake and metabolism in a subterranean habitat. Adaptations include: anatomical changes, such as improved sense of smell and detection of preys; behavioral changes, such as increased locomotion and a characteristic feeding posture; as well as physiological changes, such as a modified metabolism that confers them a higher resistance to fasting. Blind cavefish appear to have modified levels of certain neurotransmitters, such as dopaminergic and serotonergic, pathways, and perhaps signaling cascades, for example, those involving TOR. At the endocrine level, cavefish seem to exhibit typical vertebrate appetite-regulating peptides that interact with each other (Table 14.1, Figure 14.1), but their expression levels and functions might have undergone modifications in order to confer the cavefish a more efficient endocrine appetite-regulating system; however, further studies examining the role of other appetite regulators—leptin, NPY, galanin—are needed to better understand the complex endocrine regulation of feeding in *Astyanax*.

REFERENCES

Abbott, M., Volkoff, H., 2011. Thyrotropin releasing hormone (TRH) in goldfish (*Carassius auratus*): role in the regulation of feeding and locomotor behaviors and interactions with the orexin system and cocaine- and amphetamine-regulated transcript (CART). Horm. Behav. 59, 236–245.

Akash, G., Kaniganti, T., Tiwari, N.K., Subhedar, N.K., Ghose, A., 2014. Differential distribution and energy status-dependent regulation of the four CART neuropeptide genes in the zebrafish brain. J. Comp. Neurol. 522, 2266–2285.

Alvarez, C.E., Sutcliffe, J.G., 2002. Hypocretin is an early member of the incretin gene family. Neurosci. Lett. 324, 169–172.

Amiya, N., Mizusawa, K., Kobayashi, Y., Yamanome, T., Amano, M., Takahashi, A., 2012. Food deprivation increases the expression of the prepro-orexin gene in the hypothalamus of the barfin flounder, *Verasper moseri*. Zoolog. Sci. 29, 43–48.

Amole, N., Unniappan, S., 2009. Fasting induces preproghrelin mRNA expression in the brain and gut of zebrafish, *Danio rerio*. Gen. Comp. Endocrinol. 161, 133–137.

Appelbaum, L., Wang, G.X., Maro, G.S., Mori, R., Tovin, A., Marin, W., Yokogawa, T., Kawakami, K., Smith, S.J., Gothilf, Y., Mignot, E., Mourrain, P., 2009. Sleep-wake regulation and hypocretin-melatonin interaction in zebrafish. Proc. Natl. Acad. Sci. U. S. A. 106, 21942–21947.

Aranda, A., Madrid, J.A., Sanchez-Vazquez, F.J., 2001. Influence of light on feeding anticipatory activity in goldfish. J. Biol. Rhythms 16, 50–57.

Babichuk, N.A., Volkoff, H., 2013. Changes in expression of appetite-regulating hormones in the cunner (*Tautogolabrus adspersus*) during short-term fasting and winter torpor. Physiol. Behav. 120, 54–63.

Bibliowicz, J., Alie, A., Espinasa, L., Yoshizawa, M., Blin, M., Hinaux, H., Legendre, L., Pere, S., Rétaux, S., 2013. Differences in chemosensory response between eyed and eyeless *Astyanax mexicanus* of the Rio Subterraneo cave. EvoDevo 4, 25.

Bilandžija, H., Ma, L., Parkhurst, A., Jeffery, W.R., 2013. A potential benefit of albinism in *Astyanax* cavefish: downregulation of the *oca2* gene increases tyrosine and catecholamine levels as an alternative to melanin synthesis. PLoS One 8, e80823.

Biswas, J., Ramteke, A.K., 2008. Timed feeding synchronizes circadian rhythm in vertical swimming activity in cave loach *Nemacheilus evezardi*. Biol. Rhythm Res. 39, 405–412.

Blevins, J.E., Chelikani, P.K., Haver, A.C., Reidelberger, R.D., 2008. PYY(3-36) induces Fos in the arcuate nucleus and in both catecholaminergic and non-catecholaminergic neurons in the nucleus tractus solitarius of rats. Peptides 29, 112–119.

Boguszewski, C.L., Paz-Filho, G., Velloso, L.A., 2010. Neuroendocrine body weight regulation: integration between fat tissue, gastrointestinal tract, and the brain. Endokrynol. Pol. 61, 194–206.

Bolliet, V., Aranda, A., Boujard, T., 2001. Demand-feeding rhythm in rainbow trout and European catfish. Synchronisation by photoperiod and food availability. Physiol. Behav. 73, 625–633.

Boujard, T., Leatherland, J.F., 1992. Circadian rhythms and feeding time in fishes. Environ. Biol. Fish 35, 109–131.

Buckley, C., MacDonald, E.E., Tuziak, S.M., Volkoff, H., 2010. Molecular cloning and characterization of two putative appetite regulators in winter flounder (*Pleuronectes americanus*): preprothyrotropin-releasing hormone (TRH) and preproorexin (OX). Peptides 31, 1737–1747.

Castan-Laurell, I., Dray, C.D., Attane, C., Duparc, T., Knauf, C., Valet, P., 2011. Apelin, diabetes, and obesity. Endocrine 40, 1–9.

Cavallari, N., Frigato, E., Vallone, D., Frohlich, N., Lopez-Olmeda, J.F., Foa, A., Berti, R., Sanchez-Vazquez, F.J., Bertolucci, C., Foulkes, N.S., 2011. A blind circadian clock in cavefish reveals that opsins mediate peripheral clock photoreception. PLoS Biol. 9, e1001142.

Cerdá-Reverter, J.M., 2009. Control Neural de la Ingesta en Peces. In: La Nutrición y Alimentación en Piscicultura. Libros del Observatorio Español de Acuicultura del CSIC (OESA). Madrid, Spain, pp. 585–634.

Cerda-Reverter, J.M., Larhammar, D., 2000. Neuropeptide Y family of peptides: structure, anatomical expression, function, and molecular evolution. Biochem. Cell Biol. 78, 371–392.

Chen, G., Feng, L., Kuang, S., Liu, Y., Jiang, J., Hu, K., Jiang, W., Li, S., Tang, L., Zhou, X., 2012. Effect of dietary arginine on growth, intestinal enzyme activities and gene expression in muscle, hepatopancreas and intestine of juvenile Jian carp (*Cyprinus carpio* var. Jian). Br. J. Nutr. 108, 195–207.

Chen, Y., Shen, Y., Pandit, N.P., Fu, J., Li, D., Li, J., 2013. Molecular cloning, expression analysis, and potential food intake attenuation effect of peptide YY in grass carp (*Ctenopharyngodon idellus*). Gen. Comp. Endocrinol. 187, 66–73.

Chen, Y., Pandit, N.P., Fu, J., Li, D., Li, J., 2014. Identification, characterization and feeding response of peptide YYb (PYYb) gene in grass carp (*Ctenopharyngodon idellus*). Fish Physiol. Biochem. 40, 45–55.

Craig, P.M., Moon, T.W., 2011. Fasted zebrafish mimic genetic and physiological responses in mammals: a model for obesity and diabetes? Zebrafish 8, 109–117.

Crespo, C.S., Cachero, A.P., Jimenez, L.P., Barrios, V., Arilla, E., 2014. Peptides and food intake. Front. Endocrinol. (Lausanne) 5, 58. http://dx.doi.org/10.3389/fendo.2014.00058.

Culver, D.C., Pipan, T., 2009. The Biology of Caves and Other Subterranean Habitats (Biology of Habitats). Oxford University Press, Oxford, UK.

Daubner, S.C., Le, T., Wang, S., 2011. Tyrosine hydroxylase and regulation of dopamine synthesis. Arch. Biochem. Biophys. 508, 1–12.

de La Serre, C.B., Moran, T.H., 2013. Chapter 144—CCK. In: Kastin, A.J. (Ed.), Handbook of Biologically Active Peptides, second ed. Academic Press, Boston, MA, pp. 1077–1083.

de Lecea, L., Sutcliffe, J.G., 2013. Chapter 108—Hypocretins (Orexins). In: Kastin, A.J. (Ed.), Handbook of Biologically Active Peptides. second ed.. Academic Press, Boston, MA, pp. 812–818.

de Perera, T.B., 2004. Fish can encode order in their spatial map. Proc. R. Soc. Lond. B 271, 2131–2134.

Dockray, G.J., 2012. Cholecystokinin. Curr. Opin. Endocrinol. Diabetes Obes. 19, 8–12.

Dockray, G.J., de Lartigue, G., 2013. Chapter 143—CART. In: Kastin, A.J. (Ed.), Handbook of Biologically Active Peptides, second ed. Academic Press, Boston, MA, pp. 1071–1076.

Duboué, E.R., Keene, A.C., Borowsky, R.L., 2011. Evolutionary convergence on sleep loss in cavefish populations. Curr. Biol. 21, 671–676.

Duboué, E.R., Borowsky, R.L., Keene, A.C., 2012. Beta-adrenergic signaling regulates evolutionarily derived sleep loss in the Mexican cavefish. Brain Behav. Evol. 80, 233–243.

Elipot, Y., Hinaux, H., Callebert, J., Rétaux, S., 2013. Evolutionary shift from fighting to foraging in blind cavefish through changes in the serotonin network. Curr. Biol. 23, 1–10.

Elipot, Y., Hinaux, H., Callebert, J., Launay, J.M., Blin, M., Rétaux, S., 2014. A mutation in the enzyme monoamine oxidase explains part of the *Astyanax* cavefish behavioural syndrome. Nat. Commun. 5, 3647. http://dx.doi.org/10.1038/ncomms4647. 2041–1723.

Espinasa, L., Yamamoto, Y., Jeffery, W.R., 2005. Non-optical releasers for aggressive behavior in blind and blinded *Astyanax* (Teleostei, Characidae). Behav. Process. 70, 144–148.

Facciolo, R.M., Crudo, M., Giusi, G., Alo, R., Canonaco, M., 2009. Light- and dark-dependent orexinergic neuronal signals promote neurodegenerative phenomena accounting for distinct behavioral responses in the teleost *Thalassoma pavo*. J. Neurosci. Res. 87, 748–757.

Facciolo, R.M., Crudo, M., Zizza, M., Giusi, G., Canonaco, M., 2012. αGABA(A) subunit-orexin receptor interactions activate learning/motivational pathways in the goldfish. Behav. Brain Res. 234, 349–356.

Filby, A., Paull, G., Hickmore, T., Tyler, C., 2010. Unravelling the neurophysiological basis of aggression in a fish model. BMC Genomics 11, 498.

Fox, B.K., Breves, J.P., Hirano, T., Grau, E.G., 2009. Effects of short- and long-term fasting on plasma and stomach ghrelin, and the growth hormone/insulin-like growth factor I axis in the tilapia, *Oreochromis mossambicus*. Domest. Anim. Endocrinol. 37, 1–11.

Fricke, D., 1987. Reaction to alarm substance in cave populations of *Astyanax fasciatus* (Characidae, Pisces). Ethology 76, 305–308.

Fuentes, E.N., Safian, D., Einarsdottir, I., Valdes, J.A., Elorza, A.A., Molina, A., Bjornsson, B.T., 2013. Nutritional status modulates plasma leptin, AMPK and TOR activation, and mitochondrial biogenesis: implications for cell metabolism and growth in skeletal muscle of the fine flounder. Gen. Comp. Endocrinol. 186, 172–180.

Gélineau, A., Boujard, T., 2001. Oral administration of cholecystokinin receptor antagonists increase feed intake in rainbow trout. J. Fish Biol. 58, 716–724.

Gonzalez, R., Unniappan, S., 2010. Molecular characterization, appetite regulatory effects and feeding related changes of peptide YY in goldfish. Gen. Comp. Endocrinol. 166, 273–279.

Gregson, J.N.S., Burt de Perera, T., 2007. Shoaling in eyed and blind morphs of the characin *Astyanax fasciatus* under light and dark conditions. J. Fish Biol. 70, 1615–1619.

Gross, J.B., Furterer, A., Carlson, B.M., Stahl, B.A., 2013. An integrated transcriptome-wide analysis of cave and surface dwelling *Astyanax mexicanus*. PLoS One 8, e55659.

Hagan, J.J., Leslie, R.A., Patel, S., Evans, M.L., Wattam, T.A., Holmes, S., Benham, C.D., Taylor, S.G., Routledge, C., Hemmati, P., Munton, R.P., Ashmeade, T.E., Shah, A.S., Hatcher, J.P., Hatcher, P.D., Jones, D.N., Smith, M.I., Piper, D.C., Hunter, A.J., Porter, R.A., Upton, N., 1999. Orexin A activates locus coeruleus cell firing and increases arousal in the rat. Proc. Natl. Acad. Sci. U. S. A. 96, 10911–10916.

Hall, M.N., 2008. mTOR. What does it do? Transplant. Proc. 40, S5–S8.

Hervant, F., Malard, F., 2012. Responses to low oxygen. In: White, W.B., Culver, D.C. (Eds.), Encyclopedia of Caves, second ed. Academic Press, Amsterdam, pp. 651–658.

Hill, J.E., Yanong, R.P.E., 2010. Freshwater Ornamental Fish Commonly Cultured in Florida. University of Florida/Institute of Food and Agricultural Sciences (IFAS), Gainesville, FL.

Himick, B.A., Peter, R.E., 1994. CCK/gastrin-like immunoreactivity in brain and gut, and CCK suppression of feeding in goldfish. Am. J. Physiol. 267, R841–R851.

Hnasko, T.S., Szczypka, M.S., Alaynick, W.A., During, M.J., Palmiter, R.D., 2004. A role for dopamine in feeding responses produced by orexigenic agents. Brain Res. 1023, 309–318.

Hokfelt, T., Ogren, S.O., Xu, Z.-Q.D., 2013. Chapter 251—classical neurotransmitters and neuropeptides. In: Kastin, A.J. (Ed.), Handbook of Biologically Active Peptides, second ed. Academic Press, Boston, MA, pp. 1835–1841.

Holzman, R., Perkol-Finkel, S., Zilman, G., 2014. Mexican blind cavefish use mouth suction to detect obstacles. J. Exp. Biol. 217, 1955–1962.

Hoskins, L.J., Volkoff, H., 2012a. The comparative endocrinology of feeding in fish: insights and challenges. Gen. Comp. Endocrinol. 176, 327–335.

Hoskins, L.J., Volkoff, H., 2012b. Daily patterns of mRNA expression of two core circadian regulatory proteins, Clock2 and Per1, and two appetite-regulating peptides, OX and NPY, in goldfish (*Carassius auratus*, Linnaeus). Comp. Biochem. Physiol. A Mol. Integr. Physiol. 163, 127–136.

Hoskins, L.J., Xu, M., Volkoff, H., 2008. Interactions between gonadotropin-releasing hormone (GnRH) and orexin in the regulation of feeding and reproduction in goldfish (*Carassius auratus*). Horm. Behav. 54, 379–385.

Hüppop, K., 1986. Oxygen consumption of *Astyanax fasciatus* (Characidae, Pisces): a comparison of epigean and hypogean populations. Environ. Biol. Fish 17, 299–308.

Hüppop, K., 1987. Food-finding ability in cave fish (*Astyanax fasciatus*). Int. J. Speleol. 16, 59–66.

Inhoff, T., Stengel, A., Peter, L., Goebel, M., Tache, Y., Bannert, N., Wiedenmann, B., Klapp, B.F., Monnikes, H., Kobelt, P., 2010. Novel insight in distribution of nesfatin-1 and phospho-mTOR in the arcuate nucleus of the hypothalamus of rats. Peptides 31, 257–262.

Ito, H., Ishikawa, Y., Yoshimoto, M., Yamamoto, N., 2007. Diversity of brain morphology in teleosts: brain and ecological niche. Brain Behav. Evol. 69, 76–86.

Javonillo, R., Malabarba, L.R., Weitzman, S.H., Burns, J.R., 2010. Relationships among major lineages of characid fishes (Teleostei: Ostariophysi: Characiformes), based on molecular sequence data. Mol. Phylogenet. Evol. 54, 498–511.

Jeffery, W.R., 2012. *Astyanax mexicanus*: a model organism for evolution and adaptation. In: Culver, W.B.W.C. (Ed.), Encyclopedia of Caves, second ed. Academic Press, Amsterdam, pp. 36–43.

Jerzemowska, G., Plucinska, K., Kulikowski, M., Trojniar, W., Wrona, D., 2012. Locomotor response to novelty correlates with the number of midbrain tyrosine hydroxylase positive cells in rats. Brain Res. Bull. 87, 94–102.

Jiang, J., Feng, L., Liu, Y., Jiang, W.D., Hu, K., Li, S.H., Zhou, X.Q., 2013. Mechanistic target of rapamycin in common carp: cDNA cloning, characterization, and tissue expression. Gene 512, 566–572.

Jonsson, E., 2013. The role of ghrelin in energy balance regulation in fish. Gen. Comp. Endocrinol. 187, 79–85.

Karra, E., Chandarana, K., Batterham, R.L., 2009. The role of peptide YY in appetite regulation and obesity. J. Physiol. 587, 19–25.

Kiris, G.A., Kumlu, M., Dikel, S., 2007. Stimulatory effects of neuropeptide Y on food intake and growth of *Oreochromis niloticus*. Aquaculture 264, 383–389.

Knauf, C., Drougard, A., Fournel, A., Duparc, T., Valet, P., 2013. Hypothalamic actions of apelin on energy metabolism: new insight on glucose homeostasis and metabolic disorders. Horm. Metab. Res. 45, 928–934.

Koven, W., Schulte, P., 2012. The effect of fasting and refeeding on mRNA expression of PepT1 and gastrointestinal hormones regulating digestion and food intake in zebrafish (*Danio rerio*). Fish Physiol. Biochem. 38, 1565–1575.

Kowalko, J.E., Rohner, N., Linden, T.A., Rompani, S.B., Warren, W.C., Borowsky, R., Tabin, C.J., Jeffery, W.R., Yoshizawa, M., 2013. Convergence in feeding posture occurs through different genetic loci in independently evolved cave populations of *Astyanax mexicanus*. Proc. Natl. Acad. Sci. U. S. A. 110, 16933–16938.

Langecker, T.G., Neumann, B., Hausberg, C., Parzefall, J., 1995. Evolution of the optical releasers for aggressive behavior in cave-dwelling *Astyanax fasciatus* (Teleostei, Characidae). Behav. Processes 34, 161–167.

Lembke, V., Goebel, M., Frommelt, L., Inhoff, T., Lommel, R., Stengel, A., Tache, Y., Grotzinger, C., Bannert, N., Wiedenmann, B., Klapp, B.F., Kobelt, P., 2011. Sulfated cholecystokinin-8 activates phospho-mTOR immunoreactive neurons of the paraventricular nucleus in rats. Peptides 32, 65–70.

Lin, F., Wu, H., Chen, H., Xin, Z., Yuan, D., Wang, T., Liu, J., Gao, Y., Zhang, X., Zhou, C., Wei, R., Chen, D., Yang, S., Wang, Y., Pu, Y., Li, Z., 2014. Molecular and physiological evidences for the role in appetite regulation of apelin and its receptor APJ in Ya-fish (*Schizothorax prenanti*). Mol. Cell. Endocrinol. 396, 46–57.

Lopez, J.M., Dominguez, L., Moreno, N., Morona, R., Joven, A., Gonzalez, A., 2009. Distribution of orexin/hypocretin immunoreactivity in the brain of the lungfishes *Protopterus dolloi* and *Neoceratodus forsteri*. Brain Behav. Evol. 74, 302–322.

Lopez-Patino, M.A., Guijarro, A.I., Isorna, E., Delgado, M.J., Alonso-Bedate, M., de Pedro, N., 1999. Neuropeptide Y has a stimulatory action on feeding behavior in goldfish (*Carassius auratus*). Eur. J. Pharmacol. 377, 147–153.

MacDonald, E., Volkoff, H., 2009. Cloning, distribution and effects of season and nutritional status on the expression of neuropeptide Y (NPY), cocaine and amphetamine regulated transcript (CART) and cholecystokinin (CCK) in winter flounder (*Pseudopleuronectes americanus*). Horm. Behav. 56, 58–65.

Martins, L., Fernandez-Mallo, D., Novelle, M.G., Vazquez, M.J., Tena-Sempere, M., Nogueiras, R., Lopez, M., Dieguez, C., 2012. Hypothalamic mTOR signaling mediates the orexigenic action of ghrelin. PLoS One 7, e46923.

Matsuda, K., Miura, T., Kaiya, H., Maruyama, K., Uchiyama, M., Kangawa, K., Shioda, S., 2006. Stimulatory effect of n-octanoylated ghrelin on locomotor activity in the goldfish, *Carassius auratus*. Peptides 27, 1335–1340.

Menuet, A., Alunni, A., Joly, J.S., Jeffery, W.R., Rétaux, S., 2007. Expanded expression of sonic hedgehog in *Astyanax* cavefish: multiple consequences on forebrain development and evolution. Development 134, 845–855.

Micale, V., Campo, S., D'Ascola, A., Guerrera, M.C., Levanti, M.B., Germana, A., Muglia, U., 2014. Cholecystokinin: how many functions? Observations in seabreams. Gen. Comp. Endocrinol. 205, 166–167.

Mitchell, R.W., Russell, W.H., Elliott, W.R., 1977. Mexican Eyeless Characin Fishes, Genus *Astyanax*: Environment, Distribution, and Evolution. Texas Tech Press, Lubbock, TX.

Miura, T., Maruyama, K., Shimakura, S.-I., Kaiya, H., Uchiyama, M., Kangawa, K., Shioda, S., Matsuda, K., 2007. Regulation of food intake in the goldfish by interaction between ghrelin and orexin. Peptides 28, 1207–1213.

Montpetit, C.J., Chatalov, V., Yuk, J., Rasaratnam, I., Youson, J.H., 2005. Expression of Neuropeptide Y family peptides in the brain and gut during stages of the life cycle of a parasitic lamprey (*Petromyzon marinus*) and a nonparasitic lamprey (*Ichthyomyzon gagei*). Ann. N. Y. Acad. Sci. 1040, 140–149.

Murashita, K., Kurokawa, T., 2011. Multiple cocaine- and amphetamine-regulated transcript (CART) genes in medaka, *Oryzias latipes*: cloning, tissue distribution and effect of starvation. Gen. Comp. Endocrinol. 170, 494–500.

Murashita, K., Fukada, H., Hosokawa, H., Masumoto, T., 2006. Cholecystokinin and peptide Y in yellowtail (*Seriola quinqueradiata*): molecular cloning, real-time quantitative RT-PCR, and response to feeding and fasting. Gen. Comp. Endocrinol. 145, 287–297.

Murashita, K., Kurokawa, T., Nilsen, T.O., Ronnestad, I., 2009. Ghrelin, cholecystokinin, and peptide YY in Atlantic salmon (*Salmo salar*): molecular cloning and tissue expression. Gen. Comp. Endocrinol. 160, 223–235.

Nakamachi, T., Matsuda, K., Maruyama, K., Miura, T., Uchiyama, M., Funahashi, H., Sakurai, T., Shioda, S., 2006. Regulation by orexin of feeding behaviour and locomotor activity in the goldfish. J. Neuroendocrinol. 18, 290–297.

Naumann, E.A., Kampff, A.R., Prober, D.A., Schier, A.F., Engert, F., 2010. Monitoring neural activity with bioluminescence during natural behavior. Nat. Neurosci. 13, 513–520.

Nelson, L.E., Sheridan, M.A., 2006. Gastroenteropancreatic hormones and metabolism in fish. Gen. Comp. Endocrinol. 148, 116–124.

Nisembaum, L.G., de Pedro, N., Delgado, M.J., Sanchez-Bretano, A., Isorna, E., 2014. Orexin as an input of circadian system in goldfish: effects on clock gene expression and locomotor activity rhythms. Peptides 52, 29–37.

Nishino, S., Mignot, E., 2011. Narcolepsy and cataplexy. Handb. Clin. Neurol. 99, 783–814.

Nishio, S., Gibert, Y., Berekelya, L., Bernard, L., Brunet, F., Guillot, E., Le Bail, J.C., Sanchez, J.A., Galzin, A.M., Triqueneaux, G., Laudet, V., 2012. Fasting induces CART down-regulation in the zebrafish nervous system in a cannabinoid receptor 1-dependent manner. Mol. Endocrinol. 26, 1316–1326.

Novak, C.M., Jiang, X.L., Wang, C.F., Teske, J.A., Kotz, C.M., Levine, J.A., 2005. Caloric restriction and physical activity in zebrafish (*Danio rerio*). Neurosci. Lett. 383, 99–104.

Oganesyan, G.A., Romanova, I.V., Aristakesyan, E.A., Kuzik, V.V., Makina, D.M., Morina, I.Y., Khramenkova, A.E., Artamokhina, I.V., Belova, V.A., 2009. The dopaminergic system of the telencephalo-diencephalic areas of the vertebrate brain in the organization of the sleep-waking cycle. Neurosci. Behav. Physiol. 39, 805–817.

Omura, Y., 1975. Influence of light and darkness on the ultrastructure of the pineal organ in the blind cave fish *Astyanax mexicanus*. Cell Tissue Res. 160, 99–112.

Panula, P., 2010. Hypocretin/orexin in fish physiology with emphasis on zebrafish. Acta Physiol. (Oxf.) 198, 381–386.

Parzefall, J., Trajano, E., 2010. Behavioral patterns in subterranean fishes. In: Trajano, E., Bichuette, M.E., Kapoor, B.G. (Eds.), Biology of Subterranean Fishes. Science Publishers, Enfield, NH, pp. 81–114.

Pavlidis, M., Sundvik, M., Chen, Y.C., Panula, P., 2011. Adaptive changes in zebrafish brain in dominant-subordinate behavioral context. Behav. Brain Res. 225, 529–537.

Penney, C.C., Volkoff, H., 2014. Peripheral injections of cholecystokinin, apelin, ghrelin, and orexin in cavefish (*Astyanax fasciatus mexicanus*): effects on feeding and on the brain expression levels of tyrosine hydroxylase, mechanistic target of rapamycin and appetite-related hormones. Gen. Comp. Endocrinol. 196, 34–40.

Perboni, S., Vignoni, M., Inui, A., 2013. Chapter 154—NPY. In: Kastin, A.J. (Ed.), Handbook of Biologically Active Peptides second ed. Academic Press, Boston, MA, pp. 1143–1148.

Peters, N., 1993. Cerebral proportions and degree of development of sense-organs in *Astyanax mexicanus* (Pisces, Characinidae)—comparison between river fish and its cave-living forms *Anoptichthys*. Zeitschrift Fur Zoologische Systematik Und Evolutionsforschung 31, 144–159.

Peterson, B.C., Waldbieser, G.C., Riley Jr., L.G., Upton, K.R., Kobayashi, Y., Small, B.C., 2012. Pre- and postprandial changes in orexigenic and anorexigenic factors in channel catfish (*Ictalurus punctatus*). Gen. Comp. Endocrinol. 176, 231–239.

Peyon, P., Saied, H., Lin, X., Peter, R.E., 1999. Postprandial, seasonal and sexual variations in cholecystokinin gene expression in goldfish brain. Brain Res. Mol. Brain Res. 74, 190–196.

Prober, D.A., Rihel, J., Onah, A.A., Sung, R.J., Schier, A.F., 2006. Hypocretin/orexin overexpression induces an insomnia-like phenotype in zebrafish. J. Neurosci. 26, 13400–13410.

Protas, M., Jeffery, W.R., 2012. Evolution and development in cave animals: from fish to crustaceans. Wiley Interdiscip. Rev. Dev. Biol. 1, 823–845. http://dx.doi.org/10.1002/wdev.61.

Purves, D., George, J., Augustine, G.J., Fitzpatrick, D., Katz, L.C., LaMantia, A.-S., McNamara, J.O., Williams, S.M., 2001. Neuroscience, second ed. Sinauer Associates, Sunderland, MA.

Puskas, N., Papp, R.S., Gallatz, K., Palkovits, M., 2010. Interactions between orexin-immunoreactive fibers and adrenaline or noradrenaline-expressing neurons of the lower brainstem in rats and mice. Peptides 31, 1589–1597.

Rasquin, P., 1949. The influence of light and darkness on thyroid and pituitary activity of the characin *Astyanax mexicanus* and its cave derivatives. Bull. Am. Mus. Nat. Hist. 93, 501–531.

Rasquin, P., Rosenbloom, L., 1954. Endocrine imbalance and tissue hyperplasia in teleosts maintained in darkness. Bull. Am. Mus. Nat. Hist. 104, 359–426.

Rétaux, S., Elipot, Y., 2013. Feed or fight: a behavioral shift in blind cavefish. Commun. Integr. Biol. 6, e23166.

Riley, L.G., Fox, B.K., Kaiya, H., Hirano, T., Grau, E.G., 2005. Long-term treatment of ghrelin stimulates feeding, fat deposition, and alters the GH/IGF-I axis in the tilapia, *Oreochromis mossambicus*. Gen. Comp. Endocrinol 142, 234–240.

Rubio, V.C., Sanchez-Vazquez, F.J., Madrid, J.A., 2008. Role of cholecystokinin and its antagonist proglumide on macronutrient selection in European sea bass *Dicentrarchus labrax*, L. Physiol. Behav. 93, 862–869.

Sakurai, T., Amemiya, A., Ishii, M., Matsuzaki, I., Chemelli, R.M., Tanaka, H., Williams, S.C., Richarson, J.A., Kozlowski, G.P., Wilson, S., Arch, J.R., Buckingham, R.E., Haynes, A.C., Carr, S.A., Annan, R.S., McNulty, D.E., Liu, W.S., Terrett, J.A., Elshourbagy, N.A., Bergsma, D.J., Yanagisawa, M., 1998. Orexins and orexin receptors: a family of hypothalamic neuropeptides and G protein-coupled receptors that regulate feeding behavior. Cell 92, 573–585.

Salin, K., Voituron, Y., Mourin, J., Hervant, F., 2010. Cave colonization without fasting capacities: an example with the fish *Astyanax fasciatus mexicanus*. Comp. Biochem. Physiol. A Mol. Integr. Physiol. 156, 451–457.

Sanchez, J.A., Sanchez-Vazquez, F.J., 2009. Feeding entrainment of daily rhythms of locomotor activity and clock gene expression in zebrafish brain. Chronobiol. Int. 26, 1120–1135.

Sapp, V., Gaffney, L., EauClaire, S.F., Matthews, R.P., 2014. Fructose leads to hepatic steatosis in zebrafish that is reversed by mTOR inhibition. Hepatology 60, 1581–1592. http://dx.doi.org/10.1002/hep.27284.

Schemmel, C., 1980. Studies on the genetics of feeding behaviour in the cave fish *Astyanax mexicanus* f. *anoptichthys*. An example of apparent monofactorial inheritance by polygenes. Z. Tierpsychol. 53, 9–22.

Seiliez, I., Gabillard, J.C., Skiba-Cassy, S., Garcia-Serrana, D., Gutierrez, J., Kaushik, S., Panserat, S., Tesseraud, S., 2008. An *in vivo* and in vitro assessment of TOR signaling cascade in rainbow trout (*Oncorhynchus mykiss*). Am. J. Physiol. 295, R329–R335.

Shahjahan, M., Kitahashi, T., Parhar, I.S., 2014. Central pathways integrating metabolism and reproduction in teleosts. Front. Endocrinol. (Lausanne) 5, 36.

Shepherd, B.S., Johnson, J.K., Silverstein, J.T., Parhar, I.S., Vijayan, M.M., McGuire, A., Weber, G.M., 2007. Endocrine and orexigenic actions of growth hormone secretagogues in rainbow trout (*Oncorhynchus mykiss*). Comp. Biochem. Physiol. A Mol. Integr. Physiol. 146, 390–399.

Skirboll, L.R., Grace, A.A., Hommer, D.W., Rehfeld, J., Goldstein, M., Hokfelt, T., Bunney, B.S., 1981. Peptide-monoamine coexistence: studies of the actions of cholecystokinin-like peptide on the electrical activity of midbrain dopamine neurons. Neuroscience 6, 2111–2124.

Soares, D., Niemiller, M.L., 2013. Sensory adaptations of fishes to subterranean environments. BioScience 63, 274–283.

Sohn, J.W., Elmquist, J.K., Williams, K.W., 2013. Neuronal circuits that regulate feeding behavior and metabolism. Trends Neurosci. 36, 504–512.

Stefater, M.A., Seeley, R.J., 2010. Central nervous system nutrient signaling: the regulation of energy balance and the future of dietary therapies. Annu. Rev. Nutr. 30, 219–235.

Strickler, A.G., Soares, D., 2011. Comparative genetics of the central nervous system in epigean and hypogean *Astyanax mexicanus*. Genetica 139, 383–391.

Sundstrom, G., Larsson, T.A., Xu, B., Heldin, J., Larhammar, D., 2013. Interactions of zebrafish peptide YYb with the neuropeptide Y-family receptors Y4, Y7, Y8a, and Y8b. Front. Neurosci. 7, 29.

Sylvester, J.B., Pottin, K., Streelman, J.T., 2011. Integrated brain diversification along the early neuraxes. Brain Behav. Evol. 78, 237–247.

Tachibana, T., Takagi, T., Tomonaga, S., Ohgushi, A., Ando, R., Denbow, D.M., Furuse, M., 2003. Central administration of cocaine- and amphetamine-regulated transcript inhibits food intake in chicks. Neurosci. Lett. 337, 131–134.

Takei, Y., Loretz, C.A., 2010. The gastrointestinal tract as an endocrine/neuroendocrine/paracrine organ: organization, chemical messengers and physiological targets. In: Martin Grosell, A.P.F., Colin, J.B. (Eds.), Fish Physiology. Academic Press, New York, pp. 261–317.

Tinoco, A.B., Naslund, J., Delgado, M.J., de Pedro, N., Johnsson, J.I., Jonsson, E., 2014. Ghrelin increases food intake, swimming activity and growth in juvenile brown trout (*Salmo trutta*). Physiol. Behav. 124, 15–22.

Troke, R., Tan, T.M., Bloom, S.R., 2013. Chapter 157—PYY. In: Kastin, A.J. (Ed.), Handbook of Biologically Active Peptides, second ed. Academic Press, Boston, MA, pp. 1160–1165.

Tuziak, S.M., Rise, M.L., Volkoff, H.L.N., 2014. An investigation of appetite-related peptide transcript expression in Atlantic cod (*Gadus morhua*) brain following a *Camelina sativa* meal-supplemented feeding trial. Gene 550, 253–263.

Unniappan, S., Canosa, L.F., Peter, R.E., 2004. Orexigenic actions of ghrelin in goldfish: feeding-induced changes in brain and gut mRNA expression and serum levels, and responses to central and peripheral injections. Neuroendocrinology 79, 100–108.

Valen, R., Jordal, A.E., Murashita, K., Ronnestad, I., 2011. Postprandial effects on appetite-related neuropeptide expression in the brain of Atlantic salmon, *Salmo salar*. Gen. Comp. Endocrinol. 171, 359–366.

Verhulst, P.J., Depoortere, I., 2012. Ghrelin's second life: from appetite stimulator to glucose regulator. World J. Gastroenterol. 18, 3183–3195.

Volkoff, H., 2014a. Appetite regulating peptides in red-bellied piranha, *Pygocentrus nattereri*: cloning, tissue distribution and effect of fasting on mRNA expression levels. Peptides 56, 116–124.

Volkoff, H., 2014b. *In vitro* assessment of interactions between appetite-regulating peptides in brain of goldfish (*Carassius auratus*). Peptides 61, 61–68.

Volkoff, H., Peter, R.E., 2000. Effects of CART peptides on food consumption, feeding and associated behaviors in the goldfish *Carassius auratus*: actions on neuropeptide Y- and orexin A-induced feeding. Brain Res. 887, 125–133.

Volkoff, H., Peter, R.E., 2001a. Characterization of two forms of cocaine- and amphetamine-regulated transcript (CART) peptide precursors in goldfish: molecular cloning and distribution, modulation of expression by nutritional status, and interactions with leptin. Endocrinology 142, 5076–5088.

Volkoff, H., Peter, R.E., 2001b. Interactions between orexin A, NPY and galanin in the control of food intake of the goldfish, *Carassius auratus*. Regul. Pept. 101, 59–72.

Volkoff, H., Wyatt, J.L., 2009. Apelin in goldfish (*Carassius auratus*): cloning, distribution and role in appetite regulation. Peptides 30, 1434–1440.

Volkoff, H., Bjorklund, J.M., Peter, R.E., 1999. Stimulation of feeding behavior and food consumption in the goldfish, *Carassius auratus*, by orexin-A and orexin-B. Brain Res. 846, 204–209.

Volkoff, H., Canosa, L.F., Unniappan, S., Cerda-Reverter, J.M., Bernier, N.J., Kelly, S.P., Peter, R.E., 2005. Neuropeptides and the control of food intake in fish. Gen. Comp. Endocrinol. 142, 3–19.

Wahlestedt, C., Reis, D.J., 1993. Neuropeptide Y-related peptides and their receptors—are the receptors potential therapeutic drug targets? Annu. Rev. Pharmacol. Toxicol. 33, 309–352.

Wall, A., Volkoff, H., 2013. Effects of fasting and feeding on the brain mRNA expressions of orexin, tyrosine hydroxylase (TH), PYY and CCK in the Mexican blind cavefish (*Astyanax fasciatus mexicanus*). Gen. Comp. Endocrinol. 183, 44–52.

Wan, Y., Zhang, Y., Ji, P., Li, Y., Xu, P., Sun, X., 2012. Molecular characterization of CART, AgRP, and MC4R genes and their expression with fasting and re-feeding in common carp (*Cyprinus carpio*). Mol. Biol. Rep. 39, 2215–2223.

Wong, M.K., Sze, K.H., Chen, T., Cho, C.K., Law, H.C., Chu, I.K., Wong, A.O., 2013. Goldfish spexin: solution structure and novel function as a satiety factor in feeding control. Am. J. Physiol. Endocrinol. Metab. 3015, E348–E366.

Woods, I.G., Schoppik, D., Shi, V.J., Zimmerman, S., Coleman, H.A., Greenwood, J., Soucy, E.R., Schier, A.F., 2014. Neuropeptidergic signaling partitions arousal behaviors in zebrafish. J. Neurosci. 34, 3142–3160.

Wullschleger, S., Loewith, R., Hall, M.N., 2006. TOR signaling in growth and metabolism. Cell 124, 471–484.

Xu, M.Y., Volkoff, H., 2007. Molecular characterization of prepro-orexin in Atlantic cod (*Gadus morhua*): cloning, localization, developmental profile and role in food intake regulation. Mol. Cell. Endocrinol. 271, 28–37.

Xu, M., Volkoff, H., 2009. Molecular characterization of ghrelin and gastrin-releasing peptide in Atlantic cod (*Gadus morhua*): cloning, localization, developmental profile and role in food intake regulation. Gen. Comp. Endocrinol. 160, 250–258.

Xu, G., Li, Y., An, W., Zhao, J., Xiang, X., Ding, L., Li, Z., Guan, Y., Wang, X., Tang, C., Zhu, Y., Wang, N., Li, X., Mulholland, M., Zhang, W., 2010. Regulation of gastric hormones by systemic rapamycin. Peptides 31, 2185–2192.

Yamamoto, Y., Byerly, M.S., Jackman, W.R., Jeffery, W.R., 2009. Pleiotropic functions of embryonic sonic hedgehog expression link jaw and taste bud amplification with eye loss during cavefish evolution. Dev. Biol. 330, 200–211.

Yan, A., Zhang, L., Tang, Z., Zhang, Y., Qin, C., Li, B., Li, W., Lin, H., 2011. Orange-spotted grouper (*Epinephelus coioides*) orexin: molecular cloning, tissue expression, ontogeny, daily rhythm and regulation of NPY gene expression. Peptides 32, 1363–1370.

Yokobori, E., Kojima, K., Azuma, M., Kang, K.S., Maejima, S., Uchiyama, M., Matsuda, K., 2011. Stimulatory effect of intracerebroventricular administration of orexin A on food intake in the zebrafish, *Danio rerio*. Peptides 32, 1357–1362.

Yokobori, E., Azuma, M., Nishiguchi, R., Kang, K.S., Kamijo, M., Uchiyama, M., Matsuda, K., 2012. Neuropeptide Y stimulates food intake in the zebrafish, *Danio rerio*. J. Neuroendocrinol. 24, 766–773.

Yoshizawa, M., Goriçki, S., Soares, D., Jeffery, W.R., 2010. Evolution of a behavioral shift mediated by superficial neuromasts helps cavefish find food in darkness. Curr. Biol. 20, 1631–1636.

Yoshizawa, M., Jeffery, W.R., van Netten, S.M., McHenry, M.J., 2014. The sensitivity of lateral line receptors and their role in the behavior of Mexican blind cavefish (*Astyanax mexicanus*). J. Exp. Biol. 217, 886–895.

Yu, L., Tucci, V., Kishi, S., Zhdanova, I.V., 2006. Cognitive aging in zebrafish. PLoS One 1, e14.

Zhang, X.Y., Yang, H.D., Zhang, Q., Wang, Z., Wang, D.H., 2011. Increased feeding and food hoarding following food deprivation are associated with activation of dopamine and orexin neurons in male Brandt's voles. PLoS One 6, e26408.

Zhang, L., Nguyen, A.D., Lee, I.C.J., Yulyaningsih, E., Riepler, S.J., Stehrer, B., Enriquez, R.F., Lin, S., Shi, Y.C., Baldock, P.A., Sainsbury, A., Herzog, H., 2012. NPY modulates PYY function in the regulation of energy balance and glucose homeostasis. Diabetes Obes. Metab. 14, 727–736.

Zhou, Q.Y., Palmiter, R.D., 1995. Dopamine-deficient mice are severely hypoactive, adipsic, and aphagic. Cell 83, 1197–1209.

Zhou, C., Zhang, X., Liu, T., Wei, R., Yuan, D., Wang, T., Lin, F., Wu, H., Chen, F., Yang, S., Chen, D., Wang, Y., Li, Z., 2014. *Schizothorax davidi* ghrelin: cDNA cloning, tissue distribution and indication for its stimulatory character in food intake. Gene 534, 72–77.

Chapter 15

Investigating the Evolution of Sleep in the Mexican Cavefish

Erik R. Duboué[1] **and Alex C. Keene**[2]
[1]*Department of Embryology, Carnegie Institute for Science, Baltimore, Maryland, USA*
[2]*Department of Biology, University of Nevada, Reno, Nevada, USA*

INTRODUCTION

Sleep is a fundamental to animal life. In diverse animal taxa, poor sleep can potently affect cognitive abilities and health (Cappuccio et al., 2010; Rechtschaffen et al., 1983; Shaw et al., 2002). Sleep is characterized by distinct behavioral and electrophysiological properties. Historically, sleep has been identified using changes in brain activity as measured by the electroencephalogram (EEG). In awake mammals, EEG patterns recording electrical activity in cortical areas are characterized as low-voltage, high-frequency (in humans, 10-30 uV and 16-25 Hz) waves. The electrophysiological stages include four stages of slow-wave sleep, characterized by a progressive transition to low-frequency, high-voltage activity, and one stage of rapid eye movement (REM) sleep (Dement and Kleitman, 1957; Kandel et al., 2013). REM sleep, also called paradoxical sleep, is correlated with cortical activity similar to awake animal, though muscle tone is reduced. While the EEG is considered a hallmark in sleep medicine, it has been primarily utilized in mammalian species, and its relevance in non-mammalian systems is unclear.

Electrophysiological methods for measuring sleep are comparatively difficult to perform in smaller, genetically amenable animals, such as fish, flies, and worms, and therefore, studies in these systems have primarily focused on behavioral markers of sleep. Behaviorally, sleep can be defined as a homeostatically regulated process marked by prolonged periods of behavioral quiescence and decreased sensory responsiveness. Collectively, the behavioral hallmarks of sleep include: (1) prolonged periods of behavioral quiescence that occur in a circadian manner; (2) reversibility of the behavioral state to distinguish from comatose or lethargy; (3) a species-specific posture; (4) an elevated arousal threshold to external stimuli; and (5) rebound following sleep deprivation. As humans, we intuitively understand these behavioral correlates: (1) we fall into states of prolonged inactivity lasting approximately 8 h that begin at around 10 p.m. to midnight; (2) we can be woken up to a state of full vigilance; (3) when we sleep,

Biology and Evolution of the Mexican Cavefish. http://dx.doi.org/10.1016/B978-0-12-802148-4.00015-3
© 2016 Elsevier Inc. All rights reserved.

we close our eyes and lie down; (4) if we are asleep, we are less in tune to our environment than when awake; and (5) if we stay up for a full night, we spend a greater portion of the following day asleep. First described over 40 years ago (Flanigan et al., 1973), the behavioral definition of sleep has opened up the field of sleep medicine and has successfully been used to characterize sleep in both vertebrates (Campbell and Tobler, 1984; Zhdanova et al., 2001) and invertebrates (Hendricks et al., 2000; Raizen et al., 2008; Shaw et al., 2000).

More recently, the introduction of genetically amenable model systems, including the nematode worm, *Caenorhabditis elegans* (Raizen et al., 2008), the fruit fly, *Drosophila melanogaster* (Hendricks et al., 2000; Shaw et al., 2000), and the zebrafish, *Danio rerio* (Prober et al., 2006; Yokogawa et al., 2007; Zhdanova et al., 2001), into sleep medicine has allowed for a powerful analysis of the genes and neural circuits driving the behavior. In addition to identifying genes regulating the sleep/wake cycle, forward genetics screens have also highlighted the conservation of molecular and neural principles underlying sleep-wake regulation. For example, the mutant mini-sleep, which was identified through forward genetic screen for short-sleeping mutants, results from mutation in the potassium channel Shaker (Cirelli et al., 2005). Reverse genetics and overexpression studies have also been a powerful tool for the analysis of genetic components modulating sleep. Early studies in zebrafish identified the role for the hypocretin system, previously implicated in human and canine narcolepsy, in fish wakefulness (Lin et al., 1999; Nishino et al., 2000; Prober et al., 2006; Yokogawa et al., 2007). Taken together, these studies highlight the powerful ability of genetically amenable model systems to identify conserved regulators of the sleep/wake cycle, and place sleep medicine in a new era where model systems that include fish and invertebrates can be used to identify regulators of mammalian sleep.

Sleep duration and architecture are highly variable, both between and within species. Several meta-analysis studies examining sleep duration among phylogenetic taxa and other parameters, such as diet, brain size, social hierarchy, and body mass index have identified correlations between sleep traits and environment (Campbell and Tobler, 1984; Lesku et al., 2006; Siegel, 2005). These studies have concluded that there is a great deal of variation between animal phyla, with animals such as the African Elephant sleeping only 3-4 h a day and animals such as the three-toed sloth, sleeping upwards of 20 h a day (De Moura Filho et al., 1983). Even among humans, sleep times can range from 5 to 10 h a night (Hartmann, 1973), and a number of genes have been identified that are implicated in these differences (He et al., 2009). Despite a widespread appreciation for the diversity in sleep between and within species, the relationship between sleep and ecological or evolutionary history remains unclear (although see Lesku et al., 2006).

Fully understanding the function of sleep will require investigation of both the molecular and neural properties regulating the behavior and ecological influences that contribute to the evolution intraspecies variation. The era of model systems in sleep research has provided a rapidly improving landscape of genes and neural principles regulating sleep, thereby providing a framework for understanding how variation in these mechanisms drive the evolution of sleep. Here we describe studies in the Mexican blind cavefish, *Astyanax mexicanus*, which shed light upon the evolution of sleep loss following the transition into a cave environment. *A. mexicanus* is unique among vertebrate model systems; the species comprises an extant, surface-dwelling morph and 29 independently evolved cave-dwelling morphs. The cave environment differs from surface rivers in many environmental aspects linked to sleep regulation, including the relative absence of light and reduced temperature, as well as reduced food availability and predation. There is a rich history in describing the morphological evolution of these fish in response to dramatic changes in environment. The evolutionary transition from surface to cave is correlated with dramatic morphologic changes, such as lack of eyes and pigmentation, both of which are understood at the genetic molecular level (Protas et al., 2006, 2008; Gross et al., 2008, 2009), yet much less is know about how this shift affects behavior. These fish provide a unique system to examine how sleep evolves in response to environmental perturbation, and how animals compensate for sleep loss, as well as the genetic regulation of sleep. We describe recent genetic and pharmacological studies that highlight the ability of this system to elucidate novel mechanisms of sleep regulation. Furthermore, we describe technology widely used in classical model systems, and discuss how its application in *A. mexicanus* may further our understanding into the genetics and evolution of sleep/wake cycles.

FISH AS A VERTEBRATE MODEL FOR SLEEP

The Zebrafish as a Model System for Sleep Studies

Sleep in fish was first described and characterized in the zebrafish, *Danio rerio*. The larval zebrafish is a particularly attractive model system due to its small size, ease of spawning, the vast number of genetic and transgenic tools available, and optical clarity allowing for *in vivo*, whole-brain imaging (Ahrens et al., 2012). In addition, the neural circuitry of the small vertebrate shares a large degree of homology to mammalian species at the circuit, neural-transmitter, and neural-peptide levels (Zhdanova, 2006).

Larval zebrafish exhibit all of the behavioral hallmarks of a sleep-like state. By 5 days post-fertilization (dpf), the larvae are largely active, exhibiting robust diurnal rhythms (Cahill et al., 1998). The nighttime phase is characterized by prolonged periods of behavioral quiescence, which can easily be reversed by a gentle mechanical disturbance (Zhdanova et al., 2001). During the night, prolonged behavioral quiescence is correlated with a change in behavioral posture—either with the larval head positioned in a downward direction, or horizontal at the bottom of the recording dish (Zhdanova et al., 2001). The arousal threshold to external sensory stimuli increases at night (Zhdanova et al., 2001; Prober et al., 2006). Zhdanova et al. (2001) first showed this using gentle mechanical stimulation on the side of a dish (i.e., physically tapping the side of the dish). The number of taps needed to elicit a behavioral reaction was scored as the arousal threshold. Consistent with an elevated arousal threshold during sleep-like states, Zhdanova and colleges found a significant increase in the number of mechanical taps needed to elicit a response during the nighttime phase (Zhdanova et al., 2001), though the length of inactivity preceding mechanical stimulation was not scored. In a similar set of experiments, Prober et al. (2006) also demonstrated diurnal rhythms in early larvae, revealing a link between sleep and the circadian clock. Importantly, this study also showed an elevated arousal threshold to external stimuli when larvae were in a period of prolonged behavioral quiescence. Zebrafish larvae exposed to sudden darkness quickly become motile (Prober et al., 2006), a response that has also been reported for both the African clawed frog, *Xenopus laevis* (Jamieson and Roberts, 2000) and *A. mexicanus* (Yoshizawa and Jeffery, 2008). The length of inactivity immediately preceding the transition to darkness is significantly negatively correlated with the number of larvae that responded, and those that did respond showed an increase in response latency (Prober et al., 2006). Importantly, only larvae that exhibited a 1-min period of inactivity before dark transition were significantly different in response latency from those immediately active before, suggesting that a 1-min period of inactivity in the larval zebrafish is indicative of a sleep-like state (Prober et al., 2006). Lastly, sleep is homeostatically regulated in larvae. When larvae are sleep-deprived by delivering a constant mechanical vibration during nighttime phases, immobility on the following day is significantly increased (Zhdanova et al., 2001). Therefore, the wake-rest patterns of larval zebrafish meet the established behavioral criteria used to define sleep.

Sleep has also been investigated in the adult zebrafish (Appelbaum et al., 2009; Yokogawa et al., 2007). Adult zebrafish are also highly diurnal, with prolonged behavioral quiescence during the nighttime phase correlated with a change in behavioral posture; the caudal fin dips in a downward direction, and adults typically prefer the bottom of the recording tank, perhaps because adults have extremely high sensitivity to mechanical stimuli in both the wake and sleep-like states (Yokogawa et al., 2007). Arousal threshold in adults has been studied using weak electrical stimulation. At low voltages, adults that are active become motile following a weak stimulus, yet those that are inactive are less likely to respond; however, a period of 6 s of immobility is correlated with a change in the probability of behavioral response, indicating increased arousal threshold following this relatively short period of immobility (Yokogawa et al., 2007). Using an elegant tracking system combined with computerized electrical stimulation, Yokogawa et al. (2007) investigated the effects of sleep deprivation by shocking the adult in the last 6 h of the night phase every time the adult exhibited 6 s of immobility. The following day, adults showed a significant increase in sleep, confirming a homeostatic rebound following deprivation in adult fish (Yokogawa et al., 2007).

Taken together, these studies suggest that both adult and larval zebrafish exhibit all the behavioral hallmarks of a sleep-like state, including prolonged periods of behavioral quiescence that occur in a circadian manner, stereotyped behavioral postures, an elevated arousal to sensory stimuli, and rebound following sleep deprivation; however, while both larvae and adults show hallmarks of a sleep-like state, there are subtle differences in sleep between the two developmental stages. Most notably, while a 1-min period of inactivity is indicative of a sleep-like state in larval stages (Prober et al., 2006), a 6-s period is used in adults (Yokogawa et al., 2007). Therefore, these studies highlight the importance of fully defining the behavioral characteristics of sleep in each species and developmental stage.

Conservation of Sleep Systems in Zebrafish

The neural systems regulating the transition between sleep and wake are complex and appear to be conserved across phyla. For example, many drugs that act either to promote sleep or wakefulness in humans or other mammals have similar effects on fly and fish models (Zhdanova, 2006; Cirelli, 2009). Consistent with this, much work in model systems has gone into understanding how neural circuits modulate either sleep or wakefulness, and how these processes interact with each other (Figure 15.1).

In mammalian systems, the brainstem has long been known to hold wake-promoting neurons, and subsequent studies have shown that noradrenergic neurons in the locus coeruleus, located near the ascending reticular activating system of the reticular formation, act to promote cortical arousal and wakefulness (Maruzzi and Magoun, 1949; Berridge and Foote, 1991). In addition to norepinephrine, other neurotransmitters such as dopamine (Wisor et al., 2001; Kume et al., 2005; Andretic et al., 2005), serotonin (Jouvet, 1968; Yuan et al., 2006), and acetylcholine (Jouvet, 1972) have all been shown to be wake-promoting.

Gamma-Aminobutyric acid (GABA) is the primary sleep-promoting neurotransmitter, and GABAergic agonists have similar effects in flies (Agosto et al., 2008) and fish (Zhdanova et al., 2001; Rihel et al., 2010) as they do in mammals (Saper et al., 2010). The activation of GABAergic neurons in the ventrolateral preoptic (VLPO) area, a subregion of the hypothalamus, acts to promote sleep (Sherin et al., 1996), and can be inhibited by the arousal-inducing neurotransmitters noradrenaline, acetylcholine, and serotonin (Gallopin et al., 2000). It has been proposed that a relatively simple sleep circuit composed of monoamine regulation of the VLPO regulates the transition from sleep to wake states. According to this model, during times of wakefulness, monoaminergic and cholinergic systems emanating from the hindbrain promote cortical arousal and inhibit GABAergic neurons in the VLPO. As wakefulness progresses, GABAergic neurons in the VLPO fire to inhibit the monoaminergic and cholinergic hindbrain systems, known as the "switch state" theory of sleep, (Saper

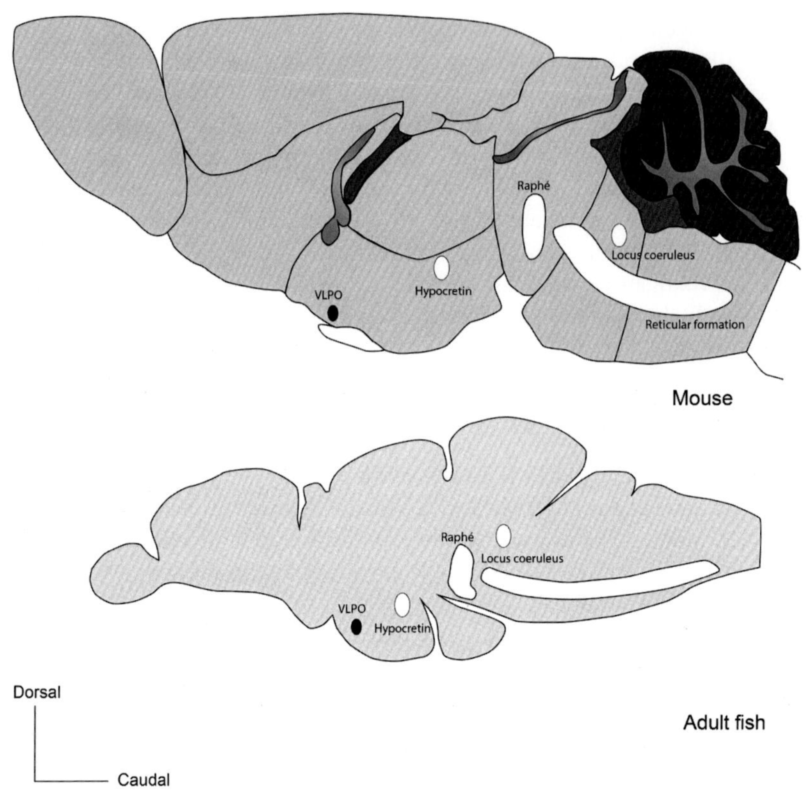

FIGURE 15.1 Sleep-regulating areas in mammalian and fish brains. Depiction of the adult mouse brain (top) and zebrafish brain (bottom). The ventrolateral preoptic (VLPO) nucleus (black circle) promotes sleep. The hypocretin-producing neurons, Raphé, and Locus coeruleus (white regions) promote wakefulness.

et al., 2010). Interestingly, VLPO neurons begin firing before sleep; thus it is unknown what the molecule(s) activate within the VLPO to begin firing.

To better understand the sleep circuit and the relationship of wake- and sleep-promoting systems, groups have begun looking into other neurotransmitters and neuropeptide systems. One neuropeptide that has received a lot of attention in both the fish and mammalian communities is hypocretin (*Hcrt*). *Hcrt* was first identified as a neuropeptide involved in sleep when the gene causing canine narcolepsy was cloned as the hypocretin receptor-2 (*hcrtr2*) (Lin et al., 1999). Since then, the loss of *Hcrt*-expressing neurons has been associated with some forms of human narcolepsy (Nishino et al., 2000), and overexpression of *Hcrt* has been shown to be wake-promoting in zebrafish larvae (Prober et al., 2006). It remains unclear how *Hcrt* acts (Chiu and Prober, 2013). Using transgenic zebrafish larvae, at least one group reported *Hcrt* neurons projecting to areas near the locus coeruleus and hindbrain dopaminergic neurons, and *in situ* hybridization showed *hcrtr2* expression colocalizing with noradrenergic cells expressing dopamine β-hydroxylase (Prober et al., 2006). Yet, another study suggested *hcrtr2* expression localized to neurons producing the neurotransmitters GABA, acetylcholine, and glycine, indicating broad classes of neurons are sensitive to *Hcrt* (Yokogawa et al., 2007). Therefore, the identification of sleep-regulating genes and neurons in fish provides a framework for understanding the neural basis for evolutionarily derived changes in sleep.

SLEEP LOSS IN CAVEFISH

To gain a full understanding of sleep, and to understand it in both a genetic and evolutionary perspective, a model system that incorporates both—a system that spawn frequently, possesses a wide genetics toolbox, is easily amenable to genetic manipulation, and has multiple well documented ecologies—is needed. *A. mexicanus* provides an excellent system to study how ecological history regulates sleep, because surface and cave populations evolve in drastically different ecologies, compete for consumable resources, and are at the top of the social hierarchy. More importantly, the 29 populations of *A. mexicanus* cavefish have evolved independently ~3-5 times from surface-dwelling form (Bradic et al., 2013; Coghill et al., 2014). Systems for measuring sleep in *A. mexicanus* larvae and adults have been established and are similar to those previously used in zebrafish (Duboué et al., 2011; Yoshizawa et al., 2015). Sleep of surface fish larvae is diurnal and similar in duration to zebrafish (Duboué et al., 2011); however, sleep is dramatically reduced in larvae of multiple, independently derived populations of cavefish, including fish from the Tinaja, Molino, and Pachón caves, demonstrating the parallel and convergent evolution of sleep loss (Duboué et al., 2011; Figure 15.2). Interestingly, sleep is reduced in adult Molino and Pachón populations compared to surface fish, but no differences are detected in Tinaja adults, raising the possibility that distinct genetic mechanisms regulate

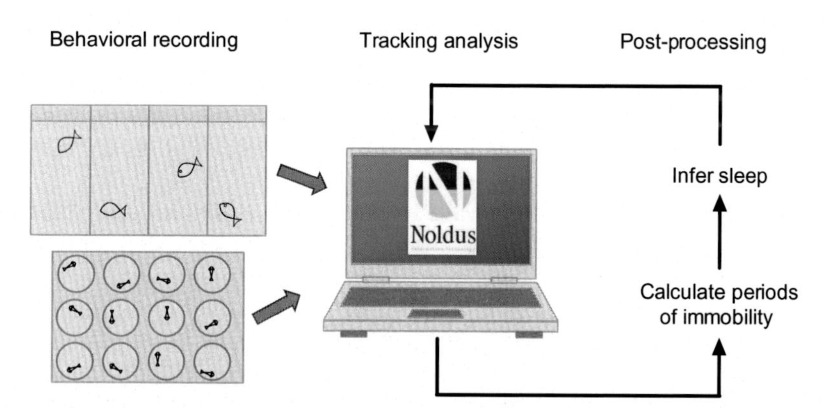

FIGURE 15.2 Recording systems for cavefish. Juvenile fish aged 20-30 dpf (bottom) are recorded in 12-well tissue culture plates. Adult fish (top) are recorded individually housed in two-gallon tanks with dividers. Video data is analyzed using Noldus Ethovision XT automated tracking software.

sleep loss in larvae and adult populations (Yoshizawa et al., 2015; Figure 15.3). Furthermore, adult cavefish from Cueva Los Sabinos, which derives from the El Abra population, do not display a reduction in sleep. Therefore, it is unclear if sleep loss has evolved in some cave populations and not others, or if these differences simply reflect a difference between the larval and adult forms.

The 29 populations of cavefish have evolved three to five independent times, though the ecology of each cave (and presumably the selective forces driving evolution) is significantly different (Bradic et al., 2013). For example, while Pachón, Tinaja, and most caves in the El Abra region have an abundance of Chiroptera (bat) life, which provides a source of defecated material for the fish to feed upon, deeper caves such as Molino are devoid of bat life (see Chapter 3). While only a few cave populations have been examined for their sleep phenotype, examining sleep in additional populations of cavefish and gaining a better understanding of the historical and present ecological differences among caves may help shed light upon the relationship between environment and evolutionarily derived sleep loss (Culver and Pipan, 2009 Figure 15.3).

The ability to generate surface-cave hybrids in *A. mexicanus* provides the powerful opportunity to analyze the multilocus traits that drive the behaviors and distinguish them. Interestingly, while the genetic mechanisms leading to morphological and behavioral cave phenotypes are typically recessive, sleep is reduced in F_1 surface-cave hybrids in a similar manner as in pure cavefish, revealing that the sleep phenotype is dominant and independent of morphological changes commonly associated with cavefish. In juvenile fish, no significant correlation was observed between sleep and several morphological traits including eye and pupil size, body length, and pigmentation in second-generation hybrids, suggesting that the evolution of sleep loss and morphological trait variation were driven by distinct genetic mechanisms (Duboué et al., 2011).

Astyanax also possess an extensive genetic toolkit that is expanding exponentially. Quantitative Trait Locus (QTL) mapping of surface/cave hybrids provides a powerful method for unraveling genes, driving sleep as well as their interactions at a genomic level. The number of genomic loci regulating sleep changes can be predicted using the Castle-Wright statistical estimator. This analysis in backcross individuals predicts that six genomic loci are associated with sleep loss in larval Pachón cavefish (Duboué et al., 2011). While the genes lying in the specific genomic loci regulating sleep have not been identified, a recently annotated genome and genomic approaches, including genotype-by-sequencing, may allow for the identification of these loci (McGaugh et al., 2014). Therefore, *A. mexicanus* may provide a system for identifying novel sleep genes. The identification of genes regulating sleep loss in Pachón fish will allow for targeted investigation of whether similar genetic mechanisms underlie sleep loss in other cave populations, providing critical insight into the molecular evolution of sleep loss.

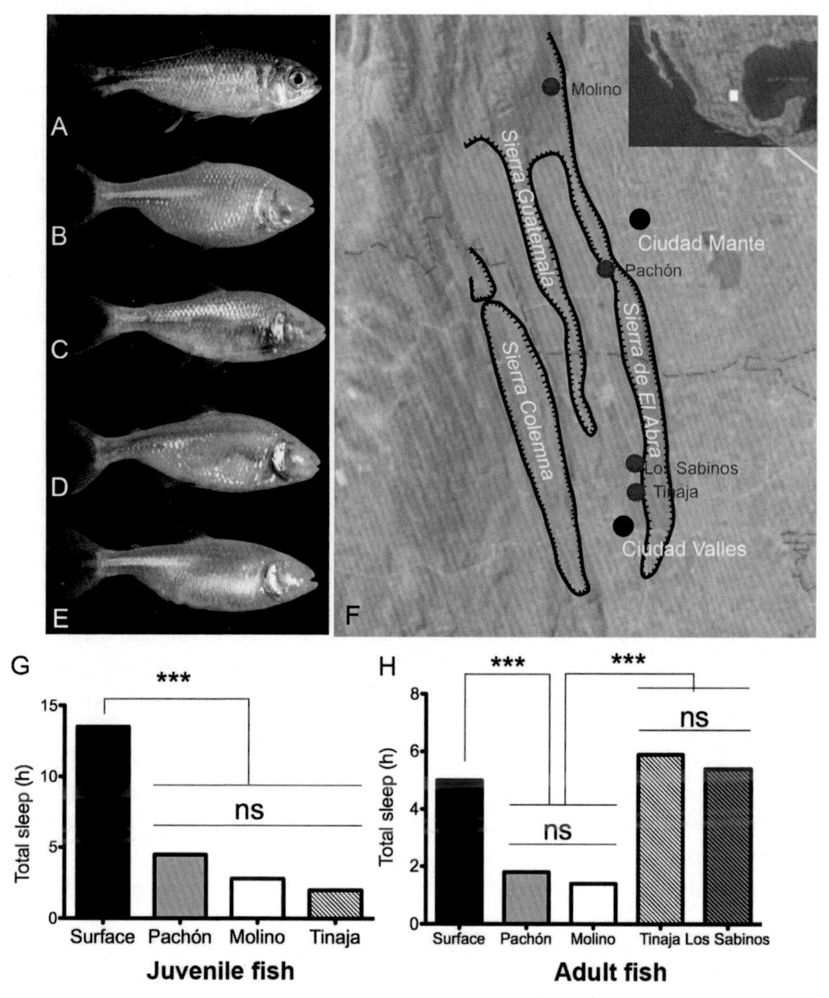

FIGURE 15.3 Convergent evolution of sleep loss in Mexican cavefishSleep has been tested in surface fish (A), Pachón (B), Tinaja (C), Los Sabinos (D), and Molino (E) fish populations. (F) Map of the Sierra de El Abra region and location of caves (red circles) of fish assayed for sleep. Black circles represent nearby towns. Surface fish are found in rivers throughout the region. Sleep levels in larval (G) and adult populations (H) over a 24 hour period. *** denotes significant differences of $p<0.001$. ns denotes non-significant differences.

Differences in Sleep Between Adults and Larvae

While the study of sleep in cavefish larvae is often preferable due to short generation, the behavior of adult fish is comparatively richer, providing the opportunity to examine interactions between sleep and other complex traits. In addition to sleep differences, adult cavefish display altered rheotaxis and vibration attraction behavior (VAB), a prey-seeking behavior where cavefish approach vibration in the water, that are dependent upon enhanced sensitivity of lateral-line sensory neurons (Yoshizawa et al., 2010, see Chapter 14). Both of these behaviors appear to be critical evolutionary changes that compensate for a loss of the visual system or changes in food availability. A hallmark of sleep is the gating of sensory information resulting in reduced responsiveness to external stimuli (Campbell and Tobler, 1984). Therefore, it is possible that sleep loss in cavefish is related to an increase in sensory information from the lateral line. Examining the relationship between sleep loss and VAB in surface-cave hybrids, or using a naturally occurring behavioral variation in pure Pachón cavefish, revealed no correlation between these behaviors, suggesting they are regulated by independent genetic and neural mechanisms (Yoshizawa et al., unpublished). The finding that VAB and sleep loss have coevolved, in some, but not all, populations tested, raises the possibility that independent selective pressures underlie the evolution of each of these behaviors, and further study of how the ecology of each cave is linked to these behaviors may provide insight into their evolution.

Why is Sleep Reduced in Cavefish

There are a number of possible explanations for reduced sleep in cavefish populations. The cave environment can be described as a harsh yet stable environment characterized by perpetual darkness, low food abundance, and year-round stable temperatures (Poulson and White, 1969; Culver, 1982). The complete absence of light consequently leads to a low abundance of photosynthetic life. While there are reports of some chemosynthetic bacteria, there is a near absence of primary producers (Culver and Pipan, 2009). This leads to two trends that are reproducible from cave to cave: (1) cave animals are generally higher on the food chain than their surface conspecifics; and (2) evolution to cave life is accompanied by significant changes in metabolic activity (Culver and Pipan, 2009).

It has been proposed that reduced sleep duration provides more time for foraging, thus, compensating for the reduction in food availability in a cave environment (Duboué et al., 2011). This theory is supported by a number of metabolic and physiological changes in cave populations that are associated with altered response to nutrient deprivation or foraging (Moran et al., 2014; Aspiras et al., 2015). The locomotor responses to starvation are blunted, and triglyceride stores are elevated in cave populations, indicating robust behavioral and physiological changes associated with changes in nutrient availability (Salin et al., 2010). Convergent evolution in feeding-associated sensory processes is also present in cavefish populations. Multiple, independently derived populations of cavefish display enhanced sensitivity to 35 Hz vibration (VAB), a similar frequency to water droplets or insects falling into stagnant water (Yoshizawa et al., 2010). This behavior (see Chapter 14) is associated with enhanced foraging capabilities, and may be adaptive in the nutrient-poor cave environment. Taken together, these findings reveal multiple foraging and feeding-state associated changes in *A. mexicanus* cave populations' behavior and physiology, which raise the possibility that sleep loss provides an additional mechanism for enhanced survival in a nutrient-poor environment.

Alternatively, reduced sleep could be a consequence of the loss of light-entrainable circadian rhythms. The circadian clock in many organisms, including flies and mammals, is associated with sleep regulation. Activity rhythms are present under light/dark conditions in both cave and surface fish (see Chapter 17); however, in constant darkness, 24-h activity rhythms persist in surface fish, but are absent in Pachón and Chica cave populations of cavefish (Beale et al., 2013). Further, behavioral activity rhythms are absent in the Somalian cavefish, *P. andruzzi* (Cavallari et al., 2011), as well as the hillstream loach of Thailand and Laos (Duboué et al., 2012), suggesting loss of circadian activity is a common response to constant darkness; however, in at least of these studies, the authors showed that in the absence of entrainable rhythms, sleep loss still persists, presumably through convergent evolution. Many molecular regulators of circadian function, including period and cryptochrome, have been shown to regulate sleep across species, and analysis of the molecular clock reveals these genes are dysregulated in cavefish populations (Beale et al., 2013). Therefore, it is possible that sleep loss results from a decoupling of neural systems regulating sleep and the circadian clock.

It has also been proposed that sleep represents a method for avoiding predation (Lesku et al., 2006). To date, no documented predators of adult cavefish have been reported. Therefore, it is possible that sleep loss has occurred due reduced need for predatory avoidance.

While the regressive explanations of evolutionarily derived sleep loss may contribute to the altered sleep observed in cavefish, we believe it is highly unlikely that these are primary factors. Sleep fulfills a number of critical physiological functions, including neuronal restoration, promotion of immune function, and behavioral and synaptic plasticity (Cirelli and Tononi, 2008; Everson, 1993; Hartmann, 1973). Therefore, while there are many possible contributors to sleep loss, we speculate that reduced sleep represents a constructive and adaptive behavior for survival in a cave environment. While the evolutionary factors driving sleep loss are unclear, it is possible that increased foraging time and reduced need for predatory avoidance may likely underlie the convergent evolution of sleep loss.

PHARMACOLOGICAL INTERROGATION OF SLEEP

The small size of larvae make them ideally suited for large-scale pharmacological screens that may be used to identify molecular regulators of behavior and development (Murphey et al., 2006; Peterson and Fishman, 2011). For example, in zebrafish, 5648 compounds from a small molecular library were screened for sleep-associated traits, including activity, sleep bout number, sleep bout length, and total sleep duration (Rihel et al., 2010). These findings provide a "pharmacological fingerprint" through which the complex effects that each compound has upon behavior can be examined. This approach also yielded a number of novel candidate molecular regulators of sleep and activity, including regulators of the potassium channel ether-a-go-go that increase waking activity without affecting rest, and immune-regulating drugs that promote sleep (Rihel et al., 2010).

We have used this approach, on a smaller scale, to examine the role of neurotransmitter systems in evolutionarily derived sleep loss. Antagonists targeting glutamate, GABA, serotonin, acetylcholine, dopamine, and norepinephrine neurotransmitter receptors were tested for their effects on sleep in surface fish and Pachón cavefish. We sought inhibitors that do not affect sleep in surface fish, but promote sleep in cavefish, to reveal a pathway that may be more active in cavefish, leading to sleep loss. Inhibition of noradrenergic signaling with the β-adrenergic antagonist propranolol and glutamate signaling with the NMDA-antagonist MK-801 enhanced sleep in cavefish, but not surface fish, raising the possibility that enhanced signaling through glutamate and β-adrenergic receptors contribute to sleep loss in cavefish (Duboué et al., 2012). The zebrafish genome encodes for four β-adrenergic receptors that are conserved in Pachón cavefish and are categorized into β1, β2, and β3 subtypes (McGaugh et al., 2014; Ruuskanen et al., 2004). Propranolol has previously been shown to increase sleep in zebrafish and mammals supporting the notion that β-adrenergic receptors have a conserved role in arousal (Mendelson et al., 1980; Rihel et al., 2010; Whitehurst et al., 1999).

Targeting β-adrenergic receptors with subtype-specific antagonists reveal that blockade of β1 adrenergic receptors alone increases sleep in Pachón cavefish. It is possible that a change in β-adrenergic receptors is due to enhanced norepinephrine release or altered localization and expression of β-adrenergic receptors. No gross anatomical differences were observed in dopamine and norepinephrine expression between surface fish and cavefish. Interestingly, the pigmentation gene ocular and cutaneous albinism-2 (*oca2*) has been implicated in the cellular transport of tyrosine in melanocytes that localize to the brain and body, revealing that loss of *oca2* correlates with higher catecholaminergic activity. Importantly, a mutation in *oca2* is also responsible for the albinism in Pachón cavefish. Mutations in *oca2* in Pachón cavefish also result in elevated norepinephrine and dopamine levels as a consequence of reduced tyrosine transport in melanocytes (Bilandžija et al., 2013). Therefore, these findings raise the possibility of a functional interaction between the evolution of albinism and sleep loss in Pachón cavefish.

The finding that norepinephrine levels are elevated in *oca2*-deficient Pachón cavefish leads to the predication that sleep loss is associated with albinism. In Pachón cave-surface fish hybrids, there was no significant correlation between sleep and pigmentation, yet there was a trend of reduced sleep in albino hybrid larvae (Duboué et al., 2011). This analysis used binary quantification of pigmentation, and a more detailed quantification of melanocyte number or total pigmentation may reveal a stronger relationship between pigmentation and sleep. Mutations in *oca2* also underlie albinism in many cave populations, including Pachón, Molino, and Micos (Gross and Wilkens, 2013; Protas et al., 2006). While many of these have not been tested for sleep, changes in sleep loss have been documented in Pachón, Tinaja, and Molino populations. Testing the effects of propranolol on sleep in additional cave populations may provide insight into whether noradrenergic modulation of sleep loss is conserved in additional cave populations. Taken together, pharmacological investigation of sleep in cavefish has implicated noradrenergic function as a regulator of sleep loss. These experiments pave the way for TALENs (Bedell et al., 2012; Dahlem et al., 2012) or CRISPR-Cas9 (Hwang et al., 2013) mediated targeted genetic disruption of noradrenergic signaling and *oca2* function to directly examine the contributions of these systems to sleep.

In addition to changes in noradrenergic signaling, it is possible that previously identified modulators of sleep, including hypocretin and melatonin, contribute to sleep loss in cave populations. While Pachón genome encodes for a single *Hcrt* receptor and *hcr* gene, the expression levels and localization have not been studied (McGaugh et al., 2014). The sleep-promoting hormone melatonin is also present in cavefish populations, but its functional role is not well understood. Interestingly, multiple cave populations are diurnal in light/dark conditions, suggesting light responsiveness in the absence of eyes (Beale et al., 2013; Yoshizawa and Jeffery, 2008). Therefore, it remains a possibility that a functional pineal gland regulates diurnal sleep-activity rhythms in cavefish and contributes to sleep loss. Future investigations of the contributions of the pineal gland and melatonin may provide insight into the role of melatonin in sleep.

Recent advances in genomic technology, including a sequenced genome for Pachón cave population, restriction site associated DNA sequencing (RADseq) based methods for high throughput sequencing, and the generation of genetic knockouts provide the opportunity to isolate novel regulators of sleep in *A. mexicanus*. Castle-Wright estimators of the number of sleep-regulating loci in Pachón cavefish indicated a small number of loci underlie sleep loss, raising the possibility that phenotypes can be verified with single gene manipulations. Full genome sequencing on hybrid fish, or pure outbred populations that have been tested for sleep behavior, may provide specific candidate genes that can then be targeted using CRIPER or TALEN technology. Therefore, *A. mexicanus* provides a powerful alternative system to identifying novel genetic regulators of sleep.

CONCLUDING REMARKS

The study of sleep in *A. mexicanus* presents a model to understand the evolution of sleep and neural regulation of sleep in other species. The convergent or parallel evolution of sleep loss has now been documented in both larvae and surface fish. A well defined genetic history of cave populations, and the ability to investigate the evolution of sleep loss in these distinct populations, may shed light upon the biological and ecological processes that underlie sleep differences throughout the metazoa. Further, new genomic and genetic technology may allow for the isolation of novel sleep regulators. This system has the potential to provide critical insights into the evolution of sleep and novel sleep-regulating genes that may improve our understanding of mammalian sleep. Therefore, *A. mexicanus* represent an emergent species for both genetic and evolutionary investigation of sleep with the unique ability to utilize naturally occurring variation and population differences to understand the biological basis for this complex behavior.

REFERENCES

Agosto, J., Choi, J.C., Parisky, K.M., Stilwell, G., Rosbash, M., Griffith, L.C., 2008. Modulation of GABAA receptor desensitization uncouples sleep onset and maintenance in Drosophila. Nat. Neurosci. 11 (3), 354–359. http://dx.doi.org/10.1038/nn2046. Epub 2008 Jan 27.

Ahrens, M.B., Li, J.M., Orger, M.B., Robson, D.N., Schier, A.F., Engert, F., Portugues, R., 2012. Brain-wide neuronal dynamics during motor adaptation in zebrafish. Nature 485, 471–477. http://dx.doi.org/10.1038/nature11057.

Allada, R., Siegel, J.M., 2008. Unearthing the phylogenetic roots of sleep. Curr. Biol. 18, R670–R679. http://dx.doi.org/10.1016/j.cub.2008.06.033.

Andretic, R., van Swinderen, B., Greenspan, R.J., 2005. Dopaminergic modulation of arousal in Drosophila. Curr. Biol. 15 (13), 1165–1175.

Appelbaum, L., Wang, G.X., Maro, G.S., Mori, R., Tovin, A., Marin, W., Yokogawa, T., Kawakami, K., Smith, S.J., Gothilf, Y., Mignot, E., Mourrain, P., 2009. Sleep-wake regulation and hypocretin-melatonin interaction in zebrafish. Proc. Natl. Acad. Sci. U. S. A., 106, 21942–21947. http://dx.doi.org/10.1073/pnas.906637106.

Aspiras, A.C., Rohner, N., Martineau, B., Borowsky, R.L., Tabin, C.J., 2015. Melanocortin 4 receptor mutations contribute to the adaptation of cavefish to nutrient-poor conditions. Proc. Natl. Acad. Sci. U. S. A. 112 (31), 9668–9673. http://dx.doi.org/10.1073/pnas.1510802112. Epub 2015 Jul 13.

Beale, A., Guibal, C., Tamai, T.K., Klotz, L., Cowen, S., Peyric, E., Reynoso, V.H., Yamamoto, Y., Whitmore, D., 2013. Circadian rhythms in Mexican blind cavefish *Astyanax mexicanus* in the lab and in the field. Nat. Commun. 4, 2769. http://dx.doi.org/10.1038/ncomms3769.

Bedell, V.M., Wang, Y., Campbell, J.M., Poshusta, T.L., Starker, C.G., Krug II, R.G., Tan, W., Penheiter, S.G., Ma, A.C., Leung, A.Y., Fahrenkrug, S.C., Carlson, D.F., Voytas, D.F., Clark, K.J., Essner, J.J., Ekker, S.C., 2012. In vivo genome editing using a high-efficiency TALEN system. Nature 491 (7422), 114–118. http://dx.doi.org/10.1038/nature11537.

Berridge, C.W., Foote, S.L., 1991. Effects of locus coeruleus activation on electroencephalographic activity in neocortex and hippocampus. J. Neurosci. 11 (10), 3135–3145.

Bilandžija, H., Ma, L., Parkhurst, A., Jeffery, W.R., 2013. A potential benefit of albinism in Astyanax cavefish: downregulation of the oca2 gene increases tyrosine and catecholamine levels as an alternative to melanin synthesis. PLoS One 8 (11), e80823.

Bradic, M., Teotónio, H., Borowsky, R.L., 2013. The population genomics of repeated evolution in the blind cavefish *Astyanax mexicanus*. Mol. Biol. Evol. 30, 2383–2400. http://dx.doi.org/10.1093/molbev/mst136.

Cahill, G.M., Hurd, M.W., Batchelor, M.M., 1998. Circadian rhythmicity in the locomotor activity of larval zebrafish. Neuroreport 9, 3445–3449.

Campbell, S.S., Tobler, I., 1984. Animal sleep: a review of sleep duration across phylogeny. Neurosci. Biobehav. Rev. 8, 269–300. http://dx.doi.org/10.1016/0149-7634(84)90054-X.

Cappuccio, F.P., D'Elia, L., Strazzullo, P., Miller, M.A., 2010. Sleep duration and all-cause mortality: a systematic review and meta-analysis of prospective studies. Sleep 33, 585–592.

Cavallari, N., Frigato, E., Vallone, D., Fröhlich, N., Lopez-Olmeda, J.F., Foà, A., Berti, R., Sánchez-Vázquez, F.J., Bertolucci, C., Foulkes, N.S., 2011. A blind circadian clock in cavefish reveals that opsins mediate peripheral clock photoreception. PLoS Biol. 9 (9), e1001142. http://dx.doi.org/10.1371/journal.pbio.1001142.

Chiu, C.N., Prober, D.A., 2013. Regulation of zebrafish sleep and arousal states: current and prospective approaches. Front Neural Circuits 7 (58). http://dx.doi.org/10.3389/fncir.2013.00058.

Cirelli, C., 2009. The genetic and molecular regulation of sleep: from fruit flies to humans. Nat. Rev. Neurosci. 10 (8), 549–560. http://dx.doi.org/10.1038/nrn2683. Review.

Cirelli, C., Bushey, D., Hill, S., Huber, R., Kreber, R., Ganetzky, B., Tononi, G., 2005. Reduced sleep in Drosophila Shaker mutants. Nature 434, 1087–1092. http://dx.doi.org/10.1038/nature03486.

Cirelli, C., Tononi, G., 2008. Is sleep essential? PLoS Biol. 6, e216. http://dx.doi.org/10.1371/journal.pbio.0060216.

Coghill, L., Darrin Hulsey, C., Chaves-Campos, J., Garcia de Leon, F., Johnson, S., 2014. Next generation phylogeography of cave and surface *Astyanax mexicanus*. Heredity 79, 368–374.

Culver, D.C., 1982. Cave life, evolution and ecology. Cambridge, Mass.

Culver, D.C., Pipan, T., 2009. The Biology of Caves and Other Subterranean Habitats. Oxford University Press, New York, USA.

Dahlem, T.J., Hoshijima, K., Jurynec, M.J., Gunther, D., Starker, C.G., Locke, A.S., Weis, A.M., Voytas, D.F., Grunwald, D.J., 2012. Simple methods for generating and detecting locus-specific mutations induced with TALENs in the zebrafish genome. PLoS Genet. 8, e1002861. http://dx.doi.org/10.1371/journal.pgen.1002861.

De Moura Filho, A.G., Huggins, S.E., Lines, S.G., 1983. Sleep and waking in the three-toed sloth, *Bradypus tridactylus*. Comp. Biochem. Physiol. A Comp. Physiol. 76, 345–355. http://dx.doi.org/10.1016/0300-9629(83)90336-5.

Dement, W., Kleitman, N., 1957. Cyclic variations in EEG during sleep and their relation to eye movements, body motility, and dreaming. Electroencephalogr. Clin. Neurophysiol. 9, 673–690. http://dx.doi.org/10.1016/0013-4694(57)90088-3.

Duboué, E.R., Borowsky, R.L., Keene, A.C., 2012. β-Adrenergic signaling regulates evolutionarily derived sleep loss in the Mexican cavefish. Brain Behav. Evol. 80, 233–243. http://dx.doi.org/10.1159/000341403.

Duboué, E.R., Keene, A.C., Borowsky, R.L., 2011. Evolutionary convergence on sleep loss in cavefish populations. Curr. Biol. 21, 671–676. http://dx.doi.org/10.1016/j.cub.2011.03.020.

Everson, C.A., 1993. Sustained sleep deprivation impairs host defense. Am. J. Physiol. 265, R1148–R1154.

Flanigan, W.F., Wilcox, R.H., Rechtschaffen, A., 1973. The EEG and behavioral continuum of the crocodilian, *Caiman sclerops*. Electroencephalogr. Clin. Neurophysiol. 34, 521–538. http://dx.doi.org/10.1016/0013-4694(73)90069-2.

Gallopin, T., Fort, P., Eggermann, E., Cauli, B., Luppi, P.H., Rossier, J., Audinat, E., Mühlethaler, M., Serafin, M., 2000. Identification of sleep-promoting neurons in vitro. Nature 404 (6781), 992–995.

Gross, J.B., Borowsky, R., Tabin, C.J., 2009. A novel role for Mc1r in the parallel evolution of depigmentation in independent populations of the cavefish Astyanax mexicanus. PLoS Genet. 5 (1), e1000326. http://dx.doi.org/10.1371/journal.pgen.1000326. Epub 2009 Jan 2.

Gross, J.B., Protas, M., Conrad, M., Scheid, P.E., Vidal, O., Jeffery, W.R., Borowsky, R., Tabin, C.J., 2008. Synteny and candidate gene prediction using an anchored linkage map of Astyanax mexicanus. Proc. Natl. Acad. Sci. U. S. A. 105 (51), 20106–20111. http://dx.doi.org/10.1073/pnas.0806238105. Epub 2008 Dec 22.

Gross, J.B., Wilkens, H., 2013. Albinism in phylogenetically and geographically distinct populations of *Astyanax* cavefish arises through the same loss-of-function Oca2 allele. Heredity 111, 122–130. http://dx.doi.org/10.1038/hdy.2013.26.

Hartmann, E.L., 1973. The Functions of Sleep. Yale University Press, New Haven, CT.

He, Y., Jones, C.R., Fujiki, N., Xu, Y., Guo, B., Holder, J.L., Rossner, M.J., Nishino, S., Fu, Y.-H., 2009. The transcriptional repressor DEC2 regulates sleep length in mammals. Science 325, 866–870. http://dx.doi.org/10.1126/science.1174443.

Hendricks, J.C., Finn, S.M., Panckeri, K.A., Chavkin, J., Williams, J.A., Sehgal, A., Pack, A.I., 2000. Rest in Drosophila is a sleep-like state. Neuron 25, 129–138. http://dx.doi.org/10.1016/S0896-6273(00)80877-6.

Hwang, W.Y., Fu, Y., Reyon, D., Maeder, M.L., Tsai, S.Q., Sander, J.D., Peterson, R.T., Yeh, J.R., Joung, J.K., 2013. Efficient genome editing in zebrafish using a CRISPR-Cas system. Nat. Biotechnol. 31, 227–229. http://dx.doi.org/10.1038/nbt.2501.

Jamieson, D., Roberts, A., 2000. Responses of young *Xenopus laevis* tadpoles to light dimming: possible roles for the pineal eye. J. Exp. Biol. 203, 1857–1867.

Jouvet, M., 1968. Insomnia and decrease of cerebral 5-hydroxytryptamine after destruction of the raphe system in the cat. Adv. Pharmacol. 6 (Pt B), 265–279.

Jouvet, M., 1972. The role of monoamines and acetylcholine-containing neurons in the regulation of the sleep-waking cycle. Ergeb Physiol 64, 166–307. No abstract available.

Kandel, E.R., Schwartz, J.H., Jessell, T.M., 2013. Principles of neural science. Neurology 4, 1414. http://dx.doi.org/10.1036/0838577016.

Kume, K., Kume, S., Park, S.K., Hirsh, J., Jackson, F.R., 2005. Dopamine is a regulator of arousal in the fruit fly. J. Neurosci. 25 (32), 7377–7384.

Lesku, J.A., Roth, T.C., Amlaner, C.J., Lima, S.L., 2006. A phylogenetic analysis of sleep architecture in mammals: the integration of anatomy, physiology, and ecology. Am. Nat. 168, 441–453. http://dx.doi.org/10.1086/506973.

Lin, L., Faraco, J., Li, R., Kadotani, H., Rogers, W., Lin, X., Qiu, X., de Jong, P.J., Nishino, S., Mignot, E., 1999. The sleep disorder canine narcolepsy is caused by a mutation in the hypocretin (orexin) receptor 2 gene. Cell 98, 365–376. http://dx.doi.org/10.1016/S0092-8674(00)81965-0.

McGaugh, S., Gross, J.B., Aken, B., Blin, M., Borowsky, R., Chalopin, D., Hinaux, H., Jeffery, W.R., Keene, A., Ma, L., Minx, P., Murphy, D., O'Quin, K.E., Rétaux, S., Rohner, N., Searle, S.M., Stahl, B.A., Tabin, C., Volff, J.N., Yoshizawa, M., Warren, W., 2014. The cavefish genome reveals candidate genes for eye loss. Nat. Commun. 5, 5307.

Mendelson, W.B., Gillin, J.C., Dawson, S.D., Lewy, A.J., Wyatt, R.J., 1980. Effects of melatonin and propranolol on sleep of the rat. Brain Res. 201 (1), 240–244.

Moran, D., Softley, R., Warrant, E.J., 2014. Eyeless Mexican cavefish save energy by eliminating the circadian rhythm in metabolism. PLoS One 9 (9), e107877. http://dx.doi.org/10.1371/journal.pone.0107877. eCollection 2014.

Moruzzi, G., Magoun, H.W., 1949. Brain stem reticular formation and activation of the EEG. Electroencephalogr. Clin. Neurophysiol. 1 (4), 455–473.

Murphey, R.D., Stern, H.M., Straub, C.T., Zon, L.I., 2006. A chemical genetic screen for cell cycle inhibitors in zebrafish embryos. Chem. Biol. Drug Des. 68, 213–219. http://dx.doi.org/10.1111/j.1747-0285.2006.00439.x.

Nishino, S., Ripley, B., Overeem, S., Lammers, G.J., Mignot, E., 2000. Hypocretin (orexin) deficiency in human narcolepsy. Lancet 355, 39–40. http://dx.doi.org/10.1016/S0140-6736(99)05582-8.

Peterson, R.T., Fishman, M.C., 2011. Designing zebrafish chemical screens. Methods Cell Biol. 105, 525–541. http://dx.doi.org/10.1016/B978-0-12-381320-6.00023-0.

Poulson, T.L., White, W.B., 1969. The cave environment. Science 165 (3897), 971–981.

Prober, D.A., Rihel, J., Onah, A.A., Sung, R.-J., Schier, A.F., 2006. Hypocretin/orexin overexpression induces an insomnia-like phenotype in zebrafish. J. Neurosci. 26, 13400–13410. http://dx.doi.org/10.1523/JNEUROSCI.4332-06.2006.

Protas, M., Tabansky, I., Conrad, M., Gross, J.B., Vidal, O., Tabin, C.J., Borowsky, R., 2008. Multi-trait evolution in a cave fish, Astyanax mexicanus. Evol. Dev. 10 (2), 196–209. http://dx.doi.org/10.1111/j.1525-142X.2008.00227.x.

Protas, M.E., Hersey, C., Kochanek, D., Zhou, Y., Wilkens, H., Jeffery, W.R., Zon, L.I., Borowsky, R., Tabin, C.J., 2006. Genetic analysis of cavefish reveals molecular convergence in the evolution of albinism. Nat. Genet. 38, 107–111. http://dx.doi.org/10.1038/ng1700.

Raizen, D.M., Zimmerman, J.E., Maycock, M.H., Ta, U.D., You, Y., Sundaram, M.V., Pack, A.I., 2008. Lethargus is a *Caenorhabditis elegans* sleep-like state. Nature 451, 569–572. http://dx.doi.org/10.1038/nature07044.

Rechtschaffen, A., Gilliland, M.A., Bergmann, B.M., Winter, J.B., 1983. Physiological correlates of prolonged sleep deprivation in rats. Science 221, 182–184. http://dx.doi.org/10.1126/science.6857280.

Rihel, J., Prober, D.A., Arvanites, A., Lam, K., Zimmerman, S., Jang, S., Haggarty, S.J., Kokel, D., Rubin, L.L., Peterson, R.T., Schier, A.F., 2010. Zebrafish behavioral profiling links drugs to biological targets and rest/wake regulation. Science 327, 348–351. http://dx.doi.org/10.1126/science.1183090.

Ruuskanen, J.O., Xhaard, H., Marjamäki, A., Salaneck, E., Salminen, T., Yan, Y.L., Postlethwait, J.H., Johnson, M.S., Larhammar, D., Scheinin, M., 2004. Identification of duplicated fourth α2-adrenergic receptor subtype by cloning and mapping of five receptor genes in zebrafish. Mol. Biol. Evol. 21, 14–28. http://dx.doi.org/10.1093/molbev/msg224.

Salin, K., Voituron, Y., Mourin, J., Hervant, F., 2010. Cave colonization without fasting capacities: an example with the fish *Astyanax fasciatus mexicanus*. Comp. Biochem. Physiol. A Mol. Integr. Physiol. 156, 451–457. http://dx.doi.org/10.1016/j.cbpa.2010.03.030.

Saper, C.B., Fuller, P.M., Pedersen, N.P., Lu, J., Scammell, T.E., 2010. Sleep state switching. Neuron 68, 1023–1042. http://dx.doi.org/10.1016/j.neuron.2010.11.032.

Shaw, P.J., Cirelli, C., Greenspan, R.J., Tononi, G., 2000. Correlates of sleep and waking in *Drosophila melanogaster*. Science 287, 1834–1837. http://dx.doi.org/10.1126/science.287.5459.1834.

Shaw, P.J., Tononi, G., Greenspan, R.J., Robinson, D.F., 2002. Stress response genes protect against lethal effects of sleep deprivation in Drosophila. Nature 417, 287–291. http://dx.doi.org/10.1038/417287a.

Sherin, J.E., Shiromani, P.J., McCarley, R.W., Saper, C.B., 1996. Activation of ventrolateral preoptic neurons during sleep. Science 271 (5246), 216–219.

Siegel, J.M., 2005. Clues to the functions of mammalian sleep. Nature 437, 1264–1271. http://dx.doi.org/10.1038/nature04285.

Whitehurst, V.E., Vick, J.A., Alleva, F.R., Zhang, J., Joseph, X., Balazs, T., 1999. Reversal of propranolol blockade of adrenergic receptors and related toxicity with drugs that increase cyclic AMP. Proc. Soc. Exp. Biol. Med. 221 (4), 382–385.

Wisor, J.P., Nishino, S., Sora, I., Uhl, G.H., Mignot, E., Edgar, D.M., 2001. Dopaminergic role in stimulant-induced wakefulness. J. Neurosci. 21 (5), 1787–1794.

Yokogawa, T., Marin, W., Faraco, J., Pézeron, G., Appelbaum, L., Zhang, J., Rosa, F., Mourrain, P., Mignot, E., 2007. Characterization of sleep in zebrafish and insomnia in hypocretin receptor mutants. PLoS Biol. 5, 2379–2397. http://dx.doi.org/10.1371/journal.pbio.0050277.

Yoshizawa, M., Gorički, Š., Soares, D., Jeffery, W.R., 2010. Evolution of a behavioral shift mediated by superficial neuromasts helps cavefish find food in darkness. Curr. Biol. 20, 1631–1636. http://dx.doi.org/10.1016/j.cub.2010.07.017.

Yoshizawa, M., Jeffery, W.R., 2008. Shadow response in the blind cavefish *Astyanax* reveals conservation of a functional pineal eye. J. Exp. Biol. 211, 292–299. http://dx.doi.org/10.1242/jeb.012864.

Yoshizawa, M., Robinson, B.G., Duboué, E.R., Masek, P., Jaggard, J.B., O'Quin, K.E., Borowsky, R.L., Jeffery, W.R., Keene, A.C., 2015. Distinct genetic architecture underlies the emergence of sleep loss and prey-seeking behavior in the Mexican cavefish. BMC Biol. 13 (15).

Yuan, Q., Joiner, W.J., Sehgal, A., 2006. A sleep-promoting role for the Drosophila serotonin receptor 1A. Curr. Biol. 16 (11), 1051–1062.

Zhdanova, I.V., 2006. Sleep in zebrafish. Zebrafish 3, 215–226. http://dx.doi.org/10.1089/zeb.2006.3.215.

Zhdanova, I.V., Wang, S.Y., Leclair, O.U., Danilova, N.P., 2001. Melatonin promotes sleep-like state in zebrafish. Brain Res. 903, 263–268. http://dx.doi.org/10.1016/S0006-8993(01)02444-1.

Chapter 16

Daily Rhythms in a Timeless Environment: Circadian Clocks in *Astyanax mexicanus*

A.D. Beale and D. Whitmore
Centre for Cell and Molecular Dynamics, University College London, London, United Kingdom

A GENERAL INTRODUCTION TO THE CIRCADIAN CLOCK

Most physiological, cellular, and behavioral processes show oscillations in a daily fashion. These oscillations can continue even in the absence of environmental time cues, and were first described nearly 300 years ago in plants (De Mairan, 1729). Behind such oscillations is an endogenous, self-sustaining time-keeper, known as the circadian clock (Pittendrigh, 1960). Most, if not all, animals and plants have evolved this endogenous timing system (Pittendrigh, 1993); this ubiquity has led to the proposal that temporally coordinating biological activity has provided a selective advantage for survival (Hut and Beersma, 2011; Pittendrigh, 1993; Sharma, 2003), which can be conferred in two ways: (1) coordination of the internal with the external world through synchronization to external cycles, the "day without" and (2) provision of temporal order within an organism, the "day within."

However, for these clocks to serve a useful biological purpose, whether internally or externally, they need to be set or entrained. Perhaps the most fundamental and palpable external cyclic event to which clocks entrain is the daily cycle of light and dark (LD), and it is this signal that serves as a primary *zeitgeber* (from the German, "time-giver") for most organisms on the planet. Constant dark environments, such as caves or the depths of the ocean, are not subject to such fundamental periodic cycles of LD. Animals that exist here provide an interesting opportunity to study the origin, evolution, and adaptive value of circadian clocks. The primary entraining signal, light, is not present in these environments, and thus, the clocks may struggle to serve a useful biological purpose.

In this chapter, we will consider what is known in the literature on clocks in cave animals before focusing on emerging stories of the circadian system of the cavefish, *Astyanax mexicanus*, to discuss what cave animals can tell us about the function and evolution of the circadian clock.

Biology and Evolution of the Mexican Cavefish. http://dx.doi.org/10.1016/B978-0-12-802148-4.00016-5
© 2016 Elsevier Inc. All rights reserved.

CLOCKS IN ZEBRAFISH

A great deal of circadian research in animals has been performed in just two organisms: the mouse and fruit fly, *Drosophila melanogaster*. Their strengths as powerful genetic model systems have significantly enhanced our fundamental understanding of how the clock works; however, we will consider here the knowledge drawn on the circadian clock from another important model in circadian biology, the zebrafish, which has significantly broadened our understanding of vertebrate circadian clock organization.

Zebrafish came onto the clock scene in 1996 as circadian researchers looked for new models to dissect the genetic basis of the circadian clock, particularly in relation to early events in development (Cahill, 1996). In this regard, zebrafish has a large advantage over the traditional vertebrate circadian model, the mouse, as fish produce large numbers of transparent and readily accessible embryos. It was this attribute that was taken advantage of in one of the very first circadian studies on zebrafish, as Cahill and colleagues demonstrated the presence of circadian rhythms of activity in early larvae (Hurd et al., 1998).

Much of what is now known about molecular clocks in teleosts comes from further studies in zebrafish. Key to this progress was the discovery of self-sustaining, light-responsive circadian clocks in nearly all zebrafish tissues, which represented the first demonstration of peripheral clocks in a vertebrate (Kaneko et al., 2006; Whitmore et al., 1998, 2000). Though peripheral clocks exist in mice, studies in zebrafish showed that these tissues are themselves directly light-responsive, drawing this vertebrate closer to *Drosophila*; the zebrafish circadian system is highly decentralized, and rhythms in clock gene expression can be generated within most cells and entrained to external stimuli, without the need of a neuronal pacemaker (Whitmore et al., 1998).

The clock itself consists of transcription-translation feedback loops involving positive and negative elements that interact to produce a cycle time of approximately 24 h. At its simplest, the positive element consists of basic helix-loop-helix (bHLH) per-arnt-sim (PAS) domain transcription factors, *clk* and *bmal*, which bind as heterodimers to E-box elements in the promoters of period (*per*) and cryptochrome (*cry*) genes to activate their expression. In turn, the translated PER and CRY inhibit the transcriptional activation of *clk-bmal* within the nucleus, and thus inhibit their own transcription, completing the feedback loop. In fish, light feeds directly into the feedback loop via specialized period and cryptochrome genes, inhibiting the clock during the light phase and resetting the clock (Carr and Whitmore, 2005; Tamai et al., 2007; Vallone et al., 2004).

It is now clear that light has a very strong impact on teleost biology, not only on resetting the circadian clock, but also on other aspects of basic cell biology, such as the activation of DNA repair processes (Hirayama et al., 2009; Tamai et al., 2004) and the regulation of cell cycle events (Dekens et al., 2003; Dickmeis et al., 2007; Idda et al., 2012; Laranjeiro et al., 2013; Tamai et al., 2012). While the identity of the photopigment or photopigments that coordinate this cell-autonomous response to light remains unconfirmed in zebrafish, in recent years components of the light-input pathway have begun to be identified. Evidence is far from conclusive, but there is a suggestion that the MAPK pathway is involved in the immediate response to light, at least in cell lines (Cermakian et al., 2002; Hirayama et al., 2007, 2009; Mracek et al., 2013; Ramos et al., 2014). Downstream of this, D-box binding factors, in particular thyrotroph embryonic factor (tef), have been proposed to mediate this light response (Vatine et al., 2009). Tef is a member of the PAR (proline and acidic amino acid-rich) subfamily of basic region/leucine zipper (bZIP) transcription factors and regulates the downstream clock genes, *cry1a* and *per2*, forming the early response of the circadian clock to light, and interacting with the core clock mechanism through their repressive function (Hirayama et al., 2003; Pando et al., 2001; Tamai et al., 2007; Ziv et al., 2005). Further studies discovered that the regulation via tef is important not just for the circadian clock, but also in mediating light activation of several other light-responsive genes, suggesting it broadly influences light-regulated zebrafish biology (Gavriouchkina et al., 2010; Mracek et al., 2012; Vatine et al., 2009; Weger et al., 2011).

If light is so crucial for biological timing and cellular physiology, what happens in a cave?

CLOCKS IN A CAVE

Cave Animals and Clocks

Organisms that live under unusual lighting conditions are of particular interest to circadian biologists. They provide the opportunity to study the endogenous circadian clock in new ways and allow questions such as what fitness advantage (if any) is offered by the circadian mechanism. Despite the ubiquitous nature of circadian clocks, implying that they provide a fitness advantage to the organism, only a few studies have been performed that provide evidence for the adaptive value of the clock. Important examples are life-span studies in SCN-lesioned chipmunks (DeCoursey et al., 2000), plant fitness studies in Arabidopsis (Green et al., 2002), and competition studies between different clock mutants on matched external LD cycles (Dodd et al., 2005), which have shown that possession of a clock is advantageous compared to "clockless" organisms.

Caves represent an unusually nonilluminated habitat that could further our understanding into the value of circadian clocks. In addition to the absence of LD cycles, temperature is extremely stable in caves, being approximately equal to the mean annual temperature of the surrounding region. Humidity is high, as evaporation rates are low, and water quality is unchanging with a steady concentration of dissolved inorganic compounds and constant low organic matter content (Poulson and White, 1969). If circadian clocks can be found in such arrhythmic environments as caves, circadian biologists will have much to say about the physiological role of the circadian clock, its evolutionary origins, and its fitness value.

Both synchrony with the external world and temporal order within an organism provide pressures for the evolution of endogenous clocks (Pittendrigh, 1993). The study of clocks in caves allows circadian biologists to test these hypotheses; since there are no external cycles in the caves, if circadian clocks are found, it would support the second hypothesis, that is, the provision of internal temporal order, is the predominant reason for the presence or retention of circadian clock machinery in cave animals. If, as has been suggested, animals that exist in an aperiodic environment gain no selective advantage from a functional circadian clock, it is a likely target for regression, as in the case of eyes and pigmentation in *Astyanax*, and we should find no evidence of a functional clock mechanism (Sharma, 2003). Of course, clock gene expression could be retained in these animals, even in the absence of a functional or synchronized circadian pacemaker, as it is likely that clock components/transcription factors may play additional essential roles in the cell, other than just as part of the clock mechanism itself. Circadian clock studies in these animals are therefore of great importance to address these questions.

Unfortunately, very few circadian studies have been performed on cave animals, and most have only examined activity or locomotion (Table 16.1).

TABLE 16.1 Historical Studies on Daily Locomotor Activity of Cave Animals

Cave Animal	Species	Study
Crayfish	*Orconectes pellucidus*	Park et al. (1941), Brown (1961) and Jegla and Poulson (1968)
Amphipod	*Niphargus puteanus*	Blume et al. (1962)
Cricket	*Hadenoecus subterraneus*	Reichle et al. (1965)
Millipede	*Glyphiulus cavernicolus*	Mead and Gilhodes (1974) and Koilraj et al. (2000)
Beetle	*Aphaenops cerberus;* *Geotrechus orpheus;* *Speonomus diecki*	Lamprecht and Weber (1978)
Fish	*Nemacheilus evezardi*	Pati (2001)
	Schistura spesei *S. jaruthanini*	Duboué and Borowsky (2012)
	Trichomycterus sp.	Trajano and Menna-Barreto (1996)
	Astyanax mexicanus	Erckens and Martin (1982) and Zafar and Morgan (1992)
Salamander	*Proteus anguinus*	Hervant et al. (2001)

These few studies reveal an array of circadian phenotypes in cave-adapted animals. Some animals can retain at least partially functioning oscillators (Duboué and Borowsky, 2012; Erckens and Martin, 1982; Pati, 2001; Reichle et al., 1965; Zafar and Morgan, 1992), some show highly variable rhythms between individuals within populations (Brown, 1961; Jegla and Poulson, 1968), and some animals show an absence of, or highly irregular, circadian rhythms (Blume et al., 1962; Hervant et al., 2001; Koilraj et al., 2000; Lamprecht and Weber, 1978; Mead and Gilhodes, 1974; Trajano and Menna-Barreto, 1996). A full comparison of the published work is difficult, due to differences in experimental conditions and animal maintenance, such as whether the animals were entrained to LD cycles before the experiment, or the use of LD cycles of varying photoperiods, or experiments under different daylengths (T-cycles). Furthermore, conclusions are highly subjected to the method of analysis; for example, though evidence suggests that the cave crayfish is able to show circadian rhythms of activity, interestingly it was first thought not to, and only upon reanalysis of the data were rhythms apparent (Brown, 1961; Park et al., 1941).

In fact, Brown's (1961) paper actually plays an interesting role in circadian history. The discovery of rhythms in a cave animal was used by Brown as evidence that circadian rhythms are not due to an internal oscillator, but rather due to extrinsic rhythmicity. While it is not clear what those external drivers could be in a cave, Brown suggested that it could be due to "geophysical factors" (Brown, 1961). This theory has long been proved incorrect; however, it demonstrates how cave animals have played, and can play, such an important role in our understanding of the value of circadian rhythms.

In summary, the persistence or absence of circadian rhythms in troglobites as a whole is not clear; conclusions are highly subjective to experimental technique and subsequent analysis. The major assay of circadian functionality, locomotor activity, is sensitive to the type of analysis performed, and perhaps a clearer idea of the presence of circadian rhythms and functional circadian clock can be gained from an analysis of their molecular components. The molecular clock is the core of the system, its transcription-translation feedback loop being the source of all output rhythms. Molecular studies on cave animals should eliminate problems in assay type and produce more consistent data, especially as many of the tools developed in zebrafish to study clock function can also be applied to study the clock in *Astyanax*.

Developing *Astyanax* as a Clock Model

Much of the advantage of *Astyanax* as a cave model for the investigation of circadian rhythms comes from its close relationship to the well studied zebrafish. As teleosts, the approaches that make zebrafish a powerful system in circadian research can also be applied to *Astyanax*, such as the ability to perform extensive genetic screens, but it is the high degree of similarity in their genomic sequence that promoted the eyeless *Astyanax* cavefish to be developed as a clock model.

Beale et al. (2013) took advantage of the close phylogenetic relationship between zebrafish and cavefish, together with available genomic data from other teleosts, to design degenerate cloning approaches to obtain coding sequence for circadian genes—a key step in the molecular characterization of the circadian system (Beale et al., 2013). With gene sequence, one can study the very characteristic circadian gene expression pattern in LD cycles and deduce the extent of clock function in a similar way to the first forays into the molecular circadian clock in zebrafish by Whitmore et al. (1998).

With the publication of the *Astyanax* genome and its release on Ensembl in 2013, many more circadian genes can be found (McGaugh et al., 2014; see Chapter 7). It now appears that cavefish have the full complement of teleost circadian genes, including two *per1* genes, two *per2* genes and two *cry1* genes (Ensembl 77).

Another key characteristic for the development of *Astyanax* as a clock model is the highly decentralized nature of teleost clocks. Decentralized clocks have been found in a range of teleosts, including zebrafish (Whitmore et al., 2000), reef fish (Park et al., 2007), goldfish (Velarde et al., 2009), and Senegalese sole (Martín-Robles et al., 2011, 2012). At this time, there is no evidence for a central or "master" pacemaker in fish, comparable to the suprachiasmatic nucleus (SCN) in mammals. In *Astyanax*, the existence of peripheral tissue clocks allows the exploration of circadian function using the "nonlethal" collection of tissue samples, such as of the caudal fin, taken at different times of day, under both lab and natural conditions. This fact has proven essential due, in part, to the scarcity of these animals and the difficulty in obtaining samples, especially under natural conditions in the wild.

THE CIRCADIAN CLOCK OF *A. MEXICANUS*

Rhythms in the Lab

Initial efforts focused on characterizing the expression pattern of key circadian genes in surface fish and cavefish under rhythmic conditions in the laboratory. These efforts looked at *per1*, a core clock gene and central to the generation of circadian rhythms within the cell and whole organism. As expected, surface fish show a robust circadian oscillation in *per1* gene expression, as has been previously reported for zebrafish and other teleost species (Martín-Robles et al., 2012; Park et al., 2007; Vallone et al., 2004; Velarde et al., 2009). This rhythm continues as the animals free-run into constant darkness (Figure 16.1(A)), indicative of a true circadian clock gene.

Interestingly, *per1* levels also oscillate in two cave populations studied, showing that cavefish retain the ability to detect light and generate molecular circadian oscillations (Figure 16.1(A)); however, there are clear and consistent differences between the rhythms seen in surface and cave populations,

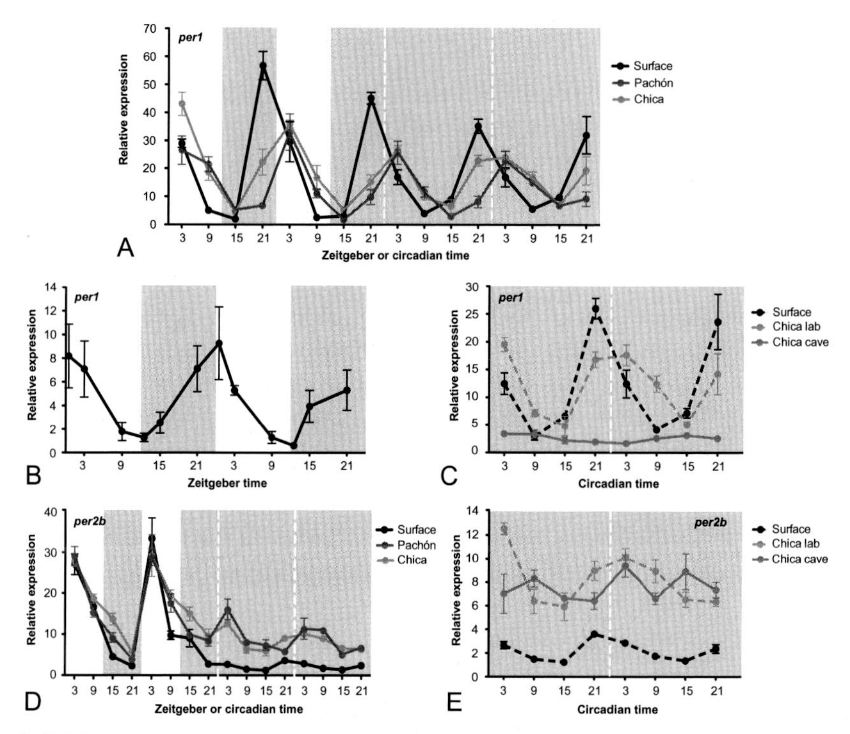

FIGURE 16.1 Molecular clock rhythms of *Astyanax* in the lab and field. (A) When entrained to a 12:12 LD cycle in the laboratory, surface fish have a high amplitude rhythm in the expression of the core clock gene, *per1*, measured by qPCR, with a peak in the late night. *Per1* levels also oscillate in adult cavefish, showing that cavefish retain the ability to detect light and generate molecular oscillations; however, consistent differences between cave and surface populations in amplitude and phase of the rhythm indicate substantial alterations in the cavefish's circadian system. (B) Surface fish in the wild show a similar rhythm to that seen in the laboratory, though no circadian oscillations are seen in wild cavefish. (C) The core clock mechanism appears to be substantially repressed, in a manner reminiscent of constant light treatment or overexpression of light-induced clock repressors in zebrafish. (D) *Per2b* expression is significantly raised in cavefish over surface fish in constant darkness in the laboratory, and is also seen in the field (E). White and grey shaded areas indicate light and dark periods, respectively. *Taken from Beale et al. (2013).*

which suggest that the circadian system in cavefish is in some way altered and point us in the direction of that alteration. First, the cavefish *per1* rhythm is lower in amplitude, with a relative peak to trough ratio in LD of 7.71- and 8.26-fold in Pachón and Chica, respectively, compared to 20.61-fold in surface fish. Secondly, the timing or entrained phase of the *per1* rhythm is clearly different, with peak expression occurring 6h later in both cave populations. As these differences persist in DD (constant darkness), these data suggest a core clock alteration rather than a change in the acute light input (Beale et al., 2013; Figure 16.1(A)). There are a number of potential mechanisms that could lead to this phenotype, including alterations in the transcriptional control of *per1* and alterations in the characteristics of the core oscillator, including that of *clk* and *bmal*, which control circadian transcriptional activation through E-boxes in promoters. Unpublished data suggest that the expression pattern of one *clk* gene, *clk1*, is similar in cave and surface fish, suggesting this transcriptional activator is not the source of the expression pattern alteration in *per1*, though as genomic duplication events in teleosts have led to multiple *clk* and *bmal* genes, the situation is likely to be complex.

One possibility is that changes in gene regulatory regions (promoters) underlie differences in circadian gene expression in cave populations. Alterations in *per1* promoter sequences between surface fish and cavefish could result in the observed phase delay. A 6-h phase delay is present in rhythms controlled by a 4xE-box heterologous promoter construct compared to a *per1b* minimal promoter in zebrafish cells, proving that the context of the E-box elements in the promoter is able to affect phase (Vallone et al., 2004). If *Astyanax* cavefish and surface fish differed in this regard, the timing of the clock would be altered. Certainly the publication of the *Astyanax* genome will help in resolving this issue.

What Happens in the Wild?

In surface fish samples collected in the Micos River in March, when the sunrise and sunset times create an almost perfect 12-h LD cycle, a high-amplitude *per1* rhythm was also observed (Figure 16.1(B)). The timing of the oscillation in natural conditions is very similar to that seen in the lab, but unsurprisingly there is greater variation. There are several possible explanations for this, including greater genetic variation in the wild compared to the lab population, as well as differences in animal health and age, light intensity, and spectral quality at the water surface of the Micos River. Together, these factors will affect gene expression patterns under natural conditions.

For samples collected within Chica cave, the dampened *per1* oscillation seen in the laboratory is even more exaggerated, as no *per1* rhythm was detected at all (Beale et al., 2013; Figure 16.1(C)). Chica cave is one of the few cave sites that contains a rhythmic bat population, giving the possibility that their dawn and dusk migration provides a time reference for the cavefish. Nevertheless, there appears to be no detectable circadian rhythm within this native cavefish population. A closer examination of *per1* expression in the cave field samples reveals that the actual levels are, in fact, much lower than those seen the lab in either surface or cavefish (Figure 16.1(C)). This very low transcript level likely reflects actual repression of core clock function rather than being the consequence of simple asynchrony or lack of clock entrainment.

Like the results from laboratory-kept fish, this situation draws parallels with constant light treatment of zebrafish cells, which leads to *per1* repression through the raised expression of a light-induced gene, *cry1a* (Tamai et al., 2007). Though *cry1a* is not significantly raised in cavefish, another light-induced clock repressor important in clock entrainment, *per2b*, is (Vatine et al., 2009; Ziv et al., 2005). *Per2b* is expressed at significantly raised levels in cave strains compared to surface fish in DD in both lab and wild samples (Beale et al., 2013; Figure 16.1(D) and (E)). This level of increased expression could explain, at least in part, the reduced amplitude and altered phase of molecular clock rhythms in cave populations. Significantly, overexpression of *per2* in zebrafish cell lines has a significant damping effect on *per1* rhythms, which is also true for overexpression of *cry1a* (Tamai et al., 2007). Thus, there is a clear perturbation of the core clock in cave populations of *A. mexicanus*, with the constitutively high expression of the light-responsive *per2b* gene, a likely candidate for causing this repressed amplitude.

In general, the cavefish molecular clock appears less robust than that found in equivalent surface individuals, with substantial modifications to the core clock and the light-input pathway.

CLOCK OUTPUTS IN *ASTYANAX*

Activity

Locomotor activity has been the assay of choice in many circadian studies, and continues to be an important assay in mice and *Drosophila*; however, it is an output of the circadian clock rather than a core component. Since *Astyanax* cavefish can clearly show a functional, but altered, molecular rhythm, we can ask the question: is this reflected in circadian clock outputs, such as activity?

Rhythms of locomotor activity have been demonstrated in *Astyanax* cavefish at juvenile and adult stages, with some suggestion that these rhythms are controlled by the circadian clock (Beale et al., 2013; Duboué et al., 2011; Erckens and Martin, 1982); however, these rhythms must persist under DD conditions if they are truly clock-controlled, which is not always the case here. Surface fish show a significant 24-h rhythm in LD, with high activity in the day, consistent with a diurnal animal. When the DD data is analyzed, a significant autocorrelation is present, suggesting that activity in surface fish is under clock control. In cavefish, the situation is slightly different. Though weakly rhythmic in LD, clock-controlled locomotor rhythms are absent in Pachón and Chica cavefish in DD (Beale et al., 2013). This corroborates previous work on locomotor rhythms in *Astyanax*, which suggests a significant reduction or lack of rhythmic behavior in cavefish (Duboué et al., 2011; Erckens and Martin, 1982). The LD activity is therefore likely due to the masking influence of light, where light itself acutely drives a behavior rather than acting through entrainment of a circadian pacemaker.

This situation is similar to that observed in the blind mole rat, *Spalax ehrenberg*, in which there is a large variability in clock-controlled activity rhythms, despite evidence for a complete and functional molecular clock (Avivi et al., 2002; Tobler et al., 1998). The rhythms have therefore become "uncoupled" from their outputs. One explanation for this uncoupling is that selection has relaxed the control of activity by the circadian clock in environments where there is little or no rhythmicity, perhaps to increase survival by improving chances for foraging in a food-restricted environment, such as a cave, a suggestion that agrees with further data proposing a general trend toward increased activity throughout the day and reduced sleep compared to surface fish (Beale et al., 2013; Duboué et al., 2011; Erckens and Martin, 1982).

These data raise the interesting possibility that the molecular circadian rhythm detected in cavefish may be uncoupled from other outputs of the clock. An analysis of additional clock-regulated outputs in this system is required to confirm if this phenomenon is specific to just locomotor behavior or extends more widely to other clock-controlled processes. Such conclusions would indeed benefit from a greater understanding of how the core clock mechanism links to the regulation of outputs in fish.

Sleep

Sleep is one such clock-influenced output that may be affected by the alterations in the molecular clock in cavefish. Sleep is under homeostatic control, as deprivation of sleep builds up the subsequent duration of sleep (sleep rebound); however, it is also regulated in a circadian manner. Sleep is typically associated with behavioral markers, such as a reduced responsiveness to external stimuli, and these occur over a circadian time period (Borbély and Tobler, 1996). Is sleep one of the characteristics that has evolved during adaptation to cave life?

Duboué et al. demonstrated a reduced total sleep phenotype in *Astyanax* cavefish fry, characterized by a significant reduction in sleep-bout duration in Pachón, Tinaja, and Molino cavefish with a greatly attenuated circadian component (Duboué et al., 2011; Figure 16.2(A) and (B); and this chapter). As suggested for *Spalax ehrenberg*, clock-uncoupled activity, together with reduced sleep and increased overall activity in cavefish, leads to the hypothesis that this will improve chances for foraging in a food-restricted environment, and agrees with other morphological and behavioral work showing numerous adaptations for improving feeding in caves, such as increases in jaw size and tastebud number, an enhanced vibration attraction behavior (VAB), and reduction in aggression (Elipot et al., 2013; Yamamoto et al., 2009; Yoshizawa et al., 2010).

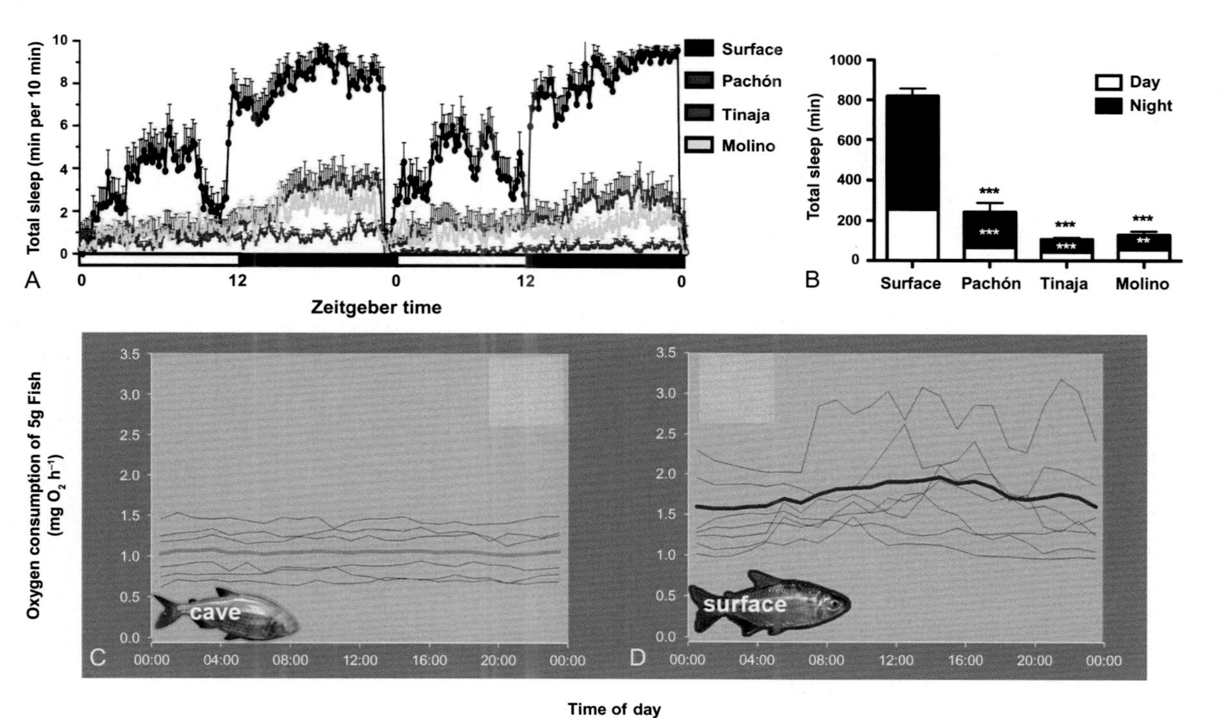

FIGURE 16.2 Circadian outputs of *Astyanax*. (A) Surface fish fry are highly diurnal, being largely active during the day and having most of the sleep duration in the night. In contrast, cavefish fry do not appear to preferentially sleep during the night or day, particularly evident for Tinaja fish. (B) Quantification of total sleep indicates that surface fish sleep significantly more than any of the cavefish populations. The route of this reduced sleep phenotype in cavefish is a significant reduction in sleep bout duration in both day and night phases. (C and D) The daily variation metabolic rate is controlled by a circadian clock in surface fish, as the oscillation continues in constant darkness. Cavefish do not show any evidence of a circadian rhythm in metabolism; by eliminating the daytime rise in metabolic rate seen in surface fish, cavefish make daily energy savings of 27% compared to surface fish in their respective natural photoperiods (LD for surface fish, DD for cavefish). *(A) and (B) Reprinted from Duboué et al. (2011b), with permission from Elsevier. (C) and (D) Reprinted from Moran et al. (2014).*

Metabolism

Another characteristic links the circadian clock to the reduced food availability in caves—metabolism. The theory that cavefish may exhibit a reduced metabolism compared to surface fish seems like common sense—reduce your physiological energy need and you reduce your need to find as much food. Unfortunately, discrepancies among the few studies regarding this issue have made good conclusions hard to come by. An early study by Hüppop suggested that Pachón cavefish have a reduced resting metabolic rate in comparison to surface fish, only for a reanalysis of the data to rule that out (Hüppop, 1986, 2000); however, a more recent study by Salin et al., also reports a reduction in metabolic rate in Pachón cavefish, agreeing with the first Hüppop study (Salin et al., 2010).

Moran and colleagues aimed to iron out these discrepancies by performing longer-duration experiments on a circadian timescale while controlling for the potential masking effect of activity, by controlling swimming speed of fish in a flume respirometer (Moran et al., 2014). They noticed that the photic conditions have a significant effect on the minimum oxygen consumption rate (and by extension, the minimal metabolism), and when the fish are in their natural photoperiod (LD for surface fish, DD for cavefish) the minimum metabolic rate is no different. Instead, it is in the daily metabolic profile that Moran et al. see the energy savings made by cavefish.

Metabolism itself shows circadian rhythms independent of the metabolic costs from daily activity and feeding. As expected, the circadian rise and fall in oxygen demand is present in surface fish, with approximately 20% higher demand during the day, even when activity and feeding is controlled for (Figure 16.2(C)). Cavefish do not show this pattern; there is no rise in oxygen demand during the day (on an LD cycle) or during the prospective day (when in constant darkness) (Figure 16.2(D)). By eliminating the clock-controlled increase in metabolic rate seen in surface fish during the day, Pachón cavefish use 27% less daily energy in its natural lighting conditions (Moran et al., 2014). For *Astyanax*, it seems, therefore, that interactions among feeding efficiency and physiological, behavioral, and morphological characteristics are a common theme of cavefish evolution; however, the mechanism by which the circadian rhythm in metabolism is lost is still unclear. One possibility suggested by Moran and colleagues is that the effective loss is due in some way to the altered molecular clock. Decoupling from the molecular clock is another possibility, and is a potential common feature of clock outputs.

Previous studies on cave animals, such as cave amphipods and salamanders, have made conclusions on circadian rhythms based upon behavioral data, with some suggesting they are lost (Blume et al., 1962; Hervant et al., 2001), and some proposing they are kept (Jegla and Poulson, 1968). Jegla and Poulson also measured oxygen consumption, which again led them to conclude that rhythms are present in the cave crayfish. Interestingly, in their study, circadian activity was more variable among animals than oxygen consumption, perhaps reflecting different strength-coupling between the clock and different outputs; however, as is clear from *Astyanax* cavefish, behavioral and metabolic rhythms may not always reflect the underlying molecular oscillator. It would, therefore, be interesting to re-examine these previously studied species at the molecular clock level.

Retinomotor Movements

Teleosts exhibit a series of physiological changes to the eye in response to the daily LD cycle, some of which are controlled by the endogenous circadian clock. While the mechanism of the degeneration of eyes in cavefish is beyond the scope of this chapter (see Chapter 10), the circadian control of some of the eye's physiological adaptations to LD before and during the structural degeneration may tell us something more about circadian system outputs.

One such adaptation is the coordinated changes in photoreceptors and retinal pigmented epithelium (RPE) cells, known as retinomotor movements. In surface-dwelling teleosts, retinomotor movements protect and tune the eye during the bright conditions of the day and are linked to the circadian clock (Zaunreiter et al., 1998); RPE granules disperse within the RPE cells to shield the outer segments of the photoreceptors, preventing bleaching (King-Smith et al., 1996), and cones move within the retina to a closer position to incoming light to improve the collection of photons (Cavallaro and Burnside, 1988).

Though weak, retinomotor rhythms are conserved in the cavefish even during the extensive eye degeneration at 5 weeks post-fertilization (Espinasa and Jeffery, 2006). Espinasa and Jeffery measured the position of the RPE pigment granules in histological sections of eyes taken from fish kept in LD, LD into DD, LD into LL, and LD into DL conditions. They showed that retinomotor rhythms are under circadian clock control, as the movement of the granules perseveres after the fish have been transferred to constant darkness and can be re-entrained when LD cycles are reversed. This unusual result confirms that cavefish do possess a functional clock that can be entrained to light and direct clock outputs; however, since movements of granules are not needed to reduce bleaching of photoreceptors, as cavefish live in darkness and their photoreceptors are essentially absent, these retinomotor rhythms appear to be a truly vestigial character.

THE ROLE OF LIGHT INPUT

Melatonin and the Pineal Gland

In addition to functional retinomotor rhythms, cavefish also appear to have a somewhat functional pineal gland and melatonin system (Wilkens et al., 1993). In nonmammalian vertebrates, a combination of circadian clock regulation and direct-light input results in rhythms of melatonin synthesis, with highest levels at night. As such, it serves as a useful marker of circadian function. Indeed, one of the earliest apparent outputs of the circadian clock during development in zebrafish is that of the pineal gland and melatonin secretion (Kazimi and Cahill, 1999; Ziv et al., 2005).

In cavefish, melatonin levels oscillate in response to LD cycles, indicating a functional pineal system; however, the amplitude of the oscillation is reduced in comparison to the surface fish, echoing the difference in core clock gene expression. Structural analysis of the pineal gland suggests it is only slightly degenerate relative to the surface fish, and while able to secrete melatonin, it is clear that the pineal system is less functional (Wilkens et al., 1993); however, a more extensive examination of melatonin signaling activity, including receptor function and tissue expression is certainly warranted at this time.

It is a different story in larvae. A functional pineal gland in cavefish mediates a true larval shadow response (Yoshizawa and Jeffery, 2008). The shadow response is a behavioral reaction to an acute loss of light; when a shadow is caused by blocking the illumination on tanks of larvae, the larvae react by swimming either toward or away from the shadow. As in *Xenopus*, the shadow response in *Astyanax* diminishes over time (Foster and Roberts, 1982), and does so at the same rate in surface fish and cavefish, perhaps due to increased opacity in the cranium or minimal inhibitory input from an embryonically organized retina (Yoshizawa and Jeffery, 2008). Yoshizawa and Jeffery speculate that the conservation of a functional pineal gland, at least with respect to the shadow response, may be a developmental constraint due to its dual physiological role, light detection (shadow response) and neurosecretion (including melatonin), a possibility supported by the conservation of *opsin* gene expression in the mammalian pineal, even though it serves no light-detecting purpose (Blackshaw and Snyder, 1997).

Interestingly, Micos cavefish secrete far less melatonin than surface fish. While still showing a rhythmic change in melatonin, melatonin synthesis is significantly reduced; daytime levels are very low, and nighttime values approach that of the daytime level in surface fish. Is this indicative of a day-like or constant light-like phenotype in cavefish, an alteration in the light-input pathway?

DNA Repair

Rhythms in activity, sleep, metabolism, retinomotor movement, and melatonin release serve to temporally coordinate an organism and its biology with the outside world. If many of the output rhythms are lost, or decoupled from the internal oscillator, why does the oscillator persist? Here we should consider physiological roles unrelated to external day and night cycles. After all, hypotheses into the selective advantage of the circadian clock pertain to temporal organization of internal physiology, as well as synchronization with the outside world.

Given that light has a profound effect on gene expression and physiology in fish (Gavriouchkina et al., 2010; Weger et al., 2011), the perturbed light-input pathway in *Astyanax* cavefish may not only affect the circadian clock through the repressive input of *per2*, but also other light-dependent biological processes. This hypothesis was tested by examining the expression of genes involved in DNA repair. Interestingly, two genes known to be involved in separate DNA repair pathways, *CPD photolyase* and *ddb2*, were significantly more highly expressed in cavefish in the dark compared to surface fish (Figure 16.3(A)). Remarkably, this expression difference was even more dramatic in cavefish sampled in the wild (Beale et al., 2013; Figure 16.3(B)).

At first, this result seems rather counterintuitive. As a photolyase, *CPD photolyase* is a directly light-activated protein involved in DNA repair. What use does a light-activated protein have in the dark? This is satisfied by CPD photolyase's secondary function; it is known to bind to sites of DNA damage even in the absence of light, serving an accessory role to enhance the effectiveness of other DNA repair pathways (Ozer et al., 1995; Sancar and Smith, 1989; Wagner et al., 2011). Indeed, this role was demonstrated to be significant in *Astyanax* cavefish as, even in the darkness, the increased expression of these DNA repair genes correlates with an improved ability to repair DNA damage in the dark (Beale et al., 2013). Thus, in an analogous situation to the enhanced DNA repair activity observed in transgenic mice overexpressing marsupial *CPD photolyase* (Schul et al., 2002), *Astyanax* cavefish can be considered a natural overexpressing condition with enhanced DNA repair activity. It is a reasonable hypothesis that the even greater expression of DNA repair genes in the wild cave populations would also lead to further enhanced DNA repair activity. Presumably an enhanced level of DNA repair activity provides some selective advantage to those wild cave populations.

Many of the physiological and behavioral changes seen in cavefish have obvious benefits in the cave; an increased number of tastebuds (Varatharasan et al., 2009; Yamamoto et al., 2009), reduction in sleep (Duboué et al., 2011), and enhanced VAB (Yoshizawa et al., 2010) all have clear benefits to fish living in the dark, especially with regard to improving feeding efficiency. Even so, it is less clear how enhanced DNA repair activity is advantageous in the dark, especially since ultraviolet (UV) light (a major DNA damage agent and the type of damage repaired by photolyases) is not present in the dark cave. What may be responsible for DNA damage in the cave? One possibility is the conditions in the cave themselves.

Levels of DNA damage are significant in cells even without UV exposure, including the hydrolysis, oxidation and alkylation of DNA bases (reviewed in Barnes and Lindahl, 2004). These processes are enhanced by certain environmental factors. It is possible that the conditions of the cave pools, hypoxic (dissolved oxygen below 0.5 mg/l) and slightly acidic, are harsh enough environments to increase instances of DNA damage. Consistent with this are interesting studies that have found DNA damage and oxidative stress are caused by

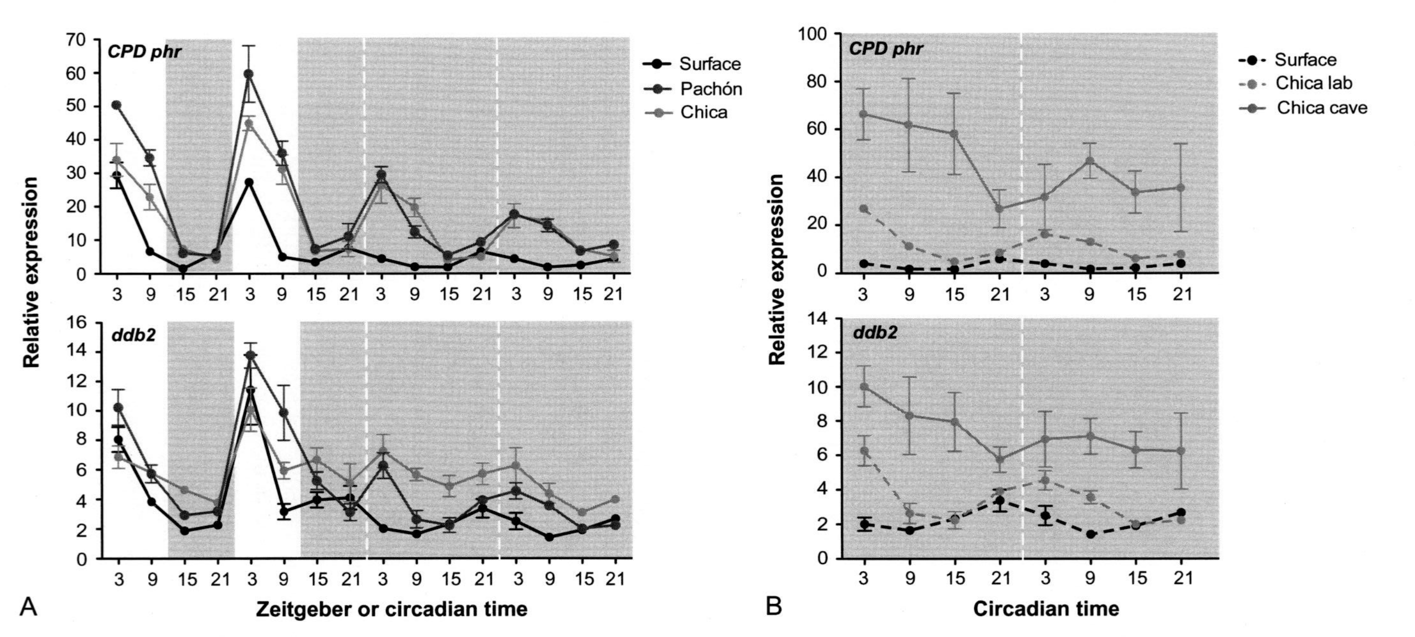

FIGURE 16.3 Hyperactivated DNA repair in *Astyanax* cavefish. In addition to the enhanced expression of light-dependent circadian genes, cavefish show an increased expression of two light-dependent DNA repair genes, *CPD phr* and *ddb2*, as measured by qPCR (A). In the lab, this difference is most clear in adult fish that have been transferred into constant darkness after having been entrained on an LD cycle, with *ddb2* in Chica cavefish being significantly raised at 7 of the 8 timepoints in DD (A). (B) This raised expression is even more exaggerated in the cave itself, with the expression of both genes being dramatically increased in wild Chica cavefish compared with surface fish or even Chica fish within the laboratory. *Taken from Beale et al. (2013).*

the hypoxic conditions in fish via indirect generation of reactive oxygen species (Lushchak and Bagnyukova, 2007; Mustafa et al., 2011).

The majority of the processing of lesions associated with oxidative damage is performed by DNA repair pathways, such as base excision repair (BER) (Cooke et al., 2003; Zharkov, 2008). Intriguingly, genes associated with the BER pathway, such as *neil1* and *xrcc1*, are also upregulated by light in zebrafish (Gavriouchkina et al., 2010; Weger et al., 2011). Although these genes are yet to be examined in cavefish, it is possible that the increased expression of *CPD photolyase* and *ddb2* represent a global increase in activation and expression of DNA repair pathway components, which may have a protective function in the hypoxic caves. It is also likely that the secondary functions of photolyases and NER components, such as accessory DNA binding in the dark (Ozer et al., 1995; Wagner et al., 2011) and in repairing oxidative and chemical DNA damage (Menoni et al., 2012) may instead be their primary roles in cavefish in the dark. Therefore, by tonically activating light-dependent signaling pathways and increasing DNA repair gene expression (and resultant DNA repair activity), individuals in the cave would reduce the frequency of deleterious mutational events. This may be a further example of an evolutionary and developmental constraint, analogous to the retention of both secretory and photosensitive functions of the pineal gland, between the light-input pathway to the circadian clock and DNA repair systems.

In a further twist, studies have shown that *CPD photolyase* protein can physically interact with the *Clk* protein to reduce *Clk/Bmal*-dependent transactivation and repress the mammalian circadian oscillator (Chaves et al., 2011). An attractive possibility in this story of the circadian clock and DNA repair in cavefish is that the high levels of *CPD photolyase* could have a similar effect and be, at least in part, responsible for the repressed levels of clock function we observe in the wild.

WHAT CAN *ASTYANAX* TELL US ABOUT OTHER CAVE SPECIES?

Astyanax cavefish have a complex circadian phenotype. While the molecular clock is largely intact and functional, it is reduced in amplitude in the lab and repressed in the wild. Rhythms in melatonin are greatly attenuated, and rhythms in metabolism are abolished altogether. Sleep and activity appear to be uncoupled from the core clock and occur arrhythmically. Finally, increased expression of light-dependent components is a possible mechanism for the repression of circadian rhythms, and results in a developmentally constrained upregulation of the DNA repair system, which is advantageous in the caves. No other species of cave animal has a picture this complete.

The only other study on molecular clocks in cave animals is a report for the Somalian cavefish, *Phreatichthys andruzzii* (Cavallari et al., 2011). The

Astyanax phenotypes are less dramatic than those reported for *P. andruzzii*, which have completely lost the capacity to entrain to an LD cycle. *P. andruzzii* possess mutations in candidate circadian photoreceptors, *TMT opsin* and *melanopsin*, a fact that potentially contributes to their "blind" circadian phenotype (Cavallari et al., 2011). Interestingly, the Somalian cavefish are able to partially entrain their molecular clock to a scheduled feeding regime, indicating that they do retain some level of clock function; however, the clock appears to have a variable infradian period of between 38 and 47 h, indicating that the core clock has undergone significant alterations, though the mechanism is not known. Furthermore, it is unknown to what extent the cavefish expresses circadian outputs, such as melatonin synthesis or metabolism, though lessons learned from *Astyanax* cavefish suggest that these are very likely to be suppressed in this cavefish. Cave species are likely to differ in the precise details of their circadian clock function, though it seems elements of the circadian light-input pathway are likely "targets" for evolutionary change.

CONCLUSION

The conservation of circadian rhythms appears to be a general phenomenon in cave-dwelling animals. Data from a broad range of cave animals, including cave crayfish, cave crickets and cavefish, suggests that the core clock mechanism is to some extent preserved in cave animals despite the constant darkness of their environments. It is unlikely that there simply has not been enough time or selective pressure to lose the clock mechanism in caves, even when taking into account the buffer of substantial redundancy in the network from gene duplication, which is especially significant in teleosts. It is clear from the loss of characteristics such as pigmentation that relaxed selection for this amount of time is able to result in significant trait loss due to neutral mutation and genetic drift (Gross et al., 2009; Protas et al., 2007, 2006). There are vast differences in the evolutionary histories and environmental pressures of the various cave animals. Secondly, as circadian rhythms are in some form retained across many cave species, despite vast differences in their evolutionary histories and environmental pressures, it strongly suggests that circadian rhythms are conserved for a purpose.

The persistence of molecular oscillators in *Astyanax* and *P. andruzzii* suggests that this purpose is to retain a "day within" (Pittendrigh, 1993; Sharma, 2003). That the molecular oscillations often are not expressed in circadian outputs, like activity, further suggests that the retention of the molecular clock is not for external synchronization. Instead, the clock's adaptive value in the absence of environmental cycles may come from providing temporal organization to certain cellular processes, such as the cell cycle, as demonstrated in zebrafish (Dekens et al., 2003; Dickmeis et al., 2007; Idda et al., 2012; Laranjeiro et al., 2013; Tamai et al., 2012).

Importance of Data from the Real World

An important feature of *Astyanax* as a model system is the potential to perform observations and experiments in the field. Though we do not have a complete answer as to the function and significance of the retained circadian clock in *Astyanax*, we do have a better picture of the situation, thanks to field studies that have proven key in the discussion of the adaptive significance of the clock. Somewhat surprisingly, given the added value of data from the field, circadian field studies in other animals are rare. One famous study was an observation of lower survival rates of SCN-lesioned chipmunks compared to SCN-intact controls (DeCoursey et al., 2000). This single, powerful study demonstrates the value of studying animals in natural habitats. Furthermore, as recently demonstrated in *Drosophila* and golden hamsters, expression of the circadian clock is often different in the field, exhibiting significant alterations in activity patterns and importance of different zeitgebers in *Drosophila* and a complete switch from nocturnal to diurnal behavior in hamsters (Gattermann et al., 2008; Vanin et al., 2012). These facts serve as an incentive (and a warning) to study circadian clock function in the field, as well as the lab.

Even so, there have been few reported field studies on circadian clock function in fish in rivers and certainly not within cave complexes. In *Astyanax* cavefish, the clock appears to be significantly repressed in the wild. While we do not fully know the reason, it is clearly remarkable that the mechanism remains largely intact over the millions of years of isolation in the caves. Unfortunately, there is no field data for the other cavefish for which molecular clock data exists, the Somalian cavefish, to add to this result. Do other cave animals retain molecular clock function? To what extent is it responsive to light? Is it expressed in the wild cave environment? More studies on cave animals in the light of data from *Astyanax* will extend the discussion further.

REFERENCES

Avivi, A., Oster, H., Joel, A., Beiles, A., Albrecht, U., Nevo, E., 2002. Circadian genes in a blind subterranean mammal II: conservation and uniqueness of the three *Period* homologs in the blind subterranean mole rat, *Spalax ehrenbergi* superspecies. Proc. Natl. Acad. Sci. U. S. A. 99, 11718–11723. http://dx.doi.org/10.1073/pnas.182423299.

Barnes, D.E., Lindahl, T., 2004. Repair and genetic consequences of endogenous DNA base damage in mammalian cells. Annu. Rev. Genet. 38, 445–476. http://dx.doi.org/10.1146/annurev.genet.38.072902.092448.

Beale, A., Guibal, C., Tamai, T.K., Klotz, L., Cowen, S., Peyric, E., Reynoso, V.H., Yamamoto, Y., Whitmore, D., 2013. Circadian rhythms in Mexican blind cavefish *Astyanax mexicanus* in the lab and in the field. Nat. Commun. 4, 2769. http://dx.doi.org/10.1038/ncomms3769.

Blackshaw, S., Snyder, S.H., 1997. Developmental expression pattern of phototransduction components in mammalian pineal implies a light-sensing function. J. Neurosci. 17, 8074–8082.

Blume, J., Günzler, E., Bünning, E., 1962. Zur Aktivitatsperiodik Bei Hohlentieren. Naturwissenschaften 49, 525.

Borbély, A.A., Tobler, I., 1996. Sleep regulation: relation to photoperiod, sleep duration, waking activity, and torpor. Prog. Brain Res. 111, 343–348.

Brown, F.A., 1961. Diurnal rhythm in cave crayfish. Nature 191, 929–930. http://dx.doi.org/10.1038/191929b0.

Cahill, G.M., 1996. Circadian regulation of melatonin production in cultured zebrafish pineal and retina. Brain Res. 708, 177–181.

Carr, A.-J.F., Whitmore, D., 2005. Imaging of single light-responsive clock cells reveals fluctuating free-running periods. Nat. Cell Biol. 7, 319–321. http://dx.doi.org/10.1038/ncb1232.

Cavallari, N., Frigato, E., Vallone, D., Fröhlich, N., Lopez-Olmeda, J.F., Foà, A., Berti, R., Sánchez-Vázquez, F.J., Bertolucci, C., Foulkes, N.S., 2011. A blind circadian clock in cavefish reveals that opsins mediate peripheral clock photoreception. PLoS Biol. 9, e1001142. http://dx.doi.org/10.1371/journal.pbio.1001142.

Cavallaro, B., Burnside, B., 1988. Prostaglandins E1, E2, and D2 induce dark-adaptive retinomotor movements in teleost retinal cones and RPE. Invest. Ophthalmol. Vis. Sci. 29, 882–891.

Cermakian, N., Pando, M.P., Thompson, C.L., Pinchak, A.B., Selby, C.P., Gutierrez, L., Wells, D.E., Cahill, G.M., Sancar, A., Sassone-Corsi, P., 2002. Light induction of a vertebrate clock gene involves signaling through blue-light receptors and MAP kinases. Curr. Biol. 12, 844–848.

Chaves, I., Nijman, R.M., Biernat, M.A., Bajek, M.I., Brand, K., da Silva, A.C., Saito, S., Yagita, K., Eker, A.P.M., van der Horst, G.T.J., 2011. The *Potorous* CPD photolyase rescues a cryptochrome-deficient mammalian circadian clock. PLoS One 6, e23447. http://dx.doi.org/10.1371/journal.pone.0023447.

Cooke, M.S., Evans, M.D., Dizdaroglu, M., Lunec, J., 2003. Oxidative DNA damage: mechanisms, mutation, and disease. FASEB J. 17, 1195–1214. http://dx.doi.org/10.1096/fj.02-0752rev.

De Mairan, J.D., 1729. Observation botanique. In: Histoire de l'Academie Royale des Science, pp. 35–36.

DeCoursey, P.J., Walker, J.K., Smith, S.A., 2000. A circadian pacemaker in free-living chipmunks: essential for survival? J. Comp. Physiol. A. 186, 169–180.

Dekens, M.P.S., Santoriello, C., Vallone, D., Grassi, G., Whitmore, D., Foulkes, N.S., 2003. Light regulates the cell cycle in zebrafish. Curr. Biol. 13, 2051–2057.

Dickmeis, T., Lahiri, K., Nica, G., Vallone, D., Santoriello, C., Neumann, C.J., Hammerschmidt, M., Foulkes, N.S., 2007. Glucocorticoids play a key role in circadian cell cycle rhythms. PLoS Biol. 5, e78. http://dx.doi.org/10.1371/journal.pbio.0050078.

Dodd, A.N., Salathia, N., Hall, A., Kévei, E., Tóth, R., Nagy, F., Hibberd, J.M., Millar, A.J., Webb, A.A.R., 2005. Plant circadian clocks increase photosynthesis, growth, survival, and competitive advantage. Science 309, 630–633. http://dx.doi.org/10.1126/science.1115581.

Duboué, E.R., Borowsky, R.L., 2012. Altered rest-activity patterns evolve via circadian independent mechanisms in cave adapted balitorid loaches. PLoS One 7, e30868. http://dx.doi.org/10.1371/journal.pone.0030868.

Duboué, E.R., Keene, A.C., Borowsky, R.L., 2011a. Evolutionary convergence on sleep loss in cavefish populations. Curr. Biol. 21, 671–676. http://dx.doi.org/10.1016/j.cub.2011.03.020.

Duboué, E.R., Keene, A.C., Borowsky, R.L., 2011b. Evolutionary convergence on sleep loss in cavefish populations. Curr. Biol. 21 (8), 671–676.

Elipot, Y., Hinaux, H., Callebert, J., Rétaux, S., 2013. Evolutionary shift from fighting to foraging in blind cavefish through changes in the serotonin network. Curr. Biol. 23, 1–10. http://dx.doi.org/10.1016/j.cub.2012.10.044.

Erckens, W., Martin, W., 1982. Exogenous and endogenous control of swimming activity in *Astyanax mexicanus* (Characidae, Pisces) by direct light response and by a circadian oscillator. 2. Features of time-controlled behavior of a cave population and their comparison to an epigean ancestral form. Z. Naturforsch. C 37, 1266–1273.

Espinasa, L., Jeffery, W.R., 2006. Conservation of retinal circadian rhythms during cavefish eye degeneration. Evol. Dev. 8, 16–22. http://dx.doi.org/10.1111/j.1525-142X.2006.05071.x.

Foster, R.G., Roberts, A., 1982. The pineal eye in *Xenopus laevis* embryos and larvae: a photoreceptor with a direct excitatory effect on behaviour. J. Comp. Physiol. 145, 413–419. http://dx.doi.org/10.1007/BF00619346.

Gattermann, R., Johnston, R.E., Yigit, N., Fritzsche, P., Larimer, S., Ozkurt, S., Neumann, K., Song, Z., Colak, E., Johnston, J., McPhee, M.E., 2008. Golden hamsters are nocturnal in captivity but diurnal in nature. Biol. Lett. 4, 253–255. http://dx.doi.org/10.1098/rsbl.2008.0066.

Gavriouchkina, D., Fischer, S., Ivacevic, T., Stolte, J., Benes, V., Dekens, M.P.S., 2010. Thyrotroph embryonic factor regulates light-induced transcription of repair genes in zebrafish embryonic cells. PLoS One 5, e12542. http://dx.doi.org/10.1371/journal.pone.0012542.

Green, R.M., Tingay, S., Wang, Z.-Y., Tobin, E.M., 2002. Circadian rhythms confer a higher level of fitness to *Arabidopsis* plants. Plant Physiol. 129, 576–584. http://dx.doi.org/10.1104/pp.004374.

Gross, J.B., Borowsky, R., Tabin, C.J., 2009. A novel role for *Mc1r* in the parallel evolution of depigmentation in independent populations of the cavefish *Astyanax mexicanus*. PLoS Genet. 5, e1000326. http://dx.doi.org/10.1371/journal.pgen.1000326.

Hervant, F., Mathieu, J., Durand, J.-P., 2001. Circadian rhythmicity, respiration and behavior in hypogean and epigean salamanders. Nat. Croat. 10, 141–152.

Hirayama, J., Fukuda, I., Ishikawa, T., Kobayashi, Y., Todo, T., 2003. New role of zCRY and zPER2 as regulators of sub-cellular distributions of zCLOCK and zBMAL proteins. Nucleic Acids Res. 31, 935–943.

Hirayama, J., Cho, S., Sassone-Corsi, P., 2007. Circadian control by the reduction/oxidation pathway: catalase represses light-dependent clock gene expression in the zebrafish. Proc. Natl. Acad. Sci. U. S. A. 104, 15747–15752. http://dx.doi.org/10.1073/pnas.0705614104.

Hirayama, J., Miyamura, N., Uchida, Y., Asaoka, Y., Honda, R., Sawanobori, K., Todo, T., Yamamoto, T., Sassone-Corsi, P., Nishina, H., 2009. Common light signaling pathways controlling DNA repair and circadian clock entrainment in zebrafish. Cell Cycle 8, 2794–2801.

Hüppop, K., 1986. Oxygen consumption of Astyanax fasciatus (Characidae, Pisces): a comparison of epigean and hypogean populations. Environ. Biol. Fishes 17, 299–308. http://dx.doi.org/10.1007/BF00001496.

Hüppop, K., 2000. How do cave animals cope with the food scarcity in caves? In: Wilkens, H., Culver, D.C., Humphries, W.F. (Eds.), Ecosystems of the World 30: Subterranean Ecosystems. Elsevier, Amsterdam, pp. 159–188.

Hurd, M.W., Debruyne, J., Straume, M., Cahill, G.M., 1998. Circadian rhythms of locomotor activity in zebrafish. Physiol. Behav. 65, 465–472.

Hut, R.A., Beersma, D.G.M., 2011. Evolution of time-keeping mechanisms: early emergence and adaptation to photoperiod. Philos. Trans. R. Soc. B 366, 2141–2154. http://dx.doi.org/10.1098/rstb.2010.0409.

Idda, M.L., Kage, E., Lopez-Olmeda, J.F., Mracek, P., Foulkes, N.S., Vallone, D., 2012. Circadian timing of injury-induced cell proliferation in zebrafish. PLoS One 7, e34203. http://dx.doi.org/10.1371/journal.pone.0034203.

Jegla, T.C., Poulson, T.L., 1968. Evidence of circadian rhythms in a cave crayfish. J. Exp. Zool. 168, 273–282. http://dx.doi.org/10.1002/jez.1401680213.

Kaneko, M., Hernandez-Borsetti, N., Cahill, G.M., 2006. Diversity of zebrafish peripheral oscillators revealed by luciferase reporting. Proc. Natl. Acad. Sci. U. S. A. 103, 14614–14619. http://dx.doi.org/10.1073/pnas.0606563103.

Kazimi, N., Cahill, G.M., 1999. Development of a circadian melatonin rhythm in embryonic zebrafish. Dev. Brain Res. 117, 47–52.

King-Smith, C., Chen, P., Garcia, D., Rey, H., Burnside, B., 1996. Calcium-independent regulation of pigment granule aggregation and dispersion in teleost retinal pigment epithelial cells. J. Cell Sci. 109 (Pt 1), 33–43.

Koilraj, A.J., Sharma, V.K., Marimuthu, G., Chandrashekaran, M.K., 2000. Presence of circadian rhythms in the locomotor activity of a cave-dwelling millipede *Glyphiulus cavernicolus sulu* (Cambalidae, Spirostreptida). Chronobiol. Int. 17, 757–765.

Lamprecht, G., Weber, F., 1978. Activity patterns of cave-dwelling beetles. Int. J. Speleol. 10, 351–379.

Laranjeiro, R., Tamai, T.K., Peyric, E., Krusche, P., Ott, S., Whitmore, D., 2013. Cyclin-dependent kinase inhibitor p20 controls circadian cell-cycle timing. Proc. Natl. Acad. Sci. U. S. A. 110, 6835–6840. http://dx.doi.org/10.1073/pnas.1217912110.

Lushchak, V.I., Bagnyukova, T.V., 2007. Hypoxia induces oxidative stress in tissues of a goby, the rotan *Perccottus glenii*. Comp. Biochem. Physiol. B, Biochem. Mol. Biol. 148, 390–397. http://dx.doi.org/10.1016/j.cbpb.2007.07.007.

Martín-Robles, A.J., Isorna, E., Whitmore, D., Muñoz-Cueto, J.A., Pendón, C., 2011. The clock gene *Period3* in the nocturnal flatfish *Solea senegalensis*: molecular cloning, tissue expression and daily rhythms in central areas. Comp. Biochem. Physiol. A, Mol. Integr. Physiol. 159, 7–15. http://dx.doi.org/10.1016/j.cbpa.2011.01.015.

Martín-Robles, A.J., Whitmore, D., Sánchez-Vázquez, F.J., Pendón, C., Muñoz-Cueto, J.A., 2012. Cloning, tissue expression pattern and daily rhythms of *Period1*, *Period2*, and *Clock* transcripts in the flatfish Senegalese sole, *Solea senegalensis*. J. Comp. Physiol. B. 182, 673–685. http://dx.doi.org/10.1007/s00360-012-0653-z.

McGaugh, S.E., Gross, J.B., Aken, B., Blin, M., Borowsky, R., Chalopin, D., Hinaux, H., Jeffery, W.R., Keene, A., Ma, L., Minx, P., Murphy, D., O'Quin, K.E., Rétaux, S., Rohner, N., Searle, S.M.J., Stahl, B.A., Tabin, C., Volff, J.-N., Yoshizawa, M., Warren, W.C., 2014. The cavefish genome reveals candidate genes for eye loss. Nat. Commun. 5, 5307. http://dx.doi.org/10.1038/ncomms6307.

Mead, M., Gilhodes, J.C., 1974. Organisation temporelle de l'activité locomotrice chez un animal cavernicole *Blaniulus lichtensteini Bröl.* (Diplopoda). J. Comp. Physiol. A. 90, 47–52.

Menoni, H., Hoeijmakers, J.H.J., Vermeulen, W., 2012. Nucleotide excision repair-initiating proteins bind to oxidative DNA lesions in vivo. J. Cell Biol. 199, 1037–1046. http://dx.doi.org/10.1083/jcb.201205149.

Moran, D., Softley, R., Warrant, E.J., 2014. Eyeless Mexican cavefish save energy by eliminating the circadian rhythm in metabolism. PLoS One 9, e107877. http://dx.doi.org/10.1371/journal.pone.0107877.

Mracek, P., Santoriello, C., Idda, M.L., Pagano, C., Ben-Moshe, Z., Gothilf, Y., Vallone, D., Foulkes, N.S., 2012. Regulation of *per* and *cry* genes reveals a central role for the D-box enhancer in light-dependent gene expression. PLoS One 7, e51278. http://dx.doi.org/10.1371/journal.pone.0051278.

Mracek, P., Pagano, C., Fröhlich, N., Idda, M.L., Cuesta, I.H., Lopez-Olmeda, J.F., Sánchez-Vázquez, F.J., Vallone, D., Foulkes, N.S., 2013. ERK signaling regulates light-induced gene expression via D-box enhancers in a differential, wavelength-dependent manner. PLoS One 8, e67858. http://dx.doi.org/10.1371/journal.pone.0067858.

Mustafa, S.A., Al-Subiai, S.N., Davies, S.J., Jha, A.N., 2011. Hypoxia-induced oxidative DNA damage links with higher level biological effects including specific growth rate in common carp, *Cyprinus carpio L.* Ecotoxicology 20, 1455–1466. http://dx.doi.org/10.1007/s10646-011-0702-5.

Ozer, Z., Reardon, J.T., Hsu, D.S., Malhotra, K., Sancar, A., 1995. The other function of DNA photolyase: stimulation of excision repair of chemical damage to DNA. Biochemistry 34, 15886–15889.

Pando, M.P., Pinchak, A.B., Cermakian, N., Sassone-Corsi, P., 2001. A cell-based system that recapitulates the dynamic light-dependent regulation of the vertebrate clock. Proc. Natl. Acad. Sci. U. S. A. 98, 10178–10183. http://dx.doi.org/10.1073/pnas.181228598.

Park, O., Roberts, T., Harris, S., 1941. Preliminary analysis of activity of the cave crayfish, *Cambarus pellucidus*. Am. Nat. 45, 154–171.

Park, J.-G., Park, Y.-J., Sugama, N., Kim, S.-J., Takemura, A., 2007. Molecular cloning and daily variations of the Period gene in a reef fish *Siganus guttatus*. J. Comp. Physiol. A. 193, 403–411. http://dx.doi.org/10.1007/s00359-006-0194-6.

Pati, A., 2001. Temporal organization in locomotor activity of the hypogean loach, *Nemacheilus evezardi*, and its epigean ancestor. Environ. Biol. Fishes 62, 119–129.

Pittendrigh, C.S., 1960. Circadian rhythms and the circadian organization of living systems. Cold Spring Harb. Symp. Quant. Biol. 25, 159–184.

Pittendrigh, C.S., 1993. Temporal organization: reflections of a Darwinian clock-watcher. Annu. Rev. Physiol. 55, 16–54. http://dx.doi.org/10.1146/annurev.ph.55.030193.000313.

Poulson, T.L., White, W.B., 1969. The cave environment. Science 165, 971–981. http://dx.doi.org/10.1126/science.165.3897.971.

Protas, M.E., Hersey, C., Kochanek, D., Zhou, Y., Wilkens, H., Jeffery, W.R., Zon, L.I., Borowsky, R., Tabin, C.J., 2006. Genetic analysis of cavefish reveals molecular convergence in the evolution of albinism. Nat. Genet. 38, 107–111. http://dx.doi.org/10.1038/ng1700.

Protas, M., Conrad, M., Gross, J.B., Tabin, C., Borowsky, R., 2007. Regressive evolution in the Mexican cave tetra, *Astyanax mexicanus*. Curr. Biol. 17, 452–454. http://dx.doi.org/10.1016/j.cub.2007.01.051.

Ramos, B.C.R., Moraes, M.N.C.M., Poletini, M.O., Lima, L.H.R.G., Castrucci, A.M.L., 2014. From blue light to clock genes in zebrafish ZEM-2S cells. PLoS One 9, e106252. http://dx.doi.org/10.1371/journal.pone.0106252.

Reichle, D., Palmer, J., Park, O., 1965. Persistent rhythmic locomotor activity in cave cricket *Hadenoecus subterraneus* and its ecological significance. Am. Midl. Nat. 74, 57–66.

Salin, K., Voituron, Y., Mourin, J., Hervant, F., 2010. Cave colonization without fasting capacities: an example with the fish *Astyanax fasciatus mexicanus*. Comp. Biochem. Physiol. A, Mol. Integr. Physiol. 156, 451–457. http://dx.doi.org/10.1016/j.cbpa.2010.03.030.

Sancar, G.B., Smith, F.W., 1989. Interactions between yeast photolyase and nucleotide excision repair proteins in *Saccharomyces cerevisiae* and *Escherichia coli*. Mol. Cell. Biol. 9, 4767–4776.

Schul, W., Jans, J., Rijksen, Y.M.A., Klemann, K.H.M., Eker, A.P.M., de Wit, J., Nikaido, O., Nakajima, S., Yasui, A., Hoeijmakers, J.H.J., van der Horst, G.T.J., 2002. Enhanced repair of cyclobutane pyrimidine dimers and improved UV resistance in photolyase transgenic mice. EMBO J. 21, 4719–4729.

Sharma, V.K., 2003. Adaptive significance of circadian clocks. Chronobiol. Int. 20, 901–919.

Tamai, T.K., Vardhanabhuti, V., Foulkes, N.S., Whitmore, D., 2004. Early embryonic light detection improves survival. Curr. Biol. 14, R104–R105.

Tamai, T.K., Young, L.C., Whitmore, D., 2007. Light signaling to the zebrafish circadian clock by Cryptochrome 1a. Proc. Natl. Acad. Sci. U. S. A. 104, 14712–14717. http://dx.doi.org/10.1073/pnas.0704588104.

Tamai, T.K., Young, L.C., Cox, C.A., Whitmore, D., 2012. Light acts on the zebrafish circadian clock to suppress rhythmic mitosis and cell proliferation. J. Biol. Rhythms 27, 226–236. http://dx.doi.org/10.1177/0748730412440861.

Tobler, I., Herrmann, M., Cooper, H.M., Negroni, J., Nevo, E., Achermann, P., 1998. Rest-activity rhythm of the blind mole rat *Spalax ehrenbergi* under different lighting conditions. Behav. Brain Res. 96, 173–183.

Trajano, E., Menna-Barreto, L., 1996. Free-running locomotor activity rhythms in cave-dwelling catfishes, *Trichomycterus sp.*, from Brazil (Teleostei, Siluriformes). Biol. Rhythm. Res. 27, 329–335. http://dx.doi.org/10.1076/brhm.27.3.329.12958.

Vallone, D., Gondi, S.B., Whitmore, D., Foulkes, N.S., 2004. E-box function in a *period* gene repressed by light. Proc. Natl. Acad. Sci. U. S. A. 101, 4106–4111. http://dx.doi.org/10.1073/pnas.0305436101.

Vanin, S., Bhutani, S., Montelli, S., Menegazzi, P., Green, E.W., Pegoraro, M., Sandrelli, F., Costa, R., Kyriacou, C.P., 2012. Unexpected features of *Drosophila* circadian behavioural rhythms under natural conditions. Nature 484, 371–375. http://dx.doi.org/10.1038/nature10991.

Varatharasan, N., Croll, R.P., Franz-Odendaal, T., 2009. Taste bud development and patterning in sighted and blind morphs of *Astyanax mexicanus*. Dev. Dyn. 238, 3056–3064. http://dx.doi.org/10.1002/dvdy.22144.

Vatine, G., Vallone, D., Appelbaum, L., Mracek, P., Ben-Moshe, Z., Lahiri, K., Gothilf, Y., Foulkes, N.S., 2009. Light directs zebrafish *period2* expression via conserved D and E boxes. PLoS Biol. 7, e1000223. http://dx.doi.org/10.1371/journal.pbio.1000223.

Velarde, E., Haque, R., Iuvone, P.M., Azpeleta, C., Alonso-Gómez, A.L., Delgado, M.J., 2009. Circadian clock genes of goldfish, *Carassius auratus*: cDNA cloning and rhythmic expression of *period* and *cryptochrome* transcripts in retina, liver, and gut. J. Biol. Rhythms 24, 104–113. http://dx.doi.org/10.1177/0748730408329901.

Wagner, K., Moolenaar, G.F., Goosen, N., 2011. Role of the insertion domain and the zinc-finger motif of *Escherichia coli* UvrA in damage recognition and ATP hydrolysis. DNA Repair 10, 483–496. http://dx.doi.org/10.1016/j.dnarep.2011.02.002.

Weger, B.D., Sahinbas, M., Otto, G.W., Mracek, P., Armant, O., Dolle, D., Lahiri, K., Vallone, D., Ettwiller, L., Geisler, R., Foulkes, N.S., Dickmeis, T., 2011. The light responsive transcriptome of the zebrafish: function and regulation. PLoS One 6, e17080. http://dx.doi.org/10.1371/journal.pone.0017080.

Whitmore, D., Foulkes, N.S., Strähle, U., Sassone-Corsi, P., 1998. Zebrafish *Clock* rhythmic expression reveals independent peripheral circadian oscillators. Nat. Neurosci. 1, 701–707. http://dx.doi.org/10.1038/3703.

Whitmore, D., Foulkes, N.S., Sassone-Corsi, P., 2000. Light acts directly on organs and cells in culture to set the vertebrate circadian clock. Nature 404, 87–91. http://dx.doi.org/10.1038/35003589.

Wilkens, H., Langecker, T.G., Olcese, J., 1993. Circadian rhythms of melatonin synthesis in the pineal organ of cave-dwelling *Astyanax fasciatus* (Teleostei: Characidae). Mém. Biospéol. 20, 279–282.

Yamamoto, Y., Byerly, M.S., Jackman, W.R., Jeffery, W.R., 2009. Pleiotropic functions of embryonic *sonic hedgehog* expression link jaw and taste bud amplification with eye loss during cavefish evolution. Dev. Biol. 330, 200–211. http://dx.doi.org/10.1016/j.ydbio.2009.03.003.

Yoshizawa, M., Jeffery, W.R., 2008. Shadow response in the blind cavefish *Astyanax* reveals conservation of a functional pineal eye. J. Exp. Biol. 211, 292–299. http://dx.doi.org/10.1242/jeb.012864.

Yoshizawa, M., Goricki, S., Soares, D., Jeffery, W.R., 2010. Evolution of a behavioral shift mediated by superficial neuromasts helps cavefish find food in darkness. Curr. Biol. 20, 1631–1636. http://dx.doi.org/10.1016/j.cub.2010.07.017.

Zafar, N., Morgan, E., 1992. Feeding entrains an endogenous rhythm of swimming activity in the blind Mexican cave fish. J. Interdisc. Cycle Res. 23, 165–166.

Zaunreiter, M., Brandstätter, R., Goldschmid, A., 1998. Evidence for an endogenous clock in the retina of rainbow trout: I. Retinomotor movements, dopamine and melatonin. NeuroReport 9, 1205–1209.

Zharkov, D.O., 2008. Base excision DNA repair. Cell. Mol. Life Sci. 65, 1544–1565. http://dx.doi.org/10.1007/s00018-008-7543-2.

Ziv, L., Levkovitz, S., Toyama, R., Falcon, J., Gothilf, Y., 2005. Functional development of the zebrafish pineal gland: light-induced expression of *period2* is required for onset of the circadian clock. J. Neuroendocrinol. 17, 314–320. http://dx.doi.org/10.1111/j.1365-2826.2005.01315.x.

Chapter 17

Social Behavior and Aggressiveness in *Astyanax*

Hélène Hinaux, Sylvie Rétaux and Yannick Elipot
Development and Evolution of the Forebrain, DECA Group, Neuro-Paris Saclay Institute, Gif sur Yvette, France

Astyanax mexicanus is a teleost fish model used for evolution studies; indeed, within the same species, there are several populations of sighted surface fish (SF) that live in Mexican rivers and at least 29 populations of blind cave-adapted fish (CF) that live in perpetual darkness. All cavefish populations are derived from surface fish ancestors, some of them independently (Bradic et al., 2012; Jeffery, 2001; Sadoglu, 1956; Strecker et al., 2004; see also Chapter 4). CF have undergone morphological differentiation, the most impressive changes being the loss of their eyes and pigmentation (see also Chapters 8, 9, and 12). They have also evolved a number of behavioral alteration, some of which are now well studied, including vibration attraction behavior (VAB), rheotaxis, and sleep.

Here, we focus on social behaviors, which encompass all interactions within groups of animals, including cooperation, aggression, and reproduction (Helfman, 2009). They also include all forms of communications (visual, acoustic, chemical, tactile, etc.). To date, this communicative aspect of social behavior is poorly understood in *Astyanax*. Which modalities do SF and CF use to communicate? As explained below, it seems that SF use mainly the visual modality while, for CF, it is unknown whether they rely on olfaction, audition, mechanosensation, or another modality to communicate.

Here we review recent studies on genetic, behavioral, evolutionary aspects of *Astyanax* social behavior, specifically reproductive behavior, schooling, alarm reaction and multiple forms of aggression.

SOCIAL BEHAVIOR

Reproductive Behavior

Reproductive behavior is by definition a major social behavior; it involves several animals, requires interactions, and usually some communication. It includes the choice of the mating partner, the courtship, the spawning act, and parental

Biology and Evolution of the Mexican Cavefish. http://dx.doi.org/10.1016/B978-0-12-802148-4.00017-7
© 2016 Elsevier Inc. All rights reserved.

cares (Helfman, 2009). *A. mexicanus* spawning behavior was first described in the lab 40 years ago (Wilkens, 1972): Upon reaching sexual maturity, *Astyanax* females swim in a reduced area along the tank walls. If they come in contact with a male, the latter becomes "activated" and starts swimming rapidly looking for females. When the male encounters a mature female, the two fish swim close together and release sperm and eggs (Wilkens, 1972). The "activation" of males (i.e., they start looking for females) is dependent on the olfactory system, because olfactory nerve transection abolishes female-induced "stimulation," and males do not change behavior when put in contact with spawning females (Wilkens, 1972).

In the original laboratory study, no major difference in this behavior could be found between SF and CF (Wilkens, 1972); however, personal observations indicate some differences between the two morphs. Locomotor activity is increased in groups of both SF and Pachón CF during specific times of the night, usually in the middle of the night or just before the light turns on. This circadian control of reproductive behavior is relatively common in fish: for example the Senegal sole spawns after dusk (Oliveira et al., 2009), the zebrafish 3 h after dawn (Blanco-Vives and Sanchez-Vazquez, 2009) and the sea bream a few hours prior to dusk (Mesequer et al., 2008). This could increase the reproductive success in a species by synchronizing male and female excitation. The rhythmicity of reproductive behavior is probably controlled by several neuropeptides whose expression oscillates (Ando et al., 2013, 2014; Kitano et al., 2011; Shahjahan et al., 2011). In contrast to SF that swim in the water column, CF swim upwards along the walls of the tank, pressed against each other (Elipot et al., 2014b). This upward movement is reminiscent of water column spawners in coral reefs (Helfman, 2009), and could reflect a final synchronizing event in the courtship sequence.

Interactions between males and females could be facilitated by the presence of small denticles on the male's anal fin that likely helps the male hook the spawning female (Borowsky, 2008; Elipot et al., 2014b); however, it is unknown whether this behavior really takes place in the natural cave environment, where the water is shallow, at least during the dry season. SF, on the other hand, swim above the breeding substrate provided in laboratory conditions and lay eggs specifically in this substrate (Elipot et al., 2014b).

Of note, it seems that reproduction is more frequent in the lab after a change of tank water (Wilkens, 1972). For CF, water changes may mimic seasonal flooding of the cave, a period of the year during which food is probably most abundant and reproductive success most likely. Females are able to breed every 2 weeks, and can spawn up to 1000 eggs (Borowsky, 2008). *A. mexicanus* show no parental care; on the contrary, after spawning, both SF and CF adults tend to eat their own eggs (Elipot et al., 2014b). Filial cannibalism (eating one's own offspring) is a common behavior, especially in teleost fishes (Manica, 2002). The offspring would serve as an alternative food source and improve future reproductive success.

In many fish species, competition for females is a key courtship component. For example, in the three-spined stickleback, males are aggressive during the breeding season, especially toward males that seem more likely to be chosen by females (Rick and Bakker, 2008). Competition can also be observed in cave animals; in *Proteus anguinus*, during the reproductive season, males defend territories next to their hiding place, attacking every other animal except sexually active females (Parzefall, 1976); however, no competition for females has been observed in either of the two *Astyanax* morphs (Wilkens, 1972).

How do blind CF find each other for reproduction independently of visual cues? It was shown that Pachón CF have a tendency to be attracted by areas previously populated by conspecifics, suggesting a reliance on chemical signals that allow fish to find each other (de Fraipont, 1992; Quinn, 1980). This type of signal is widespread in schooling species (herring, minnow, plotosid catfish, salmonids) (Helfman, 2009; Pfeiffer, 1982); therefore, this CF ability could be shared with SF, although this has never been tested. The nature of the chemical signal is still undetermined for most species, even if phosphatidylcholine was proposed to participate in this gathering behavior in the catfish *Plotosus lineatus* (Matsumura et al., 2004).

Another important question is whether mating partners are selected or random. The selection of mating partners according to certain traits (sexual selection) is common in the animal kingdom (Amcoff et al., 2013; Archer et al., 2012; Chen et al., 2012; Hendry et al., 2014; House et al., 2013; Pérez I de Lanuza et al., 2013; Tobias et al., 2012). In SF, females apparently prefer large males (Figure 17.1), relying only on visual cues in the light (Plath et al., 2006). It is also the case in eyed individuals of the Micos population (for explanations on the origins and history of various cave populations mentioned, see Chapter 4). This cave population is unusual, as it comprises eyed and blind morphs, because of

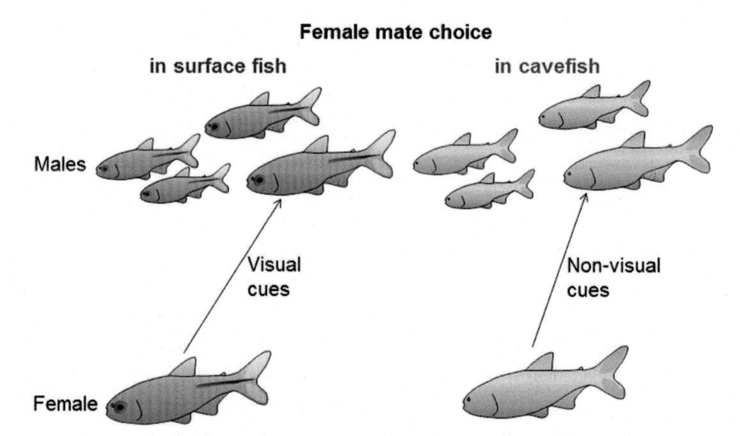

FIGURE 17.1 *Astyanax* reproductive behavior and female mate choice. Females of the *Astyanax mexicanus* species prefer large males. SF rely on visual cues to choose their mate, while CF use nonvisual cues.

recurrent hybridizations with SF (Mitchell et al., 1977). The visual mate choice could rely on sexually dimorphic phenotypes, like the rounder belly of females, and the presence of small denticles on the male's anal fin (Elipot et al., 2014b), that must help sighted *Astyanax* discriminate males and females. This visual mate choice is lost in CF; however, a novel trait evolved, as females of Micos and Yerbaníz caves are able to choose large males in the dark. In these populations, nonvisual mate choice has evolved (Figure 17.1). The choice here likely involves olfaction or mechanoreception (Plath et al., 2006). The same probably holds true for *Poecilia mexicana*; in this cave species too, females reportedly chose large males in the dark (Arndt et al., 2003).

There is an important lack of knowledge regarding *Astyanax* reproductive behavior, especially in the wild. In particular, it would be interesting to know if there is seasonality in this behavior, as it is the case of Amblyopsid CF, a family of freshwater fishes found in the southern and eastern United States, in which reproduction occurs after spring floods, when food is abundant (Poulson and Smith, 1969; Romero and Paulson, 2001). In the lab, *Astyanax* reproduction occurs at the same rate throughout the year (Elipot et al., 2014b). Taken together, the study of reproduction in CF is underexplored and presents the opportunity to investigate how fish compensate for the loss of visual cues in complex behaviors.

Schooling and Shoaling Behaviors

Schooling can be defined as the tendency of fish to synchronize their behavior and swim in an oriented manner relative to one another, while shoaling is the tendency of fish to aggregate with other fish of the same species (Helfman, 2009; Pitcher, 1983). Schooling can be observed simply by tapping on the glass of any fish tank and observing whether the stimulus induces fish to move together. These two behaviors are "distinct, but overlapping"; all schools are shoals. Usually studies on schooling assess three main parameters to describe the school, with these parameters being correlated: the number of fishes in the school; their swimming speed; and the distance between two neighboring fishes (Partridge, 1980; Partridge et al., 1980).

Several laboratory studies have established that *A. mexicanus* eyed morphs form shoals and schools, whereas CF do not (Gregson and Burt de Perera, 2007; John, 1964; Parzefall, 1983; Romero and Paulson, 2001) (Figure 17.2). In the wild, field observations confirm that SF from different rivers school, while only isolated swim has been observed in Pachón, Chica, Micos (Parzefall, 1983), Tinaja, and Los Sabinos cave populations (personal observations). *Astyanax* SF school during the day, but not at night. This change seems to be triggered by the change in illumination (John, 1964; Parzefall, 1993), indirectly suggesting that schooling in *Astyanax* is at least in part visually driven.

In fact, the schooling usually relies upon two sensory modalities: vision and mechanoreception. Indeed, in pollock, visual function acts as an

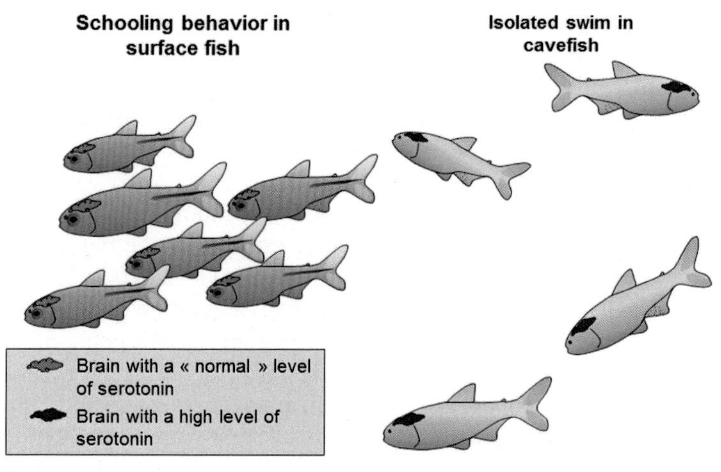

FIGURE 17.2 Evolution of schooling behavior. SF form schools in rivers and in large tanks. CF perform only isolated swims. This difference seems to be associated to changes in serotonin brain levels.

attractive force for schooling behavior (blinded fish swim farther away from their neighbors), and lateral-line signals have an opposite repulsive effect (fish whose lateral line was sectioned swim closer to each other) (Partridge and Pitcher, 1980). It is also known that the more numerous fish are in a school, the closer they will swim (Partridge, 1980; Partridge et al., 1980). The author's interpretation is that lateral-line system is a short-range sensory system, and that fish relying only on this cue underestimate the number of neighbors in the group.

A simple model of schooling behavior evolution in *Astyanax* CF could thus be that the loss of vision (Alvarez, 1946; Hubbs and Innes, 1936; Mitchell et al., 1977) and the enhancement of the lateral-line system (Jeffery et al., 2000; Schemmel, 1967; Teyke, 1990) led directly to a change in the strength of the attractive and repulsive forces, and thus to a loss of schooling behavior; however, while pollock are able to school even when blinded (Pitcher et al., 1976), SF are unable to school in the dark or when blinded by lens enucleation, showing that vision is required for schooling and shoaling behavior in *Astyanax* (Gregson and Burt de Perera, 2007; Kowalko et al., 2013). Other species need visual stimuli to school; for example, *P. mexicana* SF are unable to school in darkness, and a SF population inhabiting milky sulfur waters near a cave entrance also lost the ability to school (Parzefall, 1993).

Genomic regions that affect the propensity to schooling behavior (called QTL, for Quantitative Trait Loci) have been uncovered. One of these regions is also involved in vision (Kowalko et al., 2013). A simple explanation of the loss of schooling behavior could thus be that CF lost vision, and as a consequence, became unable to school; however, both Micos and Piedras CF with eyes do not school, suggesting that these traits (vision and schooling) are under independent genetic determinism (Parzefall and Fricke, 1991). Further, F2 hybrids resulting from crosses between SF and Tinaja CF or SF and Piedras CF do not school, fortifying the notion that schooling loss is independent from eye loss (Kowalko et al., 2013; Parzefall, 1993; Senkel, 1983). Therefore, multiple lines of evidence indicate that the loss of vision does not entirely explain the loss of schooling in CF.

What other mechanism could lead to loss of schooling? CF have an expanded lateral line, the sensory system along the body of the fish that senses water flow and vibration (see Chapter 14 and Jeffery et al., 2000; Schemmel, 1967; Teyke, 1990). Kowalko et al. showed that lateral-line expansion in CF is unlikely to have driven the evolution of schooling. In particular, SF do not show a significant difference in schooling behavior in the absence of lateral-line function (Kowalko et al., 2013). Significant neuroanatomical differences have been documented among surface and cave populations; the hypothalamus and olfactory bulbs are enlarged in CF, while the retina and optic tectum are smaller (Menuet et al., 2007; Pottin et al., 2011; see also Chapter 12 on neural evolution in this book). Pachón CF brains have elevated levels of monoamines and catecholamines (Bilandžija et al., 2013; Elipot et al., 2013, 2014a). There may also be functional differences in these systems in Tinaja CF (Duboué et al., 2012). Evolutionary loss of schooling behavior could thus result from brain differences rather than from sensory modifications. Kowalko performed deprenyl pharmacological treatments on SF, which specifically increase serotonin levels, but not other monoamines (Elipot et al., 2014a). This resulted in a significant decrease in schooling behavior; thus the elevated serotonin levels in the CF brain (Elipot et al., 2013, 2014a) are very likely to contribute to the loss of schooling behavior (Kowalko et al., 2013) (Figure 17.2).

Evolutionarily, it seems that schooling and shoaling are not advantageous in a cave. They protect fish from predation (Ioannou et al., 2012), and most fish school/shoal only during daytime when predators are present (Croft et al., 2005; Ryer and Olla, 1998). Of note, there exist other collective strategies to avoid predation, including vertical avoidance (fish diving in response to a predator [Rieucau et al., 2014]), escape through "flash expansion" (radial outward movement of the fish) or "fountain effect" (fish split, swim outward, and regroup behind the predator) (Partridge, 1982). These have not been studied in *Astyanax*.

Schooling and shoaling are less pronounced in habitats with less predation risk (Plath and Schlupp, 2008). *Astyanax* SF face strong predation pressure by a number of bird species in their natural habitat (Parzefall, 1983). On the contrary, there seems to be very few predators in cave environments in general (Barr, 1968). Even more so, it is also possible that schooling is a disadvantage in caves; CF probably use chemical gradients to find food (Bibliowicz et al., 2013), and a large group of fishes (a school or a shoal) would disrupt these gradients much more quickly than isolated fishes.

Alarm Reaction

Alarm reaction is another way to detect and escape from predators. It has long been known that when stressed, fish tend to group to avoid predation (Thinès and Legrain, 1973), or an established school can lose its coherence and flee (Frisch, 1941). This reaction can be the consequence of the emission of alarm substance in the water. This substance was discovered long ago, after a casual observation that a school of minnows became entirely frightened after only one member of the school was injured or killed (Frisch, 1941). This alarm reaction can be triggered by a filtered extract of a dead fish. On the contrary, the simple sight of a dead fish through glass does not provoke alarm reaction, implying that a chemical substance is emitted by the injured fish and detected by conspecifics. Skin extracts from an injured fish can trigger the alarm reaction, suggesting that the organ producing the alarm substance is the skin (Frisch, 1941).

Alarm substance and alarm reaction are present in Ostariophysi (Pfeiffer, 1967, 1977), including in the widely used model species *Danio rerio* (Speedie and Gerlai, 2008). Alarm substance is produced by specialized "piston cells" that are oval, with a central nucleus, and distributed uniformly throughout the epidermis (Pfeiffer, 1967). Fish react much more strongly to the alarm substance produced by conspecifics (Frisch, 1941), suggesting that the substance varies between species. The best candidates for the alarm substance were for long hypoxanthine-3 N-oxide and other molecules with a nitrogen oxide functional group (Argentini, 1976; Brown et al., 2000; Pfeiffer et al., 1985), but such molecule was never detected in the skin (Smith, 1999). More recently, it was suggested that alarm substance in zebrafish is a mixture, in which the active cue is chondroitin, a glycosaminoglycan present in the mucus (Mathuru et al., 2012). Concerning the detection, olfactory nerve transection experiments showed that olfaction is required for alarm reaction (Hamdani et al., 2000; Frisch, 1941).

In *Astyanax* SF, the detection of alarm substance leads to a stereotypic reaction lasting a few minutes: zigzag swimming, reduced swimming activity, less food intake, and more time spent at the bottom of the tank (Figure 17.3). In Piedras CF, the only behavioral changes observed are a reduction of food intake, and an increased swimming activity (Parzefall and Fricke, 1991). In Pachón and Chica CF, the alarm reaction seems to be lost (Pfeiffer, 1966). In particular, no zigzag swimming is observed in Pachón and Piedras CF, and it is only rarely observed in Chica and in the hybrid Micos population (Figure 17.3); however, these CF seem to avoid the place of substance release, as this behavior lasts for a few days (Fricke, 1987). Thus, alarm reaction seems lost in various caves with diverse types of habitats and population histories, some being from independent origin.

Early genetic studies on Chica and Pachón CF suggested that alarm reaction inheritance is due to two independent dominant alleles, only one of which was sufficient to allow alarm reaction (Pfeiffer, 1966). It seems that change in swimming speed, zigzag swimming, and feeding rate are not genetically linked (Fricke and Parzefall, 1989).

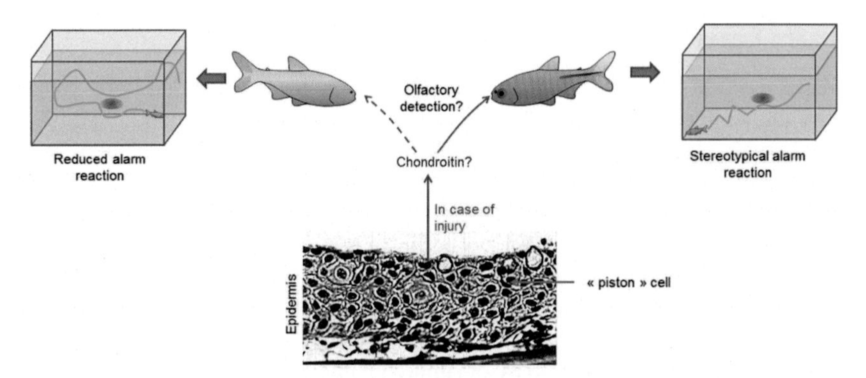

FIGURE 17.3 Evolution of alarm reaction. When an *Astyanax* fish is injured, its skin (Pfeiffer, 1963)—possibly the "piston" cells—releases an alarm substance (which could be chondroitin). This substance is smelled by other fish, which leads to different reactions in SF and CF: SF swim in zigzag, and then swim less, and at the bottom of the tank; while CF only avoid the place where the substance was released. The reduced alarm reaction in CF could be due to changes in various steps—detection of the substance, brain integration in the olfactory bulbs and other structures—but the production of the alarm substance seems to be unaffected.

Alarm reaction and alarm substance production are not correlated; Chica CF (which have a reduced alarm reaction) probably still produce alarm substance as their skin extracts are sufficient to induce an alarm reaction in other species of Characids and their epidermis contains "piston cells" (Pfeiffer, 1963). The cyprinid CF *Caecobarbus geertsi* also retained alarm substance-producing cells while it lost alarm reaction (Pfeiffer, 1963; Thinès and Legrain, 1973). This suggests that alarm reaction has been repeatedly lost or reduced in caves, probably because it is not advantageous. This also suggests that alarm reaction can be lost more easily than alarm substance production.

Territoriality and Hierarchy

Aggressive interactions usually result from competition. Their outcome can produce dominance hierarchies in groups of aggregating fishes, or else territoriality (defense of a personal space) in solitary species. In addition, a hierarchy can arise between neighboring territory holders (Helfman, 2009). *Astyanax* SF can show both types of behavior.

In *Astyanax* SF, fish show territoriality that is dependent upon the aquarium space available. In smaller tanks (<250 L), *Astyanax* SF stop schooling and start defending territories (Burchards et al., 1985; Parzefall, 1985). Territoriality appears at 10 months of age (Parzefall and Hausberg, 2001). It is apparently not associated with reproductive behavior (Wilkens, 1972), contrarily to *P. anguinus* (Parzefall, 1976) or to the cichlid fish *Astotilapia burtoni* (Fernald and Hirata, 1979). In this latter species, territorial behavior is associated with larger GnRH (Gonadotropin Releasing Hormone) neurons in the preoptic area, expressing a higher level of GnRH1 messenger RNA (mRNA) (Davis and Fernald, 1990; White et al., 2002). GnRH is a hypothalamic hormone involved in the control of the reproductive function (Roa, 2013). It could be interesting to test whether in *Astyanax* SF, too, the shift between schooling and territoriality is correlated with changes in GnRH1 levels.

In the wild, this behavior has also been observed. In shallow passages of rivers, some territorial individuals can be found, usually quite large, defending a 1 m^2 territory, particularly during the dry season. They spread fins when conspecifics approach. As in the lab, no territoriality was witnessed in Pachón, Chica, and Micos caves (Parzefall, 1983).

Hierarchy and territoriality are linked; individuals that occupy the largest part of the tank are also the ones who attack most often, i.e., the dominant individuals (Burchards et al., 1985; personal observations). It is also the case in cichlids, where only dominant males are territorial. The dominance is correlated with higher levels of GnRH-R1 mRNA in the pituitary (Maruska et al., 2011) and with higher levels of androgen and estrogen receptor mRNAs in the anterior brain (Burmeister et al., 2007).

Hierarchy in *Astyanax* SF is already present in 7-week-old fish, with a dominant fish that can be recognized in each tank (Parzefall and Hausberg, 2001). Dominance is not clearly correlated with body size (Burchards et al., 1985), as is the case in other fish species, like cave molly, angelfish, clown fish, or platyfish (Aldenhoven, 1984; Buston, 2003; Culumber and Monks, 2014; Parzefall, 1969); however, dominant SF individuals have a pigmentation pattern that is more intense (personal observations), which could be a visual signal for the group. Our observations suggest that hierarchy within SF groups is pyramidal, meaning that fishes having a dominant rank are less numerous than subordinate fishes. Dominant *Astyanax* are systematically females (Elipot et al., 2013). The position within the hierarchy might be determined by the level of serotonin in the brain of the fishes, the dominant fish having lower levels of serotonin in the posterior brain (the hindbrain raphe nucleus; see also next section). Raphe serotonin neurons are involved in the control of mood and anxiety in mammals (An et al., 2013; Donaldson et al., 2014; Rood et al., 2014). Dominant SF undergo an experience-dependent down-regulation of their raphe serotonin levels, a mechanism that would be lost in CF Pachón, explaining the absence of hierarchy (Elipot et al., 2013; Rétaux and Elipot, 2013). It would be interesting to investigate whether dominant individuals can be recognized by conspecifics on the basis of chemical cues, as in Bullhead catfishes (Helfman, 2009; Todd et al., 1967).

The loss of hierarchy in CF could be a sign that this social organization is disadvantageous in caves. Indeed, in another CF species, *P. mexicana*, epigean male individuals develop a hierarchy according to body size, and this hierarchy is also lost in cave animals (Parzefall, 1969, 1974, 1979); however, some subterranean catfishes retain aggressive behaviors allowing the establishment of a hierarchy (Ercolini et al., 1981; Trajano, 1991). It is thus unclear if hierarchical organization of the group confers any disadvantage in cave environments.

AGGRESSIVENESS

Aggressiveness is a trait found in most parts of the animal kingdom. In some cases, it is linked to reproduction, and thus limited to specific periods of time (some weeks or days). For example, stickleback males protect their territory during the nesting period, or couples of cichlids protect their larvae during parental care. In other cases, aggressiveness is not restricted to certain periods; one or more individuals protect a territory against intruders all year long, or are aggressive when they hunt—for example, packs of wolves. One can also distinguish interspecies from intraspecies aggressiveness. In most groups of animals in which a hierarchical order exists, dominant(s) use intimidation and/or aggressiveness to submit subordinate(s). Aggressiveness characterizes an animal that is naturally inclined to attack for reasons that include territoriality, hierarchy, predation, defense, as well as many others. Here, we review aggressiveness within the same species, in *Astyanax*, as a social behavior.

Loss of Aggressiveness in Cave *Astyanax*

To analyze aggressiveness in vertebrates, the classical and widely used test is the "mirror test" (Dunham and Wilczynski, 2014; Norton and Bally-Cuif, 2010). It consists of placing an animal next to a mirror and analyzing its reaction to its own reflection, by measuring the number of aggressions, the time of latency before attack, or the frequency of attacks (Figure 17.4(A)); however, it is impossible to

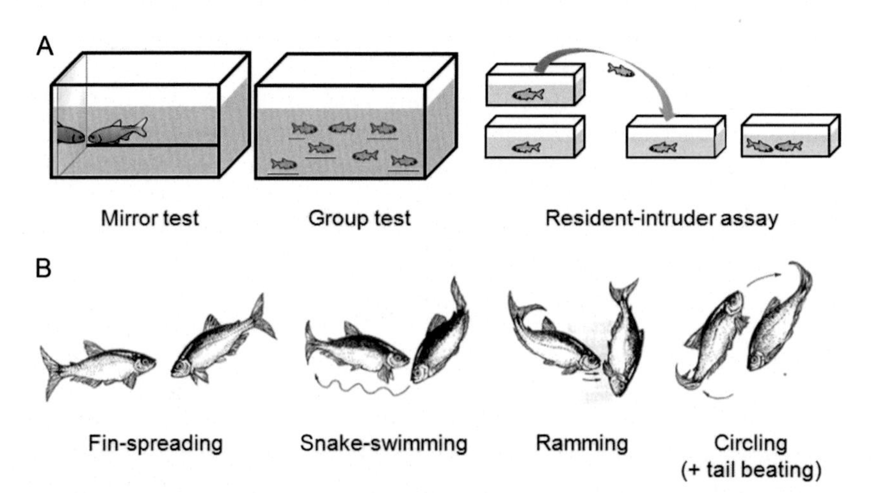

FIGURE 17.4 Measuring aggressiveness in *Astyanax*. (A) Possible tests for aggressiveness analysis. The mirror test is easier (no complex interaction between animals), but compatible with sighted animals only. The group test and resident-intruder assay are compatible with both blind and sighted fish, but introduce complex interactions between individuals. (B) Steps for ritualization of aggressiveness in *Astyanax* SF. The chronology is from left to right. *Panel (A) modified from Elipot et al. (2013) and panel (B) from Parzefall and Hausberg (2001).*

use this assay to compare *Astyanax* sighted SF and blind CF, because the latter do not see. To circumvent the problem of working with blind animals, alternative protocols have been used, which assess aggressiveness between individuals in groups of two or more fish (Figure 17.4(A)). The simplest type of test is a "resident-intruder assay," in which two animals are put in contact with one another (Figure 17.4(A)). Such tests can be more difficult in terms of interpretation than the mirror test, due to complex interactions between fish, but they have been successfully used in surface and blind *Astyanax* CF populations.

Breder was the first to describe aggressiveness in *A. mexicanus*. He explained that surface populations show "erratic viciousness" (Breder, 1943). "Fin-spreading" is the first movement of the assailant; its head is lowered and its body takes on an oblique position. In addition, "snake-swimming" and "circling" may be performed. The highest degree of aggressive manifestation is characterized by "tail-beating," "ramming," and "biting" (Figure 17.4(B)) (Burchards et al., 1985). In larval/juvenile SF, aggressiveness is first observable 16 days post-fertilization (dpf), when the size of the larvae is approximatively 12 mm (Parzefall and Hausberg, 2001). This aggressiveness is not related to territoriality, which appears after 10-12 months of age (Parzefall and Hausberg, 2001), nor with population densities during growth (Elipot et al., 2013). In adult SF, subordinate individuals with no opportunity to hide or escape are eventually killed (Elipot et al., 2013; Espinasa et al., 2005).

No aggressiveness is observed in Pachón CF, although some fin injuries are visible after 6-7 weeks of age (23 mm) (Parzefall and Hausberg, 2001). Globally, lab studies have revealed cave morphs to have lost aggressiveness, although some "attacks" can be observed between Micos, Piedras, and Pachón individuals (Burchards et al., 1985; Elipot et al., 2013; Espinasa et al., 2005; Hofmann and Hausberg, 1993; Parzefall, 1985). It was proposed that CF may have evolved new aggressive patterns responding to nonvisual signals (Langecker et al., 1995) or after body contact between CF (Parzefall and Hausberg, 2001). Even though *Astyanax* CF can display some level of aggressiveness, it is quantitatively "negligible" compared to SF; during an hour-long resident-intruder assay, SF perform 10 times more attacks than Pachón or Molino CF, i.e., more than 1000 versus 100 (Elipot et al., 2013).

One could think that the loss of aggressiveness in cave populations is simply due to their absence of vision; they would be peaceful, because they would not be able to find and target their congeners. In fact, the link between vision and aggressiveness is controversial in *Astyanax*. Burchards found that SF do not attack in the dark, suggesting a visually driven behavior (Burchards et al., 1985). Other studies describe SF attacks as more frequent in the light than the dark (Brust-Burchards, 1980). In the case of Micos CF, which naturally present different sizes of eyes due to recurrent hybridization with SF, individuals with large eyes show "intermediate" aggressiveness compared to SF and Pachón (Brust-Burchards, 1980), while those with small eyes perform few attacks and do not attack their reflect in a mirror (Dölle, 1981). Other studies performed in the dark show that SF aggressiveness is similar when compared to lighted conditions, suggesting vision is not involved in this process (Brust-Burchards, 1980; Elipot et al., 2013). Finally, in SF blinded at early larval stage by a lensectomy surgical procedure that results in eye degeneration like in CF (Yamamoto and Jeffery, 2000), aggressiveness is conserved (Elipot et al., 2013; Espinasa et al., 2005). In sum, it is probable that SF aggressiveness is not strictly visually driven, but is indeed facilitated by vision. It also indirectly suggests that loss of aggressiveness in CF is not due to their absence of eyes. A definitive response to this question would come from analyses on CF with restored eyes and restored visual function.

Genetic Basis of Aggressiveness

Aggressiveness appears to be genetically encoded in *A. mexicanus*. This idea is supported by experiments where hybrids are generated between different cave populations and SF, and by the findings that SF that have been reared in total isolation are aggressive at their first contact with a congener, suggesting that aggressiveness is an "innate" behavior (Burchards et al., 1985; Elipot et al., 2013; Fricke and Parzefall, 1989; Langecker et al., 1995; Parzefall and Hausberg, 2001; Wilkens, 1988).

In hybrids resulting from a cross between SF and Pachón, the onset of aggressive behavior occurs later than in SF. In contrast to the moderately aggressive Fl generation, fish from the F2 generation display an uncoordinated aggressive behavior (Parzefall and Hausberg, 2001) and their level of aggressiveness is comparable to SF (Parzefall, 1985). In addition, fish resulting from a backcross between F1 hybrids and CF show no aggression, while those resulting from a backcross with SF present an aggressiveness similar to SF (Parzefall, 1985). This suggests that cave alleles reduce aggressive behavior. Finally, hybrids resulting from a Pachón x Molino cross are nonaggressive (Elipot et al., 2013), suggesting that the same gene(s) may be involved in the loss of aggressiveness in these two independently evolved cave populations. In another study, the extreme aggressiveness reported in F1 hybrids after an SF x Pachón cross, or even greater in crosses of SF x Molino cross remains enigmatic and may involve a heterosis effect (Elipot et al., 2013). Overall, these data support a genetic basis for the reduction of aggressiveness in blind populations of *Astyanax*. Moreover, *Astyanax* aggressive behavior is a complex trait with polygenic determinism (Elipot et al., 2013; Fricke and Parzefall, 1989; Wilkens, 1988). This is also the case in *P. mexicana*, in which aggressiveness is encoded by several genes (Parzefall, 1989, 1979).

Neural Basis of Aggressiveness

The neural circuits mediating the behavioral program of aggression are the subject of intense investigation, including in humans, but are not yet understood. It is known, however, that some neurotransmitter systems are involved in the control of this behavior. For example, the hypothalamic arginine vasopressin neuropeptide-containing neurons control aggressiveness in the golden hamster, *Mesocricetus auratus* (Ferris et al., 1997), in the common clownfish, *Amphiprion ocellaris* (Yaeger et al., 2014), and in the medaka, *Oryzias latipes* (Kagawa, 2014). Also, testosterone is known to be involved in the control of sexual, social, and aggressive behaviors of males in many vertebrates, such as the bearded capuchin monkeys, *Sapajus libidinosus* (Mendonça-Furtado et al., 2014), the mouse, *Mus musculus* (Marie-Luce et al., 2013), or the mangrove rivulus fish, *Kryptolebias marmoratus* (Li et al., 2014).

Another important player in agonistic behavior in vertebrates including fish is serotonin, or 5-HT for 5-hydroxytryptamine (Dennis et al., 2008; Forsatkar et al., 2014; Heiming et al., 2013; Höglund et al., 2001; Jansen et al., 2011; Larson and Summers, 2001; Lynn et al., 2007; Naumenko et al., 2013; Norton and Bally-Cuif, 2010; Saudou et al., 1994; Winberg et al., 2001). Serotonin (5-HT) neurotransmission intensity and aggressiveness are generally inversely correlated. For example, in the fighting fish, *Betta splendens*, increasing brain 5-HT level by pharmacological treatments decreases aggressiveness and abolishes death struggles between males (Lynn et al., 2007). Or in *Brycon amazonicus*, a characid fish phylogenetically close to *Astyanax*, a diet enriched in L-tryptophan, the precursor of 5-HT, modifies the aggressive pattern (Wolkers et al., 2014).

Astyanax Pachón CF carry a partial loss-of-function point mutation in the coding sequence of MAO (monoamine oxidase), the serotonin-degrading enzyme. Because there is only one form of MAO in *Astyanax* as in other teleosts (Anichtchik et al., 2006; Chen et al., 1994; Ekström and Van Veen, 1984; Elipot et al., 2014a; Kumazawa et al., 1998; Nicotra and Senatori, 1989; Senatori et al., 2009; Setini et al., 2005), and because *Astyanax* MAO selectively degrades serotonin, but not dopamine nor norepinephrine (Elipot et al., 2014a), this mutation leads to elevated 5-HT levels and low metabolite (5HIAA) level in the brains of CF. This low 5HIAA/5-HT ratio is a sign of high 5-HT neurotransmission; CF show a "hyper-serotonergic" phenotype (Elipot et al., 2014a). Such a condition is at least partially responsible for their loss of aggressiveness. Along this line, treatment of *Astyanax* SF with the MAO inhibitor deprenyl strongly decreases the number of attacks performed in a resident-intruder assay (Elipot et al., 2013).

Which serotonin neural circuit could be involved in the control of aggressiveness? In teleosts, this neurotransmitter system is organized into several neuronal groups located in the hindbrain raphe (as in mammals), but also in the hypothalamus (Bellipanni et al., 2002; Ekström et al., 1985; Ekström and Van Veen, 1984; Elipot et al., 2013; Lillesaar, 2011; Sallinen et al., 2009). As described below, different parts of this serotonin network are involved in distinct aspects of aggressive behavior in CF and SF.

Different Types of Aggressive Behavior Between SF and CF

Careful qualitative and quantitative observations have led several authors to propose that CF have evolved a new type of "aggressiveness," potentially as an adaptation to life in darkness (Elipot et al., 2013; Hausberg, 1995; Langecker et al., 1995; Parzefall and Hausberg, 2001).

Foraging-Related Aggressiveness in CF

CF are more efficient than SF to find food in darkness (Hüppop, 1987), even as larvae (Espinasa et al., 2014). Such an improvement of the feeding abilities probably has several behavioral bases. The most striking is the modification of the feeding posture, with a precise feeding angle of 45° to the substrate that allows them to be four times more efficient than SF to find food in the dark (Schemmel, 1980). VAB facilitates food finding (Yoshizawa et al., 2010). The dramatic decrease in sleep duration is thought to increase the time available for foraging (Duboué et al., 2011). Besides, an increase in chemosensory capabilities could also improve food-finding abilities (Bibliowicz et al., 2013; Jeffery et al., 2000; Schemmel, 1967; Yamamoto et al., 2009).

We have recently proposed that CF "aggressiveness" corresponds to foraging (Elipot et al., 2013). This hypothesis is supported by several lines of evidences. First, analyses of aggressiveness patterns during a resident-intruder assay show that CF perform their few attacks exclusively at the beginning of the test—contrarily to SF, who do not attack at the beginning, but progressively become highly aggressive toward the second half of the 1-h test. This was interpreted as blind CF searching for food when the resident and intruder are put together. Second, CF attack more after 3 weeks of starvation, and their aggressive pattern is not changed (Elipot et al., 2013). Contrarily, SF aggressiveness does not seem to have a foraging component. Indeed, after 3 weeks of starvation, SF show reduced aggressiveness, an opposite reaction compared to CF (Elipot et al., 2013).

Serotonin is likely also playing a role in the foraging-related aggressiveness in CF. Indeed, serotonin is known to modulate food intake (Lam et al., 2010; Pérez Maceira et al., 2014), serotonin levels are globally increased in CF due to the MAO mutation (Elipot et al., 2014a), and one specific 5-HT neuronal group in the anterior hypothalamus of CF is increased in size (Elipot et al., 2013). Developmental and pharmacological manipulations suggest that this particular 5-HT group could control foraging intensity. This may occur directly or indirectly, through the neighboring hypothalamic neuronal networks controlling hunger and satiety. Therefore, the change in the size of the CF anterior hypothalamic nucleus could potentially underlie a regulatory change in motivation for food or in mechanisms controlling food intake and feeding behavior (Elipot et al., 2013; Rétaux and Elipot, 2013).

Hierarchical Aggressiveness in SF

In the wild, SF show gregarious instincts and live in large schools, but to our knowledge, detailed behavior has not been studied. As discussed earlier, in the lab some SF individuals are dominant compared to others. In large tanks, they school as in the river, while in small tanks, they are also gregarious, but develop territoriality (Burchards et al., 1985). Our personal observations witnessed that one or more animals often showing intense fin coloration have a particular status within the school, suggesting that a hierarchical order exists in the group.

The resident-intruder aggressiveness assay may provide insight into the relationship between hierarchy and aggressiveness. In the case of SF, at the beginning of the test, the resident and the intruder observe each other and do not attack. Then, one fish, probably the most dominant, starts striking at the other. It is likely that the attack and the flight of the subordinate improve the dominance status of the assailant and consequently increase the attack frequency with time (Elipot et al., 2013; Rétaux and Elipot, 2013). We have proposed that aggressiveness in SF is a demonstration of hierarchy. This idea is also supported by decreased aggressiveness and loss of aggressive pattern after starvation (Elipot et al., 2013). In fact, in starvation condition, SF decrease their metabolism (Salin et al., 2010). In such condition, establishment of hierarchy could probably be considered as deleterious, because the dominant fish consumes energy to chase the others, and because the subordinates need energy to flee and to heal their injuries. Interestingly, metabolic state and aggressiveness are also linked in the mosquitofish, *Gambusia holbrooki* (Seebacher et al., 2013), illustrating the impact of body energy homeostasis on the control of motivated behaviors. In short, *Astyanax* SF would not lose energy being aggressive when they starve.

The hindbrain raphe nucleus is likely involved in the control of SF hierarchical aggressiveness. Indeed, dominant SF have low 5-HT raphe levels, while subordinate SF show high raphe 5-HT levels (while their hypothalamic 5-HT level is not dependent upon hierarchical order). High position in the hierarchy is also inversely correlated to 5-HT level in the hindbrain raphe neurons of the hamster (Cooper et al., 2009) or the cichlid fish *Astatotilapia burtoni* (Loveland et al., 2014). Of note, it is possible to elicit some (moderate) "hierarchical" aggressiveness in CF after manipulation of their raphe nucleus. Interestingly, the 5-HT level in the raphe of a CF is similar to a subordinate SF, suggesting that dominant SF down-regulate their raphe 5-HT level, while the 5-HT level of subordinate SF is "basal" (Elipot et al., 2013). This down-regulation is probably achieved by the MAO enzyme. As MAO activity in CF is affected by a genetic mutation, CF may not be able to down-regulate 5-HT level in their raphe and acquire the dominant status (Elipot et al., 2014a). Thus, the CF MAO mutation may well underlie the loss of "hierarchical aggressiveness," and consequently of global aggressiveness, of CF.

Loss of Aggressiveness in CF—An Adaptive Change?

Does a dark environment systematically promote the loss of aggressiveness? As mentioned above, both *A. mexicanus* and *P. mexicana* CF lost aggressiveness (Parzefall, 1974). On the contrary, the blind Somalian catfish, *Uegitglanis zammaranoi*, (Ercolini et al., 1981) or the blind cave salamander,

P. anguinus, (Parzefall, 1976) are considered as aggressive animals. Thus, loss of aggressiveness is not a general rule in darkness, and each case/species should be analyzed for its own sake. Burchards alleged that the loss of aggressiveness could be explained by the absence of stabilizing selection for this trait in darkness; the probability that two individuals meet is lower in caves, because of a presumed low density of population, rendering this trait "non-useful" (Burchards et al., 1985).

On the other hand, the same MAO mutation (the same allele) is found in several populations of CF, some being independently evolved (Elipot et al., 2014a). These findings suggest that the MAO mutation was present at low frequency in the ancestral SF populations and was selected from this standing genetic variation in the various caves. This mutated allele was not detected in the extant SF population, suggesting either that it has disappeared, or that it is very rare. The phenotype that was selected in the caves could have been the high serotonin levels in the brains. As discussed above, 5-HT is not only involved in aggressiveness control, but also in other behaviors, such as schooling and foraging, and the loss of schooling in cave environment could be adaptive to finding food in the dark. Moreover, the hyperserotonergic brain phenotype may help the CF solve the "feed or fight" conflict; in conditions where food is scarce, it seems more advantageous to spend time looking for food (Rétaux and Elipot, 2013).

CONCLUSION

A. mexicanus proves a fantastic model to study social behaviors and their molecular and neural basis, notably because of the many behavioral differences between the two morphs, and the possibility to cross them to investigate the genetic basis of these trait differences. After having reviewed the current knowledge on *A. mexicanus* social behaviors, it clearly appears that social interactions between CF are strongly reduced. These fish have lost schooling behavior, they've lost most of the stereotypic reaction to alarm substance release, and they've lost all forms of aggressiveness, including the ability to form hierarchies or establish territories. Blindness only plays a minor role in all these behavioral losses.

However, CF retained (at least in the laboratory) social interactions for reproductive behavior, a behavior that is crucial for survival. Notably, a courtship sequence seems to be present in CF. It even seems that new features evolved in the caves—the ability to choose mates using nonvisual cues is an innovation of the cave populations. Besides, CF have also evolved VAB, a trait that helps them find food in the dark (see Chapter 14). Because this behavior relies on the attraction for vibrations at 35 Hz, and because this is about the same frequency that is produced by moving *Astyanax* (Bleckmann et al., 1991), it is possible that this behavior is a kind of communication among CF that would help them gather at food sources (Yoshizawa and Jeffery, 2011) and even, maybe, breed. After all, finding food and finding mates are two primary challenges for CF to survive in the dark. It is thus possible that social interactions are not lacking in *Astyanax* CF, but they are entirely different from those known among river fishes, and yet to be discovered.

ACKNOWLEDGMENTS

Work in the group was supported by ANR grant [Astyco] and [Blindtest], Fondation pour la Recherche Médicale and CNRS. Many thanks to all members of the group (Alexandre Alié, Maryline Blin, Lucie Devos) and particularly to Stéphane Père, Magalie Bouvet, and Diane Denis for taking care of our *Astyanax* colony, and Laurent Legendre for collaborative help on husbandry methods.

REFERENCES

Aldenhoven, J.M., 1984. Social Organization and Sex Change in an Angelfish Centropyge Bicolor on the Great Barrier Reef. Macquarie University, North Ryde, New South Wales.

Alvarez, J., 1946. Revision del genero Anoptichthys con descripcion de una especie nueva (Pisc., Characidae). An. Esc. Nac. Cienc. Biol. Mex. 4, 263–282.

Amcoff, M., Gonzalez-Voyer, A., Kolm, N., 2013. Evolution of egg dummies in Tanganyikan cichlid fishes: the roles of parental care and sexual selection. J. Evol. Biol. 26, 2369–2382. http://dx.doi.org/10.1111/jeb.12231.

An, Y., Inoue, T., Kitaichi, Y., Izumi, T., Nakagawa, S., Song, N., Chen, C., Li, X., Koyama, T., Kusumi, I., 2013. Anxiolytic-like effect of mirtazapine mediates its effect in the median raphe nucleus. Eur. J. Pharmacol. 720, 192–197.

Ando, H., Shahjahan, M., Hattori, A., 2013. Molecular neuroendocrine basis of lunar-related spawning in grass puffer. Gen. Comp. Endocrinol. 181, 211–214. http://dx.doi.org/10.1016/j.ygcen.2012.07.027.

Ando, H., Ogawa, S., Shahjahan, M., Ikegami, T., Doi, H., Hattori, A., Parhar, I., 2014. Diurnal and circadian oscillations in expression of kisspeptin, kisspeptin receptor and gonadotrophin-releasing hormone 2 genes in the grass puffer, a semilunar-synchronised spawner. J. Neuroendocrinol. 26, 459–467.

Anichtchik, O., Sallinen, V., Peitsaro, N., Panula, P., 2006. Distinct structure and activity of monoamine oxidase in the brain of zebrafish (Danio rerio). J. Comp. Neurol. 498, 593–610. http://dx.doi.org/10.1002/cne.21057.

Archer, C.R., Zajitschek, F., Sakaluk, S.K., Royle, N.J., Hunt, J., 2012. Sexual selection affects the evolution of lifespan and ageing in the decorated cricket Gryllodes sigillatus. Evolution 66, 3088–3100. http://dx.doi.org/10.1111/j.1558-5646.2012.01673.x.

Argentini, M., 1976. Isolierung des Schreckstoffes aus der Haut der Elritze Phoxinus phoxinus (L). University of Zurich, Zurich.

Arndt, M., Parzefall, J., Plath, M., 2003. Does sexual experience influence mate choice decisions in cave molly females (Poecilia mexicana, Poeciliidae, Teleostei). Subterran Biol. 2, 53–58.

Barr, T.C., 1968. Cave ecology and the evolution of troglobites. In: Dobzhansky, T., Hecht, M.K., Steere, W.C. (Eds.), Evolutionary Biology. Appleton-Century-Crofts, New York, pp. 35–102.

Bellipanni, G., Rink, E., Bally-Cuif, L., 2002. Cloning of two tryptophan hydroxylase genes expressed in the diencephalon of the developing zebrafish brain. Mech. Dev. 119, S215–S220. http://dx.doi.org/10.1016/S0925-4773(03)00119-9.

Bibliowicz, J., Alié, A., Espinasa, L., Yoshizawa, M., Blin, M., Hinaux, H., Legendre, L., Père, S., Rétaux, S., 2013. Differences in chemosensory response between eyed and eyeless *Astyanax mexicanus* of the Rio Subterráneo cave. EvoDevo 4, 25. http://dx.doi.org/10.1186/2041-9139-4-25.

Bilandžija, H., Ma, L., Parkhurst, A., Jeffery, W.R., 2013. A potential benefit of albinism in Astyanax cavefish: downregulation of the oca2 gene increases tyrosine and catecholamine levels as an alternative to melanin synthesis. PLoS One 8, e80823. http://dx.doi.org/10.1371/journal.pone.0080823.

Blanco-Vives, B., Sanchez-Vazquez, F., 2009. Synchronisation to light and feeding time of circadian rhythms of spawning and locomotor activity in zebrafish. Physiol. Behav. 98, 268–275.

Bleckmann, H., Breithaupt, T., Blickhan, R., Tautz, J., 1991. The time course and frequency content of hydrodynamic events caused by moving fish, frogs, and crustaceans. J. Comp. Physiol. A. 168, 749–757.

Borowsky, R., 2008. *Astyanax mexicanus*, the blind Mexican cave fish: a model for studies in development and morphology. In: Emerging Model Organisms: A Laboratory Manual. CSH Press, Cold Spring Harbor, NY, pp. 469–480.

Bradic, M., Beerli, P., García-de León, F.J., Esquivel-Bobadilla, S., Borowsky, R.L., 2012. Gene flow and population structure in the Mexican blind cavefish complex (*Astyanax mexicanus*). BMC Evol. Biol. 12, 9. http://dx.doi.org/10.1186/1471-2148-12-9.

Breder, C.M., 1943. A note on erratic viciousness in *Astyanax mexicanus* (Phillipi). Copeia 1943, 82. http://dx.doi.org/10.2307/1437770.

Brown, G.E., Adrian, J.C., Smyth, E., Leet, H., Brennan, S., 2000. Ostariophysan alarm pheromones: laboratory and field tests of the functional significance of nitrogen oxides. J. Chem. Ecol. 26, 139–154.

Brust-Burchards, H., 1980. Das Aggressionsverhalten von Fischen. Eine vergleichende Betrachtung unter besonderer Berücksichtigung von *Astyanax mexicanus*. University of Hamburg, Hamburg.

Burchards, H., Dölle, A., Parzefall, J., 1985. Aggressive behaviour of an epigean population of *Astyanax mexicanus* (Characidae, Pisces) and some observations of three subterranean populations. Behav. Processes 11, 225–235.

Burmeister, S.S., Kailasanath, V., Fernald, R.D., 2007. Social dominance regulates androgen and estrogen receptor gene expression. Horm. Behav. 51, 164–170. http://dx.doi.org/10.1016/j.yhbeh.2006.09.008.

Buston, P., 2003. Social hierarchies: size and growth modification in clownfish. Nature 424, 145–146. http://dx.doi.org/10.1038/424145a.

Chen, K., Wu, H.F., Grimsby, J., Shih, J.C., 1994. Cloning of a novel monoamine oxidase cDNA from trout liver. Mol. Pharmacol. 46, 1226–1233.

Chen, I.-P., Stuart-Fox, D., Hugall, A.F., Symonds, M.R.E., 2012. Sexual selection and the evolution of complex color patterns in dragon lizards. Evolution 66, 3605–3614. http://dx.doi.org/10.1111/j.1558-5646.2012.01698.x.

Cooper, M.A., Grober, M.S., Nicholas, C.R., Huhman, K.L., 2009. Aggressive encounters alter the activation of serotonergic neurons and the expression of 5-HT1A mRNA in the hamster dorsal raphe nucleus. Neuroscience 161, 680–690. http://dx.doi.org/10.1016/j.neuroscience.2009.03.084.

Croft, D.P., James, R., Ward, A.J.W., Botham, M.S., Mawdsley, D., Krause, J., 2005. Assortative interactions and social networks in fish. Oecologia 143, 211–219. http://dx.doi.org/10.1007/s00442-004-1796-8.

Culumber, Z.W., Monks, S., 2014. Does fin coloration signal social status in a dominance hierarchy of the livebearing fish Xiphophorus variatus? Behav. Process. 107, 158–162. http://dx.doi.org/10.1016/j.beproc.2014.08.010.

Davis, M.R., Fernald, R.D., 1990. Social control of neuronal soma size. J. Neurobiol. 21, 1180–1188.

De Fraipont, M., 1992. Réponse d'*Astyanax mexicanus* aux stimulations chimiques provenant de groupes de congénères à différents stades du développement. Mém. Biospél. 19, 209–213.

Pérez I de Lanuza, G., Font, E., Monterde, J.L., 2013. Using visual modelling to study the evolution of lizard coloration: sexual selection drives the evolution of sexual dichromatism in lacertids. J. Evol. Biol. 26, 1826–1835. http://dx.doi.org/10.1111/jeb.12185.

Dennis, R.L., Chen, Z.Q., Cheng, H.W., 2008. Serotonergic mediation of aggression in high and low aggressive chicken strains. Poult. Sci. 87, 612–620. http://dx.doi.org/10.3382/ps.2007-00389.

Dölle, A., 1981. Über Ablauf und Funktion des Aggressionsverhaltens von *Astyanax mexicanus* (Characidae, Pisces) unter Berücksichtigung zweier Höhlenpopulationen. University of Hamburg, Hamburg.

Donaldson, Z.R., Piel, D.A., Santos, T.L., Richardson-Jones, J., Leonardo, E.D., Beck, S.G., Champagne, F.A., Hen, R., 2014. Developmental effects of serotonin 1A autoreceptors on anxiety and social behavior. Neuropsychopharmacology 39, 291–302. http://dx.doi.org/10.1038/npp.2013.185.

Duboué, E.R., Keene, A.C., Borowsky, R.L., 2011. Evolutionary convergence on sleep loss in cavefish populations. Curr. Biol. 21, 671–676. http://dx.doi.org/10.1016/j.cub.2011.03.020.

Duboué, E.R., Borowsky, R.L., Keene, A.C., 2012. β-Adrenergic signaling regulates evolutionarily derived sleep loss in the Mexican cavefish. Brain Behav. Evol. 80, 233–243. http://dx.doi.org/10.1159/000341403.

Dunham, L.A., Wilczynski, W., 2014. Arginine vasotocin, steroid hormones and social behavior in the green anole lizard (Anolis carolinensis). J. Exp. Biol. 217, 3670–3676. http://dx.doi.org/10.1242/jeb.107854.

Ekström, P., Van Veen, T., 1984. Distribution of 5-hydroxytryptamine (serotonin) in the brain of the teleostGasterosteus aculeatus L. J. Comp. Neurol. 226, 307–320. http://dx.doi.org/10.1002/cne.902260302.

Ekström, P., Nyberg, L., van Veen, T., 1985. Ontogenetic development of serotoninergic neurons in the brain of a teleost, the three-spined stickleback. An immunohistochemical analysis. Brain Res. 349, 209–224.

Elipot, Y., Hinaux, H., Callebert, J., Rétaux, S., 2013. Evolutionary shift from fighting to foraging in blind cavefish through changes in the serotonin network. Curr. Biol. 23, 1–10. http://dx.doi.org/10.1016/j.cub.2012.10.044.

Elipot, Y., Hinaux, H., Callebert, J., Launay, J.-M., Blin, M., Rétaux, S., 2014a. A mutation in the enzyme monoamine oxidase explains part of the *Astyanax* cavefish behavioural syndrome. Nat. Commun. 5(4647), 1–11. http://dx.doi.org/10.1038/ncomms4647.

Elipot, Y., Legendre, L., Père, S., Sohm, F., Rétaux, S., 2014b. *Astyanax* transgenesis and husbandry: how cavefish enters the laboratory. Zebrafish 11, 291–299. http://dx.doi.org/10.1089/zeb.2014.1005.

Ercolini, A., Berti, R., Cianfanelli, A., 1981. Aggressive behavior in Uegitglanis zammaranoi Gianferrari (Clariidae: Siluriformes), anophtalmic phreatic fish from Somalia. Monit. Zool. Ital. Suppl. 14, 39–56.

Espinasa, L., Yamamoto, Y., Jeffery, W.R., 2005. Non-optical releasers for aggressive behavior in blind and blinded Astyanax (Teleostei, Characidae). Behav. Process. 70, 144–148. http://dx.doi.org/10.1016/j.beproc.2005.06.003.

Espinasa, L., Bibliowicz, J., Jeffery, W.R., Rétaux, S., 2014. Enhanced prey capture skills in Astyanax cavefish larvae are independent from eye loss. EvoDevo 5, 35. http://dx.doi.org/10.1186/2041-9139-5-35.

Fernald, R.D., Hirata, N.R., 1979. The ontogeny of social behavior and body coloration in the African cichlid fish Haplochromis burtoni. Z. Für Tierpsychol. 50, 180–187.

Ferris, C.F., Melloni, R.H., Koppel, G., Perry, K.W., Fuller, R.W., Delville, Y., 1997. Vasopressin/serotonin interactions in the anterior hypothalamus control aggressive behavior in golden hamsters. J. Neurosci. Off. J. Soc. Neurosci. 17, 4331–4340.

Forsatkar, M.N., Nematollahi, M.A., Amiri, B.M., Huang, W.-B., 2014. Fluoxetine inhibits aggressive behaviour during parental care in male fighting fish (Betta splendens, Regan). Ecotoxicology 23, 1794–1802. http://dx.doi.org/10.1007/s10646-014-1345-0.

Fricke, D., 1987. Reaction to alarm substance in cave populations of Astyanax fasciatus (Characidae, Pisces). Ethology 76, 305–308. http://dx.doi.org/10.1111/j.1439-0310.1987.tb00691.x.

Fricke, D., Parzefall, J., 1989. Alarm reaction, aggression and schooling in cave and river populations of Astyanax fasciatus and their hybrids. Mém. Biospél. 26, 177–182.

Frisch, K.V., 1941. Die Bedeutung des Geruchsinnes im Leben der Fische. In: Süffert, F. (Ed.), Die Naturwissenschaften. Springer, Berlin, Heidelberg, pp. 321–333.

Gregson, J.N.S., Burt de Perera, T., 2007. Shoaling in eyed and blind morphs of the characin Astyanax fasciatus under light and dark conditions. J. Fish Biol. 70, 1615–1619. http://dx.doi.org/10.1111/j.1095-8649.2007.01430.x.

Hamdani, E.H., Stabell, O.B., Alexander, G., Døving, K.B., 2000. Alarm reaction in the crucian carp is mediated by the medial bundle of the medial olfactory tract. Chem. Senses 25, 103–109.

Hausberg, C., 1995. Das aggressionsverhalten von Astyanax fasciatus (Characidae, Teleostei): Zur Ontogenie, Genetik und Evolution der epigäischen und hypogäischen Form. University of Hamburg, Hamburg.

Heiming, R.S., Mönning, A., Jansen, F., Kloke, V., Lesch, K.-P., Sachser, N., 2013. To attack, or not to attack? The role of serotonin transporter genotype in the display of maternal aggression. Behav. Brain Res. 242, 135–141. http://dx.doi.org/10.1016/j.bbr.2012.12.045.

Helfman, G.S., 2009. The Diversity of Fishes. Wiley-Blackwell, Oxford.

Hendry, C.R., Guiher, T.J., Pyron, R.A., 2014. Ecological divergence and sexual selection drive sexual size dimorphism in New World pitvipers (Serpentes: Viperidae). J. Evol. Biol. 27, 760–771. http://dx.doi.org/10.1111/jeb.12349.

Hofmann, S., Hausberg, C., 1993. The aggressive behavior of the Micos cave population (Astyanax fasciatus, Characidae, Teleostei) after selection for functional eyes in comparison to an epigean one. Mém. Biospél. 20, 101–103.

Höglund, E., Kolm, N., Winberg, S., 2001. Stress-induced changes in brain serotonergic activity, plasma cortisol and aggressive behavior in Arctic charr (Salvelinus alpinus) is counteracted by L-DOPA. Physiol. Behav. 74, 381–389. http://dx.doi.org/10.1016/S0031-9384(01)00571-6.

House, C.M., Lewis, Z., Hodgson, D.J., Wedell, N., Sharma, M.D., Hunt, J., Hosken, D.J., 2013. Sexual and natural selection both influence male genital evolution. PLoS One 8, e63807. http://dx.doi.org/10.1371/journal.pone.0063807.

Hubbs, C.L., Innes, W.T., 1936. The first known blind fish of the family Characidae: a new genus from Mexico. Occas. Paper Mus. Zool. Univ. Mich. 342, 1–9.

Hüppop, K., 1987. Food-finding ability in cave fish (Astyanax fasciatus). Int. J. Speleol. 16, 59–66. http://dx.doi.org/10.5038/1827-806X.16.1.4.

Ioannou, C.C., Guttal, V., Couzin, I.D., 2012. Predatory fish select for coordinated collective motion in virtual prey. Science 337, 1212–1215. http://dx.doi.org/10.1126/science.1218919.

Jansen, F., Heiming, R.S., Kloke, V., Kaiser, S., Palme, R., Lesch, K.-P., Sachser, N., 2011. Away game or home match: the influence of venue and serotonin transporter genotype on the display of offensive aggression. Behav. Brain Res. 219, 291–301. http://dx.doi.org/10.1016/j.bbr.2011.01.029.

Jeffery, W.R., 2001. Cavefish as a model system in evolutionary developmental biology. Dev. Biol. 231, 1–12. http://dx.doi.org/10.1006/dbio.2000.0121.

Jeffery, W., Strickler, A., Guiney, S., Heyser, D., Tomarev, S., 2000. Prox 1 in eye degeneration and sensory organ compensation during development and evolution of the cavefish Astyanax. Dev. Genes Evol. 210, 223–230. http://dx.doi.org/10.1007/s004270050308.

John, K.R., 1964. Illumination, vision, and schooling of *Astyanax mexicanus* (Fillipi). J. Fish. Res. Board Can. 21, 1453–1473.

Kagawa, N., 2014. Comparison of aggressive behaviors between two wild populations of Japanese Medaka, *Oryzias latipes* and *O. sakaizumii*. Zool. Sci. 31, 116–121. http://dx.doi.org/10.2108/zsj.31.116.

Kitano, H., Irie, S., Ohta, K., Hirai, T., Yamaguchi, A., Matsuyama, M., 2011. Molecular cloning of two gonadotropin receptors and their distinct mRNA expression profiles in daily oogenesis of the wrasse Pseudolabrus sieboldi. Gen. Comp. Endocrinol. 172, 268–276. http://dx.doi.org/10.1016/j.ygcen.2011.03.012.

Kowalko, J.E., Rohner, N., Rompani, S.B., Peterson, B.K., Linden, T.A., Yoshizawa, M., Kay, E.H., Weber, J., Hoekstra, H.E., Jeffery, W.R., Borowsky, R., Tabin, C.J., 2013. Loss of schooling behavior in cavefish through sight-dependent and sight-independent mechanisms. Curr. Biol. 23, 1874–1883. http://dx.doi.org/10.1016/j.cub.2013.07.056.

Kumazawa, T., Seno, H., Ishii, A., Suzuki, O., Sato, K., 1998. Monoamine oxidase activities in catfish (Parasilurus asotus) tissues. J. Enzym. Inhib. 13, 377–384.

Lam, D.D., Garfield, A.S., Marston, O.J., Shaw, J., Heisler, L.K., 2010. Brain serotonin system in the coordination of food intake and body weight. Pharmacol. Biochem. Behav. 97, 84–91. http://dx.doi.org/10.1016/j.pbb.2010.09.003.

Langecker, T.G., Neumann, B., Hausberg, C., Parzefall, J., 1995. Evolution of the optical releasers for aggressive behavior in cave-dwelling Astyanax fasciatus (Teleostei, Characidae). Behav. Processes 34, 161–167.

Larson, E.T., Summers, C.H., 2001. Serotonin reverses dominant social status. Behav. Brain Res. 121, 95–102.

Li, C.-Y., Earley, R.L., Huang, S.-P., Hsu, Y., 2014. Fighting experience alters brain androgen receptor expression dependent on testosterone status. Proc. Biol. Sci. 281, 20141532. http://dx.doi.org/10.1098/rspb.2014.1532.

Lillesaar, C., 2011. The serotonergic system in fish. J. Chem. Neuroanat. 41, 294–308. http://dx.doi.org/10.1016/j.jchemneu.2011.05.009.

Loveland, J.L., Uy, N., Maruska, K.P., Carpenter, R.E., Fernald, R.D., 2014. Social status differences regulate the serotonergic system of a cichlid fish, Astatotilapia burtoni. J. Exp. Biol. 217, 2680–2690. http://dx.doi.org/10.1242/jeb.100685.

Lynn, S.E., Egar, J.M., Walker, B.G., Sperry, T.S., Ramenofsky, M., 2007. Fish on Prozac: a simple, noninvasive physiology laboratory investigating the mechanisms of aggressive behavior in Betta splendens. Adv. Physiol. Educ. 31, 358–363. http://dx.doi.org/10.1152/advan.00024.2007.

Manica, A., 2002. Filial cannibalism in teleost fish. Biol. Rev. Camb. Philos. Soc. 77, 261–277. http://dx.doi.org/10.1017/S1464793101005905.

Marie-Luce, C., Raskin, K., Bolborea, M., Monin, M., Picot, M., Mhaouty-Kodja, S., 2013. Effects of neural androgen receptor disruption on aggressive behavior, arginine vasopressin and galanin systems in the bed nucleus of stria terminalis and lateral septum. Gen. Comp. Endocrinol. 188, 218–225. http://dx.doi.org/10.1016/j.ygcen.2013.03.031.

Maruska, K.P., Levavi-Sivan, B., Biran, J., Fernald, R.D., 2011. Plasticity of the reproductive axis caused by social status change in an African cichlid fish: I. Pituitary gonadotropins. Endocrinology 152, 281–290. http://dx.doi.org/10.1210/en.2010-0875.

Mathuru, A.S., Kibat, C., Cheong, W.F., Shui, G., Wenk, M.R., Friedrich, R.W., Jesuthasan, S., 2012. Chondroitin fragments are odorants that trigger fear behavior in fish. Curr. Biol. 22, 538–544. http://dx.doi.org/10.1016/j.cub.2012.01.061.

Matsumura, K., Matsunaga, S., Fusetani, N., 2004. Possible involvement of phosphatidylcholine in school recognition in the catfish, Plotosus lineatus. Zool. Sci. 21, 257–264. http://dx.doi.org/10.2108/zsj.21.257.

Mendonça-Furtado, O., Edaes, M., Palme, R., Rodrigues, A., Siqueira, J., Izar, P., 2014. Does hierarchy stability influence testosterone and cortisol levels of bearded capuchin monkeys (Sapajus libidinosus) adult males? A comparison between two wild groups. Behav. Process. 109 (Pt A), 79–88. http://dx.doi.org/10.1016/j.beproc.2014.09.010.

Menuet, A., Alunni, A., Joly, J.-S., Jeffery, W.R., Retaux, S., 2007. Expanded expression of Sonic Hedgehog in Astyanax cavefish: multiple consequences on forebrain development and evolution. Development 134, 845–855. http://dx.doi.org/10.1242/dev.02780.

Mesequer, C., Ramos, J., Bayarri, M., Oliveira, C., Sanchez-Vazquez, F., 2008. Light synchronization of the daily spawning rhythms of gilthead sea bream (Sparus aurata L) kept under different photoperiod and after shifting the LD cycle. Chronobiol. Int. 25, 666–679.

Mitchell, R.W., Russell, W.H., Elliott, W.R., 1977. Mexican Eyeless Characin Fishes, Genus Astyanax: Environment, Distribution, and Evolution, Special publications (Texas Tech University). Texas Press, Lubbock.

Naumenko, V.S., Kozhemyakina, R.V., Plyusnina, I.F., Kulikov, A.V., Popova, N.K., 2013. Serotonin 5-HT1A receptor in infancy-onset aggression: comparison with genetically defined aggression in adult rats. Behav. Brain Res. 243, 97–101. http://dx.doi.org/10.1016/j.bbr.2012.12.059.

Nicotra, A., Senatori, O., 1989. Some characteristics of mitochondrial monoamine oxidase activity in eggs of carp (Cyprinus carpio) and rainbow trout (Salmo gairdneri). Comp. Biochem. Physiol. C 92, 401–404.

Norton, W., Bally-Cuif, L., 2010. Adult zebrafish as a model organism for behavioural genetics. BMC Neurosci. 11, 90. http://dx.doi.org/10.1186/1471-2202-11-90.

Oliveira, C., Dinis, M., Soares, F., Cabrita, E., Pousão-Ferreira, P., Sanchez-Vazquez, F., 2009. Lunar and daily spawning rhythms of Senegal sole Solea senegalensis. J. Fish Biol. 75, 61–74.

Partridge, B.L., 1980. The effect of school size on the structure and dynamics of minnow schools. Anim. Behav. 28, 68–77. http://dx.doi.org/10.1016/S0003-3472(80)80009-1.

Partridge, B.L., 1982. Structure and function of fish schools. Sci. Am. 246, 114–123.

Partridge, B.L., Pitcher, T.J., 1980. The sensory basis of fish schools: relative roles of lateral line and vision. J. Comp. Physiol. A. 135, 315–325. http://dx.doi.org/10.1007/BF00657647.

Partridge, B.L., Pitcher, T., Cullen, J.M., Wilson, J., 1980. The three-dimensional structure of fish schools. Behav. Ecol. Sociobiol. 6, 277–288. http://dx.doi.org/10.1007/BF00292770.

Parzefall, J., 1969. Zur Vergleichenden Ethologie Verschiedener Mollienesia-Arten Einschliesslich Einer Höhlenform Von M. Sphenops. Behaviour 33, 1–37. http://dx.doi.org/10.1163/156853969X00297.

Parzefall, J., 1974. Rückbildung aggressiver Verhaltensweisen bei einer Höhlenform von Poecilia sphenops (Pisces, Poeciliidae). Z. Für Tierpsychol. 35, 66–84. http://dx.doi.org/10.1111/j.1439-0310.1974.tb00433.x.

Parzefall, J., 1976. Die Rolle der chemischen Information im Verhalten des Grottenolms Proteus anguineus Laur. (Proteidae, Urodela). Z. Für Tierpsychol. 42, 29–49. http://dx.doi.org/10.1111/j.1439-0310.1976.tb00954.x.

Parzefall, J., 1979. Zur Genetik und biologischen Bedeutung des Aggressionsverhaltens von Poecilia sphenops (Pisces, Poecilidae) Untersuchung an Bastarden ober- und unterirdisch lebender Populationen. Z. Für Tierpsychol. 50, 399–422.

Parzefall, J., 1983. Field observations in epigean and cave populations of the Mexican characid *Astyanax mexicanus* (Pisces, Characidae). Mém. Biospéol. 10, 171–176.

Parzefall, J., 1985. On the heredity of behavior patterns in cave animals and their epigean relatives. NSS Bull. 47, 128–135.

Parzefall, J., 1989. Sexual and aggressive behaviour in species hybrids of Poecilia mexicana and Poecilia velifera (Pisces, Poeciliidae). Ethology 82, 101–115. http://dx.doi.org/10.1111/j.1439-0310.1989.tb00491.x.

Parzefall, J., 1993. Schooling behavior in population hybrids of Astyanax fasciatus and Poecilia Mexicana (Pisces, Characidae, Poecilidae). In: Schröder, H., Bauer, J., Schartl, M. (Eds.), Trends in Ichthyolgy, pp. 297–303.

Parzefall, J., Fricke, D., 1991. Alarm reaction and schooling in population hybrids of Astyanax fasciatus (Pisces, Characidae). Mém. Biospél. 18, 29–32.

Parzefall, J., Hausberg, C., 2001. Ontogeny of the aggressive behaviour in epigean and hypogean populations of Astyanax fasciatus (Characidae, Teleostei) and their hybrids. Mém. Biospél. 28, 157–161.

Pérez Maceira, J.J., Mancebo, M.J., Aldegunde, M., 2014. The involvement of 5-HT-like receptors in the regulation of food intake in rainbow trout (Oncorhynchus mykiss). Comp. Biochem. Physiol. C Toxicol. Pharmacol. 161, 1–6. http://dx.doi.org/10.1016/j.cbpc.2013.12.003.

Pfeiffer, W., 1963. Vergleichende Untersuchungen über die Schreckreaktion und den Schreckstoff bei Ostariophysen. Z. Vergl. Physiol. 47, 111–147.

Pfeiffer, W., 1966. Über die Vererbung der Schreckreaktion bei Astyanax (Characidae, Pisces). Z. Vererbungsl. 98, 97–105. http://dx.doi.org/10.1007/BF00897181.

Pfeiffer, W., 1967. Schreckreaktion und Schreckstoffzellen bei Ostariophysi und Gonorhynchiformes. Z. Vergl. Physiol. 56, 380–396. http://dx.doi.org/10.1007/BF00298056.

Pfeiffer, W., 1977. The distribution of fright reaction and alarm substance cells in fishes. Copeia 1977, 653–665.

Pfeiffer, W., 1982. Chemical signals in communication. In: Hara, T.J. (Ed.), Chemoreception in Fishes. Elsevier, Amsterdam, pp. 307–326.

Pfeiffer, W., Riegelbauer, G., Meier, G., Scheibler, B., 1985. Effect of hypoxanthine-3(N)-oxide and hypoxanthine-1(N)-oxide on central nervous excitation of the black tetraGymnocorymbus ternetzi (Characidae, Ostariophysi, Pisces) indicated by dorsal light response. J. Chem. Ecol. 11, 507–523. http://dx.doi.org/10.1007/BF00989562.

Pitcher, T.J., 1983. Heuristic definitions of fish shoaling behaviour. Anim. Behav. 31, 611–613.

Pitcher, T., Partridge, B., Wardle, C., 1976. A blind fish can school. Science 194, 963–965. http://dx.doi.org/10.1126/science.982056.

Plath, M., Schlupp, I., 2008. Parallel evolution leads to reduced shoaling behavior in two cave dwelling populations of Atlantic mollies (Poecilia mexicana, Poeciliidae, Teleostei). Environ. Biol. Fishes 82, 289–297. http://dx.doi.org/10.1007/s10641-007-9291-9.

Plath, M., Rohde, M., Schröder, T., Taebel-Hellwig, A., Schlupp, I., 2006. Female mating preferences in blind cave tetras Astyanax fasciatus (Characidae, Teleostei). Behaviour 143, 15–32.

Pottin, K., Hinaux, H., Retaux, S., 2011. Restoring eye size in *Astyanax mexicanus* blind cavefish embryos through modulation of the Shh and Fgf8 forebrain organising centres. Development 138, 2467–2476. http://dx.doi.org/10.1242/dev.054106.

Poulson, T.L., Smith, P.M., 1969. The basis for seasonal growth and reproduction in aquatic cave organisms. In: Actes 4th Int. Congr. Speleol. Ljubl. Yugosl. 4-5, pp. 197–201.

Quinn, T.P., 1980. Locomotor responses of juvenile blind cave fish, Astyanax jordani, to the odors of conspecifics. Behav. Neural Biol. 29, 123–127.

Rétaux, S., Elipot, Y., 2013. Feed or fight: a behavioral shift in blind cavefish. Commun. Integr. Biol. 6, e23166. http://dx.doi.org/10.4161/cib.23166.

Rick, I.P., Bakker, T.C.M., 2008. Males do not see only red: UV wavelengths and male territorial aggression in the three-spined stickleback (Gasterosteus aculeatus). Naturwissenschaften 95, 631–638. http://dx.doi.org/10.1007/s00114-008-0365-0.

Rieucau, G., Boswell, K.M., De Robertis, A., Macaulay, G.J., Handegard, N.O., 2014. Experimental evidence of threat-sensitive collective avoidance responses in a large wild-caught herring school. PLoS One 9, e86726. http://dx.doi.org/10.1371/journal.pone.0086726.

Roa, J., 2013. Role of GnRH neurons and their neuronal afferents as key integrators between food intake regulatory signals and the control of reproduction. Int. J. Endocrinol. 2013, 518046. http://dx.doi.org/10.1155/2013/518046.

Romero, A., Paulson, K.M., 2001. It's a wonderful hypogean life: a guide to the troglomorphic fishes of the world. Environ. Biol. Fishes 62, 13–41.

Rood, B.D., Calizo, L.H., Piel, D., Spangler, Z.P., Campbell, K., Beck, S.G., 2014. Dorsal raphe serotonin neurons in mice: immature hyperexcitability transitions to adult state during first three postnatal weeks suggesting sensitive period for environmental perturbation. J. Neurosci. 34, 4809–4821. http://dx.doi.org/10.1523/JNEUROSCI.1498-13.2014.

Ryer, C., Olla, B., 1998. Effect of light on juvenile walleye pollock shoaling and their interaction with predators. Mar. Ecol. Prog. Ser. 167, 215–226. http://dx.doi.org/10.3354/meps167215.

Sadoglu, P., 1956. A preliminary report on the genetics of the Mexican cave characins. Copeia 2, 113–114.

Salin, K., Voituron, Y., Mourin, J., Hervant, F., 2010. Cave colonization without fasting capacities: an example with the fish Astyanax fasciatus mexicanus. Comp. Biochem. Physiol. A Mol. Integr. Physiol. 156, 451–457. http://dx.doi.org/10.1016/j.cbpa.2010.03.030.

Sallinen, V., Sundvik, M., Reenilä, I., Peitsaro, N., Khrustalyov, D., Anichtchik, O., Toleikyte, G., Kaslin, J., Panula, P., 2009. Hyperserotonergic phenotype after monoamine oxidase inhibition in larval zebrafish. J. Neurochem. 109, 403–415. http://dx.doi.org/10.1111/j.1471-4159.2009.05986.x.

Saudou, F., Amara, D.A., Dierich, A., LeMeur, M., Ramboz, S., Segu, L., Buhot, M.C., Hen, R., 1994. Enhanced aggressive behavior in mice lacking 5-HT1B receptor. Science 265, 1875–1878.

Schemmel, C., 1967. Vergleichende Untersuchungen an den Hautsinnesorganen ober- und unterirdisch lebender Astyanax-Formen: Ein Beitrag zur Evolution der Cavernicolen. Z. Morphol. Tiere 61, 255–316. http://dx.doi.org/10.1007/BF00400988.

Schemmel, C., 1980. Studies on the genetics of feeding behaviour in the cave fish *Astyanax mexicanus* f. anoptichthys. Z. Tierpsychol. 53, 9–22.

Seebacher, F., Ward, A.J.W., Wilson, R.S., 2013. Increased aggression during pregnancy comes at a higher metabolic cost. J. Exp. Biol. 216, 771–776. http://dx.doi.org/10.1242/jeb.079756.

Senatori, O., Setini, A., Scirocco, A., Nicotra, A., 2009. Effect of short-time exposures to nickel and lead on brain monoamine oxidase from *Danio rerio* and *Poecilia reticulata*. Environ. Toxicol. 24, 309–313. http://dx.doi.org/10.1002/tox.20431.

Senkel, S., 1983. Zum Schwarmverhalten von Bastarden zwischen Fluss- und Höhlen populationen bei *Astyanax mexicanus* (Pisces, Characidae). University of Hamburg, Hamburg.

Setini, A., Pierucci, F., Senatori, O., Nicotra, A., 2005. Molecular characterization of monoamine oxidase in zebrafish (Danio rerio). Comp. Biochem. Physiol. B Biochem. Mol. Biol. 140, 153–161. http://dx.doi.org/10.1016/j.cbpc.2004.10.002.

Shahjahan, M., Ikegami, T., Osugi, T., Ukena, K., Doi, H., Hattori, A., Tsutsui, K., Ando, H., 2011. Synchronised expressions of LPXRFamide peptide and its receptor genes: seasonal, diurnal and circadian changes during spawning period in grass puffer: LPXRFa and its receptor gene expression in pufferfish. J. Neuroendocrinol. 23, 39–51. http://dx.doi.org/10.1111/j.1365-2826.2010.02081.x.

Smith, R.J.F., 1999. What good is smelly stuff in the skin? cross function and cross taxa effects in fish "alarm substances". In: Johnston, R.E., Müller-Schwarze, D., Sorensen, P.W. (Eds.), Advances in Chemical Signals in Vertebrates. Springer US, Boston, MA, pp. 475–487.

Speedie, N., Gerlai, R., 2008. Alarm substance induced behavioral responses in zebrafish (Danio rerio). Behav. Brain Res. 188, 168–177. http://dx.doi.org/10.1016/j.bbr.2007.10.031.

Strecker, U., Faúndez, V.H., Wilkens, H., 2004. Phylogeography of surface and cave Astyanax (Teleostei) from Central and North America based on cytochrome b sequence data. Mol. Phylogenet. Evol. 33, 469–481. http://dx.doi.org/10.1016/j.ympev.2004.07.001.

Teyke, T., 1990. Morphological differences in neuromasts of the blind cave fish Astyanax hubbsi and the sighted river fish *Astyanax mexicanus*. Brain Behav. Evol. 35, 23–30.

Thinès, G., Legrain, J.M., 1973. Effects of alarm substance on the behaviour of the cavefish Anoptichthys and Caecobarbus geertsi. Ann. Spéleol. 28, 291–297.

Tobias, J.A., Montgomerie, R., Lyon, B.E., 2012. The evolution of female ornaments and weaponry: social selection, sexual selection and ecological competition. Philos. Trans. R. Soc. Lond. B Biol. Sci. 367, 2274–2293. http://dx.doi.org/10.1098/rstb.2011.0280.

Todd, J.H., Atema, J., Bardach, J.E., 1967. Chemical communication in social behavior of a fish, the yellow bullhead (Ictalurus natalis). Science 158, 672–673. http://dx.doi.org/10.1126/science.158.3801.672.

Trajano, E., 1991. Agonistic behavior of Pimelodella kronei, a troglobitic catfish from southeastern Brazil (Siluriformes, Pimelodidae). Behav. Process. 23, 113–124.

White, S.A., Nguyen, T., Fernald, R.D., 2002. Social regulation of gonadotropin-releasing hormone. J. Exp. Biol. 205, 2567–2581.

Wilkens, H., 1972. Über Präadaptationen für das Höhlenleben, untersucht am Laichverhalten ober- und unterirdischer Populationen des *Astyanax mexicanus* (Pisces). Zool. Anz. 188, 1–11.

Wilkens, H., 1988. Evolution and genetics of epigean and cave Astyanax fasciatus (Characidae, Pisces). Support for the Neutral Mutation Theory. Evol. Biol. 23, 271–367.

Winberg, S., Øverli, Ø., Lepage, O., 2001. Suppression of aggression in rainbow trout (Oncorhynchus mykiss) by dietary L-tryptophan. J. Exp. Biol. 204, 3867–3876.

Wolkers, C.P.B., Serra, M., Szawka, R.E., Urbinati, E.C., 2014. The time course of aggressive behaviour in juvenile matrinxã *Brycon amazonicus* fed with dietary L-tryptophan supplementation: l-tryptophan reduces *Brycon amazonicus* aggression. J. Fish Biol. 84, 45–57. http://dx.doi.org/10.1111/jfb.12252.

Yaeger, C., Ros, A.M., Cross, V., Deangelis, R.S., Stobaugh, D.J., Rhodes, J.S., 2014. Blockade of arginine vasotocin signaling reduces aggressive behavior and c-Fos expression in the preoptic area and periventricular nucleus of the posterior tuberculum in male *Amphiprion ocellaris*. Neuroscience 267, 205–218. http://dx.doi.org/10.1016/j.neuroscience.2014.02.045.

Yamamoto, Y., Jeffery, W.R., 2000. Central role for the lens in cave fish eye degeneration. Science 289, 631–633. http://dx.doi.org/10.1126/science.289.5479.631.

Yamamoto, Y., Byerly, M.S., Jackman, W.R., Jeffery, W.R., 2009. Pleiotropic functions of embryonic sonic hedgehog expression link jaw and taste bud amplification with eye loss during cavefish evolution. Dev. Biol. 330, 200–211. http://dx.doi.org/10.1016/j.ydbio.2009.03.003.

Yoshizawa, M., Jeffery, W.R., 2011. Evolutionary tuning of an adaptive behavior requires enhancement of the neuromast sensory system. Commun. Integr. Biol. 4, 89.

Yoshizawa, M., Gorički, Š., Soares, D., Jeffery, W.R., 2010. Evolution of a behavioral shift mediated by superficial neuromasts helps cavefish find food in darkness. Curr. Biol. 20, 1631–1636. http://dx.doi.org/10.1016/j.cub.2010.07.017.

Chapter 18

Spatial Mapping in Perpetual Darkness: EvoDevo of Behavior in *Astyanax mexicanus* Cavefish

Ana Santacruz, Oscar M. Garcia, Maryana Tinoco-Cuellar, Emma Rangel-Huerta and Ernesto Maldonado

EvoDevo Lab, Unidad Académica de Sistemas Arrecifales, Instituto de Ciencias del Mar y Limnología, Universidad Nacional Autónoma de México, Puerto Morelos, Mexico

EVOLUTION AND DEVELOPMENT OF BEHAVIOR

A current focus of evolutionary developmental biology (EvoDevo) is animal form. Central to this topic is the idea that the shape of every animal is the product of two processes: development and evolution (Carroll, 2005). In other words, changes in development can create a diverse variety of forms, which natural selection then acts upon. Animal behavior is also a product of developmental and evolutionary forces; behavioral plasticity allows for adaptation to a changing environment, and it is possible that variation in behavior is as rich as variation in shape and form. In principle, a combination of analytical tools used to study "Developmental Biology" as molecular and cellular biology or imaging, "Behavior" like conditional learning and maze experiments and "Evolution" such as the construction of phylogenetic trees of genes related to behavior, could contribute greatly to the field of EvoDevo.

Astyanax mexicanus (Mexican Tetra) is a model organism with two morphotypes of the same species: fish that live underground in cave ponds, and fish that inhabit surface rivers. There are obvious morphological and physiological differences between these morphotypes. Whereas the surface type has large eyes and is silver with a black band extending along each side to the tail, cavefish have neither eyes nor pigmentation (Jeffery, 2009; Gross, 2012). The origin of these morphological differences has been the subject of numerous studies (Borowsky, 2008; Rétaux et al., 2008; Jeffery, 2009). Additionally, there is growing interest in differences in behaviors—including schooling (Kowalko et al., 2013a,b), foraging/feeding (Yoshizawa et al., 2010; Bibliowicz et al., 2013; Kowalko et al., 2013a,b), social behavior and aggressiveness (Rétaux and Elipot, 2013; Elipot et al., 2014), and sleep behavior (Duboué et al., 2011)—between surface fish and cavefish.

Biology and Evolution of the Mexican Cavefish. http://dx.doi.org/10.1016/B978-0-12-802148-4.00018-9
© 2016 Elsevier Inc. All rights reserved.

Here, we discuss how the sensory systems and brain of *A. mexicanus* cavefish interpret its environment in the absence of a visual system. We focus on how the cavefish brain creates a spatial map of its surroundings, how other senses are utilized in the absence of sight, and how findings pertaining to the cavefish can be extrapolated to understand aspects of spatial map formation in blind people.

We begin with a summary of how spatial maps are formed using the visual system. Then, we comment on several experiments investigating how *Astyanax* cavefish navigate their habitat, and how observations from these behavioral assays have shaped our current model of spatial mapping in perpetual darkness.

BUILDING SPATIAL MAPS FROM THE VISUAL SENSORY SYSTEM

A puzzling question in behavioral neurobiology is how the brain makes sense of spaces. When you climb a stairwell for the first time or move around corners, which neurons are responsible for the integration of sensory information and the mapping of space? In addition, at what point in development do you learn to recognize the landmarks that help you successfully navigate a given space? When an animal encounters a previously experienced environment, how is orientation memory retrieved and used to recognize the correct path? Questions like this could be answered in organisms that have experienced a radical change in their environment. A particularly interesting model is the *Astyanax* cavefish that evolved to process information from an environment of perpetual darkness. Although the main theme of this chapter is spatial mapping in the absence of vision, we first present an overview of how spatial representations of the world that correspond to our rich visual experience are created. This question has been studied in several organisms, including mammals and fishes, by observing how the organism, often guided by visual cues, navigates a maze.

Mammals

In mice and rats, neurons called "place cells" are only active when the animal is at a specific location. When the animal moves, other place cells are activated. These differential activity patterns are believed to be the basis by which spatial contexts are distinguished. Place cells were identified in 1971 by O'Keefe and Dostrovsky, who recorded neuronal activity in the hippocampus to study spatial navigation. The authors found that approximately 6% of the neurons from which they were able to record only responded when the rat was situated in a particular part of the testing platform and facing in a particular direction (O'Keefe and Dostrovsky, 1971). It is now clear that these place cells in the hippocampus provide the rest of the brain with a spatial reference map. This process has been confirmed under different conditions, including a 48-meter maze with visual cues along the track (Rich et al., 2014). In addition to place cells, there are "head cells," which process information about the current heading of the animal as determined by exteroceptive (external visual, olfactory, and auditory sensory landmarks) and interoceptive cues (internal self-motion signals, vestibular signals, etc.) (Taube, 2007). Furthermore, "grid cells" encode the distance traveled by the animal and provide an internal representation of the speed and direction of movement. Grid cells are divided into coherent units arranged in functional modules, and each module is composed of a set of grid cells that share the same wavelength and orientation (Sargolini et al., 2006). Spatially tuned neurons have been described for other mammalian organisms, such as primates (Georges-Francois et al., 1999) and bats (Yartsev et al., 2011) as well as for nonmammalian vertebrates, such as birds (Siegel et al., 2006) and fish (Canfield and Mizumori, 2004).

Zebrafish

Zebrafish (*Danio rerio*) has been the model of choice for the study of the cellular and molecular underpinnings of visually guided behavior. Zebrafish are not only amenable to behavioral assay, but also a genetically tractable model organism for the study of development. Zebrafish have excellent tetrachromatic vision, vestibular sensation, and chemosensation, and can easily detect vibratory stimuli with both auditory and lateral-line neuromasts (Blaser and Vira, 2014). Studies have been carried out in both larvae and adults of this cyprinid fish to investigate its optomotor, vestibule-ocular and optokinetic reflexes; scototaxis (dark/light preference); circadian rhythmicity; thigmotaxis (exploratory wall-hugging); prey capture; stress; depth preference; startle and escape responses; and social behaviors, such as shoaling (Maximino et al., 2010; Bianco et al., 2011; Ahrens et al., 2012; Schnorr et al., 2012; Cameron et al., 2013; Blaser and Vira, 2014; Chen et al., 2014; Gerlai, 2014; Giacomini et al., 2014; Moore and Whitmore, 2014).

Many sophisticated methods have been actively developed to enable the neuroimaging of zebrafish larvae during noninvasive assays of visually guided behavior (Feierstein et al., 2014), this work could be used as reference for nonvisual behavior research in *Astyanax* cavefish. Because mutagenesis and transgenesis are feasible in zebrafish, the intersection of these approaches has facilitated a unique set of experiments in this model organism. Akira Muto at the Kawakami Laboratory in Japan, was able to record the activity of single neurons in the neuropil of the tectum (where the dendrites of tectal neurons form synapses with the axons of retinal afferents) while the fish were hunting for a free-moving paramecium (a natural prey of zebrafish larvae) (Muto et al., 2013). Because Muto observed that the anterior tectum of the zebrafish was more active immediately before prey capture than during approach, he proposed that the anterior tectum links the visual and motor pathways during prey capture. This work utilized a transgenic zebrafish expressing a very sensitive GFP-modified calcium sensor protein that allowed for the visualization of the neural activity underlying the organism's perception of its world (Muto et al., 2013).

Fish respond to whole-field visual motion with compensatory eye and body movements to stabilize their gaze and position with respect to their surroundings. Therefore, rotational and translational stimuli and horizontal and vertical stimuli must be distinguished. Kubo et al. (2014) used a combination of optogenetic activation and silencing (optogenetics is a transgenic method allowing excitable cells to be switched on or off by a light beam) and *in vivo* calcium imaging during optic flow stimulation to describe four tectum clusters that process horizontal visually guided whole-field motion. For example, they found that a specific arborization field in the tectum (AF9) is necessary and sufficient to drive smooth eye movements (Kubo et al., 2014).

We believe the use of powerful imaging techniques such as light-field and light-sheet microscopy, whether the larvae is in a fixed position (Prevedel et al., 2014; Vladimirov et al., 2014), or in motion (Muto et al., 2013), will lead to important advances in our understanding of spatial mapping in zebrafish. These same methods could be seamlessly applied to *Astyanax* cavefish larvae as well, because of their transparent body and similar size to the zebrafish larvae.

EXPERIMENTS IN *ASTYANAX MEXICANUS* CAVEFISH NAVIGATION

Experiments have been designed to elucidate how cavefish navigate without vision and to determine which other senses are used during swimming to acquire information about the surroundings (Figure 18.1). *Astyanax* cavefish repeatedly pass by novel objects at a short distance, thereby detecting nearby objects by sensing distortions in the flow field of water and avoiding obstacles without touching them (von Campenhausen et al., 1981; Hassan, 1989; Windsor et al., 2011). This flow field is produced by the cavefish's own swimming movements and is distorted by stationary objects within range. Such distortions are then detected by the lateral-line system, which is formed by superficial neuromasts that are embedded in subdermal canals. Since the layer that surrounds the fish has damping properties, it will attenuate or increase the amplitude of the hydrodynamic stimulus, which cavefish might use to generate a form of hydrodynamic imaging and to build a spatial map of their surroundings (Windsor et al., 2008; Sutherland et al., 2009).

For this detection system to work, the cavefish must maintain constant motion. When an *Astyanax* cavefish is placed in a new tank, there is an increase in swimming velocity during the habituation period (Teyke, 1988; Burt de Perera, 2004a,b). According to a computer-generated model, increasing increments in swimming speed would result in increases in the amplitude of the self-generated flow signal (Hassan, 1993). Furthermore, faster swimming would aid in encountering novel features for the purpose of gaining spatial knowledge. During the habituation phase, cavefish swim at the periphery of their enclosures, and the number of contacts with objects in the environment decreases over time (Teyke, 1985). This wall-following behavior, or thigmotaxis, is an exploratory

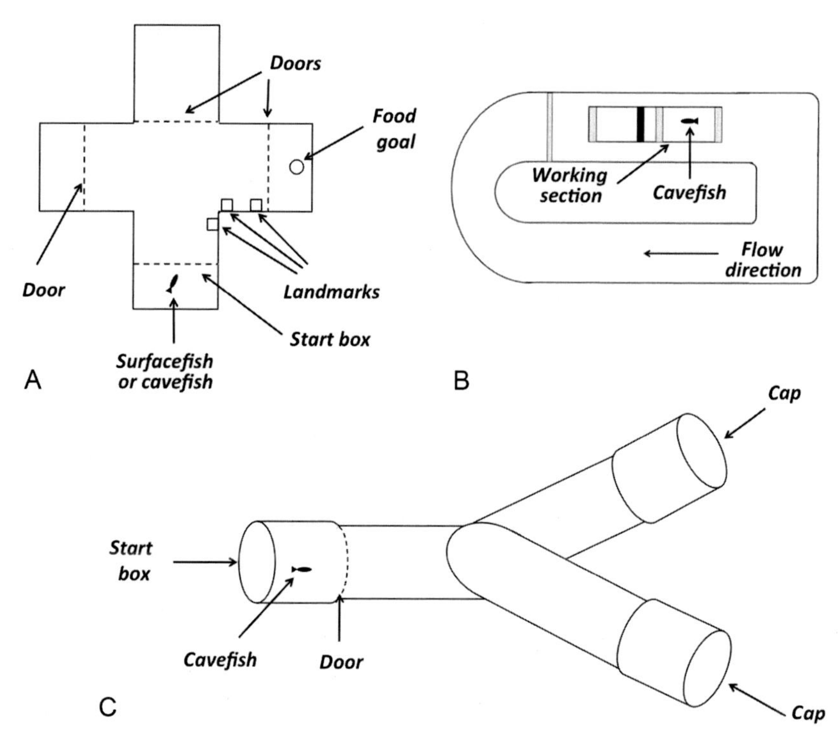

FIGURE 18.1 Maze designs that have been used by different authors to learn about spatial mapping in the *Astyanax* surface fish and cavefish. (A) A T-maze in which, both surface fish or cavefish *Astyanax* initiate in a start box and then could choose to turn either left or right to find a food goal along the route landmarks could be used as a reference. After training, the landmarks were moved to a different place and while *Astyanax* surface fish increase the latency and the distance traveled to reach the food goal, the cavefish *Astyanax* only increased the traveled distance. From these experiments, it was concluded that different mechanisms for orientation are used by both morphs of *Astyanax* (Sutherland et al., 2009). (B) Recirculating flow chamber used to study if the lateral line is required for rheotaxis, *Astyanax* cavefish were placed in the working section so they could swim against the flow direction at different flow velocities. The authors observed that gentamicin-treated cavefish (which loss the lateral line) performed equally to untreated fish during rheotaxis experiments (Van Trump and McHenry, 2013). (C) Three-dimensional Y-maze that can be rotated around its axis. *Astyanax* surface fish was tested in this maze comparing its ability to form spatial maps using landmarks along the horizontal and vertical axes to determine if the learned axes (horizontal and vertical) were in conflict. As a result, the fish showed a strong preference to use the vertical axis (Holbrook and Burt de Perera, 2009).

behavior that has been observed in several organisms (Besson and Martin, 2005; Yaski et al., 2009), including *Astyanax* cavefish and surface fish (Sharma et al., 2009). It has been demonstrated that different wall shapes (concave or convex) and different arena (tank) sizes do not affect *Astyanax* wall-following behavior, as cavefish continued to follow the wall at frequencies significantly above chance levels (Patton et al., 2010).

The increases in swimming velocity and wall-hugging behavior both seem to enhance information-gathering in novel environments. During exploration, cavefish alternate burst and coast movements using coordinated body bends and beats of the caudal fin to accelerate during the burst phase and then passively glide as the drag slows them prior to generating another burst (Windsor et al., 2008). Surface *Astyanax* fish also exhibit these swimming cycle sequences (Tan et al., 2011). Coombs and collaborators reasoned that flow sensing would be difficult to achieve during the burst phase, because body undulations generate turbulent flow and shed vortices in the wake of the fish. Instead, flow sensing might take place during the coasting phase, when the body is held straight and the flow field is more stable (Figure 18.2) (Tan et al., 2011). This model implies that cavefish might modulate the duration of the coasting phase for sensing purposes.

Active nonvisual sensing strategies, such as echolocation and electrolocation, are utilized by other organisms, including whales, bats, and electric fishes. Because locomotion occurs during signal production, active flow sensing must differ in cavefish. It is therefore difficult to imagine that locomotion is modulated for adaptive control of the outgoing signal (Tan et al., 2011). The lateral-line system must be present for flow sensing in *Astyanax*, since removal of neuromasts by cobalt/calcium treatment increases collision rates (Windsor et al., 2008). In addition, cranial neuromasts are more numerous in *Astyanax* cavefish than in surface fish, and the cupula covering each neuromast is larger (300 μm) and more sensitive in cavefish than in sighted fish (40 μm) (Teyke, 1990; Yoshizawa et al., 2014). Besides, cranial neuromasts are specially important for vibration attraction behavior (VAB) (Yoshizawa et al., 2010), which is the ability of the cavefish to swim toward the source of a water disturbance, which could potentially be food. This relation is a particularly interesting link between a constructive morphological trait (i.e., more numerous and larger neuromasts) and the evolution of a behavioral change (VAB).

Most of the experiments described thus far have been carried out in *Astyanax* fish moving horizontally in water that is 5 cm deep, thus ensuring that the fish will remain in focus during filming. In consequence, the vertical component of the space has been largely overlooked. It is important to acknowledge this problem, as fish, in their natural habitats, move within multiple spatial dimensions as they pitch and roll to the left or right and up or down while moving forward or backward. Because of the buoyancy of the swim bladder, vertical and horizontal movements have a very similar energetic cost (Holbrook and Burt de Perera, 2013). Burt de Perera and coworkers reported that if visual cues are not present, *Astyanax* surface fish encode spatial information from both horizontal and vertical axes with similar resolution and use information from both axes with similar accuracy (Holbrook and Burt de Perera, 2009, 2013); however, in the presence of both horizontal and vertical visual landmarks, fish prefer the previously learned vertical component, suggesting that in sighted fish (Figure 18.1), the vertical axis is learned and robustly remembered (Holbrook and Burt de

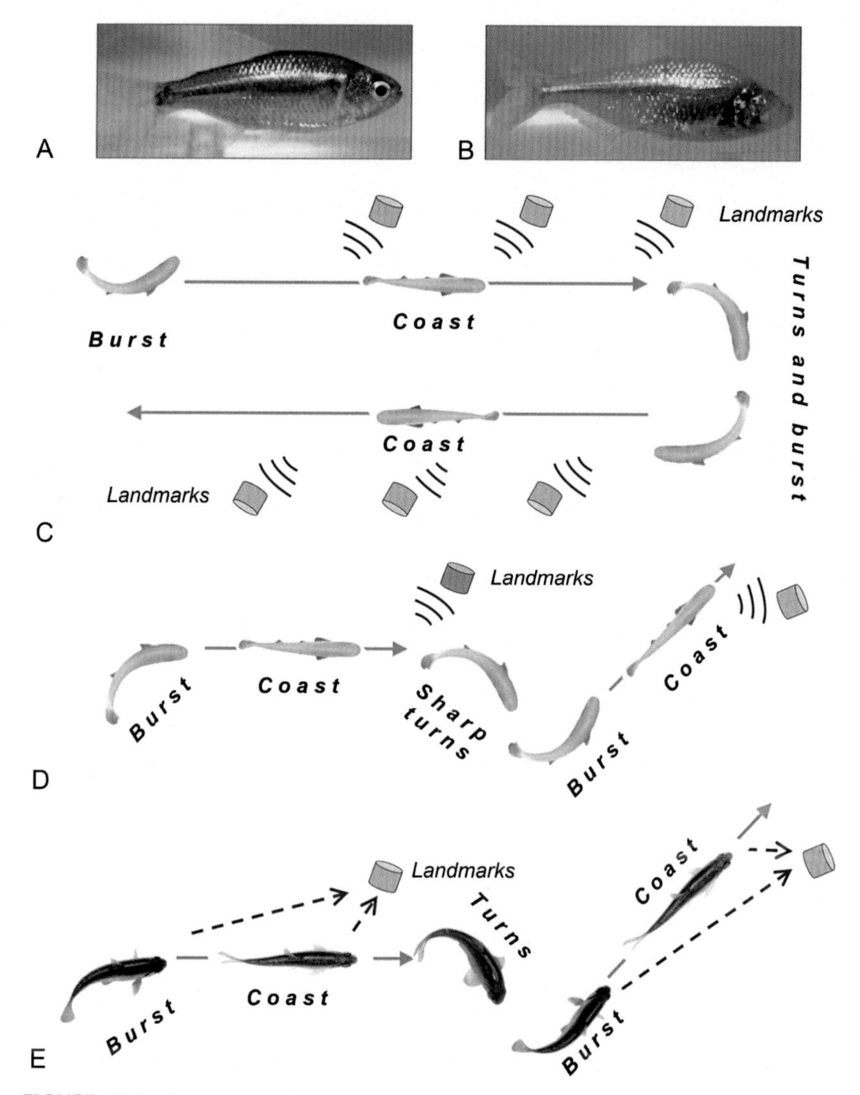

FIGURE 18.2 *Astyanax mexicanus* space mapping behavior. (A) *A. mexicanus* surface river sighted morphotype and (B) *A. mexicanus* cavefish blind morphotype. Both specimens were captured in the wild. (C) When *Astyanax* cavefish is placed in a new environment, there is an increase in swimming velocity, and then it moves in a straight line (orange arrow) devoid of sharp turns, alternating short burst with long coast movements; during the coast phase, flow sensing takes places, which is when the cavefish learns sequentially environmental landmarks. (D) When in a familiar environment, *Astyanax* cavefish swimming cycles are different; the length of the burst and coast phases are both short, and it is most common to observe sharp turns. (E) *Astyanax* surface fish in a new environment do not have long coast movements during swim cycles, since the fish can determine the spatial relationship guided by visual cues from any vantage point (arrow dotted blue lines). Swimming cycles become less frequent as the surface fish gets familiar with its new environment.

Perera, 2011, 2013). This preference is also true of benthic fish, such as the "Bronze corydoras" (*Corydoras aeneus*), which spend much of their time moving horizontally over the substrate (Davis et al., 2014). At present, there is no clear information regarding a preference in *Astyanax* cavefish for the horizontal versus the vertical axis.

It is clear, however, that other senses are enhanced in blind *Astyanax* cavefish; they have more tastebuds (Varatharasan et al., 2009), better tuned mechanosensation (more neuromasts and higher sensitive lateral line) (Yoshizawa et al., 2014), larger nostril openings and more acute chemosensation (in particular, olfaction) than surface fish (Bibliowicz et al., 2013). Surprisingly, hearing is equivalent between *Astyanax* cavefish and surface fish (Popper, 1970) and has even been reported to be poorer in some cavefishes of the *Amblyopsidae* family (Niemiller et al., 2013); however, some sensory systems, such as somatosensation (Wang et al., 2013) (which is carried out by a variety of skin sensory neurons that detect different types of thermal, chemical, and mechanical touch stimuli) and proprioception (the perception of body position), have not yet been fully tested. Proprioception has been studied in the bluegill sunfish (*Lepornis macrochirus*), in which afferent nerve fibers responding specifically to fin ray bending were detected (Williams et al., 2013). Any model of spatial cognition in the absence of vision must consider the role of multisensory integration over all senses, and because early experience determines how senses interact (Wallace and Stein, 2007), behavioral experiments performed during blind cavefish development will be essential.

NAVIGATING AND CREATING SPATIAL MAPS IN THE COMPLETE ABSENCE OF VISION

In 1985, after systematically tracking and observing blind *Astyanax* cavefish swimming around a new tank, Thomas Teyke formulated a series of interesting questions (Teyke, 1985), some of which remain unanswered today. When a familiar space is altered, is new exploration required? For how long is a spatial map stored? Does the creation of a new spatial map replace older space maps? Is there a correlation between the time required for exploration and space configuration?

The experiments described above have shown that active flow sensing, using locomotion and the lateral line, is utilized during cavefish navigation. One particularly keen observation stands out: *Astyanax* cavefish (blind) exhibit a higher frequency of swim cycles (burst and coast) in straight trajectories than *Astyanax* surface fish (sighted), during the initial exploration of a novel environment (Figure 18.2). Coombs and collaborators have proposed that sighted fish can determine the spatial relationship of distant landmark features while swimming from a single vantage point, whereas blind cavefish must reconstruct spatial relationships from sequential encounters with different features at close range (Tan et al., 2011). This difference suggests that three-dimensional space is learned and remembered differently in *Astyanax* surface fish and cavefish. Because *Astyanax* cavefish evolved from an ancestral form of the *Astyanax* surface fish (Gross, 2012), modifications to the sensory, processing, and motor neural systems of the cavefish must have occurred such that it is completely equipped for navigation in perpetual darkness.

The spatial cognitive map is anisotropic, and there are clear differences in visually guided spatial mapping among animals that mainly navigate horizontally (such as humans and rats) and organisms that navigate horizontally and vertically (such as bats, birds, and fishes). As noted above, in mammals, place cells encode focal position, and grid cells encode distance. Interestingly, place cells respond less accurately, and grid cells do not respond at all in rats during vertical navigation (climbing a wall) (Hayman et al., 2011); however, grid cells in the Egyptian fruit bat, *Rousettus aegyptiacus*, fire with equal spatial resolution in the horizontal and vertical dimensions (Yartsev and Ulanovsky, 2013). Therefore, when designing behavioral experiments for *Astyanax* fishes, caution must be taken when considering information garnered from rat studies regarding the neural encoding of position. Zebrafish behavioral studies are wide-ranging and sophisticated (Huntingford, 2012; Kalueff et al., 2013; Feierstein et al., 2014); in particular, studies mapping neural circuits during fictive navigation in transgenic fluorescent zebrafish larvae address fundamental questions concerning the neural circuit basis of behavior. In zebrafish, neurons that show place cell-like features have only recently been identified in the forebrain (Ahrens et al., 2013), and a grid cell equivalent has yet to be discovered. The recent whole-genome sequencing of the *Astyanax* cavefish (McGaugh et al., 2014) will certainly launch behavioral studies at the genetic and cellular levels that may yield information, through screening the mutations among the genes, which are known to be involved in neural development and/or wiring. Some of these mutations could be related to the differences between cavefish and surface fish neural circuits used in navigation, and may be specifically associated with its ability to form spatial maps.

Once a sighted or blind *Astyanax* has explored and learned a new space, an internal memory model of the surroundings is created. In humans, this working representation of a space has been named "egocentric memory" (Land, 2014). In humans, the motor cortex seems to be organized in such a way that the neurons responsible for movements are tuned to particular directions. There are two important sensory systems required for the formation of egocentric memories: the visual and the vestibular systems. In other words, visualizing a landmark and simultaneously registering the position of the cue with respect to the body is essential for spatial mapping. There is strong evidence that visual and egocentric memory representations overlap throughout the visual field, and that both can be used to shift body position (Land, 2014). It is clear that *Astyanax* surface fish, during exploration, can learn visual cues that can later be retrieved as a representation of directions that maps onto actions or movements (Holbrook and Burt de Perera, 2011, 2013; Tan et al., 2011).

Zebrafish, medaka (*Oryzias latipes*), and goldfish (*Carassius auratus*) all have vestibulo-ocular reflexes (Beck et al., 2004). An interesting unanswered question is how the *Astyanax* cavefish relates its body position to the perception of its surroundings. Studies of the vestibular and proprioceptive systems of *Astyanax* cavefish and how these systems are integrated with spatial mapping might shed light upon how this fish has successfully adapted to perpetual darkness.

HUMAN SENSORY DEPRIVATION AND SPACE MAPPING

One-third of the cortical surface in primates is involved with visual function. When vision is lost at birth or later in life, plastic rearrangements take place in the brain, the so-called "compensatory plasticity" that affects the auditory, tactile, olfactory, gustatory, and thermal sensory systems (Kupers and Ptito, 2014). Blind persons show a reduction in the volume of the posterior part of the hippocampus, an area important in spatial navigation (Chebat et al., 2007), as well as an increase in the thickness of certain cortical occipital areas that are involved in vision in sighted individuals (Qin et al., 2013).

Of these rearrangements, enhanced auditory perception has been most thoroughly tested in blind individuals, who show an increased sensitivity in absolute pitch discrimination tasks matched only by sighted subjects with musical training. In particular, auditory spatial localization is enhanced in blind individuals (Kupers and Ptito, 2014). The comparison of brain activity by functional magnetic resonance imaging (fMRI) in blind and sighted persons during auditory stimulation has shown the recruitment of typically visual occipital areas (i.e., striate and extrastriate) for general auditory processing in blind subjects (Gougoux et al., 2009); however, whether this cortical reorganization only occurs in early-onset and not late-onset blind individuals is controversial (Voss et al., 2004; Wan et al., 2010). Cross-modal activation of the visual cortex has also been observed during different tactile tasks in visually impaired persons (Ptito et al., 2012; Kupers and Ptito, 2014). Just five days of eyesight deprivation (blindfolding) is sufficient to stimulate colonization of the primary visual cortex by touch (Merabet et al., 2008). Similarly, a larger olfactory bulb volume is correlated with higher olfactory performance in blind individuals (Rombaux et al., 2010). Some individuals, like many animals, are able to use reflected sound waves (echolocation) to perceive close objects or obstacles. Studies comparing echolocation abilities in blind people and sighted individuals show that blind subjects perform better than sighted controls (Kupers and Ptito, 2014). When two blind subjects—one with early-onset and one with late-onset blindness—were studied by fMRI, echolocation was found to activate occipital and nonauditory cortical areas more strongly in the early-blind individual (Thaler et al., 2011).

ASTYANAX CAVEFISH AND EYESIGHT LOSS IN HUMANS

Can we use what we know about the adaptations of *Astyanax* cavefish to perpetual darkness to better understand sensory substitution and compensatory plasticity in blind human individuals?

The basic ionic mechanisms of the action potential were first described in classical experiments on the giant axon of the squid (Hodgkin and Huxley, 1952), and the molecular mechanisms of learning and memory were first elucidated in the sea slug, *Aplysia* (Kandel, 2001). It has recently been proposed that some properties of visually guided neuromotor integration are shared between zebrafish and monkeys (Joshua and Lisberger, 2014). In fact, in both species, integrator neurons show a wide diversity of time-varying responses. Mati Joshua and collaborators generated a computer model combining data from both species; this network model integrates and reproduces the time-varying firing rates of 18 neurons in the oculomotor neural integrator of monkeys and zebrafish. Previously, Joshua and colleagues had regarded the variety of response properties observed in monkey brainstem neurons as a curiosity, then they found a zebrafish integrator circuit model characterized by multiple dimensions of slow firing-rate dynamics (Miri et al., 2011). Application of the fish model to monkey data revealed that the noise became understandable a signal; in other words, the zebrafish-derived model explained the data acquired in monkeys surprisingly well, suggesting that this method of neural integration may be common across a wide range of species (Joshua and Lisberger, 2014).

Astyanax cavefish is a model organism uniquely positioned to elucidate the neural basis of perceptual learning. It is a unique case in which multisensory substitution has naturally evolved to compensate completely for the loss of sight. The space-recognition mechanisms in *Astyanax* blind cavefish must be analyzed in light of the wealth of experimental data regarding multisensory substitution in blind human subjects (Kupers and Ptito, 2014); therefore, at this moment, conditions are ideal for *Astyanax* to provide insights into the neural encoding of position in the absence of vision. In light of the recent publication of the *Astyanax* cavefish whole-genome sequence (McGaugh et al., 2014), it is very likely that neural circuit *in vivo* imaging will be performed in the near future in *Astyanax* cavefish. This is challenging, but feasible, since we have learned from zebrafish that it is possible to study fish embryos with an unprecedented level of precision, observing single neurons and molecular events, while tracing neural circuits during fictive navigation (Feierstein et al., 2014).

ACKNOWLEDGMENTS

Research in the Ernesto Maldonado Laboratory was funded by a grant from PAPPIT-UNAM IN208512 and by CONACYT grants 166046 and 170814, "Cavefish Evolution Development and Behavior. *Astyanax* International Meeting (Evolución, Desarrollo y Comportamiento de Peces de Cueva. Congreso Internacional de *Astyanax*)."

REFERENCES

Ahrens, M.B., Huang, K.H., Narayan, S., Mensh, B.D., Engert, F., 2013. Two-photon calcium imaging during fictive navigation in virtual environments. Front. Neural Circuits 7, 104.

Ahrens, M.B., Li, J.M., Orger, M.B., Robson, D.N., Schier, A.F., Engert, F., Portugues, R., 2012. Brain-wide neuronal dynamics during motor adaptation in zebrafish. Nature 485 (7399), 471–477.

Beck, J.C., Gilland, E., Tank, D.W., Baker, R., 2004. Quantifying the ontogeny of optokinetic and vestibuloocular behaviors in zebrafish, medaka, and goldfish. J. Neurophysiol. 92 (6), 3546–3561.

Besson, M., Martin, J.R., 2005. Centrophobism/thigmotaxis, a new role for the mushroom bodies in Drosophila. J. Neurobiol. 62 (3), 386–396.

Bianco, I.H., Kampff, A.R., Engert, F., 2011. Prey capture behavior evoked by simple visual stimuli in larval zebrafish. Front. Syst. Neurosci. 5, 101.

Bibliowicz, J., Alie, A., Espinasa, L., Yoshizawa, M., Blin, M., Hinaux, H., Legendre, L., Pere, S., Retaux, S., 2013. Differences in chemosensory response between eyed and eyeless *Astyanax mexicanus* of the Rio Subterraneo cave. EvoDevo 4 (1), 25.

Blaser, R.E., Vira, D.G., 2014. Experiments on learning in zebrafish (*Danio rerio*): a promising model of neurocognitive function. Neurosci. Biobehav. Rev. 42, 224–231.

Borowsky, R., 2008. *Astyanax mexicanus*, the blind Mexican cave fish: a model for studies in development and morphology. CSH Protoc. 2008. pdb.emo107.

Burt de Perera, T., 2004a. Fish can encode order in their spatial map. Proc. Biol. Sci. 271 (1553), 2131–2134.

Burt de Perera, T., 2004b. Spatial parameters encoded in the spatial map of the blind Mexican cave fish, *Astyanax fasciatus*. Anim. Behav. 68 (2), 291–295.

Cameron, D.J., Rassamdana, F., Tam, P., Dang, K., Yanez, C., Ghaemmaghami, S., Dehkordi, M.I., 2013. The optokinetic response as a quantitative measure of visual acuity in zebrafish. J. Vis. Exp. (80), e50832.

Canfield, J.G., Mizumori, S.J., 2004. Methods for chronic neural recording in the telencephalon of freely behaving fish. J. Neurosci. Methods 133 (1-2), 127–134.

Carroll, S.B., 2005. Endless Forms Most Beautiful: The New Science of EvoDevo and the Making of the Animal Kingdom. Norton, New York.

Chebat, D.R., Chen, J.K., Schneider, F., Ptito, A., Kupers, R., Ptito, M., 2007. Alterations in right posterior hippocampus in early blind individuals. Neuroreport 18 (4), 329–333.

Chen, C.C., Bockisch, C.J., Bertolini, G., Olasagasti, I., Neuhauss, S.C., Weber, K.P., Straumann, D., Ying-Yu Huang, M., 2014. Velocity storage mechanism in zebrafish larvae. J. Physiol. 592 (Pt 1), 203–214.

Davis, V.A., Holbrook, R.I., Schumacher, S., Guilford, T., Perera, T.B., 2014. Three-dimensional spatial cognition in a benthic fish, *Corydoras aeneus*. Behav. Process. 109 (Pt B), 151–156.

Duboué, E.R., Keene, A.C., Borowsky, R.L., 2011. Evolutionary convergence on sleep loss in cavefish populations. Curr. Biol. 21 (8), 671–676.

Elipot, Y., Hinaux, H., Callebert, J., Launay, J.M., Blin, M., Retaux, S., 2014. A mutation in the enzyme monoamine oxidase explains part of the *Astyanax* cavefish behavioural syndrome. Nat. Commun. 5, 3647.

Feierstein, C.E., Portugues, R., Orger, M.B., 2014. Seeing the whole picture: a comprehensive imaging approach to functional mapping of circuits in behaving zebrafish. Neuroscience 296, 26–38.

Georges-Francois, P., Rolls, E.T., Robertson, R.G., 1999. Spatial view cells in the primate hippocampus: allocentric view not head direction or eye position or place. Cereb. Cortex 9 (3), 197–212.

Gerlai, R., 2014. Social behavior of zebrafish: from synthetic images to biological mechanisms of shoaling. J. Neurosci. Methods 234, 59–65.

Giacomini, A.C., de Abreu, M.S., Koakoski, G., Idalencio, R., Kalichak, F., Oliveira, T.A., da Rosa, J.G., Gusso, D., Piato, A.L., Barcellos, L.J., 2014. My stress, our stress: blunted cortisol response to stress in isolated housed zebrafish. Physiol. Behav. 139C, 182–187.

Gougoux, F., Belin, P., Voss, P., Lepore, F., Lassonde, M., Zatorre, R.J., 2009. Voice perception in blind persons: a functional magnetic resonance imaging study. Neuropsychologia 47 (13), 2967–2974.

Gross, J.B., 2012. The complex origin of *Astyanax* cavefish. BMC Evol. Biol. 12, 105.

Hassan, E.-S., 1989. Hydrodynamic imaging of the surroundings by the lateral line of the blind cave fish *Anoptichthys jordani*. In: Coombs, S., Gomer, P., Münz, H. (Eds.), The Mechanosensory Lateral Line: Neurobiology and Evolution. Springer-Verlag, New York, pp. 217–228.

Hassan, E.-S., 1993. Mathematical description of the stimuli to the lateral line system of fish, derived from a three-dimensional flow field analysis. III. The case of an oscillating sphere near the fish. Biol. Cybern. 69 (5-6), 525–538.

Hayman, R., Verriotis, M.A., Jovalekic, A., Fenton, A.A., Jeffery, K.J., 2011. Anisotropic encoding of three-dimensional space by place cells and grid cells. Nat. Neurosci. 14 (9), 1182–1188.

Hodgkin, A.L., Huxley, A.F., 1952. The components of membrane conductance in the giant axon of Loligo. J. Physiol. 116 (4), 473–496.

Holbrook, R.I., Burt de Perera, T., 2011. Three-dimensional spatial cognition: information in the vertical dimension overrides information from the horizontal. Anim. Cogn. 14 (4), 613–619.

Holbrook, R.I., Burt de Perera, T., 2009. Separate encoding of vertical and horizontal components of space during orientation in fish. Anim. Behav. 78, 241–245.

Holbrook, R.I., Burt de Perera, T., 2013. Three-dimensional spatial cognition: freely swimming fish accurately learn and remember metric information in a volume. Anim. Behav. 86 (5), 1077–1083.

Huntingford, F.A., 2012. The physiology of fish behaviour: a selective review of developments over the past 40 years (section sign). J. Fish Biol. 81 (7), 2103–2126.

Jeffery, W.R., 2009. Regressive evolution in *Astyanax* cavefish. Annu. Rev. Genet. 43, 25–47.

Joshua, M., Lisberger, S.G., 2014. A tale of two species: neural integration in zebrafish and monkeys. Neuroscience 296, 80–91.

Kalueff, A.V., Gebhardt, M., Stewart, A.M., Cachat, J.M., Brimmer, M., Chawla, J.S., Craddock, C., Kyzar, E.J., Roth, A., Landsman, S., Gaikwad, S., Robinson, K., Baatrup, E., Tierney, K., Shamchuk, A., Norton, W., Miller, N., Nicolson, T., Braubach, O., Gilman, C.P., Pittman, J., Rosemberg, D.B., Gerlai, R., Echevarria, D., Lamb, E., Neuhauss, S.C., Weng, W., Bally-Cuif, L., Schneider, H., Zebrafish Neuroscience Research Consortium, 2013. Towards a comprehensive catalog of zebrafish behavior 1.0 and beyond. Zebrafish 10 (1), 70–86.

Kandel, E.R., 2001. The molecular biology of memory storage: a dialogue between genes and synapses. Science 294 (5544), 1030–1038.

Kowalko, J.E., Rohner, N., Linden, T.A., Rompani, S.B., Warren, W.C., Borowsky, R., Tabin, C.J., Jeffery, W.R., Yoshizawa, M., 2013a. Convergence in feeding posture occurs through different genetic loci in independently evolved cave populations of *Astyanax mexicanus*. Proc. Natl. Acad. Sci. U. S. A. 110 (42), 16933–16938.

Kowalko, J.E., Rohner, N., Rompani, S.B., Peterson, B.K., Linden, T.A., Yoshizawa, M., Kay, E.H., Weber, J., Hoekstra, H.E., Jeffery, W.R., Borowsky, R., Tabin, C.J., 2013b. Loss of schooling behavior in cavefish through sight-dependent and sight-independent mechanisms. Curr. Biol. 23 (19), 1874–1883.

Kubo, F., Hablitzel, B., Dal Maschio, M., Driever, W., Baier, H., Arrenberg, A.B., 2014. Functional architecture of an optic flow-responsive area that drives horizontal eye movements in zebrafish. Neuron 81 (6), 1344–1359.

Kupers, R., Ptito, M., 2014. Compensatory plasticity and cross-modal reorganization following early visual deprivation. Neurosci. Biobehav. Rev. 41, 36–52.

Land, M.F., 2014. Do we have an internal model of the outside world? Philos. Trans. R. Soc. Lond. B Biol. Sci. 369 (1636), 20130045.

Maximino, C., Marques de Brito, T., Dias, C.A., Gouveia Jr., A., Morato, S., 2010. Scototaxis as anxiety-like behavior in fish. Nat. Protoc. 5 (2), 209–216.

McGaugh, S.E., Gross, J.B., Aken, B., Blin, M., Borowsky, R., Chalopin, D., Hinaux, H., Jeffery, W.R., Keene, A., Ma, L., Minx, P., Murphy, D., O'Quin, K.E., Retaux, S., Rohner, N., Searle, S.M., Stahl, B.A., Tabin, C., Volff, J.N., Yoshizawa, M., Warren, W.C., 2014. The cavefish genome reveals candidate genes for eye loss. Nat. Commun. 5, 5307.

Merabet, L.B., Hamilton, R., Schlaug, G., Swisher, J.D., Kiriakopoulos, E.T., Pitskel, N.B., Kauffman, T., Pascual-Leone, A., 2008. Rapid and reversible recruitment of early visual cortex for touch. PLoS ONE 3 (8), e3046.

Miri, A., Daie, K., Arrenberg, A.B., Baier, H., Aksay, E., Tank, D.W., 2011. Spatial gradients and multidimensional dynamics in a neural integrator circuit. Nat. Neurosci. 14 (9), 1150–1159.

Moore, H.A., Whitmore, D., 2014. Circadian rhythmicity and light sensitivity of the zebrafish brain. PLoS ONE 9 (1), e86176.

Muto, A., Ohkura, M., Abe, G., Nakai, J., Kawakami, K., 2013. Real-time visualization of neuronal activity during perception. Curr. Biol. 23 (4), 307–311.

Niemiller, M.L., Higgs, D.M., Soares, D., 2013. Evidence for hearing loss in amblyopsid cavefishes. Biol. Lett. 9 (3), 20130104.

O'Keefe, J., Dostrovsky, J., 1971. The hippocampus as a spatial map. Preliminary evidence from unit activity in the freely-moving rat. Brain Res. 34 (1), 171–175.

Patton, P., Windsor, S., Coombs, S., 2010. Active wall following by Mexican blind cavefish (*Astyanax mexicanus*). J. Comp. Physiol. A Neuroethol. Sens. Neural Behav. Physiol. 196 (11), 853–867.

Popper, A.N., 1970. Auditory capacities of the Mexican blind cave fish (*Astyanax jordani*) and its eyed ancestor (*Astyanax mexicanus*). Anim. Behav. 18, 552–562.

Prevedel, R., Yoon, Y.G., Hoffmann, M., Pak, N., Wetzstein, G., Kato, S., Schrodel, T., Raskar, R., Zimmer, M., Boyden, E.S., Vaziri, A., 2014. Simultaneous whole-animal 3D imaging of neuronal activity using light-field microscopy. Nat. Methods 11 (7), 727–730.

Ptito, M., Matteau, I., Zhi Wang, A., Paulson, O.B., Siebner, H.R., Kupers, R., 2012. Crossmodal recruitment of the ventral visual stream in congenital blindness. Neural Plast. 2012, 304045.

Qin, W., Liu, Y., Jiang, T., Yu, C., 2013. The development of visual areas depends differently on visual experience. PLoS ONE 8 (1), e53784.

Rétaux, S., Elipot, Y., 2013. Feed or fight: a behavioral shift in blind cavefish. Commun. Integr. Biol. 6 (2), e23166.

Rétaux, S., Pottin, K., Alunni, A., 2008. Shh and forebrain evolution in the blind cavefish *Astyanax mexicanus*. Biol. Cell. 100 (3), 139–147.

Rich, P.D., Liaw, H.P., Lee, A.K., 2014. Place cells. Large environments reveal the statistical structure governing hippocampal representations. Science 345 (6198), 814–817.

Rombaux, P., Huart, C., De Volder, A.G., Cuevas, I., Renier, L., Duprez, T., Grandin, C., 2010. Increased olfactory bulb volume and olfactory function in early blind subjects. Neuroreport 21 (17), 1069–1073.

Sargolini, F., Fyhn, M., Hafting, T., McNaughton, B.L., Witter, M.P., Moser, M.B., Moser, E.I., 2006. Conjunctive representation of position, direction, and velocity in entorhinal cortex. Science 312 (5774), 758–762.

Schnorr, S.J., Steenbergen, P.J., Richardson, M.K., Champagne, D.L., 2012. Measuring thigmotaxis in larval zebrafish. Behav. Brain Res. 228 (2), 367–374.

Sharma, S., Coombs, S., Patton, P., Burt de Perera, T., 2009. The function of wall-following behaviors in the Mexican blind cavefish and a sighted relative, the Mexican tetra (*Astyanax*). J. Comp. Physiol. A Neuroethol. Sens. Neural Behav. Physiol. 195 (3), 225–240.

Siegel, J.J., Nitz, D., Bingman, V.P., 2006. Lateralized functional components of spatial cognition in the avian hippocampal formation: evidence from single-unit recordings in freely moving homing pigeons. Hippocampus 16 (2), 125–140.

Sutherland, L.H., Holbrook, R.I., Burt de Perera, T., 2009. Sensory system affects orientational strategy in a short-range spatial task in blind and eyed morphs of the fish, *Astyanax fasciatus*. Ethology 115 (5), 504–510.

Tan, D., Patton, P., Coombs, S., 2011. Do blind cavefish have behavioral specializations for active flow-sensing? J. Comp. Physiol. A Neuroethol. Sens. Neural Behav. Physiol. 197 (7), 743–754.

Taube, J.S., 2007. The head direction signal: origins and sensory-motor integration. Annu. Rev. Neurosci. 30, 181–207.

Teyke, T., 1985. Collision with and avoidance of obstacles by blind cave fish *Anoptichthys jordani* (Characidae). J. Comp. Physiol. A 157 (6), 837–843.

Teyke, T., 1988. Flow field, swimming velocity and boundary layer: parameters which affect the stimulus for the lateral line organ in blind fish. J. Comp. Physiol. A 163 (1), 53–61.

Teyke, T., 1990. Morphological differences in neuromasts of the blind cave fish *Astyanax hubbsi* and the sighted river fish *Astyanax mexicanus*. Brain Behav. Evol. 35 (1), 23–30.

Thaler, L., Arnott, S.R., Goodale, M.A., 2011. Neural correlates of natural human echolocation in early and late blind echolocation experts. PLoS ONE 6 (5) e20162.

Van Trump, W.J., McHenry, M.J., 2013. The lateral line system is not necessary for rheotaxis in the Mexican blind cavefish (Astyanax fasciatus). Integr. Comp. Biol. 53 (5), 799–809.

Varatharasan, N., Croll, R.P., Franz-Odendaal, T., 2009. Taste bud development and patterning in sighted and blind morphs of *Astyanax mexicanus*. Dev. Dyn. 238 (12), 3056–3064.

Vladimirov, N., Mu, Y., Kawashima, T., Bennett, D.V., Yang, C.T., Looger, L.L., Keller, P.J., Freeman, J., Ahrens, M.B., 2014. Light-sheet functional imaging in fictively behaving zebrafish. Nat. Methods 11 (9), 883–884.

von Campenhausen, C., Riess, I., Weissert, R., 1981. Detection of stationary objects by the blind cave fish *Anoptichthys jordani* (Characidae). J. Comp. Physiol. A 143 (3), 369–374.

Voss, P., Lassonde, M., Gougoux, F., Fortin, M., Guillemot, J.P., Lepore, F., 2004. Early- and late-onset blind individuals show supra-normal auditory abilities in far-space. Curr. Biol. 14 (19), 1734–1738.

Wallace, M.T., Stein, B.E., 2007. Early experience determines how the senses will interact. J. Neurophysiol. 97 (1), 921–926.

Wan, C.Y., Wood, A.G., Reutens, D.C., Wilson, S.J., 2010. Early but not late-blindness leads to enhanced auditory perception. Neuropsychologia 48 (1), 344–348.

Wang, F., Julien, D.P., Sagasti, A., 2013. Journey to the skin: somatosensory peripheral axon guidance and morphogenesis. Cell Adhes. Migr. 7 (4), 388–394.

Williams, R., Neubarth, N., Hale, M.E., 2013. The function of fin rays as proprioceptive sensors in fish. Nat. Commun. 4, 1729.

Windsor, S., Paris, J., de Perera, T.B., 2011. No role for direct touch using the pectoral fins, as an information gathering strategy in a blind fish. J. Comp. Physiol. A Neuroethol. Sens. Neural Behav. Physiol. 197 (4), 321–327.

Windsor, S.P., Tan, D., Montgomery, J.C., 2008. Swimming kinematics and hydrodynamic imaging in the blind Mexican cave fish (*Astyanax fasciatus*). J. Exp. Biol. 211 (Pt 18), 2950–2959.

Yartsev, M.M., Ulanovsky, N., 2013. Representation of three-dimensional space in the hippocampus of flying bats. Science 340 (6130), 367–372.

Yartsev, M.M., Witter, M.P., Ulanovsky, N., 2011. Grid cells without theta oscillations in the entorhinal cortex of bats. Nature 479 (7371), 103–107.

Yaski, O., Portugali, J., Eilam, D., 2009. The dynamic process of cognitive mapping in the absence of visual cues: human data compared with animal studies. J. Exp. Biol. 212 (Pt 16), 2619–2626.

Yoshizawa, M., Goricki, S., Soares, D., Jeffery, W.R., 2010. Evolution of a behavioral shift mediated by superficial neuromasts helps cavefish find food in darkness. Curr. Biol. 20 (18), 1631–1636.

Yoshizawa, M., Jeffery, W.R., van Netten, S.M., McHenry, M.J., 2014. The sensitivity of lateral line receptors and their role in the behavior of Mexican blind cavefish (*Astyanax mexicanus*). J. Exp. Biol. 217 (Pt 6), 886–895.

PART V

Future Applications

Chapter 19

Transgenesis and Future Applications for Cavefish Research

Kathryn M. Tabor and Harold A. Burgess
Program in Genomics of Differentiation, Eunice Kennedy Shriver National Institute of Child Health and Human Development, Bethesda, Maryland, USA

INTRODUCTION

High throughput sequencing and powerful computational techniques are enabling rapid progress toward associating genotypic changes with phenotypic adaptations to the subterranean habitat in *Astyanax mexicanus* (McGaugh et al., 2014). Researchers will soon confront the challenge of showing how specific genetic changes lead to shifts in development or physiology that produce these morphological and behavioral phenotypes. Modern genetic tools that have been used to study development and function in model organisms are well suited to help answer such questions. In zebrafish, a species now widely used for biomedical research, genome modification using transgenic techniques has been instrumental in revealing basic mechanisms of vertebrate development and behavior. Inspired by these applications, in this chapter we will focus on transgenic tools that offer tractable experimental approaches for connecting genotype to phenotype in cavefish.

Following the breakthrough work of Palmiter and Brinster in generating transgenic mice (Palmiter et al., 1982), gene transfer experiments in fish were pioneered by aquaculture researchers who aimed to improve phenotypic characteristics, such as growth rate and infection resistance. Goldfish were the first fish species transgenically modified, by microinjection of a minigene for human growth hormone expression into fertilized eggs (Zhu et al., 1985). Reports of successful transgenesis in other fish species rapidly followed, and the first transgenic zebrafish were reported in 1988 (Stuart et al., 1988). These experiments used injection of linear DNA, which multimerizes before integrating into the genome at relatively low efficiency; however, large transgene concatamers are targets for epigenetic silencing

Biology and Evolution of the Mexican Cavefish. http://dx.doi.org/10.1016/B978-0-12-802148-4.00019-0
2016 Published by Elsevier Inc.

mechanisms, and thus, despite stable transgene inheritance, reporter expression was often undetectable (Stuart et al., 1988). A promising advance was the development of retroviral vectors pseudotyped to infect fish embryos. These vectors transduce single-copy integrations and are highly efficient (Burns et al., 1993; Lin et al., 1994a). Unfortunately, the relatively small cargo capacity of retroviruses limited their usefulness for constructing new transgenic lines. Improved efficiency was also obtained using an I-SceI meganuclease-based system; however, transgenic fish created using this method typically contain transgene concatamers, increasing the risk of silencing (Thermes et al., 2002).

Ultimately the low efficiency of transgenesis and the problem of concatamerized insertions were solved by the serendipitous discovery of a native transposon system in medaka fish. Study of a spontaneous pigmentation mutant led to the molecular cloning of an active hAT family transposable element that had disrupted the tyrosinase gene (Koga et al., 1996). The Tol2 transposon comprised four open reading frames, one of which proved to be a functional transposase (Kawakami et al., 1998; Kawakami and Shima, 1999). Co-injection of Tol2 transposase with a transgene flanked by transposon-derived sequence resulted in very high rates of stable and heritable transgene integration (Kawakami et al., 2000). Critically, in transposon-mediated transgenesis, a circular plasmid is injected, thereby avoiding the tendency for multimerization before integration. Transposon-mediated transgenesis, currently the state-of-the-art method for generating transgenic zebrafish, has been successfully used in several other vertebrate species, and has already shown promise in experiments with cavefish (Elipot et al., 2014; Kawakami, 2007).

In mice, precise reporter gene integration into targeted genomic loci is routinely performed using homologous recombination in embryonic stem cells. Until recently, this method was not available to researchers using fish; however, the newly developed TALE nuclease and CRISPR technologies, which induce a double-stranded break at a precisely targeted site in the genome, present new opportunities for targeted transgene integration (Cong et al., 2013; Jinek et al., 2012; Mali et al., 2013; Miller et al., 2011; Zhang et al., 2011). These methods are usually employed to create loss of function alleles, but in fish, double-stranded breaks stimulate the integration of co-injected DNA fragments, facilitating the integration of transgenes into the targeted locus at high efficiency (Auer et al., 2014; Kimura et al., 2014). Moreover, this method has now been adapted to promote efficient homologous recombination in zebrafish (Shin et al., 2014). The CRISPR system appears to be effective in virtually every organism that has been tested to date, and there is every reason to think it will also enable homologous recombination in cavefish. Given the range of technical approaches, gene transfer experiments in laboratory cavefish stocks are unlikely to face substantial technical hurdles, making it feasible to implement a wide range of powerful experimental approaches.

VISUALIZING DEVELOPMENT AND ANATOMY

Cavefish share a key advantage of most bony fish species for studying development, in that fertilization is external and embryos are largely transparent. Transgenic methods can complement these features by expressing a fluorescent reporter gene in specific cell populations, allowing their development to be monitored *in vivo*. In the first such experiments in zebrafish, *lacZ* was used as a reporter so that transgene expression could be observed in live embryos treated with a fluorescein-based substrate (Lin et al., 1994b); however, after the cloning of the gene encoding green fluorescent protein (GFP) from *Aequorea victoria* (Prasher et al., 1992), it was soon discovered that GFP expression alone was sufficient to produce fluorescence in bacteria and nematodes (Chalfie et al., 1994; Inouye and Tsuji, 1994). Within a year, visible GFP fluorescence was obtained from a single-copy transgene in zebrafish (Amsterdam et al., 1995). In these experiments, reporters were expressed broadly or ubiquitously. The next advance was the creation of transgenic fish with spatially controlled expression of GFP in blood cells and muscle (Higashijima et al., 1997; Long et al., 1997). Hundreds of transgenic lines expressing fluorescent proteins in a variety of tissues have since been generated and used for a huge variety of applications. For instance, the *islet-1:GFP* transgenic line has been instrumental in many studies of cranial motor neuron development, migration, and axon pathfinding (Higashijima et al., 2000). Similarly the *fli:EGFP* line, which labels blood vessels, has facilitated hundreds of *in vivo* studies, ranging from microangiography to models of vascular disease (Lawson and Weinstein, 2002).

Transgenic experiments in cavefish will be able to build upon reagents used to drive cell-type specific expression in zebrafish. In general, the first step in generating a new transgenic line is to identify transcriptional regulatory sequences (enhancers) from an endogenous gene that is expressed in the cell type of interest. Enhancer sequences are usually located within a few hundred kilobases of the coding sequence, and are often identified by interspecific sequence conservation. In synthetic transgenes, a DNA fragment, usually between 0.5 and 15 kb in length and containing one or more enhancer elements is placed upstream of a promoter followed by a reporter-encoding complementary DNA (cDNA). For instance, one of the first tissue-specific transgenic zebrafish used a 3.9 kb enhancer/promoter fragment from the *α-actin* gene to drive GFP expression in muscle cells (Higashijima et al., 1997). Defined enhancer fragments have since been successfully used to create many zebrafish lines with tissue-specific expression; however, cell-type specific enhancer fragments cannot yet be reliably identified by computational methods alone (Fisher et al., 2006); for the *phox2b* gene, algorithms relying on sequence conservation failed to detect about half the enhancers that were experimentally defined (McGaughey et al., 2008). Thus instead of attempting to predict functional enhancer sequences, an alternate approach is to use large genomic regions that encompass multiple presumptive regulatory elements for a gene of interest. For this, genomic fragments containing enhancer elements have been cloned from the pufferfish, *Fugu rubripes*. Because of the compact size of the *Fugu* genome, fragments that are technically manageable in size may nevertheless comprise a wealth of regulatory elements, increasing the likelihood of recapitulating the endogenous gene expression (Barton et al., 2001; Kimura-Yoshida et al., 2004). Alternatively, transgenic animals can be made using large genome fragments cloned into artificial chromosomes. Artificial chromosomes may contain as much as 300kb of genomic DNA, and can be manipulated *in vitro* to introduce reporters whose expression is controlled by the ensemble of enhancer elements contained within the construct (Jessen et al., 1999; Suster et al., 2009). Notably, artificial chromosomes were used to make transgenic fish with fluorescent labeling of glutamatergic and glycinergic neurons. These lines have proven extremely useful for defining neuron identity and enabled live neurons to be targeted for tracing and electrophysiology experiments (Kinkhabwala et al., 2011; Koyama et al., 2011). Moreover, as noted above, the demonstration that CRISPR technology stimulates homologous recombination in zebrafish holds tremendous promise for facilitating the targeted expression of genetic reporters in almost any gene expression pattern.

Shifts in cell lineage may underlie conspicuous changes in eye morphology and neural patterning that occur in cave morphs (Pottin et al., 2011). Developmental fate-mapping studies in zebrafish have benefited from genetic methods that mark cell lineages. One sophisticated approach makes use of Cre recombinase from bacteriophage P1, which catalyzes the recombination between two short cognate DNA sequences ("lox" sites) (Sternberg et al., 1986). Depending upon the orientation of the lox sites, Cre can be used to excise or invert the intervening sequence. Already used for more than two decades to conditionally delete genes in mice (Gu et al., 1994), more recently Cre has been employed in fish to regulate transgene expression (Thummel et al., 2005). By irreversibly recombining target sites, transient Cre expression permanently marks cells and all their subsequent progeny. Transgenic expression of a Cre in progenitor cells during development circumvents the demanding technical manipulations required for other fate-mapping methods (Mongera et al., 2013; Wang et al., 2011). A variation of this technique is to regulate Cre activity in order to induce stochastic recombination at loxP targets. Sparse labeling of progenitor cells enables clonal analysis of cell lineages during tissue formation (Collins et al., 2010). Other active recombinases have been isolated from bacteria and yeast (Nern et al., 2011), and several shown to function in zebrafish (Boniface et al., 2009; Lister, 2010; Park and Leach, 2013). Because each recombinase acts upon a unique recognition sequence, multiple recombinases can be used in a single animal to create complex patterns of gene expression. This has been used to uniquely label only cellular lineages that expressed two recombinases, where each was separately controlled by different regulatory elements (Park and Leach, 2013; Sajgo et al., 2014).

Quantitative trait loci under selection during adaptation may produce relatively subtle changes that are difficult to reveal using histochemical methods on fixed animals that capture only a single point in time. Another way to visualize changes that affect tissue development and morphogenesis is to use live transgenic reporters of molecular signaling. In these animals, promoter elements that contain binding sites for transcription factors downstream of signaling cascades are coupled to destabilized variants of fluorescent reporter proteins that have a short intracellular half-life in the cell. Fluorescent signals thus represent the activation of signaling cascades in near real-time. Transgenic zebrafish lines based upon this method have been created that act as live reporters for major developmental signaling systems, including wnt, fibroblast growth factor (FGF), hedgehog, bone morphogenetic protein (BMP), and retinoic acid, enabling visualization of their activity during development (Dorsky et al., 2002; Laux et al., 2011; Molina et al., 2007; Pittlik and Begemann, 2012; Schwend et al., 2010). Such live reporters may be particularly valuable for identifying subtle changes in the timing or strength of developmental cues that are responsible for phenotypic shifts.

TRANSGENIC APPROACHES TO TESTING GENETIC CAUSALITY

Studies on cave morphs will continue to identify specific genetic changes that are strongly associated with phenotypic adaptations. The next step is to definitively establish a causal connection between a given genotypic modification and an associated phenotype. In the best case, genetic mapping experiments describe an interval associated with a single gene, strongly suggesting that allelic variants affecting the gene or regulatory sequences are responsible for the altered phenotype. Researchers have generated and mapped genetic mutations in zebrafish for more than 30 years (Walker and Streisinger, 1983). Most successfully mapped mutants have harbored nonsense or missense mutations, leading to a presumptive loss of protein function. Nevertheless, the "gold standard" for proof in zebrafish has called for independent evidence, due to the possibility that the true causative change is a closely linked sequence variant. A challenge ahead for the cavefish research community will be to set feasible but rigorous standards for definitively linking genotype to phenotype. In laboratory mutagenesis experiments, an independent allele with the same phenotype is often recovered, a luxury that may also occasionally be afforded in the study of natural cavefish populations (Protas et al., 2006). If not, phenocopy by creating new mutant alleles using TALE nuclease or CRISPR technologies may provide independent evidence of the causal role for simple loss of function mutations; however, many cave morph phenotypes are likely to arise from polygenic inheritance of sequence modifications that quantitatively alter the levels or spatiotemporal expression patterns of multiple genes (reviewed in Wilkens, 2015). Such changes are more challenging to validate, but once again, transgenic methods may be instrumental.

Variants that may affect expression patterns can be verified relatively easily by testing their ability to drive reporter gene expression in transgenic animals. For instance, in stickleback, the effect of a regulatory sequence change on armor plate formation was confirmed by making transgenic fish using the wildtype and altered sequence to drive GFP expression (O'Brown et al., 2015). Phenotypic changes thought to result from increases in protein expression levels may be assessed by making transgenic fish with an additional copy of the wildtype protein. Conversely, changes that appear to depress expression levels can be confirmed using transgenic rescue experiments. In zebrafish, rescue is often performed by injecting an artificial chromosome containing the wildtype gene of interest into mutant embryos. For instance, the *chock* mutant, which contains a premature stop codon in the *rx3* gene, fails to form eyes. Eye formation in mutants was restored after injection of a bacteriophage artificial chromosome containing the wildtype *rx3* gene and flanking regulatory sequence, firmly establishing a causative role for the mutation (Kennedy et al., 2004). Successful rescue has also been achieved by generating stable transgenic fish that ubiquitously express the wildtype protein. A nice example was the restoration of optokinetic responses to red light in *pob* mutants crossed to a transgenic fish ubiquitously expressing a wildtype copy of the effected gene (Taylor et al., 2005). Similarly, rescue can be achieved using a tissue-specific promoter that selectively restores expression of the wildtype protein in the targeted cell group (Low et al., 2011). Thus transgenic methods may offer a rigorous means to test whether specific genetic variants are responsible for differences between surface and cave morphs.

ANALYSIS OF NEURONAL CIRCUITS THAT CONTROL BEHAVIOR

Cave morphs show a suite of behavioral adaptations for coping with the extreme pressures of the cave environment. At least some changes in cave morph behavior, such as increased sensitivity to sensory cues, are elaborations of responses shared by surface morphs (Yoshizawa, 2015). It will be fascinating to discover the affiliated modifications to neuronal development or function that enable these adaptations. Do independent cave morphs acquire similar behavioral adaptations by modifying the same neuronal pathways? Or are similar adaptations more often acquired by modifying distinct neuronal systems? These studies hold the promise of illuminating general principles at play during the evolution of the brain. In model organisms, transgenic methods are increasingly central to experimental analyses of nervous system structure and function, and in this section, we will describe approaches that are being used in zebrafish that may help to probe the neuronal basis of behavior in cavefish: neuron tracing, targeted cell ablation, optogenetics, and live brain imaging.

At least some behavioral adaptations in cave morphs are likely to be connected to changes in brain architecture (Rétaux et al., 2015). Whereas gross brain structure can be readily analyzed using immunohistochemical methods, analysis of connectivity within the brain is enormously facilitated by transgenic lines that express reporter genes in defined neuronal cell types. As for cavefish, early larval zebrafish brains are transparent and small enough to image comprehensively using light microscopy. Reconstruction of the arborization patterns of transgenically labeled neurons has been used to provide a comprehensive projection map of dopaminergic and noradrenergic neurons (Tay et al., 2011). In this study, single-cell resolution was achieved by laboriously imaging hundreds of transient transgenic fish that showed mosaic reporter expression. A promising new approach is the use of "Brainbow," a transgenic reporter system for labeling individual neurons with unique fluorescent signals (Livet et al., 2007; Pan et al., 2011). In Brainbow, Cre-mediated recombination occurs within transgenes that comprise multiple fluorescent proteins, each flanked by different lox sites. The stochastic choice of recombination sites in each neuron leads to a unique combination of proteins being expressed so that the soma and neurites have a distinctive signature when imaged with multichannel fluorescence microscopy. In zebrafish, Brainbow has been used to simultaneously visualize multiple retinal ganglion cells *in vivo*, and follow changes in their arborization patterns in the optic tectum over several days (Robles et al., 2013). Such tools will facilitate neuronal tracing experiments that may reveal changes in connectivity responsible for adaptation to the cave environment.

Functional analyses that probe the contribution of neuronal cell groups to behaviors are an essential complement to neuroanatomical studies. Lesion studies are a fundamental technique for studying the neuronal mechanisms of behavior. In zebrafish, traditional anatomical lesions have largely been supplanted by genetic methods that offer improved cell-type specificity. In the first zebrafish ablation experiments, tissue-specific promoters were used to control expression of the diphtheria toxin A-subunit (DTA), an enzyme that blocks translation by modifying elongation factor-2, leading to apoptosis. Transient transgenic fish expressing DTA were used to selectively interfere with lens maturation during embryogenesis, demonstrating a key role for this tissue in regulating the organization of the neural retina (Kurita et al., 2003). In mammals, targeted ablation has been achieved by injection of diphtheria and related toxins coupled to cellular targeting moieties, or by use of transgenic animals that express diphtheria toxin receptor in specific tissues (Chatterjee et al., 2012; Clark et al., 2007; Saito et al., 2001; Weldon and Pastan, 2011); however, in zebrafish, it has proved difficult to develop stable lines carrying a DTA transgene, likely because of extreme toxicity from a low level of leaky expression; a single DTA molecule is sufficient to cause cell death (Yamaizumi et al., 1978). Instead, ablation studies most often use the bacterial nitroreductase *NfsB*. *NfsB* kills cells by converting exogenously applied nitroimidazole prodrugs, such as metronidazole into DNA interstrand cross-linking agents (Roberts et al., 1986). Metronidazole is cell-permeable, whereas its metabolites are not, ensuring that only cells expressing *NfsB* are ablated, while neighboring cells remain intact (Bridgewater et al., 1997). This system was originally exploited in zebrafish to selectively kill pancreatic cells (Curado et al., 2007; Pisharath et al., 2007) and has since been used to ablate a wide variety of transgenically defined neuronal cell types, and is thus an effective tool for dissecting neuronal circuits (Bergeron et al., 2014).

The development of optogenetic tools that allow neuronal activity to be controlled by light provided neurobiologists with a powerful new tool to study neuronal cell function. Light-gated cation channels, such as channelrhodopsin-2 (ChR2) and its many variants can be used in zebrafish to selectively activate neurons, allowing researchers to assess their contribution to behavior or influence on synaptically connected neurons (Douglass et al., 2008; Hong et al., 2013; Zhu et al., 2009). Conversely, neuronal activity can be inhibited by using light-activated ion pumps, such as halorhodopsin or archaerhodopsin that hyperpolarize the cell membrane. These proteins have been transgenically expressed in zebrafish and used to define the behavioral function of neuronal cell types (Bergeron et al., 2014). An alternative to the targeted expression of optogenetic proteins is to generate transgenic animals with pan-neuronal expression and to use spatially patterned light to regulate activity (Arrenberg et al., 2009, 2010).

Optical methods for monitoring neuronal activity are increasingly being used in place of, or as a complement to, traditional electrophysiological recordings. In a pioneering study almost 20 years ago, Fetcho and colleagues injected calcium green dye coupled to high molecular weight dextran into the spinal cord, where it was retrogradely transported along the axons of reticulospinal neurons to their cell bodies in the hindbrain (O'Malley et al., 1996). Calcium green fluorescence indicates intracellular calcium concentration, which is a proxy for neuronal firing. The group imaged fluorescence in multiple identified reticulospinal neurons simultaneously in semirestrained behaving larvae. This remarkable study demonstrated the power of calcium imaging for recording neural activity in fish; neurons normally inaccessible to electrode recordings can be easily monitored, multiple cells can be visualized simultaneously, and the intact animal can perform a range of normal behavioral responses. Calcium dyes, however, offer few options for selectively targeting neurons, a drawback that was overcome by the development of genetically encoded calcium indicators (GECIs), which can be transgenically expressed (Higashijima et al., 2003). In the last decade, there has been a rapid improvement in the sensitivity and dynamic range of GECIs. Pan-neuronal expression of GECIs in transgenic fish, coupled with rapid imaging technology has enabled the real-time monitoring of activity of nearly all neurons in the larval zebrafish brain (Vladimirov et al., 2014). GECIs with large changes in fluorescence signals during neuronal firing have even permitted the detection of activity changes in freely swimming fish (Muto et al., 2013).

In short, the last decade has seen the introduction of a suite of transformative genetic tools for studying brain function. Together with the genetic reagents needed to direct expression of these proteins to specific cell groups, it should be possible to compare neuronal circuitry in surface and cave morphs, and identify modifications in network architecture that gives rise to adaptive changes in behavior.

CONCLUDING REMARKS

Sydney Brenner, famed for introducing the nematode *Caenorhabditis elegans* as a model organism for the study of neural development, opened his 2002 Nobel lecture by noting that "choosing the right organism for one's research is as important as finding the right problems to work on" (Brenner, 2003). Unexpectedly, he then continued by observing that the increasing sophistication of genetic techniques and depth of genomic information had started to reduce the incentive to focus biological research on a few popular model organisms. Indeed, there is increasing recognition that the narrow set of model species widely used in biomedical research offers too limited a scope for addressing many fundamental questions about the biological world (Bolker, 2012). Extraordinary strides in genome editing techniques introduced over the last decade have made available a set of tools that can enable complex genetic manipulations in almost any animal species. The growth of a research community coalescing around *A. mexicanus* is thus well timed, with these methods opening up a wealth of exciting opportunities to reveal fundamental principles of evolution and development.

REFERENCES

Amsterdam, A., Lin, S., Hopkins, N., 1995. The *Aequorea victoria* green fluorescent protein can be used as a reporter in live zebrafish embryos. Dev. Biol. 171, 123–129.

Arrenberg, A.B., Del Bene, F., Baier, H., 2009. Optical control of zebrafish behavior with halorhodopsin. Proc. Natl. Acad. Sci. U. S. A. 106, 17968–17973.

Arrenberg, A.B., Stainier, D.Y.R., Baier, H., Huisken, J., 2010. Optogenetic control of cardiac function. Science 330, 971–974.

Auer, T.O., Duroure, K., De Cian, A., Concordet, J.P., Del Bene, F., 2014. Highly efficient CRISPR/Cas9-mediated knock-in in zebrafish by homology-independent DNA repair. Genome Res. 24, 142–153.

Barton, L.M., Göttgens, B., Gering, M., Gilbert, J.G.R., Grafham, D., Rogers, J., Bentley, D., Patient, R., Green, A.R., 2001. Regulation of the stem cell leukemia (SCL) gene: a tale of two fishes. Proc. Natl. Acad. Sci. U. S. A. 98, 6747–6752.

Bergeron, S.A., Carrier, N., Li, G.H., Ahn, S., Burgess, H.A., 2014. Gsx1 expression defines neurons required for prepulse inhibition. Mol. Psychiatry. Epublished 2014 Sep 16. http://dx.doi.org/10.1038/mp.2014.106.

Bolker, J., 2012. Model organisms: there's more to life than rats and flies. Nature 491, 31–33.

Boniface, E.J., Lu, J., Victoroff, T., Zhu, M., Chen, W., 2009. FlEx-based transgenic reporter lines for visualization of Cre and Flp activity in live zebrafish. Genesis 47, 484–491.

Brenner, S., 2003. Nobel lecture: nature's gift to science. Biosci. Rep. 23, 225–237.

Bridgewater, J.A., Knox, R.J., Pitts, J.D., Collins, M.K., Springer, C.J., 1997. The bystander effect of the nitroreductase/CB1954 enzyme/prodrug system is due to a cell-permeable metabolite. Hum. Gene Ther. 8, 709–717.

Burns, J.C., Friedmann, T., Driever, W., Burrascano, M., Yee, J.K., 1993. Vesicular stomatitis virus G glycoprotein pseudotyped retroviral vectors: concentration to very high titer and efficient gene transfer into mammalian and nonmammalian cells. Proc. Natl. Acad. Sci. U. S. A. 90, 8033–8037.

Chalfie, M., Tu, Y., Euskirchen, G., Ward, W., Prasher, D., 1994. Green fluorescent protein as a marker for gene expression. Science 263, 802–805.

Chatterjee, D., Chandran, B., Berger, E.A., 2012. Selective killing of Kaposi's sarcoma-associated herpesvirus lytically infected cells with a recombinant immunotoxin targeting the viral gpK8.1A envelope glycoprotein. MAbs 4, 233–242.

Clark, S.D., Alderson, H.L., Winn, P., Latimer, M.P., Nothacker, H.P., Civelli, O., 2007. Fusion of diphtheria toxin and urotensin II produces a neurotoxin selective for cholinergic neurons in the rat mesopontine tegmentum. J. Neurochem. 102, 112–120.

Collins, R.T., Linker, C., Lewis, J., 2010. MAZe: a tool for mosaic analysis of gene function in zebrafish. Nat. Methods 7, 219–223.

Cong, L., Ran, F.A., Cox, D., Lin, S., Barretto, R., Habib, N., Hsu, P.D., Wu, X., Jiang, W., Marraffini, L.A., Zhang, F., 2013. Multiplex genome engineering using CRISPR/Cas systems. Science 339, 819–823.

Curado, S., Anderson, R.M., Jungblut, B., Mumm, J., Schroeter, E., Stainier, D.Y., 2007. Conditional targeted cell ablation in zebrafish: a new tool for regeneration studies. Dev. Dyn. 236, 1025–1035.

Dorsky, R.I., Sheldahl, L.C., Moon, R.T., 2002. A transgenic Lef1/beta-catenin-dependent reporter is expressed in spatially restricted domains throughout zebrafish development. Dev. Biol. 241, 229–237.

Douglass, A.D., Kraves, S., Deisseroth, K., Schier, A.F., Engert, F., 2008. Escape behavior elicited by single, channelrhodopsin-2-evoked spikes in zebrafish somatosensory neurons. Curr. Biol. 18, 1133–1137.

Elipot, Y., Legendre, L., Père, S., Sohm, F., Rétaux, S., 2014. Astyanax transgenesis and husbandry: how cavefish enters the laboratory. Zebrafish 11, 291–299.

Fisher, S., Grice, E.A., Vinton, R.M., Bessling, S.L., McCallion, A.S., 2006. Conservation of RET regulatory function from human to zebrafish without sequence similarity. Science 312, 276–279.

Gu, H., Marth, J.D., Orban, P.C., Mossmann, H., Rajewsky, K., 1994. Deletion of a DNA polymerase beta gene segment in T cells using cell type-specific gene targeting. Science 265, 103–106.

Higashijima, S.-i., Okamoto, H., Ueno, N., Hotta, Y., Eguchi, G., 1997. High-frequency generation of transgenic zebrafish which reliably express GFP in whole muscles or the whole body by using promoters of zebrafish origin. Dev. Biol. 192, 289–299.

Higashijima, S., Hotta, Y., Okamoto, H., 2000. Visualization of cranial motor neurons in live transgenic zebrafish expressing green fluorescent protein under the control of the islet-1 promoter/enhancer. J. Neurosci. 20, 206–218.

Higashijima, S., Masino, M.A., Mandel, G., Fetcho, J.R., 2003. Imaging neuronal activity during zebrafish behavior with a genetically encoded calcium indicator. J. Neurophysiol. 90, 3986–3997.

Hong, E., Santhakumar, K., Akitake, C.A., Ahn, S.J., Thisse, C., Thisse, B., Wyart, C., Mangin, J.M., Halpern, M.E., 2013. Cholinergic left-right asymmetry in the habenulo-interpeduncular pathway. Proc. Natl. Acad. Sci. U. S. A. 110, 21171–21176.

Inouye, S., Tsuji, F.I., 1994. Aequorea green fluorescent protein. FEBS Lett. 341, 277–280.

Jessen, J.R., Willett, C.E., Lin, S., 1999. Artificial chromosome transgenesis reveals long-distance negative regulation of rag1 in zebrafish. Nat. Genet. 23, 15–16.

Jinek, M., Chylinski, K., Fonfara, I., Hauer, M., Doudna, J.A., Charpentier, E., 2012. A programmable dual-RNA-guided DNA endonuclease in adaptive bacterial immunity. Science 337, 816–821.

Kawakami, K., 2007. Tol2: a versatile gene transfer vector in vertebrates. Genome Biol. 8, S7.

Kawakami, K., Shima, A., 1999. Identification of the *Tol2* transposase of the medaka fish *Oryzias latipes* that catalyzes excision of a nonautonomous *Tol2* element in zebrafish *Danio rerio*. Gene 240, 239–244.

Kawakami, K., Koga, A., Hori, H., Shima, A., 1998. Excision of the *Tol2* transposable element of the medaka fish, *Oryzias latipes*, in zebrafish, *Danio rerio*. Gene 225, 17–22.

Kawakami, K., Shima, A., Kawakami, N., 2000. Identification of a functional transposase of the *Tol2* element, an Ac-like element from the Japanese medaka fish, and its transposition in the zebrafish germ lineage. Proc. Natl. Acad. Sci. U. S. A. 97, 11403–11408.

Kennedy, B.N., Stearns, G.W., Smyth, V.A., Ramamurthy, V., van Eeden, F., Ankoudinova, I., Raible, D., Hurley, J.B., Brockerhoff, S.E., 2004. Zebrafish *rx3* and *mab21l2* are required during eye morphogenesis. Dev. Biol. 270, 336–349.

Kimura, Y., Hisano, Y., Kawahara, A., Higashijima, S., 2014. Efficient generation of knock-in transgenic zebrafish carrying reporter/driver genes by CRISPR/Cas9-mediated genome engineering. Sci. Rep. 4, 6545.

Kimura-Yoshida, C., Kitajima, K., Oda-Ishii, I., Tian, E., Suzuki, M., Yamamoto, M., Suzuki, T., Kobayashi, M., Aizawa, S., Matsuo, I., 2004. Characterization of the pufferfish Otx2 cis-regulators reveals evolutionarily conserved genetic mechanisms for vertebrate head specification. Development 131, 57–71.

Kinkhabwala, A., Riley, M., Koyama, M., Monen, J., Satou, C., Kimura, Y., Higashijima, S., Fetcho, J., 2011. A structural and functional ground plan for neurons in the hindbrain of zebrafish. Proc. Natl. Acad. Sci. U. S. A. 108, 1164–1169.

Koga, A., Suzuki, M., Inagaki, H., Bessho, Y., Hori, H., 1996. Transposable element in fish. Nature 383, 30.

Koyama, M., Kinkhabwala, A., Satou, C., Higashijima, S., Fetcho, J., 2011. Mapping a sensory-motor network onto a structural and functional ground plan in the hindbrain. Proc. Natl. Acad. Sci. U. S. A. 108, 1170–1175.

Kurita, R., Sagara, H., Aoki, Y., Link, B.A., Arai, K.-i., Watanabe, S., 2003. Suppression of lens growth by αA-crystallin promoter-driven expression of diphtheria toxin results in disruption of retinal cell organization in zebrafish. Dev. Biol. 255, 113–127.

Laux, D.W., Febbo, J.A., Roman, B.L., 2011. Dynamic analysis of BMP-responsive Smad activity in live zebrafish embryos. Dev. Dyn. 240, 682–694.

Lawson, N.D., Weinstein, B.M., 2002. *In vivo* imaging of embryonic vascular development using transgenic zebrafish. Dev. Biol. 248, 307–318.

Lin, S., Gaiano, N., Culp, P., Burns, J.C., Friedmann, T., Yee, J.K., Hopkins, N., 1994a. Integration and germ-line transmission of a pseudotyped retroviral vector in zebrafish. Science 265, 666–669.

Lin, S., Yang, S., Hopkins, N., 1994b. lacZ expression in germline transgenic zebrafish can be detected in living embryos. Dev. Biol. 161, 77–83.

Lister, J.A., 2010. Transgene excision in zebrafish using the phiC31 integrase. Genesis 48, 137–143.

Livet, J., Weissman, T.A., Kang, H., Draft, R.W., Lu, J., Bennis, R.A., Sanes, J.R., Lichtman, J.W., 2007. Transgenic strategies for combinatorial expression of fluorescent proteins in the nervous system. Nature 450, 56–62.

Long, Q., Meng, A., Wang, H., Jessen, J.R., Farrell, M.J., Lin, S., 1997. GATA-1 expression pattern can be recapitulated in living transgenic zebrafish using GFP reporter gene. Development 124, 4105–4111.

Low, S.E., Amburgey, K., Horstick, E., Linsley, J., Sprague, S.M., Cui, W.W., Zhou, W., Hirata, H., Saint-Amant, L., Hume, R.I., Kuwada, J.Y., 2011. TRPM7 is required within zebrafish sensory neurons for the activation of touch-evoked escape behaviors. J. Neurosci. 31, 11633–11644.

Mali, P., Yang, L., Esvelt, K.M., Aach, J., Guell, M., DiCarlo, J.E., Norville, J.E., Church, G.M., 2013. RNA-guided human genome engineering via Cas9. Science 339, 823–826.

McGaugh, S.E., Gross, J.B., Aken, B., Blin, M., Borowsky, R., Chalopin, D., Hinaux, H., Jeffery, W.R., Keene, A., Ma, L., Minx, P., Murphy, D., O'Quin, K.E., Rétaux, S., Rohner, N., Searle,

S.M.J., Stahl, B.A., Tabin, C., Volff, J.-N., Yoshizawa, M., Warren, W.C., 2014. The cavefish genome reveals candidate genes for eye loss. Nat. Commun. 5, 5307.

McGaughey, D.M., Vinton, R.M., Huynh, J., Al-Saif, A., Beer, M.A., McCallion, A.S., 2008. Metrics of sequence constraint overlook regulatory sequences in an exhaustive analysis at *phox2b*. Genome Res. 18, 252–260.

Miller, J.C., Tan, S., Qiao, G., Barlow, K.A., Wang, J., Xia, D.F., Meng, X., Paschon, D.E., Leung, E., Hinkley, S.J., Dulay, G.P., Hua, K.L., Ankoudinova, I., Cost, G.J., Urnov, F.D., Zhang, H.S., Holmes, M.C., Zhang, L., Gregory, P.D., Rebar, E.J., 2011. A TALE nuclease architecture for efficient genome editing. Nat. Biotechnol. 29, 143–148.

Molina, G.A., Watkins, S.C., Tsang, M., 2007. Generation of FGF reporter transgenic zebrafish and their utility in chemical screens. BMC Dev. Biol. 7, 62.

Mongera, A., Singh, A.P., Levesque, M.P., Chen, Y.Y., Konstantinidis, P., Nusslein-Volhard, C., 2013. Genetic lineage labeling in zebrafish uncovers novel neural crest contributions to the head, including gill pillar cells. Development 140, 916–925.

Muto, A., Ohkura, M., Abe, G., Nakai, J., Kawakami, K., 2013. Real-time visualization of neuronal activity during perception. Curr. Biol. 23, 307–311.

Nern, A., Pfeiffer, B.D., Svoboda, K., Rubin, G.M., 2011. Multiple new site-specific recombinases for use in manipulating animal genomes. Proc. Natl. Acad. Sci. U. S. A. 108, 14198–14203.

O'Brown, N.M., Summers, B.R., Jones, F.C., Brady, S.D., Kingsley, D.M., 2015. A recurrent regulatory change underlying altered expression and Wnt response of the stickleback armor plates gene *EDA*. eLife 4. e05290.

O'Malley, D.M., Kao, Y.H., Fetcho, J.R., 1996. Imaging the functional organization of zebrafish hindbrain segments during escape behaviors. Neuron 17, 1145–1155.

Palmiter, R.D., Brinster, R.L., Hammer, R.E., Trumbauer, M.E., Rosenfeld, M.G., Birnberg, N.C., Evans, R.M., 1982. Dramatic growth of mice that develop from eggs microinjected with metallothionein-growth hormone fusion genes. Nature 300, 611–615.

Pan, Y.A., Livet, J., Sanes, J.R., Lichtman, J.W., Schier, A.F., 2011. Multicolor Brainbow imaging in zebrafish. Cold Spring Harb. Protoc. 2011. pdb.prot5546.

Park, J.T., Leach, S.D., 2013. TAILOR: transgene activation and inactivation using lox and rox in zebrafish. PLoS ONE 8, e85218.

Pisharath, H., Rhee, J.M., Swanson, M.A., Leach, S.D., Parsons, M.J., 2007. Targeted ablation of beta cells in the embryonic zebrafish pancreas using *E. coli* nitroreductase. Mech. Dev. 124, 218–229.

Pittlik, S., Begemann, G., 2012. New sources of retinoic acid synthesis revealed by live imaging of an Aldh1a2-GFP reporter fusion protein throughout zebrafish development. Dev. Dyn. 241, 1205–1216.

Pottin, K., Hinaux, H., Retaux, S., 2011. Restoring eye size in *Astyanax mexicanus* blind cavefish embryos through modulation of the Shh and Fgf8 forebrain organising centres. Development 138, 2467–2476.

Prasher, D.C., Eckenrode, V.K., Ward, W.W., Prendergast, F.G., Cormier, M.J., 1992. Primary structure of the *Aequorea victoria* green-fluorescent protein. Gene 111, 229–233.

Protas, M.E., Hersey, C., Kochanek, D., Zhou, Y., Wilkens, H., Jeffery, W.R., Zon, L.I., Borowsky, R., Tabin, C.J., 2006. Genetic analysis of cavefish reveals molecular convergence in the evolution of albinism. Nat. Genet. 38, 107–111.

Rétaux, S., Alié, A., Blin, M., Devos, L., Elipot, Y., Hinaux, H., 2015. Neural development and evolution in *Astyanax mexicanus*: comparing cavefish and surface fish brains. In: Keene, A., Yoshizawa, M., McGhaugh, S. (Eds.), Biology and Evolution of the Mexican Cavefish. Elsevier, Amsterdam.

Roberts, J.J., Friedlos, F., Knox, R.J., 1986. CB 1954 (2,4-dinitro-5-aziridinyl benzamide) becomes a DNA interstrand crosslinking agent in Walker tumour cells. Biochem. Biophys. Res. Commun. 140, 1073–1078.

Robles, E., Filosa, A., Baier, H., 2013. Precise lamination of retinal axons generates multiple parallel input pathways in the tectum. J. Neurosci. 33, 5027–5039.

Saito, M., Iwawaki, T., Taya, C., Yonekawa, H., Noda, M., Inui, Y., Mekada, E., Kimata, Y., Tsuru, A., Kohno, K., 2001. Diphtheria toxin receptor-mediated conditional and targeted cell ablation in transgenic mice. Nat. Biotechnol. 19, 746–750.

Sajgo, S., Ghinia, M.G., Shi, M., Liu, P., Dong, L., Parmhans, N., Popescu, O., Badea, T.C., 2014. Dre-Cre sequential recombination provides new tools for retinal ganglion cell labeling and manipulation in mice. PLoS ONE 9, e91435.

Schwend, T., Loucks, E.J., Ahlgren, S.C., 2010. Visualization of Gli activity in craniofacial tissues of hedgehog-pathway reporter transgenic zebrafish. PLoS ONE 5, e14396.

Shin, J., Chen, J., Solnica-Krezel, L., 2014. Efficient homologous recombination-mediated genome engineering in zebrafish using TALE nucleases. Development 141, 3807–3818.

Sternberg, N., Sauer, B., Hoess, R., Abremski, K., 1986. Bacteriophage P1 *cre* gene and its regulatory region. Evidence for multiple promoters and for regulation by DNA methylation. J. Mol. Biol. 187, 197–212.

Stuart, G.W., McMurray, J.V., Westerfield, M., 1988. Replication, integration and stable germ-line transmission of foreign sequences injected into early zebrafish embryos. Development 103, 403–412.

Suster, M.L., Sumiyama, K., Kawakami, K., 2009. Transposon-mediated BAC transgenesis in zebrafish and mice. BMC Genomics 10, 477.

Tay, T.L., Ronneberger, O., Ryu, S., Nitschke, R., Driever, W., 2011. Comprehensive catecholaminergic projectome analysis reveals single-neuron integration of zebrafish ascending and descending dopaminergic systems. Nat. Commun. 2, 171.

Taylor, M.R., Kikkawa, S., Diez-Juan, A., Ramamurthy, V., Kawakami, K., Carmeliet, P., Brockerhoff, S.E., 2005. The zebrafish *pob* gene encodes a novel protein required for survival of red cone photoreceptor cells. Genetics 170, 263–273.

Thermes, V., Grabher, C., Ristoratore, F., Bourrat, F., Choulika, A., Wittbrodt, J., Joly, J.S., 2002. *I-SceI* meganuclease mediates highly efficient transgenesis in fish. Mech. Dev. 118, 91–98.

Thummel, R., Burket, C.T., Brewer, J.L., Sarras Jr., M.P., Li, L., Perry, M., McDermott, J.P., Sauer, B., Hyde, D.R., Godwin, A.R., 2005. Cre-mediated site-specific recombination in zebrafish embryos. Dev. Dyn. 233, 1366–1377.

Vladimirov, N., Mu, Y., Kawashima, T., Bennett, D.V., Yang, C.-T., Looger, L.L., Keller, P.J., Freeman, J., Ahrens, M.B., 2014. Light-sheet functional imaging in fictively behaving zebrafish. Nat. Methods 11, 883–884.

Walker, C., Streisinger, G., 1983. Induction of mutations by gamma-rays in pregonial germ cells of zebrafish embryos. Genetics 103, 125–136.

Wang, Y., Rovira, M., Yusuff, S., Parsons, M.J., 2011. Genetic inducible fate mapping in larval zebrafish reveals origins of adult insulin-producing β-cells. Development 138, 609–617.

Weldon, J.E., Pastan, I., 2011. A guide to taming a toxin—recombinant immunotoxins constructed from Pseudomonas exotoxin A for the treatment of cancer. FEBS J. 278, 4683–4700.

Wilkens, H., 2015. Classic genetics and hybridization in *Astyanax*. In: Keene, A., Yoshizawa, M., McGhaugh, S. (Eds.), Biology and Evolution of the Mexican Cavefish. Elsevier, Amsterdam.

Yamaizumi, M., Mekada, E., Uchida, T., Okada, Y., 1978. One molecule of diphtheria toxin fragment a introduced into a cell can kill the cell. Cell 15, 245–250.

Yoshizawa, M., 2015. Evolution of neural regulation for foraging behavior in *A. mexicanus*. In: Keene, A., Yoshizawa, M., McGhaugh, S. (Eds.), Biology and Evolution of the Mexican Cavefish. Elsevier, Amsterdam.

Zhang, F., Cong, L., Lodato, S., Kosuri, S., Church, G.M., Arlotta, P., 2011. Efficient construction of sequence-specific TAL effectors for modulating mammalian transcription. Nat. Biotechnol. 29, 149–153.

Zhu, Z., He, L., Chen, S., 1985. Novel gene transfer into the fertilized eggs of gold fish (*Carassius auratus* L. 1758). J. Appl. Ichthyol. 1, 31–34.

Zhu, P., Narita, Y., Bundschuh, S.T., Fajardo, O., Zhang Schärer, Y.-P., Chattopadhyaya, B., Arn Bouldoires, E., Stepien, A.E., Deisseroth, K., Arber, S., Sprengel, R., Rijli, F.M., Friedrich, R.W., 2009. Optogenetic dissection of neuronal circuits in zebrafish using viral gene transfer and the Tet system. Front. Neural Circuits 3, 21.

Concluding Remarks: The *Astyanax* Community

William R. Jeffery
Department of Biology, University of Maryland, College Park, Maryland, USA

In concluding this book, I will emphasize a single attribute among the other many favorable reasons for conducting research on the *Astyanax* system. This is the existence of a vibrant and interactive international group of researchers and scholars—the *Astyanax* community. It is through the collaborative efforts of the *Astyanax* community that this book was conceived and executed. How and why did the *Astyanax* community evolve?

After the discovery of *Astyanax* (e.g., *Anoptichthys*) cavefish in the 1930s (Hubbs and Innes, 1936; also see Romero, 2001 for historical review), the first "wave" of research on *Astyanax* involved the Herculean efforts of single individuals. The Mexican biologist Jose Alvarez discovered two new *Astyanax* cavefish populations and described their facial bones (Alvarez, 1946), the American aquarium hobbyist C. Basil Jordan brought the first *Astyanax* cavefish to the United States (Romero, 2001), and the Turkish biologist Perihan Şadoğlu conducted the first crosses between cavefish and surface fish, discovering that albinism is controlled by a single Mendelian gene (Şadoğlu, 1957). During this period, a few small groups were also established, and these made incremental advances in understanding the basic biology of *Astyanax*. Charles Breder and his students and colleagues working at the New York Museum of Natural History were responsible for comprehensive ecological and physiological studies, as well as the first mechanistic work on eye degeneration (Breder, 1942; Breder and Rasquin, 1947). Robert Mitchell of Texas Tech University and his associates did extensive field work in the Sierra de El Abra region of Mexico, discovering more than two dozen additional *Astyanax* cavefish populations and making great strides in describing and mapping the fish caves (Mitchell et al., 1977). Horst Wilkens (1971) and his students at the University of Hamburg conducted evolutionary genetic studies showing that different genes were involved in repeated independent eye loss in different cavefish population. Jakob Parzefall (1983) and his students, also at the University of Hamburg, conducted detailed behavioral studies, which were the forerunners of some of the modern behavioral analyses described in this book. John Avis and Robert Selander of

Biology and Evolution of the Mexican Cavefish. http://dx.doi.org/10.1016/B978-0-12-802148-4.09988-6
© 2016 Elsevier Inc. All rights reserved.

the University of Texas (1972) did allozyme analysis that led to the realization that *Astyanax* surface fish and *Anoptichthys* cavefish populations are members of the same species complex, now called *Astyanax mexicanus*.

Astyanax cavefish are derived from a relatively small part of Mexico, but the individual *Astyanax* researchers worked in relative isolation in different parts of the world, and few Mexican researchers were involved. The interactions between the first *Astyanax* researchers were probably quite minimal; and it is possible that most of them had never met.

The next "wave" of *Astyanax* research included new groups attracted by the potential of this system to address fundamental questions in evolutionary genetics and EvoDevo, a new area of biology that was emerging at the time. Although some collaborative activities were carried out, for example, the first quantitative trait loci (QTL) for regressive and constructive traits were reported in a paper by Borowsky and Wilkens (2002), the new groups were also far-flung and generally not communicating with each other. An exception was a surprise meeting of the groups of H. Wilkens and W. Jeffery on a Mexican jungle trail in 1995. Without each other's knowledge, the same groups sometimes even conducted similar research projects; for instance, studies of the newly discovered *pax6* gene's role in cavefish eye degeneration (Behrens et al., 1997; Jeffery and Martasian, 1998). The small number of initial *Astyanax* groups expanded and produced their own students, who moved on to form their own research groups. Current groups at University College, London (UCL) (Y. Yamamoto), University of Cincinnati (J. Gross), and Marist University (L. Espinasa) were established during this period. New investigators also became interested in working on the *Astyanax* system, bringing expertise from diverse areas, such as neuroscience (Sylvie Rétaux, France), population genetics (Richard Borowsky, New York University), circadian rhythms (David Whitmore, UCL), and molecular systematics (Patricia Ornelas-Garcia, Mexico). The need for more effective communication became acute.

In 2007, a small party of interested individuals met at Hotel Taninul near Cuidad Valles, Mexico, to explore the possibility of establishing a rendezvous for worldwide *Astyanax* researchers (Figure 1(A)). This resulted in the first *Astyanax* International Meeting: AIM 2009 (Figure 2). The setting of the meeting was just as spectacular as the cavefish themselves. Hotel Taninul is located in a lush meadow abutting the crested limestone ridge of the Sierra de El Abra. The resurgence of a large spring on the hotel grounds has been modified into a tile-lined swimming pool, a perfect spot for discussion while the participants were immersed to the neck in soothing warm water. Many of the experiments, ideas, and the collaborative efforts described in the chapters of this book, were probably conceived in this idyllic location.

AIM has now underdone three iterations (Figure 1(B)), attracting increasing numbers of investigators from all over the world. Accordingly, J. Gross, one of the AIM 2013 organizers, provided his introductory comments in four different languages. Holding AIM at a site in Mexico, "donde cavefish ciegos nacen," is an important priority, opening the possibility for Mexican scientists, the original discoverers of cavefish, to easily return to *Astyanax* research. Three of the chapters

FIGURE 1 Participants in the 2007 planning session (A) for AIM 2009 and (B) AIM 2013. Authors of chapters in this book are identified by numbers. 1. M. Yoshizawa. 2. A. Parkhurst. 3. W. Jeffery. 4. S. Rétaux. 5. J. Gross. 6. E. Maldonado. 7. L. Espinasa. 8. A. Keene. 9. O. Garcia. 10. A. Santacruz. 11. M. Blin. 12. H. Hinaux 13. N. Rohner. 14. W. Elliott. 15. S. McGaugh. 16. R. Borowsky. 17. L. Ma. 18. H. Bilandžija. 19. K. O'Quin.

in this book are result of their recent efforts. Other important byproducts of AIM have been: the formation of the international *Astyanax* community; the entry of *Astyanax* research into the modern genomic era (see Chapter 6); and the idea to publish this book, the first dealing with *Astyanax* cavefish. As noted in a recent commentary (Gross et al., 2015), *Astyanax* research is on the rise; after humble beginnings, it has attracted widespread attention, and it is now an emerging model in biological and biomedical research (Maher, 2009; Albertson et al., 2009). The *Astyanax* community has many important and novel discoveries to make in the future, including further successes in linking the genotype to the phenotype. The chapter authors and editors are to be congratulated for their work bringing this fine system into its rightful limelight.

FIGURE 2 AIM 2009 logo. Drawing by Špela Gorički.

REFERENCES

Albertson, R.C., Cresko, W., Detrich III, H.W., Postlethwait, J.H., 2009. Evolutionary mutant models for human disease. Trends Genet. 25, 74–81.

Alvarez, J., 1946. Revisi6n del generero *Anoptichthys* con descripti6n de una especies nueva (Pisc. Characidae). An. Esc. Nac. Cien. Biol. Mex. 4, 263–282.

Avis, J.C., Selander, R.K., 1972. Evolutionary genetics of cave-dwelling fishes of the genus *Astyanax*. Evolution 26, 1–19.

Behrens, M., Langecker, G.T., Wilkens, H., Schmale, H., 1997. Comparative analysis of *Pax-6* sequence and expression in the eye development of blind cave fish *Astyanax fasciatus* and its epigean conspecific. Mol. Biol. Evol. 14, 299–308.

Borowsky, R., Wilkens, H., 2002. Mapping a cavefish genome: polygenic systems and regressive evolution. J. Hered. 93, 19–21.

Breder Jr., C.M., 1942. Descriptive ecology of La Cueva Chica with special reference to the blind fish, *Anoptichthys*. Zoologica 27, 7–15.

Breder Jr., C.M., Rasquin, P., 1947. Evidence for the lack of a growth principle in the optic cyst of Mexican cave fish. Zoologica 32, 29–33.

Gross, J.B., Meyer, B., Perkins, M., 2015. The rise of *Astyanax* cavefish. Dev. Dyn. http://dx.doi.org/10.1002/dvdy.24253.

Hubbs, C.L., Innes, W.T., 1936. The first known blind fish of the family Characidae: a new genus from Mexico. Occ. Pap. Mus. Zool. 342, 1–7.

Jeffery, W.R., Martasian, D.P., 1998. Evolution of eye regression in the cavefish *Astyanax*: apoptosis and the Pax-6 gene. Am. Zool. 38, 685–696.

Maher, B., 2009. Evolution: biology's next top model? Nature 458, 695–698.

Mitchell, R.W., Russell, W.H., Elliot, W.R., 1977. Mexican eyeless characin fishes, genus *Astyanax*: environment, distribution, and evolution. Spec. Publ. Mus. Texas Tech Univ. 12, 1–89.

Parzefall, J., 1983. Field observation in epigean and cave populations of Mexican characid *Astyanax mexicanus* (Pisces, Characidae). Mém. Biospéol. 10, 171–176.

Romero, A., 2001. Scientists prefer them blind: the history of hypogean fish research. Envron. Biol. Fishes 62, 43–71.

Şadoğlu, P., 1957. A Mendelian gene for albinism in natural cavefish population. Experientia 13, 394.

Wilkens, H., 1971. Genetic interpretation of regressive evolutionary processes: studies of hybrid eyes of two *Astyanax* cave populations (Characidae, Pisces). Evolution 25, 530–544.

Index

Note: Page numbers followed by *b* indicate boxes, *f* indicate figures, and *t* indicate tables.

Q

R

S

Printed in the United States
By Bookmasters